LONDON MATHEMATICAL SOCIETY LECTURE NOTE SERIES

The titles below are available from booksellers, or from Cambridge University Press at
www.cambridge.org/mathematics

348 Transcendental dynamics and complex analysis, P.J. RIPPON & G.M. STALLARD (eds)
349 Model theory with applications to algebra and analysis I, Z. CHATZIDAKIS, D. MACPHERSON, A. PILLAY & A. WILKIE (eds)
350 Model theory with applications to algebra and analysis II, Z. CHATZIDAKIS, D. MACPHERSON, A. PILLAY & A. WILKIE (eds)
351 Finite von Neumann algebras and masas, A.M. SINCLAIR & R.R. SMITH
352 Number theory and polynomials, J. MCKEE & C. SMYTH (eds)
353 Trends in stochastic analysis, J. BLATH, P. MÖRTERS & M. SCHEUTZOW (eds)
354 Groups and analysis, K. TENT (ed)
355 Non-equilibrium statistical mechanics and turbulence, J. CARDY, G. FALKOVICH & K. GAWEDZKI
356 Elliptic curves and big Galois representations, D. DELBOURGO
357 Algebraic theory of differential equations, M.A.H. MACCALLUM & A.V. MIKHAILOV (eds)
358 Geometric and cohomological methods in group theory, M.R. BRIDSON, P.H. KROPHOLLER & I.J. LEARY (eds)
359 Moduli spaces and vector bundles, L. BRAMBILA-PAZ, S.B. BRADLOW, O. GARCÍA-PRADA & S. RAMANAN (eds)
360 Zariski geometries, B. ZILBER
361 Words: Notes on verbal width in groups, D. SEGAL
362 Differential tensor algebras and their module categories, R. BAUTISTA, L. SALMERÓN & R. ZUAZUA
363 Foundations of computational mathematics, Hong Kong 2008, F. CUCKER, A. PINKUS & M.J. TODD (eds)
364 Partial differential equations and fluid mechanics, J.C. ROBINSON & J.L. RODRIGO (eds)
365 Surveys in combinatorics 2009, S. HUCZYNSKA, J.D. MITCHELL & C.M. RONEY-DOUGAL (eds)
366 Highly oscillatory problems, B. ENGQUIST, A. FOKAS, E. HAIRER & A. ISERLES (eds)
367 Random matrices: High dimensional phenomena, G. BLOWER
368 Geometry of Riemann surfaces, F.P. GARDINER, G. GONZÁLEZ-DIEZ & C. KOUROUNIOTIS (eds)
369 Epidemics and rumours in complex networks, M. DRAIEF & L. MASSOULIÉ
370 Theory of p-adic distributions, S. ALBEVERIO, A.YU. KHRENNIKOV & V.M. SHELKOVICH
371 Conformal fractals, F. PRZYTYCKI & M. URBAŃSKI
372 Moonshine: The first quarter century and beyond, J. LEPOWSKY, J. MCKAY & M.P. TUITE (eds)
373 Smoothness, regularity and complete intersection, J. MAJADAS & A. G. RODICIO
374 Geometric analysis of hyperbolic differential equations: An introduction, S. ALINHAC
375 Triangulated categories, T. HOLM, P. JØRGENSEN & R. ROUQUIER (eds)
376 Permutation patterns, S. LINTON, N. RUŠKUC & V. VATTER (eds)
377 An introduction to Galois cohomology and its applications, G. BERHUY
378 Probability and mathematical genetics, N. H. BINGHAM & C. M. GOLDIE (eds)
379 Finite and algorithmic model theory, J. ESPARZA, C. MICHAUX & C. STEINHORN (eds)
380 Real and complex singularities, M. MANOEL, M.C. ROMERO FUSTER & C.T.C WALL (eds)
381 Symmetries and integrability of difference equations, D. LEVI, P. OLVER, Z. THOMOVA & P. WINTERNITZ (eds)
382 Forcing with random variables and proof complexity, J. KRAJÍČEK
383 Motivic integration and its interactions with model theory and non-Archimedean geometry I, R. CLUCKERS, J. NICAISE & J. SEBAG (eds)
384 Motivic integration and its interactions with model theory and non-Archimedean geometry II, R. CLUCKERS, J. NICAISE & J. SEBAG (eds)
385 Entropy of hidden Markov processes and connections to dynamical systems, B. MARCUS, K. PETERSEN & T. WEISSMAN (eds)
386 Independence-friendly logic, A.L. MANN, G. SANDU & M. SEVENSTER
387 Groups St Andrews 2009 in Bath I, C.M. CAMPBELL *et al* (eds)
388 Groups St Andrews 2009 in Bath II, C.M. CAMPBELL *et al* (eds)
389 Random fields on the sphere, D. MARINUCCI & G. PECCATI
390 Localization in periodic potentials, D.E. PELINOVSKY
391 Fusion systems in algebra and topology, M. ASCHBACHER, R. KESSAR & B. OLIVER
392 Surveys in combinatorics 2011, R. CHAPMAN (ed)
393 Non-abelian fundamental groups and Iwasawa theory, J. COATES *et al* (eds)
394 Variational problems in differential geometry, R. BIELAWSKI, K. HOUSTON & M. SPEIGHT (eds)
395 How groups grow, A. MANN
396 Arithmetic differential operators over the p-adic integers, C.C. RALPH & S.R. SIMANCA
397 Hyperbolic geometry and applications in quantum chaos and cosmology, J. BOLTE & F. STEINER (eds)
398 Mathematical models in contact mechanics, M. SOFONEA & A. MATEI
399 Circuit double cover of graphs, C.-Q. ZHANG
400 Dense sphere packings: a blueprint for formal proofs, T. HALES
401 A double Hall algebra approach to affine quantum Schur–Weyl theory, B. DENG, J. DU & Q. FU
402 Mathematical aspects of fluid mechanics, J.C. ROBINSON, J.L. RODRIGO & W. SADOWSKI (eds)
403 Foundations of computational mathematics, Budapest 2011, F. CUCKER, T. KRICK, A. PINKUS & A. SZANTO (eds)

404 Operator methods for boundary value problems, S. HASSI, H.S.V. DE SNOO & F.H. SZAFRANIEC (eds)
405 Torsors, étale homotopy and applications to rational points, A.N. SKOROBOGATOV (ed)
406 Appalachian set theory, J. CUMMINGS & E. SCHIMMERLING (eds)
407 The maximal subgroups of the low-dimensional finite classical groups, J.N. BRAY, D.F. HOLT & C.M. RONEY-DOUGAL
408 Complexity science: the Warwick master's course, R. BALL, V. KOLOKOLTSOV & R.S. MACKAY (eds)
409 Surveys in combinatorics 2013, S.R. BLACKBURN, S. GERKE & M. WILDON (eds)
410 Representation theory and harmonic analysis of wreath products of finite groups, T. CECCHERINI-SILBERSTEIN, F. SCARABOTTI & F. TOLLI
411 Moduli spaces, L. BRAMBILA-PAZ, O. GARCÍA-PRADA, P. NEWSTEAD & R.P. THOMAS (eds)
412 Automorphisms and equivalence relations in topological dynamics, D.B. ELLIS & R. ELLIS
413 Optimal transportation, Y. OLLIVIER, H. PAJOT & C. VILLANI (eds)
414 Automorphic forms and Galois representations I, F. DIAMOND, P.L. KASSAEI & M. KIM (eds)
415 Automorphic forms and Galois representations II, F. DIAMOND, P.L. KASSAEI & M. KIM (eds)
416 Reversibility in dynamics and group theory, A.G. O'FARRELL & I. SHORT
417 Recent advances in algebraic geometry, C.D. HACON, M. MUSTAŢĂ & M. POPA (eds)
418 The Bloch–Kato conjecture for the Riemann zeta function, J. COATES, A. RAGHURAM, A. SAIKIA & R. SUJATHA (eds)
419 The Cauchy problem for non-Lipschitz semi-linear parabolic partial differential equations, J.C. MEYER & D.J. NEEDHAM
420 Arithmetic and geometry, L. DIEULEFAIT *et al* (eds)
421 O-minimality and Diophantine geometry, G.O. JONES & A.J. WILKIE (eds)
422 Groups St Andrews 2013, C.M. CAMPBELL *et al* (eds)
423 Inequalities for graph eigenvalues, Z. STANIĆ
424 Surveys in combinatorics 2015, A. CZUMAJ *et al* (eds)
425 Geometry, topology and dynamics in negative curvature, C.S. ARAVINDA, F.T. FARRELL & J.-F. LAFONT (eds)
426 Lectures on the theory of water waves, T. BRIDGES, M. GROVES & D. NICHOLLS (eds)
427 Recent advances in Hodge theory, M. KERR & G. PEARLSTEIN (eds)
428 Geometry in a Fréchet context, C.T.J. DODSON, G. GALANIS & E. VASSILIOU
429 Sheaves and functions modulo *p*, L. TAELMAN
430 Recent progress in the theory of the Euler and Navier–Stokes equations, J.C. ROBINSON, J.L. RODRIGO, W. SADOWSKI & A. VIDAL-LÓPEZ (eds)
431 Harmonic and subharmonic function theory on the real hyperbolic ball, M. STOLL
432 Topics in graph automorphisms and reconstruction (2nd Edition), J. LAURI & R. SCAPELLATO
433 Regular and irregular holonomic D-modules, M. KASHIWARA & P. SCHAPIRA
434 Analytic semigroups and semilinear initial boundary value problems (2nd Edition), K. TAIRA
435 Graded rings and graded Grothendieck groups, R. HAZRAT
436 Groups, graphs and random walks, T. CECCHERINI-SILBERSTEIN, M. SALVATORI & E. SAVA-HUSS (eds)
437 Dynamics and analytic number theory, D. BADZIAHIN, A. GORODNIK & N. PEYERIMHOFF (eds)
438 Random walks and heat kernels on graphs, M.T. BARLOW
439 Evolution equations, K. AMMARI & S. GERBI (eds)
440 Surveys in combinatorics 2017, A. CLAESSON *et al* (eds)
441 Polynomials and the mod 2 Steenrod algebra I, G. WALKER & R.M.W. WOOD
442 Polynomials and the mod 2 Steenrod algebra II, G. WALKER & R.M.W. WOOD
443 Asymptotic analysis in general relativity, T. DAUDÉ, D. HÄFNER & J.-P. NICOLAS (eds)
444 Geometric and cohomological group theory, P.H. KROPHOLLER, I.J. LEARY, C. MARTÍNEZ-PÉREZ & B.E.A. NUCINKIS (eds)
445 Introduction to hidden semi-Markov models, J. VAN DER HOEK & R.J. ELLIOTT
446 Advances in two-dimensional homotopy and combinatorial group theory, W. METZLER & S. ROSEBROCK (eds)
447 New directions in locally compact groups, P.-E. CAPRACE & N. MONOD (eds)
448 Synthetic differential topology, M.C. BUNGE, F. GAGO & A.M. SAN LUIS
449 Permutation groups and cartesian decompositions, C.E. PRAEGER & C. SCHNEIDER
450 Partial differential equations arising from physics and geometry, M. BEN AYED *et al* (eds)
451 Topological methods in group theory, N. BROADDUS, M. DAVIS, J.-F. LAFONT & I. ORTIZ (eds)
452 Partial differential equations in fluid mechanics, C.L. FEFFERMAN, J.C. ROBINSON & J.L. RODRIGO (eds)
453 Stochastic stability of differential equations in abstract spaces, K. LIU
454 Beyond hyperbolicity, M. HAGEN, R. WEBB & H. WILTON (eds)
455 Groups St Andrews 2017 in Birmingham, C.M. CAMPBELL *et al* (eds)
456 Surveys in combinatorics 2019, A. LO, R. MYCROFT, G. PERARNAU & A. TREGLOWN (eds)
457 Shimura varieties, T. HAINES & M. HARRIS (eds)

London Mathematical Society Lecture Note Series: 457

Shimura Varieties

Edited by

THOMAS HAINES
University of Maryland, College Park

MICHAEL HARRIS
Columbia University

CAMBRIDGE
UNIVERSITY PRESS

CAMBRIDGE
UNIVERSITY PRESS

University Printing House, Cambridge CB2 8BS, United Kingdom

One Liberty Plaza, 20th Floor, New York, NY 10006, USA

477 Williamstown Road, Port Melbourne, VIC 3207, Australia

314-321, 3rd Floor, Plot 3, Splendor Forum, Jasola District Centre, New Delhi - 110025, India

103 Penang Road, #05-06/07, Visioncrest Commercial, Singapore 238467

Cambridge University Press is part of the University of Cambridge.

It furthers the University's mission by disseminating knowledge in the pursuit of education, learning and research at the highest international levels of excellence.

www.cambridge.org
Information on this title: www.cambridge.org/9781108704861
DOI: 10.1017/9781108649711

This volume forms the sequel to *On the Stabilization of the Trace Formula* published by International Press of Boston, Inc., 2011. ISBN 9781571462275

First published 2020

A catalogue record for this publication is available from the British Library

Library of Congress Cataloging in Publication data
Names: Haines, Thomas J., editor. | Harris, Michael, 1954 – editor.
Title: Shimura varieties / edited by Thomas Haines, Michael Harris.
Description: Cambridge ; New York : Cambridge University Press, 2019. |
Series: London Mathematical Society lecture note series; 457 | "This volume forms the sequel to On the stabilization of the trace formula published by International Press of Boston, Inc., 2011." |
Includes bibliographical references. | Contents: Introduction to Volume II / T. J. Haines and M. Harris – Lectures on Shimura varieties / A. Genestier and B.C. Ngô – Unitary Shimura varieties / Marc-Hubert Nicole – Integral models of Shimura varieties of PEL type / Sandra Rozensztajn – Introduction to the Langlands-Kottwitz method / Yihang Zhu – Integral Canonical Models of Shimura varieties : an update / Mark Kisin – The Newton stratification / Elena Mantovan – On the geometry of the Newton stratification / Eva Viehmann – Construction of automorphic Galois representations : the self-dual case / Sug Woo Shin – The local Langlands correspondence for GL_n over p-adic fields, and the cohomology of compact unitary Shimura varieties / Peter Scholze – Une application des variétés de Hecke des groupes unitaires / Gaëtan Chenevier – A patching lemma / Claus M. Sorensen – On subquotients of the étale cohomology of Shimura varieties / Christian Johansson and Jack A. Thorne. | Text in English with one contribution in French.
Identifiers: LCCN 2019037003 (print) | LCCN 2019037004 (ebook) |
ISBN 9781108704861 (paperback) | ISBN 9781108649711 (epub)
Subjects: LCSH: Shimura varieties. | Automorphic forms.
Classification: LCC QA242.5 .S45 2019 (print) | LCC QA242.5 (ebook) | DDC 516.3/5–dc23
LC record available at https://lccn.loc.gov/2019037003
LC ebook record available at https://lccn.loc.gov/2019037004

ISBN 978-1-108-70486-1 Paperback

CONTENTS

INTRODUCTION TO VOLUME II

T. J. HAINES AND M. HARRIS

The present volume is the second in a series of collections of mainly expository articles on the arithmetic theory of automorphic forms. The books are primarily intended for two groups of readers. The first group is interested in the structure of automorphic forms on reductive groups over number fields, and specifically in qualitative information about the multiplicities of automorphic representations. The second group is interested in the problem of classifying l-adic representations of Galois groups of number fields. Langlands' conjectures elaborate on the notion that these two problems overlap to a considerable degree. The goal of this series of books is to gather into one place much of the evidence that this is the case, and to present it clearly and succinctly enough so that both groups of readers are not only convinced by the evidence but can pass with minimal effort between the two points of view.

The first volume mainly dealt with the mechanics of the stable trace formula, with special emphasis on the role of the Fundamental Lemma, whose proof had recently appeared. The present volume is largely concerned with application of the methods of arithmetic geometry to the theory of Shimura varieties. The primary motivation, in both cases, is the construction of the compatible families of ℓ-adic Galois representations attached to cohomological automorphic representations π of $GL(n)$ over totally imaginary quadratic extensions of totally real number fields – CM fields, in other words. These are the Galois representations that allow the most direct generalizations of methods developed for the study of the arithmetic of elliptic curves over $\mathbb{Q}$. This is one reason they have increasingly attracted the attention of algebraic number theorists. The latter belong to the second group of readers to whom this book is addressed.

Most of the Galois representations that can be attached to automorphic representations are realized in the ℓ-adic cohomology of Shimura varieties, with coefficients in local systems. The Shimura varieties to be considered in this book are attached to unitary similitude groups; they belong to the class of *PEL Shimura varieties*, the first large family of locally symmetric varieties studied by Shimura in his long series of papers in the early 1960s. The abbreviation stands for Polarization-Endomorphisms-Level; like the other PEL Shimura varieties, the ones studied here are moduli spaces of polarized abelian varieties of a fixed dimension, together with an algebra of endomorphisms and a fixed level structure, all of which have to satisfy certain compatibilities. If one drops the endomorphisms one is left with the moduli space of polarized abelian varieties of fixed dimension g with level structure, in other words with the familiar Siegel moduli space. This is the Shimura variety attached to the group $GSp(2g)$ of similitudes of a $2g$-dimensional vector space endowed with a non-degenerate

---

Date: March 23, 2019.

alternating form. The decision to focus instead on unitary similitude groups was dictated by the close relation between $GL(n)$ and unitary groups, which is responsible for the role of the Shimura varieties considered here in the construction of the Galois representations attached to $GL(n)$ over a totally real or CM field. A detailed treatment of Siegel modular varieties can be found in Morel's two articles [M08, M11].

The Shimura varieties considered in this book are all *proper* as complex algebraic varieties; the corresponding unitary similitude groups are anisotropic modulo center. This choice is sufficient in order to attach Galois representations to the automorphic representations that contribute to the cohomology[1] of the adèlic locally symmetric space attached to $GL(n)$. The Galois representations discussed in chapters 8-11 of this Volume, and in Chapter CHL.IV.C of Volume 1, are directly attached to the *polarized* cohomological cuspidal automorphic representations of $GL(n)$; these are the ones that admit descents to unitary groups. Galois representations can also be attached to cuspidal automorphic representations of $GL(n)$ that are cohomological but not polarized [HLTT, Sch, B]; their construction is based on techniques from p-adic geometry that are beyond the scope of the present volume.

A. Zeta functions of unitary Shimura varieties

Let $\mathcal{K}$ be a totally imaginary quadratic extension of a totally real field F. Let $(V, [\cdot,\cdot])$ be a non-degenerate hermitian space for $\mathcal{K}/F$, i.e., V is an n-dimensional $\mathcal{K}$-vector space and $[\cdot,\cdot]$ is a $\mathbf{c}$-sesquilinear form satisfying $[x, y] = \mathbf{c}([y, x])$ for all $x, y \in V$. Let $GU(V)$ denote the group of similitudes of $(V, [\cdot,\cdot])$, and let $G \subset GU(V)$ be the subgroup of similitudes with rational similitude factor, defined as in the chapters of Genestier-Ngô, Nicole, Rozensztajn, and Zhu. This is viewed as an algebraic group over $\mathbb{Q}$. Let $K_f \subset G(\mathbf{A}_f)$ be an open compact subgroup, and let $K_\infty \subset G(\mathbb{R})$ be a maximal connected subgroup that contains the center of $G(\mathbb{R})$ and is compact modulo the center. Then $X := G(\mathbb{R})/K_\infty$ is the union of finitely many copies of a hermitian symmetric domain, and the double coset space

$$G(\mathbb{Q})\backslash G(\mathbf{A})/K_\infty \times K_f$$

can be identified with the set of points of a complex quasi-projective variety $Sh(G, X)_{K_f}$. In Chapter 1 and (in more detail) in Chapter 2, it is explained that X is endowed with a canonical structure, following Deligne, that does not depend on the choice of K_∞, and the Shimura variety $Sh(G, X)_{K_f}$ has a model over the *reflex field* E, a number field determined by the pair (G, X) and contained in the Galois closure of $\mathcal{K}$ over $\mathbb{Q}$. This model identifies $Sh(G, X)_{K_f}$ with (a part of) the moduli space of quadruples $(A, \iota, \lambda, \bar{\eta})$, where A is an abelian variety of dimension n, λ is a polarization of A, ι is an action of an appropriate order in $\mathcal{K}$ on A, and $\bar{\eta}$ is a level structure depending on K_f. The quadruple $(A, \iota, \lambda, \bar{\eta})$ are moreover required to satisfy certain compatibilities that were introduced by Shimura in his definition of *PEL types* (for Polarization, Endomorphism, Level); we refer the reader to Chapter 2 for details.

We always assume that the derived subgroup $G^{der} \subset G$ is an *anisotropic*, which equivalent to the condition $[x, x] \neq 0$ for all $x \in V \setminus \{0\}$. If $n > 2$, this is equivalent

[1] Here and below, by *cohomology* of a locally symmetric space we will understand the cohomology with coefficients in local systems attached to algebraic representations of the corresponding reductive group.

to the condition that, for some complex embedding $\tau : \mathcal{K} \hookrightarrow \mathbb{C}$, the induced hermitian form on $V \otimes_{\mathcal{K},\tau} \mathbb{C}$ is positive- or negative-definite. Under this hypothesis, it is a well-known theorem of Borel and Harish-Chandra that the double coset space $G(\mathbb{Q})\backslash G(\mathbf{A})/K_\infty \times K_f$ is a compact complex variety, and thus the Shimura variety $Sh(G,X)_{K_f}$ is projective.

Suppose the group G is unramified at p; in other words, it is quasi-split over $\mathbb{Q}_p$ and splits over an unramified extension of $\mathbb{Q}_p$. Then $G(\mathbb{Q}_p)$ contains a hyperspecial maximal compact subgroup given as the stabilizer of some $\Lambda \subset V(\mathbb{Q}_p)$, a lattice that is self-dual relative to $[\cdot,\cdot]$. We suppose that $K_f = K_p \times K^p$ where K_p is such a hyperspecial maximal compact subgroup of $G(\mathbb{Q}_p)$; moreover, we suppose that K^p is a sufficiently small compact open subgroup of the adeles of $\mathbb{Q}$ away from p. In Chapter 3 it is explained that the Shimura variety $Sh(G,X)_{K_f}$ then has a *smooth* model S_{K^p} over the integers $\mathcal{O}_\mathfrak{p}$ in the completion of E at any p-adic place $\mathfrak{p}$; moreover, under our hypothesis that G^{der} is anisotropic, it is known that S_{K^p} is *projective* over $\mathcal{O}_\mathfrak{p}$.

Remark. (Contributed by Y. Zhu) For the smooth quasi-projective integral model of the PEL Shimura variety considered by Kottwitz (at a hyperspecial prime), it is known that it is projective if and only if G^{der} is anisotropic over $\mathbb{Q}$. A special case of this assertion, when $End_B(V)$ is a division algebra, is proved by Kottwitz on p. 392 of [K92a]. The general case follows from the main results of Lan's thesis [Lan13]. In fact, Lan constructs smooth projective toroidal compactifications of the integral model, whose boundary strata are also smooth, see [Lan13, Theorem 4.1.1.1, Theorem 7.3.3.4]. Thus the projectivity of Kottwitz's integral model is equivalent to the projectivity of the generic fiber. But for the generic fiber, it is well known that projectivity is equivalent to G^{der} being anisotropic (see for instance [Pink]). We mention that the same projectivity criterion is also valid for PEL integral models of more general levels. This can be deduced from the results of [Lan11], cf. the discussion on p. 7 of [Pera]. We note that many of Lan's results have been generalized by Madapusi Pera to Hodge type, in [Pera]. See especially [Pera, Corollary 4.1.7] for the same projectivity criterion for the Hodge-type integral models.

The computation of the local factor of the zeta function at $\mathfrak{p}$ comes down to a parametrization of the fixed points, on the special fiber of S_{K^p}, of a correspondence T_a obtained by composing a Hecke operator T (at a prime not dividing p) with the power $Frob_\mathfrak{p}^a$, $a >> 0$, of the (geometric) Frobenius automorphism at $\mathfrak{p}$. The points of the special fiber are partitioned into *isogeny classes* of PEL quadruples $(A, \iota, \lambda, \bar{\eta})$ – in other words, isogeny classes of abelian varieties A, together with the additional structures – and each such isogeny class is preserved by T_a. Thus the determination of the local factor can be divided into four steps. This division is artificial and does not follow the actual proof, and the account given below is a gross oversimplification; however, it works as a first approximation. Precise statements can be found in Chapter 4, specifically Theorem 5.3.1.

(i) Parametrization of the set of isogeny classes

The crucial observation is that the Honda-Tate classification of abelian varieties over finite fields, in terms of their Frobenius automorphisms, allows us to parametrize the set $\mathcal{V}_\mathcal{K}$ of isogeny classes of PEL quadruples by a certain subset of elliptic conjugacy classes in $I(\mathbb{Q})$, where I runs through a certain family of reductive group over $\mathbb{Q}$ whose conjugacy classes can be related to those of G. In this way the isogeny class supplies an elliptic conjugacy class $[\gamma_0] \subset G(\mathbb{Q})$.

(ii) Parametrization of an individual isogeny class

The next step is to parametrize the fixed points belonging to the isogeny class corresponding to the conjugacy class $[\gamma_0]$. Suppose for a moment that A is an abelian variety over a finite field k of characteristic p. Then for any prime $\ell \neq p$, the ℓ-adic Tate module $T_\ell(A)$ is canonically a lattice in $H_1(A_{\bar{k}}, \mathbb{Q}_\ell)$, defined as the dual of the ℓ-adic cohomology $H^1(A_{\bar{k}}, \mathbb{Q}_\ell)$. An A' related to A by a prime-to-p-isogeny then defines by duality a lattice in

$$H^1(A_{\bar{k}}, \mathbf{A}_f^p) := \prod_{\ell \neq p}' H^1(A_{\bar{k}}, \mathbb{Q}_\ell),$$

the restricted product being taken with respect to the integral cohomology. In this way the prime-to-p isogeny classes of PEL quadruples can be related to $G(\mathbf{A}_f^p) = \prod'_{\ell \neq p} G(\mathbb{Q}_\ell)$-orbits in the space of lattices in $H^1(A_{\bar{k}}, \mathbf{A}_f^p)$. A standard construction relates fixed points of Hecke correspondences on this set of orbits with orbital integrals in $G(\mathbf{A}_f^p)$.

Incorporating p-power isogenies is more subtle. Instead of ℓ-adic étale cohomology, one needs to classify Frobenius-stable lattices in the rational Dieudonné module of A. Since the Frobenius operator is only σ-linear, where σ is a generator of the Galois group of the maximal unramified extension $\mathbb{Q}_p^{un}$ of $\mathbb{Q}_p$, it is not surprising that this part of the classification leads to *twisted* orbital integrals over finite unramified extensions of $\mathbb{Q}_p$.

In this way, the global conjugacy class $[\gamma_0]$ of Step A is joined by a pair (γ, δ) with $\gamma \in G(\mathbf{A}_f^p)$ and $\delta \in G(\mathbb{Q}_p{}^{un})$. The triple $(\gamma_0; \gamma, \delta)$ is called a *Kottwitz triple* if it satisfies the axioms of Definition 4.1.1 of Chapter 4.

(iii) Reconciliation of the global and local data by Galois cohomology

Steps (i) and (ii) have established a map from the set of points of S_{K^p} over finite fields to the set of Kottwitz triples. Thus the Lefschetz trace of the correspondence T_a is a weighted sum over the set of triples in the image of this map. Perhaps the deepest point in the Langlands-Kottwitz method is the result of Kottwitz that asserts that the image consists precisely of those triples for which a certain Galois cohomological invariant (which is naturally known as the *Kottwitz invariant*) vanishes. A series of reductions translates this fundamental observation into an application of the comparison theorem of p-adic Hodge theory. For all this, see Chapter 1, Proposition 6.3.1, and Chapter 4, §4, especially Theorem 4.2.8 and the sketch of its proof.

The Kottwitz invariant of a triple $(\gamma_0; \gamma, \delta)$ closely resembles the cohomological invariant that is the subject of Theorem 4.5 (also due to Kottwitz) of Chapter [H.I.A] of [CHLN]. The most obvious difference – that the Kottwitz invariant attached to points on Shimura varieties involves a twisted conjugacy class at p – turns out (again thanks to Kottwitz) to be a manageable problem. Thus in the end the Lefschetz trace of T_a can be written, just as in formula (4.7) of Chapter [H.I.A] of [CHLN], as a weighted sum of adelic orbital integrals, indexed by global conjugacy classes. A similar analysis applies to compute the Lefschetz trace of the operator T_a acting on the cohomology of an ℓ-adic local system $\mathcal{W}_\rho$ defined by an algebraic representation ρ of G.

The fourth step is the comparison of the formula that results from steps (i)-(iii) with the stabilized trace of a Hecke operator T'_a acting on the space of automorphic forms $L_2(G(\mathbb{Q})\backslash G(\mathbf{A})/Z)$, where $Z \subset G(\mathbb{R})$ is a subgroup of the center chosen to make the quotient space compact (for example, one can take Z to be the subgroup $\mathfrak{A}_G$ introduced in Chapter L.IV.A of Volume 1). The operators T'_a and T_a coincide at finite places prime to p. At p the local component of T'_a is a Hecke operator determined explicitly by a and the Shimura datum, while the archimedean component of T'_a is a discrete series pseudocoefficient determined by ρ, as in Chapter CHL.IV.B of [CHLN]. In this way the characteristic polynomial of $Frob_\mathfrak{p}$ on eigenspaces of the Hecke operators away from p is determined by the traces of Hecke operators at p on these same eigenspaces, confirming the Langlands conjecture in the situations where the method applies.

The sketch given here is loosely based on two fundamental papers of Kottwitz: The idea of comparing the Lefschetz trace formula for points of Shimura varieties over finite fields with the Selberg trace formula for automorphic forms first appeared in a paper by Ihara [I67], where it was applied to modular curves. Ihara's method was extended by Langlands to the adelic setting, allowing for certain kinds of bad reduction, in [L73]. Langlands wrote a series of papers on the subject in the 1970s, and formulated a conjecture on the form of the zeta function of a general Shimura variety in [L79]. Over the next decade Kottwitz developed the techniques of Galois cohomology, in parallel with those introduced in Volume 1 for the purposes of stabilizing the trace formula, in the end obtaining a formula for the Lefschetz traces of the operators T_a with the same shape as the automorphic trace of T'_a. Readers of Volume 1 will recall that the stabilization of the elliptic part of the Arthur-Selberg trace formula breaks down into three steps analogous to (A-C) above. After the first two steps, the (elliptic part of the) geometric side of the trace formula can be written as a sum over conjugacy classes of pairs (γ, γ_v), where γ (resp. γ_v) is a global (resp. adelic) conjugacy class, satisfying some cohomological restrictions. The analogous formula was proved by Kottwitz in [K92a] for PEL type Shimura varieties attached to groups G whose Lie algebras are of type A or C.

The formula for Lefschetz traces was rewritten in [K90] as a sum over stable global conjugacy classes of purely local orbital integrals, at the cost of introducing endoscopic groups, and assuming the Fundamental Lemma and related conjectures. The resulting formula was completely analogous to the elliptic part of the stable trace formula. While everything developed in the first part of the present book is at least implicit in [K90], the latter is not so easy to use as a reference, because his formulation of the answer makes no distinction between conjectures, like the Fundamental Lemma, that have since been proved, and the most general conjectures relative to the parametrization of points on general Shimura varieties. Moreover, Kottwitz went further: he rewrote the spectral side of the trace formula in terms of Arthur parameters, defined in terms of the Langlands group whose existence is still hypothetical. In Arthur's book [A13] on classical groups, as (implicitly) in the chapters by Clozel, Harris, and Labesse in Volume 1, cuspidal automorphic representations of $GL(n)$ are used as a substitute for Arthur parameters. One of the aims of the Paris book project was to provide a usable reference for the applications of the trace formula to the Shimura varieties most relevant to the construction of Galois representations. Since these Galois representations are attached to (polarized) cohomological cuspidal automorphic representations of $GL(n)$, it is completely natural to use these representations as parameters.

Remark. There are other good introductions to this subject matter, notably Milne's three papers [Mi90, Mi92, Mi05]. The treatment here differs from that of Milne in that, as in Volume 1, we focus specifically on the case of Shimura varieties attached to unitary similitude groups. This is the most important case for the construction of Galois representations; moreover, in this case, the elliptic conjugacy classes that arise in the trace formula, as well as the transfer factors used to stabilize the trace formula, are easy to describe in terms of algebraic number theory (see §8 of the first chapter of [CHLN]).

Unitary groups vs. similitude groups

The Shimura variety attached to the (rational) similitude group $GU(V)$ is of PEL type. In particular, it is (a piece of) the solution of a moduli problem that has been well understood in characteristic zero since Shimura's work of the 1960s, and the *determinant condition* introduced by Kottwitz in [K92a] (see Chapter 3, section 2.2) provides a definition of the moduli problem that is valid in all good characteristics. This makes it possible to treat the parametrization of the points efficiently, but the reduction of the automorphic theory of $GU(V)$ to that of $U(V)$, and thus (by base change) to that of $GL(n)$, is not trivial; see §1 of Chapter CHL.IV.C of [CHLN] for an illustration of this. Even when this issue has been resolved, the parametrization of automorphic representations of $GU(V)$ requires an auxiliary Hecke character that is destined to disappear when constructing automorphic Galois representations, but that in the intermediate steps introduces additional variables that pose an ultimately useless challenge to fitting the crucial formulas in a single line.

It is also possible to attach a Shimura variety directly to the unitary group. One of us (M.H.) has written a fair number of papers about the Shimura varieties attached to similitude groups and is quite embarrassed to confess that he only learned of this possibility in 2017, after having failed to read carefully Chapter 27 of [GGP], and thus contributed significantly to the proliferation of excessively long formulas. Langlands's conjecture on the zeta functions of Shimura varieties ([L79]; see also Conjecture 0.4 below), applied in this case, yields a simple relation between the Galois representations on the ℓ-adic cohomology of the Shimura varieties attached to unitary groups, on the one hand, and the automorphic representations of $GL(n)$, on the other hand. However, these varieties are of *abelian type* but not of PEL type – they do not parametrize families of abelian varieties, but rather of motives closely related to those of abelian varieties. The proof of the Langlands conjecture – at places of good reduction – for Shimura varieties of abelian type, by the Langlands-Kottwitz method, is the subject of work in progress by Kisin, Shin, and Zhu [KSZ] that should appear in the near future.

B. Geometry of Shimura varieties in positive characteristic

For arithmetic applications, the existence and characterization of *integral* canonical models for Shimura varieties play an important role. Let us discuss this problem in a fairly general context, where the Shimura variety $Sh(G,X)_{K_f}$ is associated to Shimura data such that $G_{\mathbb{Q}_p}$ is an unramified group and K_f factorizes as $K_f = K^p K_p$ for $K_p \subset G(\mathbb{Q}_p)$ a hyperspecial maximal compact subgroup and for $K^p \subset G(\mathbb{A}_f^p)$ a sufficiently small compact open subgroup. We fix K_p but let K^p vary, and consider the pro-E-scheme

$$Sh_{K_p}(G,X) = \varprojlim_{K^p} Sh(G,X)_{K^p K_p}.$$

Here $Sh(G,X)_{K^p K_p}$ denotes the canonical model of the Shimura variety over the reflex field E. An integral model of $Sh_{K_p}(G,X)$ over $\mathcal{O}_\mathfrak{p}$ consists of a $G(\mathbb{A}_f^p)$-equivariant inverse system (S_{K^p}) of $\mathcal{O}_\mathfrak{p}$-models of the inverse system $(Sh(G,X)_{K^p K_p, E_\mathfrak{p}})_{K^p}$. We also view the pro-$\mathcal{O}_\mathfrak{p}$-scheme $S_{K_p} = \varprojlim_{K^p} S_{K^p}$ as an integral model of $Sh_{K_p}(G,X)_{E_\mathfrak{p}}$. We say such an integral model is *smooth* if for some subgroup K_0^p, $S_{K_0^p}$ is smooth over $\mathcal{O}_\mathfrak{p}$ and $S_{K_1^p}$ is étale over $S_{K_2^p}$ for all $K_1^p \subseteq K_2^p \subseteq K_0^p$. Given our assumption on K_p, a smooth integral model should always exist. In the case where the Shimura data is PEL type (assuming $p > 2$ in Type D), Kottwitz [K92a] showed that the natural PEL moduli problem attached to $(G, X, K^p K_p)$ is represented by a smooth quasi-projective scheme over $\mathcal{O}_\mathfrak{p}$, essentially by reducing to the Siegel case (GSp, $S^\pm$), which was handled earlier by Mumford using geometric invariant theory [Mu65].

If $Sh_{K_p}(G,X)$ has at least one smooth integral model, in principle it will have many. So it is important to define a favorable notion of smooth integral model and characterize it by a condition which pins it down up to unique isomorphism. One such characterization, reminiscent of the Néron extension property of Néron models, was suggested by Milne [Mi92, §2]: the pro-$\mathcal{O}_\mathfrak{p}$-scheme S_{K_p} should be termed an *integral canonical model* if it is smooth (hence regular as a scheme) and satisfies the following extension property: given any regular $\mathcal{O}_\mathfrak{p}$-scheme Y such that $Y_{E_\mathfrak{p}}$ is dense in Y, any $E_\mathfrak{p}$-morphism $Y_{E_\mathfrak{p}} \to Sh_{K_p}(G,X)_{E_\mathfrak{p}}$ extends uniquely to an $\mathcal{O}_\mathfrak{p}$-morphism $Y \to S_{K_p}$. As explained by Moonen [Mo98, §3], it is not clear such models always exist; it is not even clear that the Siegel modular scheme satisfies this extension property. Fundamentally, the problem is that the class of test schemes Y in this definition is too broad. In showing that the Siegel modular scheme satisfies Milne's extension property, one needs to know that for any closed subscheme $Z \subset Y$ disjoint from $Y_{E_\mathfrak{p}}$ and of codimension at least 2, any abelian scheme over $Y \backslash Z$ extends to an abelian scheme over all of Y. However, this statement is false in general ([Mo98, §3.4]). But the required version of it beomes true, thanks to a lemma of Faltings [Mo98, 3.6], if we require Y to also be *formally smooth* over $\mathcal{O}_\mathfrak{p}$ (see [Mo98, §3.5]). The problem with Milne's suggestion is therefore averted by modifying the extension property: we require it only of *regular* and *formally smooth* $\mathcal{O}_\mathfrak{p}$-schemes Y (see [Ki10, §2.3.7]). Then Kisin's main result in [Ki10] is that, for $p > 2$, every abelian type Shimura variety $Sh_{K_p}(G,X)$ possesses an integral canonical model S_{K_p} in this modified sense. Since S_{K_p} is itself regular and formally smooth over $\mathcal{O}_\mathfrak{p}$ by construction, the modified extension property characterizes S_{K_p} uniquely up to unique isomorphism. Vasiu proved results similar to Kisin's in his earlier works [Va99], but the characterization is stated in a different way. It is not hard to show that Kottwitz's PEL moduli problems are represented by (finite unions of) integral canonical models. In particular this applies to the PEL unitary Shimura varieties which are the main focus of this volume.

Chapter 5 is a summary of Kisin's paper [Ki10], which constructs integral canonical models for Shimura varieties of abelian type (assuming $p > 2$) according to the following outline:

(i) Hodge type case

For Hodge type data (G, X) endowed with an embedding of Shimura data $(G, X) \hookrightarrow (\mathrm{GSp}(V), S^{\pm})$, one constructs an integral canonical model S_{K_p} as the normalization of the closure of $Sh_{K_p}(G, X)$ in $S_{K'_p}(\mathrm{GSp}(V), S^{\pm})$ for $K'_p \subset \mathrm{GSp}(\mathbb{Q}_p)$ the stabilizer of a certain lattice $V_{\mathbb{Z}} \subset V$. Here, the integral model $S_{K'_p}(\mathrm{GSp}(V), S^{\pm})$ comes from a moduli problem, formulated with the choice of $V_{\mathbb{Z}}$. (Warning: the latter is not necessarily an integral canonical model associated to $(\mathrm{GSp}(V), S^{\pm})$ so in particular is not necessarily the same as Kottwitz' Siegel modular scheme, unless K'_p is hyperspecial.)

This first step relies on Kisin's Key Lemma, which asserts that certain tensors arising from certain Hodge classes via the p-adic comparision theorems are *integral*. For this Kisin uses his earlier results on crystalline representations. Then Kisin is able to make use of Faltings' deformation ring for p-divisible groups equipped with a collection of Tate cycles [Fa99, §7] to show that S_{K_p} as constructed above is smooth and satisfies the modified extension property.

(ii) Abelian type case

Let (G, X) be a Shimura variety of abelian type, and let (G_1, X_1) be a Shimura variety of Hodge type such that there is a central isogeny $G_1^{der} \to G^{der}$ inducing an isomorphism $(G_1^{ad}, X_1^{ad}) \xrightarrow{\sim} (G^{ad}, X^{ad})$. Kisin constructs the integral canonical model for $Sh_{K_p}(G, X)$ from the one already constructed for $Sh_{K_{1,p}}(G_1, X_1)$.

Having the integral canonical model S_{K_p} in hand, it now makes sense to consider its special fiber $\bar{S}_{K_p}$, which is a well-defined object depending canonically on the data (G, X, K_p). Again, for PEL type Shimura varieties like the ones we consider in this volume, this amounts to studying the Kottwitz moduli problems over the residue field of $\mathcal{O}_{\mathfrak{p}}$. Even in PEL situations, there are difficult questions revolving around the counting of points (used to understand zeta functions and to construct Galois representations in the cohomology of Shimura varieties), and also questions related to various stratifications (Newton, Ekedahl-Oort, Kottwitz-Rapoport, etc.). In this volume, the point-counting problems and their relations to automorphic forms and Galois representations are addressed in the articles (Zhu), (Shin), and (Scholze), and their contents are further described in Parts A and C of this introduction. In the rest of Part B, we discuss the articles (Mantovan) and (Viehmann), which are concerned with the geometry of the special fibers $\bar{S}_{K_p}$, most importantly with the Newton stratification, the Oort foliations, and their applications. Although some statements are currently known to extend to Hodge type Shimura varieties, for the most part we will limit our discussion to PEL Shimura varieties, for simplicity.

Let $k \cong \mathbb{F}_p$ be a finite prime field, and let L denote the fraction field of the Witt ring $W(\bar{k})$. Let σ be the Frobenius automorphism of $W(\bar{k})$ over $W(k)$ induced by $x \mapsto x^p$ on k, and use the same symbol for the automorphism of L. An *F-isocrystal* is a pair (V, Φ) consisting of a finite-dimensional L-vector space V and a σ-linear bijection $\Phi : V \to V$.

Using the Dieudonné-Manin classification of F-isocrystals, one can associate to a simple F-isocrystal its Newton polygon. In [Ko85], Kottwitz defined, for any connected reductive group $G/\mathbb{Q}_p$, the set of F-isocrystals with G-structure (called *G-isocrystals* here), and showed that when G is quasi-split this set is identified with the pointed set $B(G)$ consisting of σ-conjugacy classes in $G(L)$. In the same setting he gave a group-theoretic meaning to Newton polygons, defining the set of *Newton points*

$$\mathcal{N}(G) = (X_*(A)_{\mathbb{Q}}/\Omega)^{\Gamma}$$

where A is a maximal $\mathbb{Q}_p$-torus with Weyl group Ω and $\Gamma = \mathrm{Gal}(\bar{\mathbb{Q}}_p/\mathbb{Q}_p)$. Further, Kottwitz gave a group-theoretic generalization of the map to Newton polygons, by defining the *Newton map*

$$\bar{\nu} : B(G) \to \mathcal{N}(G) \, , \, b \mapsto \bar{\nu}_b$$

(see (Viehmann), §3). Here $\nu_b \in \mathrm{Hom}_L(\mathbb{D}, G)$ refers to the *slope homomorphism* which is the group-theoretic counterpart of slopes of an isocrystal (see (Mantovan), §2.3). In the cases of interest to us (where $G^{der} = G^{sc}$ and G is quasisplit over $\mathbb{Q}_p$), the map $\bar{\nu}$ is injective, and the natural partial order $\preceq$ on $\mathcal{N}(G)$ gives $B(G)$ the structure of a poset.

Now return to the global data (G, X), where $G_{\mathbb{Q}_p}$ is still assumed to be quasi-split, and let $\mu_{\bar{\mathbb{Q}}_p}$ denote a geometric cocharacter of $G_{\bar{\mathbb{Q}}_p}$ coming from (G, X) in the usual way. Kottwitz defined a finite subset $B(G_{\mathbb{Q}_p}, \mu_{\bar{\mathbb{Q}}_p}) \subset B(G)$ using $\preceq$. In PEL cases, a closed geometric point x of $\bar{S}_{K_p}$ gives rise to the $G_{\mathbb{Q}_p}$-isocrystal associated to the (rational) Dieudonné module $\mathbb{D}H_{x,\mathbb{Q}}$ of its associated p-divisible group H_x, and thus to an element $[b_x] \in B(G_{\mathbb{Q}_p}, \mu_{\bar{\mathbb{Q}}_p})$. The *Newton stratum* (resp. closed Newton set) associated to $[b] \in B(G_{\mathbb{Q}_p}, \mu_{\bar{\mathbb{Q}}_p})$ is the locus $\mathcal{N}_{[b]}$ (resp. $\mathcal{N}_{\preceq[b]}$) of points $x \in \bar{S}_{K_p}(\bar{k})$ such that $[b_x] = [b]$ (resp. $[b_x] \preceq [b]$). It is known thanks to Grothendieck (for p-divisible groups without additional structure) and thanks to Rapoport-Richartz [RR96] in general, that $\mathcal{N}_{\preceq[b]}$ is indeed a closed subset.

Fundamental quesions one can ask about Newton strata for $[b] \in B(G_{\mathbb{Q}_p}, \mu_{\bar{\mathbb{Q}}_p})$:

- Is every Newton stratum $\mathcal{N}_{[b]}$ non-empty?
- Does the stratification behave well, i.e., is $\mathcal{N}_{\preceq[b]}$ the union of the $\mathcal{N}_{[b']}$ for $[b'] \preceq [b]$?
- What is the geometry of $\mathcal{N}_{[b]}$, in particular, what is its dimension?
- How does $\mathcal{N}_{[b]}$ relate to other important objects, such as Rapoport-Zink spaces $RZ_{(G,b,\mu)}$ and the central leaves $C_{[b]}$ in $\mathcal{N}_{[b]}$? (See (Mantovan), §3, 4.)

A large part of the article (Viehmann) is a summary of the current knowledge about these questions. The first point has been established in many cases, including in the PEL cases of interest to us, by Viehmann-Wedhorn, see Theorem 4.1 of chapter (Viehmann); in addition (Viehmann) explains various others approaches to the non-emptiness question. The second point was proved, in both PEL and Hodge type cases, by Hamacher [Ha15b, Ha17] (see (Viehmann), Theorem 5.2). The most important part of the proof in the PEL case is to first answer the third point, in particular to prove that $\mathcal{N}_{[b]}$ is equidimensional and that $\dim \mathcal{N}_{[b]}$ is as conjectured (up to a minor correction) by Rapoport [Ra05, p. 296] – see (Viehmann), Theorem 5.7. An essential tool is the relation

$$\dim \mathcal{N}_{[b]} = \dim X_\mu(b)_{\mathbb{Q}_p} + \dim C_{[b]}, \tag{0.1}$$

where $C_{[b]}$ is the *central leaf* in the Newton stratum and $X_\mu(b)_{\mathbb{Q}_p}$ is the *affine Deligne-Lusztig variety*, which can be identified with the reduced Rapoport-Zink space $\overline{RZ}_{(G,b,\mu)}$. The equation (0.1) reflects the Oort-Mantovan structure of $\mathcal{N}_{[b]}$ as an "quasi-product" of $RZ_{(G,b,\mu)}$ and $C_{[b]}$; see (Mantovan), §5.

The affine Deligne-Lusztig variety has a counterpart in the equal characteristic world, $X_\mu(b)_{\mathbb{F}_p((t))}$, which had been intensively studied by Görtz-Haines-Kottwitz-Reuman [GHKR], Viehmann [Vi06] (for split groups G over $\mathbb{F}_p((t))$) and by Hamacher [Ha15a] (for unramified groups), and also by many others. In particular the cited works completely proved Rapoport's conjectural closed formula for $\dim X_\mu(b)$, when G is unramified but μ is arbitrary (not necessarily minuscule) – see (Viehmann) Theorem 5.13. In [Ha15b, Ha17], Hamacher transported the techniques of these papers over to the p-adic context $X_\mu(b)_{\mathbb{Q}_p}$, and deduced the required dimension formulae for PEL and Hodge type Shimura varieties.

The quasi-product structure of $\mathcal{N}_{[b]}$ is discussed in greater depth in (Mantovan), §5, and is extended to quasi-products of Igusa varieties $\mathrm{Ig}_{m,\mathbb{X}}$ and elements in the truncated Rapoport-Zink tower, denoted in (Mantovan) by $\mathcal{M}^{n,d}_{b,\mathbb{X}}$, where $\mathbb{X}$ is a certain "completely slope divisible" p-divisible group compatible in a certain sense with $\mu_{\bar{\mathbb{Q}}_p}$. The quasi-product structure refers to finite surjective morphisms

$$\mathrm{Ig}_{m,\mathbb{X}} \times \bar{\mathcal{M}}^{n,d}_{b,\mathbb{X}} \to \mathcal{N}_{[b]}(\bar{k}).$$

These structures give rise to Mantovan's formula expressing the cohomology of $\mathcal{N}_{[b]}$ in terms of the cohomology of the corresponding Igusa varieties and Rapoport-Zink spaces, see (Mantovan), §5.3.

Finally, we mention that all of the above questions can be framed and studied for more general subgroups K_p. In particular, the cases where K_p is a parahoric subgroup have attracted a lot of attention in recent years, but unfortunately this is beyond the scope of the present volume and we will not attempt to summarize the progress that has been made in this direction. Let us only mention that Kisin and Pappas [KP] have recently succeeded in constructing integral models S_{K_p} for abelian type Shimura varieties $Sh_{K_p}(G,X)$ in most cases where K_p is parahoric and $p > 2$. Furthermore, it is fully expected that these integral models, while neither regular nor formally smooth, are nevertheless characterized uniquely by a kind of valuative criterion of properness for characteristic $(0,p)$ valuation rings. A careful study of the fine structure of their reductions modulo p has borne some fruit but much work remains to be done. The reader is encouraged to consult the paper of He and Rapoport [HeRa] for a contemporary (group-theoretic) point of view unifying all the various stratifications which are considered in this subject.

C. Automorphic Galois representations

For the moment we make the following special assumptions on $\mathcal{K}$ and the hermitian space V. We assume $\mathcal{K}$ contains an imaginary quadratic field $\mathcal{K}_0$, we fix an embedding $\iota : \mathcal{K}_0 \hookrightarrow \mathbb{C}$, and we let Σ be the set of complex embeddings of $\mathcal{K}$ that coincide with ι upon restriction to $\mathcal{K}_0$. Then Σ is a CM type: it consists of one choice of each pair of complex conjugate embeddings. We choose one $\sigma_0 \in \Sigma$ and we assume the signature of the induced hermitian form on $V_{\sigma_0} = V \otimes_{\mathcal{K},\sigma_0} \mathbb{C}$ is $(1, n-1)$; at the remaining $\sigma \in \Sigma$ we assume the induced signature is $(0, n)$. Finally, we assume $\mathcal{K} \supsetneq \mathcal{K}_0$.

With these choices there is a natural Shimura datum (see Proposition 3.7 of Nicole's Chapter 2, or more explicitly formula (3.2) of Chapter CHL.IV.C of [CHLN]) that defines a complex structure on the variety $Sh(G,X)_{K_f}$ – which is projective, because $\mathcal{K} \neq \mathcal{K}_0$. Moreover, $Sh(G,X)_{K_f}$ has a *canonical model*, as in §5.5 of Chapter 1, over the field $\sigma_0(\mathcal{K}) \subset \mathbb{C}$; and this canonical model, as explained in Chapter 2, is consistent with its interpretation as a PEL moduli space. Finally, at primes p at which the level subgroup K_f is hyperspecial, the PEL problem has a natural smooth integral model, as mentioned above.

The ℓ-adic cohomology of $Sh(G,X)_{K_f}$ thus acquires a canonical action of $Gal(\overline{\mathbb{Q}}/\sigma_0(\mathcal{K})) \xrightarrow{\sim} Gal(\overline{\mathbb{Q}}/\mathcal{K})$ which is unramified at hyperspecial places for K_f. Moreover, suppose π_f is an irreducible smooth representation of $GU(V)(\mathbf{A}_f)$ such that $\pi_f^{K_f} \neq 0$, let $\mathbb{T}(K_f)$ be the convolution algebra of compactly supported functions on $K_f \backslash GU(V)(\mathbf{A}_f)/K_f$, and let

$$H_\ell[\pi_f] = Hom_{\mathbb{T}(K_f)}(\pi_f^{K_f}, H_\ell^\bullet(Sh(G,X)_{K_f,\overline{\mathbb{Q}}}, \mathcal{W}_\rho)) \tag{0.2}$$

denote the $\pi_f^{K_f}$-isotypic subspace for the action of the unramified Hecke algebra relative to K_f. This is a representation space for $Gal(\overline{\mathbb{Q}}/\mathcal{K})$, and the Langlands-Kottwitz method determines the characteristic polynomials of $Frob_{\mathfrak{p}}$, acting on $H_\ell[\pi_f]$, at primes $\mathfrak{p}$ of $\mathcal{K}$ dividing hyperspecial places for K_f. We assume $H_\ell[\pi_f]$ to be non-trivial; then π_f extends to (several) automorphic representation(s) π of $GU(V)(\mathbf{A})$. In favorable situations $H_\ell[\pi_f]$ is non-trivial, concentrated in middle-degree cohomology of $Sh(G,X)_{K_f,\overline{\mathbb{Q}}}$, and $\dim_{\mathbb{Q}_\ell} H_\ell[\pi_f] = n$ – this being the cardinality of the discrete series L-packet for $GU(V)(\mathbb{R})$ (see Chapter A.III.B of [CHLN]). We assume this to be the case for the time being. Then we have associated to π_f, or to π, an n-dimensional representation r_π of $Gal(\overline{\mathbb{Q}}/\mathcal{K})$.

In order to complete our construction of automorphic Galois representations, there remain several tasks.

(i) The representation r_π should be attached to a cuspidal automorphic representation Π of $GL(n)_\mathcal{K}$ on the cohomology of the corresponding locally symmetric space. To pass from π to Π one applies stable base change, as in Chapter L.IV.A of [CHLN]. But we really want to start with Π that satisfies the *polarization condition*

$$\Pi \circ c \xrightarrow{\sim} \Pi^\vee, \tag{0.3}$$

where $c \in Gal(\mathcal{K}/F)$ is complex conjugation, and define an n-dimensional r_Π as r_π for any π whose base change is isomorphic to Π. That is, to descend Π to (a packet of) π, one applies stable base change in reverse. This is not always possible, however. We return to this in point (iii); for the moment we restrict attention to those Π that do descend to our chosen $GU(V)$.

(0) The normalization is not as simple as that described above; one needs to twist by a Galois character to compensate the presence of the similitude factor. If we had been working with unitary groups all along, this would not be an issue; we disregard it for the purposes of this introduction.

(ii) It follows from the Chebotarev density theorem that, up to semisimplification, the isomorphism class of r_π, or r_Π is uniquely determined by what the Langlands-Kottwitz method tells us about the characteristic polynomials of $Frob_\mathfrak{p}$ at unramified places. But what about places v of $\mathcal{K}$ where π_f is ramified? Assume this only happens at v that split in the extension $\mathcal{K}$ of its maximal totally real subfield F; this allows us in particular to define a local component π_v of π_f, even though the group $GU(V)$ is only an algebraic group over $\mathbb{Q}$ (the similitude factor again); moreover π_v can be identified with an irreducible representation of $GL(n, \mathcal{K}_v)$. Then the answer is explained in Shin's Chapter 8: the representation of a decomposition group Γ_v at such a v is the representation associated to π_v by the *local Langlands correspondence*. Shin's method uses v-adic uniformization of $Sh(G, X)_{K_f}$, as in [HT01], where the local Langlands correspondence was defined by nearby cycles on Drinfel'd's tower of local Lubin-Tate moduli spaces. The local structure in Shin's work is equivalent to that in [HT01], but his global computation is considerably more elaborate, because [HT01] was written in a twisted setting for which the subtle Galois cohomological considerations of the stable trace formula are irrelevant.

(iii) Now we start with Π satisfying condition 0.3. The polarization condition guarantees the existence of a (packet of) descent(s) to π on certain groups $GU(V)$. But we want to know that Π descends to $GU(V)$ with our chosen signature. This is possible, provided there are no local obstructions to descent. One sees (as in Chapter L.IV.A of [CHLN]) that the cohomological condition guarantees that there are no local obstructions at archimedean places, and that there are no local obstructions at a finite place v at which $GU(V)$ is quasi-split. In particular, the classification of unitary groups implies that there are no local obstructions anywhere if n is odd. However, if n is even, there may be no hermitian space V with the indicated signatures that is quasi-split at all finite places. See the discussion in Lemmas 1.4.1 and 1.4.2 of Chapter CHL.IV.B of [CHLN].

The best option is then to replace the (even-dimensional) V by an (odd-dimensional) V' and realize the desired Galois representation in the part of the cohomology of $Sh(GU(V'), X')$ (in the obvious notation), obtained by endoscopic transfer from $U(n) \times U(1)$. This strategy was pioneered by Blasius and Rogawski when $n = 2$. For general even n, it was applied in Chapter CHL.IV.C of [CHLN], along with the Langlands-Kottwitz method, to obtain an n-dimensional r_Π – or more precisely, $r_{\Pi,\ell}$, on ℓ-adic cohomology – with the right components at unramified places. The same strategy was applied (independently) in a series of papers by Shin, summarized in his Chapter 8, that determine the local component of $r_{\Pi,\ell}$ at every v not dividing ℓ, using the methods already mentioned in point (ii).

(iv) However, there is another local obstruction, this time at archimedean places, to choosing an auxiliary character of $U(1)$ so that the endoscopic transfer from $U(n) \times U(1)$ contributes an n-dimensional piece in the cohomology. The strategy in point (iii) works when π (or the original local system ρ) satisfies a mild regularity condition, first identified for $n = 2$ by Blasius and Rogawski, now generally known as *Shin-regularity*.

(v) Shin-regularity nevertheless fails in some cases of particular importance for applications; it fails, for example, when Π is unramified at all finite places, when $[F : \mathbb{Q}]$ is even, and when the local system $\mathcal{W}_\rho$ is the constant sheaf. To construct the missing Galois representations, one uses a method of p-adic approximation, due to Chenevier, and explained in his Chapter 10. A similar method was first applied by Wiles to ordinary Hilbert modular forms; Chenevier's construction is based on the *eigenvariety* attached to *totally definite* unitary groups $U(V_0)$, and exploits the fact that the local obstruction to descending Π to $GU(V)$ (or rather to $U(V)$) is sufficient to guarantee its descent to $U(V_0)$. Chenevier's method has the additional merit of guaranteeing that the resulting r_Π are *de Rham* in Fontaine's sense, with the predicted Hodge-Tate weights.

(vi) Along the way we have acquired a few debts that have now come due. We have, for example, assumed that $\mathcal{K}$ contains an imaginary quadratic field, that all ramified places for π_f are split in $\mathcal{K}/F$, and (although we haven't said so) that $\mathcal{K}/F$ is unramified at all finite places. It is possible to reduce to this situation by replacing F and $\mathcal{K}$ by F' and $\mathcal{K}' := \mathcal{K}\cdot F'$ for certain well-chosen totally real quadratic extensions F'/F. Moreover, the eigenvariety argument only works at primes p such that Π_f contains an Iwahori-fixed vector at every divisor $\mathfrak{p}$ of p. We can reduce to this situation by replacing F (and $\mathcal{K}$) by certain well-chosen *solvable* extensions F'/F. In this way we have a lot of n-dimensional representations of the Galois groups $Gal(\overline{\mathbb{Q}}/\mathcal{K}')$. It remains to show that these can be patched together into a single n-dimensional representation of $Gal(\overline{\mathbb{Q}}/\mathcal{K})$. That this is possible is proved in Sorensen's Chapter 11.

The results of this construction were put together in the article [CH]. The main result of [CH] is the existence, for every cohomological Π satisfying (0.3), of a compatible family of semisimple n-dimensional λ-adic representations of $Gal(\overline{\mathbb{Q}}/\mathcal{K})$ such that, (i) at each v prime to λ, the semisimplified local representation at v is related to Π_v by the local Langlands correspondence, and (ii) at v dividing (the residue characteristic of) λ, the local representation at v is de Rham with Hodge-Tate weights determined by the local system $\mathcal{W}_\rho$. In fact, the main result of [CH] is a bit stronger than (i) and (ii): it says that the local representation at v is *crystalline* when Π is unramified at v dividing (the residue characteristic of) λ ; it determines the characteristic polynomial of the crystalline Frobenius operator at such v; and it says a bit more than indicated about the local representations at v prime to λ. The complete compatibility with the local Langlands correspondence, including at $\ell = p$, was obtained by Caraiani in the two papers cited as [Car12] and [Car14] in Chapter 8; see the discussion there.

The current status of the Langlands conjecture on zeta functions of Shimura varieties

Langlands devoted several papers in the 1970s to working out the zeta functions of Shimura varieties in special cases. He stopped short of formulating a general conjecture, however; as Kottwitz put it in [K90], "Langlands was able to predict the contribution of tempered automorphic representations to the zeta function." Here Kottwitz is apparently referring to formula (7.6) of [L79]. We will refer to this formula as "the Langlands conjecture" but, as Langlands was well aware, it was only valid under optimal circumstances. In particular, as we already mentioned, it only applies to L-packets that are tempered; moreover, as Langlands stated explicitly in [L79], it needs to be corrected to take account of endoscopy.

Kottwitz's article [K90] contains the first complete statement of a conjecture that incorporates both endoscopic representations (as in Chapter CHL.IV.C of [CHLN]) and non-tempered representations. This is the last displayed formula in §10 of [K90], and it is worth noting that Kottwitz not only refrains from stating it explicitly as a conjecture but doesn't even dignify the formula with an equation number. Moreover, as noted above, it is doubly conjectural in that it is formulated in the framework of Arthur's conjectural parameters, which are not known to exist in general. Nevertheless, it has served as the basis for the study of zeta functions of Shimura varieties over the last three decades, and the editors of this volume have no doubt that it is correct. For example, the article [Zhu] verifies Kottwitz's conjecture for Shimura varieties attached to orthogonal groups of signature $(n, 2)$. Sections (0.1)-(0.4) of [Zhu] provide an excellent account of the issues encountered in reformulating Kottwitz's conjecture in terms of automorphic representations (as opposed to parameters).

Let (G, X) be a datum defining a Shimura variety $Sh(G, X)$, as in Chapter 1, Definition 4.3.1, or Chapter 2, 2.2. Let $\mu : (\mathbb{G}_m)_{\mathbb{C}} : G_{\mathbb{C}}$ be the corresponding cocharacter; then the *reflex field* $E = E(G, X)$ is the field of definition of its G-conjugacy class. This conjugacy class defines a character of a maximal torus of the Langlands dual group $\hat{G}$. Using this character, Langlands (see also [K84, Lemma 2.1.2]) defines a representation $r_\mu : {}^LG \to GL(V_\mu)$, where ${}^LG = \hat{G} \rtimes Gal(\overline{\mathbb{Q}}/E)$ is the L-group of G over E. The Langlands conjecture is stated in terms of the L-function attached to r_μ. Namely, suppose π is a cuspidal automorphic representation of $G(\mathbf{A}_{\mathbb{Q}})$ that contributes to the cohomology of $Sh(G, X)$.[2] Let d be the dimension of X as a complex manifold. Assuming π_v is tempered for every place v of $\mathbb{Q}$, and a few other favorable hypotheses, then $H_\ell[\pi_f]$, defined as in (0.2), is concentrated in degree d:

$$H_\ell[\pi_f] = Hom_{\mathbb{T}(K_f)}(\pi_f^{K_f}, H^d_\ell(Sh(G, X)_{K_f, \overline{\mathbb{Q}}}, \mathcal{W}_\rho)).$$

Let $K_f \subset G(\mathbf{A}_f)$ be an open compact subgroup such that $\pi_f^{K_f} \neq 0$. Let

$$r_\pi : Gal(\overline{\mathbb{Q}}/E) \to GL(H_\ell[\pi_f])$$

denote the Galois representation on $H_\ell[\pi_f]$.

[2]We assume $Sh(G, X)$ is projective, as above; in general one should replace cohomology by intersection cohomology. Moreover, we recall that discussion applies as well to cohomology with coefficients in local systems.

Conjecture 0.4 (Langlands-Kottwitz). *Let p be a rational prime that is hyperspecial for K_f. The local factor at p of the L-function of r_π – that is, the product over primes $\mathfrak{p}$ of E dividing p – is given by*

$$L_p(s, r_\pi) = L(s - \frac{d}{2}, \pi_p, r_\mu)^{m(\pi)},$$

where $m(\pi)$ is a certain explicit multiplicity (depending on K_f).

When it is known, the identity in Conjecture 0.4 is derived from the formula for the Lefschetz traces of the operators T_a discussed above. The general procedure developed in [K90] for proving the identity [K90] has been applied successively in increasing generality, especially to the PEL-type Shimura varieties attached to inner forms of unitary and symplectic similitude groups: [K92b, HT01, M10, Shi11, M11]; special cases relevant to the construction of automorphic Galois representations were treated in Chapter CHL.IV.3 of [CHLN].

The Shimura varieties of *Hodge type* are those that can be identified with moduli spaces of abelian varieties endowed with certain Hodge classes. Alternatively, they are the Shimura varieties attached to data (G, X) that admit embeddings $(G, X) \hookrightarrow (GSp(2g), \mathfrak{S}_g^{\pm})$, where $\mathfrak{S}_g^{\pm}$ is the union of the Siegel upper and lower halfspaces of genus g. All Shimura varieties of PEL type are also of Hodge type. Shimura varieties of *abelian type* are, roughly speaking, attached to data (G, X) that fit into diagrams of the form

$$(G, X) \overset{b}{\leftarrow} (G_1, X_1) \hookrightarrow (GSp(2g), \mathfrak{S}_g^{\pm})$$

where b is a surjection whose kernel is an algebraic group of multiplicative type. These were classified by Deligne in [D79].

Dong Uk Lee has recently posted a proof [Lee] of Langlands's conjecture for Shimura varieties of Hodge type; as mentioned above, Kisin, Shin, and Zhu have announced a proof of the conjecture for Shimura varieties of abelian type. When they are published, these results will supersede all the special cases treated previously. The remaining cases involve groups G of type E_6 or E_7, or groups G of type D such that $G^{ad}(\mathbb{R})$ is a product of groups of type $D^{\mathbf{H}}$ and $D^{\mathbf{R}}$, in the notation of Table 1.3.9 of [D79]. No integral models have been constructed for the corresponding Shimura varieties, and for the moment Langlands's conjecture is out of reach for these varieties.

Contents of Volume II

The book is divided into three parts, corresponding to the three sections of the description above. We briefly summarize the contents of each chapter.

Part A develops the necessary theory to analyze the unramified local factors of zeta functions of Shimura varieties of PEL type, with special attention to the Shimura varieties attached to unitary similitude groups.

Chapter 1 (Genestier-Ngô) contains the notes of an introductory course on Shimura varieties, presented by the authors at the IHES in 2006. It covers much of the material contained in Milne's expository articles [Mi90, Mi92, Mi05] but with less detail. Its final section describes Kottwitz's parametrization of points on the Siegel modular variety in [K92a]. The chapter's final sentence, written in 2006, anticipates the comparison of Kottwitz's formula for the Siegel modular variety with the Arthur-Selberg formula for the symplectic similitude group; this has since been completed by Morel in [M08].

Chapter 2 (Nicole) develops the explicit theory of Shimura varieties attached to unitary similitude groups: as complex analytic spaces, as quasi-projective complex algebraic varieties, and as moduli spaces of PEL types. Alternatively, the chapter can be seen as an explicit description of the complex points of these moduli spaces as locally symmetric spaces. The chapter concludes with modular and automorphic descriptions of Hecke correspondences, and with an account of the Matsushima-Murakami computation of the cohomology of a compact Shimura variety in terms of automorphic forms.

Chapter 3 (Rozensztajn) continues the description of the PEL moduli spaces of the previous chapter, this time with attention to the integral models at places of good reduction. The moduli problem is shown to be representable by a smooth scheme over $Spec(\mathbb{Z}_{(p)})$ whose generic fiber coincides with the Shimura variety constructed in Chapter 2. Since the methods apply equally to Shimura varieties attached to reductive groups of type A and type C, Chapter 3 explains the construction of both families.

Chapter 4 (Zhu) is a detailed explanation of the Langlands-Kottwitz method, applied to compute the unramified local factors of the zeta functions of the PEL Shimura varieties constructed in the two preceding chapters. More precisely, the chapter carries out steps (i)-(iii) outlined above, introduces the Kottwitz invariant, and stabilizes the resulting expression. For the unitary Shimura varieties that are the subject of this volume, this chapter completes the construction sketched in [K90] and thus verifies the statements used in the construction of automorphic Galois representations in chapters [CHL.IV.B] and [CHL.IV.C] of [CHLN].

Part B reviews most what is currently known about integral canonical models of Shimura varieties with hyperspecial level structure at p, and about the fine structure of their reductions modulo p.

Chapter 5 (Kisin) gives an outline of Kisin's construction of integral canonical models for Hodge type Shimura varieties, and also indicates how the construction can be pushed further to encompass abelian type Shimura varieties. The article works under the assumption $p > 2$, although the article [Ki10] being summarized proves some partial results even for $p = 2$. The article states Kisin's Key Lemma about integrality of certain tensors arising from p-adic comparison isomorphisms, and gives an outline of the argument for how the Key Lemma is used to prove that the candidate integral canonical models are indeed smooth. The article ends with a sketch of the proof of the Key Lemma, using Kisin's earlier results on classification of crystalline representations and p-divisible groups.

Chapter 6 (Mantovan) reivews the classical Dieudonné-Manin theory of F-isocrystals and their Newton polygons, and then explains Kottwitz' group-theoretic generalization to G-isocrystals and the Newton and Kottwitz maps, where G is a connected reductive group over $\mathbb{Q}_p$. Newton strata and Oort foliations are described and the structure of Newton strata as the quasi-product of elements of the Igusa towers and Rapoport-Zink towers is used to explain Mantovan's formula.

Chapter 7 (Viehmann) gives an overview of the state of knowledge about the Newton stratification in PEL and Hodge type Shimura varieties with hyperspecial level structure at p. The closure relations for Newton strata (Grothendieck's conjecture on closures of

Newton strata) are explained to be a consequence of a strengthened version of Oort'a purity theorem, combined with a computation of dimensions of Newton strata. Function field analogues are also discussed. The dimensions of affine Deligne-Lusztig varieties play a key role in establishing the required dimensions of Newton strata, which combined with purity results yields the proof of Grothendieck's conjecture.

Part C brings together several articles on the Galois representations attached in [Shi11, CHLN, CH] to (polarized cohomological) automorphic representations of $GL(n)$ over a CM field.

Chapter 8 (Shin) is an updated version of Shin's notes for a lecture delivered at the Clay Summer School in 2009. It contains a complete account of the construction of the automorphic Galois representations, following the author's previous papers ([Shi11] and its companions). The notes explain how to determine the local ℓ-adic Galois representations at ramified places (prime to ℓ), by extending the methods of Part A to places of bad reduction for the PEL Shimura varieties whose connected components are locally symmetric spaces for the group $U(n-1, 1)$. In this way Shin was able to apply arguments analogous to those developed in [HT01] for the proof of the local Langlands correspondence for $GL(n)$, while allowing for non-trivial endoscopy. This completes the construction outlined in Section C above, up through point (iv) (Shin-regularity).

Chapter 9 (Scholze) contains notes for a talk on Scholze's proof of the local Langlands correspondence between representations of $GL(n, F)$, where F is a p-adic field, and Weil-Deligne parameters for F. While it was based, like that in [HT01], on a study of the nearby cycles on the PEL Shimura varieties attached to twisted inner forms of the unitary similitude groups studied in the present volume, it differs from that proof in two important respects. In the first place, Scholze managed to apply the Langlands-Kottwitz method directly to the Shimura varieties, and therefore entirely avoids the analysis of points on Igusa varieties. Scholze's second innovation – what he describes as his main innovation – was to replace reference to Henniart's numerical correspondence with a geometric proof of a relation between inertia invariants in the local Galois representation and Iwahori-invariants in the corresponding representation of $GL(n, F)$, based on Grothendieck's purity conjecture (proved by Thomason).

The last three chapters prove original results, of independent interest, that have not appeared elsewhere in the literature.

Chapter 10 (Chenevier) applies p-adic analytic geometry to study the variation of automorphic Galois representations as a function of the archimedean parameter of the associated automorphic representation. The methods are those previously developed in [BCh, Bu] to construct p-adic families of automorphic forms and Galois representations, and are quite different from the trace formula methods on which the other chapters in this book are based. The p-adic families are parametrized by *eigenvarieties* – rigid analytic spaces whose points correspond roughly to systems of eigenvalues of Hecke operators – and the methods of this chapter are used to study their p-adic continuity properties. The main result, Theorem 3.3, uses these continuity properties to construct Galois representations attached to automorphic representations that do not satisfy the regularity conditions of Shin's Chapter 8, and of [CHLN]; this is the main step used to complete point (v) in the outline of Section C.

Starting with a (polarized, cohomological) cuspidal automorphic representation of $GL(n)$ over a CM field F, the constructions of the previous chapters attach families of ℓ-adic representations $\rho_{F'}$ of Galois groups of certain finite solvable extensions F' of F. The main theorem of Chapter 11 (Sorensen) provides conditions under which the $\rho_{F'}$ can be patched together into a representation of the absolute Galois group of F itself. The methods go back to work of Blasius and Ramakrishan; they are Galois-theoretic and make no direct reference to automorphic forms. This completes point (vi) in the outline of Section C and, together with the results of the earlier chapters, it provides the basis for the main theorem of [CH].

Chapter 12 (Johansson-Thorne) contains a general characterization of the kinds of Galois representations that should occur in the intersection cohomology of Shimura varieties, assuming standard conjectures – essentially those of Buzzard and Gee in [BG14] – relating Galois representations to automorphic representations with integral infinitesimal parameter. The main conclusion is that these representations are almost always polarized (conjugate self-dual); this conclusion should also apply to the cohomology of Shimura varieties attached to exceptional groups, for which no modular description is yet known. The chapter also constructs subquotients of the ℓ-adic cohomology of Shimura varieties that are not polarized, and extends the main result to the cohomology of open Shimura varieties of abelian type.

Acknowledgements

The introduction to Volume 1 ended with a paragraph thanking the many colleagues whose contributions helped bring it into being. The editors of Volume 2 wish to reaffirm our gratitude to our colleagues and collaborators. Ten years have elapsed since the two week-long programs at the Banff International Research Station at which many of the authors of the chapters spoke, and we again thank the Banff center for hosting the programs. We thank all the authors for participating actively in the project and for making an effort to make their texts accessible to the broadest possible audience. We thank the referees – every chapter that had not previously been published was sent to a referee – for taking the project seriously and for holding the authors to a high standard. We are especially grateful to the two colleagues, one of them not a specialist in the trace formula, who took the time to read the first four chapters as a unit and to help us to maintain the volume's continuity.

Thanks are due to Alain Genestier and Ngô Bao Chau, and to Mark Kisin, for agreeing to submit texts that had already appeared elsewhere; we also thank their publishers – respectively the Société Mathématique de France and the Institut de Mathématiques de Bordeaux – for allowing us to include them here. We also thank Peter Scholze and Sug Woo Shin for accepting our invitation to publish their lecture notes. Finally, we are grateful to International Press for producing Volume 1 to the highest standards, and to Cambridge University Press for agreeing to take up the challenge of completing the project.

References

[A13] Arthur, J., *The endoscopic classification of representations: Orthogonal and symplectic groups, American Mathematical Society Colloquium Publications*, **61**, Providence, RI: American Mathematical Society (2013).

[B] Boxer, G., Torsion in the coherent cohomology of Shimura varieties and Galois representations, `https://arxiv.org/abs/1507.05922`.

[BCh] Bellaïche, J., Chenevier, G. *Families of Galois representations and Selmer groups, Astérisque*, **342** Soc. Math. France, 314 p. (2009).

[Bu] K. Buzzard, Eigenvarieties, in *L-functions and Galois Representations*, Cambridge University Press, Durham (2007), 59–120.

[BG14] Buzzard, K., Gee, T. The conjectural connections between automorphic representa-tions and Galois representations, in *Automorphic forms and Galois representations* Vol. 1, volume 414 of London Math. Soc. Lecture Note Ser. Cambridge Univ. Press, Cambridge (2014) 135–187.

[CH] Chenevier, G., Harris, M., Construction of automorphic Galois representations, II, *Camb. J. Math.*, **1** (2013) 53–73.

[CHLN] Clozel, L., Harris, M., Labesse, J.-P., Ngô, B.-C., *The stable trace formula, Shimura varieties, and arithmetic applications. Volume I: Stabilization of the trace formula*, Boston: International Press (2011) (cited as **Volume 1**).

[D79] Deligne, P. Variétés de Shimura: interprétation modulaire, et techniques de construction de modèles canoniques, in *Proc. Sympos. Pure Math.*, Vol. XXXIII, part II, AMS, Providence, R.I., (1979), pp. 247–289.

[Fa99] G. Faltings: *Integral crystalline cohomology over very ramified valuation rings*, J. Amer. Math. Soc. **12** (1999), 117-144.

[GGP] Gan, W. T., Gross, B. Prasad, D., Symplectic local root numbers, central critical L values, and restriction problems in the representation theory of classical groups. Sur les conjectures de Gross et Prasad. I. *Astérisque*, **346** (2012), 1–109.

[GHKR] U. Görtz, T. Haines, R. Kottwitz and D. Reuman, *Dimensions of some affine Deligne-Lusztig varieties*, Ann. Sci. école Norm. Sup. (4) **39** (2006), no. 3, 467-511.

[Ha15a] P. Hamacher, *The dimension of affine Deligne-Lusztig varieties in the affine Grassmannian*, IMRN **2015**, no. 23, 12804-12839.

[Ha15b] P. Hamacher, *The geometry of Newton strata in the reduction modulo p of Shimura varieties of PEL type*, Duke Math. J. **164** (2015), 2809-2895.

[Ha17] P. Hamacher, *The almost product structure of Newton strata in the deformation space of a Barsotti-Tate group with crystalline Tate tensors*, Math.Ż. **287** (2017), no. 3-4, 1255-1277.

[HT01] Harris, M., Taylor, R., *The geometry and cohomology of some simple Shimura varieties, Annals of Mathematics Studies*, **151**, , Princeton, NJ: Princeton University Press (2001).

[HLTT] Harris, M., Lan, K.W., Taylor, R., Thorne, J., *Res Math Sci* (2016) 3: 37. `https://doi.org/10.1186/s40687-016-0078-5`.

[HeRa] X. He, M. Rapoport, *Stratifications in the reduction of Shimura varieties*, manuscripta math. **152**, (2017), 317-343.

[I67] Ihara, Y., Hecke Polynomials as congruence ζ functions in elliptic modular case, *Ann. of Math.*, **85** (1967) 267–295.

[Ki10] M. Kisin, *Integral models for Shimura varieties of abelian type*, J. Amer. Math. Soc. **23**, no. 4 (2010), 967-1012.

[KP] M. Kisin, G. Pappas, *Integral models of Shimura varieties with parahoric level structure*, Publ. Math. Inst. Hautes tudes Sci. **128** (2018), 121-218.

[KSZ] Kisin, M., Shin, S. W., and Zhu, Y., The stable trace formula for certain Shimura varieties of abelian type, in preparation.

[K84] Kottwitz, R., Shimura varieties and twisted orbital integrals. *Math. Ann.*, **269** (1984) 287–300.

[Ko85] R. Kottwitz, *Isocrystals with additional structure*, Compositio Math. **56**, no. 2, (1985), 201-220.

[K90] Kottwitz, R., Shimura varieties and λ-adic representations, in L. Clozel and J. S. Milne, eds., *Automorphic forms, Shimura varieties, and L-functions, Vol. I (Ann Arbor, MI, 1988), Perspect. Math.*, **10**, Boston, MA: Academic Press (1990) 161–209.

[K92a] Kottwitz, R., Points on some Shimura varieties over finite fields. *J. Amer. Math. Soc.*, **5** (1992) 373–444.

[K92b] Kottwitz, R. On the λ-adic representations associated to some simple Shimura varieties, *Inv. Math.*, **108** (1992) 653–665.

[Lan11] Lan, K.-W. Elevators for degenerations of PEL structures, *Math. Res. Lett.*, **18** (2011) 889–907.

[Lan13] Lan, K.-W.: *Arithmetic compactifications of PEL-type Shimura varieties*, London Mathematical Society Monographs, vol. 36, Princeton University Press, Princeton, (2013).

[L73] Langlands, R. P., Modular forms and ℓ-adic representations, in *Modular functions of one variable, II* (Proc. Internat. Summer School, Univ. Antwerp, Antwerp, 1972), *Lecture Notes in Math.*, **349**, Springer, Berlin (1973) 361–500.

[L79] Langlands, R. P., Automorphic represnetations, Shimura varieties, and motives. Ein Märchen in: *Proc. Sympos. Pure Math.*, Vol. XXXIII, part II, AMS, Providence, R.I., (1979) 205–246.

[Lee] Lee, D. U., Galois gerbs and Lefschetz number formula for Shimura varieties of Hodge type, `https://arxiv.org/abs/1801.03057`.

[Mi90] Milne, J.S., Canonical models of (mixed) Shimura varieties and automorphic vector bundles, in L. Clozel and J. S. Milne, eds., *Automorphic forms, Shimura varieties, and L-functions, Vol. I (Ann Arbor, MI, 1988), Perspect. Math.*, **10**, Boston, MA: Academic Press (1990) 283–414.

[Mi92] Milne, J.S., The points on a Shimura variety modulo a prime of good reduction. in R. P. Langlands and D. Ramakrishnan, eds., *The zeta functions of Picard modular surfaces*, Montreal, QC: Univ. Montreal, (1992) 151–253.

[Mi05] Milne, J.S., Introduction to Shimura varieties. in *Harmonic analysis, the trace formula, and Shimura varieties, Clay Math. Proc., 4*, Providence, RI: Amer. Math. Soc. (2005) 265–378.

[Mo98] B. Moonen, *Models of Shimura varieties in mixed characteristics*, Galois Representations in Arithmetic Algebraic Geometry, ed. A.J. Scholl, R.L. Taylor, London Math. Soc. Lect. Notes Ser. **254**, Cambridge Univ. Press, 1998, pp. 267-350.

[M08] Morel, S., Complexes pondérés sur les compactifications de Baily-Borel. Le cas des variétés de Siegel, *J. Amer. Math. Soc.*, **21** (2008), 23–61.

[M10] Morel, S., *On the cohomology of certain non-compact Shimura varieties.* With an appendix by Robert Kottwitz, *Annals of Mathematics Studies*, **173** Princeton, NJ: Princeton University Press (2010).

[M11] Morel, S., Cohomologie d'intersection des variétés modulaires de Siegel, suite. *Compos. Math.*, **147** (2011), no. 6, 1671–1740.

[Mu65] D. Mumford, *Geometric Invariant Theory*, Springer, Heidelberg, 1965.

[Pera] Madapusi Pera, K., *Toroidal compactifications of integral models of Shimura varieties of Hodge type*, arXiv.1712.03708 [math.AG].

[Pink] Pink, R., Arithmetical compactification of mixed Shimura varieties, *Bonner Math. Schriften*, **209** (1990).

[Ra05] M. Rapoport, *A guide to the reduction modulo p of Shimura varieties*, Astérisque **298** (2005), 271-318.

[RR96] M. Rapoport, M. Richartz, *On the classification and specialization of F-isocrystals with additional structure*, Compositio Math. **103** (1996), 153-181.

[Sch] Scholze, P., On torsion in the cohomology of locally symmetric varieties, *Annals of Math.*, **182** (2015) 945–1066.

[Shi11] Shin, S. W., Galois representations arising from some compact Shimura varieties, *Ann. of Math.*, **173** (2011) 1645–1741.

[Va99] A. Vasiu, *Integral canonical models of Shimura varieties of preabelian type*, Asian J. Math. **3**(2), (1999), 401-518.

[Vi06] E. Viehmann, *The dimension of some affine Deligne-Lusztig varieties*, Ann. Sci. École Norm. Sup. (4) **39** (2006), no. 3, 513-526.

[Zhu] Zhu, Y., The stabilization of the Frobenius–Hecke traces on the intersection cohomology of orthogonal Shimura varieties, `http://front.math.ucdavis.edu/1801.09404`.

LECTURES ON SHIMURA VARIETIES

A. GENESTIER AND B.C. NGÔ

Contents

Abstract

The main goal of these lectures is to explain the representability of the moduli space of abelian varieties with polarizations, endomorphisms and level structures, due to Mumford [GIT], and the description of the set of its points over a finite field, due to Kottwitz [JAMS]. We also try to motivate the general definition of Shimura varieties and their canonical models as in the article of Deligne [Corvallis]. We will leave aside important topics like compactifications, bad reduction and the p-adic uniformization of Shimura varieties.

These are the notes of the lectures on Shimura varieties delivered by one of us at the Asia-French summer school organized at IHES in July 2006. It is basically based on the notes of a course delivered by the two of us at Université Paris-Nord in 2002.

1 Quotients of Siegel's upper half space

1.1 Review on complex tori and abelian varieties

Let V denote a complex vector space of dimension n and U a lattice in V; by definition U is a discrete subgroup of V of rank $2n$. The quotient $X = V/U$ of V by U acting on V by translation, is naturally equipped with a structure of compact complex manifold and a structure of abelian group.

Lemma 1.1.1. *We have canonical isomorphisms from* $\mathrm{H}^r(X,\mathbb{Z})$ *to the group of alternating* r*-forms* $\bigwedge^r U \to \mathbb{Z}$.

Proof. Since $X = V/U$ with V contractible, $\mathrm{H}^1(X,\mathbb{Z}) = \mathrm{Hom}(U,\mathbb{Z})$. The cup-product defines a homomorphism

$$\bigwedge^r \mathrm{H}^1(X,\mathbb{Z}) \to \mathrm{H}^r(X,\mathbb{Z})$$

which is an isomorphism since X is isomorphic to $(S_1)^{2n}$ as real manifolds (where $S_1 = \mathbb{R}/\mathbb{Z}$ is the unit circle). □

Let L be a holomorphic line bundle over the compact complex variety X. Its Chern class $c_1(L) \in \mathrm{H}^2(X,\mathbb{Z})$ is an alternating 2-form on U which can be made explicit as follows. By pulling back L to V by the quotient morphism $\pi : V \to X$, we get a trivial line bundle, since every holomorphic line bundle over a complex vector space is trivial. We choose an isomorphism $\pi^* L \to \mathcal{O}_V$. For every $u \in U$, the canonical isomorphism $u^*\pi^* L \simeq \pi^* L$ gives rise to an automorphism of $\mathcal{O}_V$ which is given by an invertible holomorphic function

$$e_u \in \Gamma(V, \mathcal{O}_V^\times).$$

The collection of these invertible holomorphic functions, for all $u \in U$, satisfies the cocycle equation

$$e_{u+u'}(z) = e_u(z+u')e_{u'}(z).$$

If we write $e_u(z) = e^{2\pi i f_u(z)}$, where $f_u(z)$ are holomorphic functions (well-defined up to a constant in $\mathbb{Z}$) the above cocycle equation is equivalent to

$$F(u_1, u_2) = f_{u_2}(z+u_1) + f_{u_1}(z) - f_{u_1+u_2}(z) \in \mathbb{Z}.$$

The Chern class

$$c_1 : \mathrm{H}^1(X, \mathcal{O}_X^\times) \to \mathrm{H}^2(X, \mathbb{Z})$$

sends the class of L in $\mathrm{H}^1(X, \mathcal{O}_X^\times)$ to the element $c_1(L) \in \mathrm{H}^2(X, \mathbb{Z})$, whose corresponding 2-form $E : \bigwedge^2 U \to \mathbb{Z}$ is given by

$$(u_1, u_2) \mapsto E(u_1, u_2) := F(u_1, u_2) - F(u_2, u_1).$$

Lemma 1.1.2. *The Neron-Severi group* $\mathrm{NS}(X)$*, defined as the image of* $c_1 : \mathrm{H}^1(X, \mathcal{O}_X^\times) \to \mathrm{H}^2(X, \mathbb{Z})$*, consists of the alternating 2-forms* $E : \bigwedge^2 U \to \mathbb{Z}$ *satisfying the equation*

$$E(iu_1, iu_2) = E(u_1, u_2);$$

here E still denotes the alternating 2-form extended to $U \otimes_{\mathbb{Z}} \mathbb{R} = V$ by $\mathbb{R}$-linearity.

Proof. The short exact sequence

$$0 \to \mathbb{Z} \to \mathcal{O}_X^\times \to \mathcal{O}_X \to 0$$

induces a long exact sequence which contains

$$\mathrm{H}^1(X, \mathcal{O}_X^\times) \to \mathrm{H}^2(X, \mathbb{Z}) \to \mathrm{H}^2(X, \mathcal{O}_X).$$

It follows that the Neron-Severi group is the kernel of the map

$$\mathrm{H}^2(X, \mathbb{Z}) \to \mathrm{H}^2(X, \mathcal{O}_X).$$

This map is the composition of the obvious maps

$$\mathrm{H}^2(X, \mathbb{Z}) \to \mathrm{H}^2(X, \mathbb{C}) \to \mathrm{H}^2(X, \mathcal{O}_X).$$

The Hodge decomposition

$$\mathrm{H}^m(X, \mathbb{C}) = \bigoplus_{p+q=m} \mathrm{H}^p(X, \Omega_X^q),$$

where Ω_X^q is the sheaf of holomorphic q-forms on X, can be made explicit [14, page 4]. For $m = 1$, we have

$$\mathrm{H}^1(X, \mathbb{C}) = V_{\mathbb{R}}^* \otimes_{\mathbb{R}} \mathbb{C} = V_{\mathbb{C}}^* \oplus \overline{V}_{\mathbb{C}}^*,$$

where $V^*_{\mathbb{C}}$ is the space of $\mathbb{C}$-linear maps $V \to \mathbb{C}$, $\overline{V}^*_{\mathbb{C}}$ is the space of conjugate $\mathbb{C}$-linear maps and $V^*_{\mathbb{R}}$ is the space of $\mathbb{R}$-linear maps $V \to \mathbb{R}$. There is a canonical isomorphism $\mathrm{H}^0(X, \Omega^1_X) = V^*_{\mathbb{C}}$ defined by evaluating a holomorphic 1-form on X on the tangent space V of X at the origin. There is also a canonical isomorphism $\mathrm{H}^1(X, \mathcal{O}_X) = \overline{V}^*_{\mathbb{C}}$.

By taking $\bigwedge^2$ on both sides, the Hodge decomposition of $\mathrm{H}^2(X, \mathbb{C})$ can also be made explicit. We have $\mathrm{H}^2(X, \mathcal{O}_X) = \bigwedge^2 \overline{V}^*_{\mathbb{C}}$, $\mathrm{H}^1(X, \Omega^1_X) = V^*_{\mathbb{C}} \otimes \overline{V}^*_{\mathbb{C}}$ and $\mathrm{H}^0(X, \Omega^2_X) = \bigwedge^2 V^*_{\mathbb{C}}$. It follows that the map $\mathrm{H}^2(X, \mathbb{Z}) \to \mathrm{H}^2(X, \mathcal{O}_X)$ is the obvious map $\bigwedge^2 U^*_{\mathbb{Z}} \to \bigwedge^2 V^*_{\mathbb{C}}$. Its kernel is precisely the set of integral 2-forms E on U which satisfy the relation $E(iu_1, iu_2) = E(u_1, u_2)$ (when they are extended to V by $\mathbb{R}$-linearity). □

Let $E : \bigwedge^2 U \to \mathbb{Z}$ be an integral alternating 2-form on U that satisfies $E(iu_1, iu_2) = E(u_1, u_2)$ after extension to V by $\mathbb{R}$-linearity. The real 2-form E on V defines a Hermitian form λ on the $\mathbb{C}$-vector space V by

$$\lambda(x, y) = E(ix, y) + iE(x, y);$$

this in turn determines E by the relation $E = \mathrm{Im}(\lambda)$. The Neron-Severi group $\mathrm{NS}(X)$ can be described in yet another way as the group of the Hermitian forms λ on the $\mathbb{C}$-vector space V having an imaginary part which takes integral values on U.

Theorem 1.1.3 (Appell-Humbert). *Isomorphy classes of holomorphic line bundles on $X = V/U$ correspond bijectively to pairs (λ, α), where*

- *$\lambda \in NS(X)$ is a Hermitian form on V such that its imaginary part takes integral values on U;*
- *$\alpha : U \to S_1$ is a map from U to the unit circle S_1 satisfying the equation*

$$\alpha(u_1 + u_2) = e^{i\pi \mathrm{Im}(\lambda)(u_1, u_2)} \alpha(u_1)\alpha(u_2).$$

For every (λ, α) as above, the line bundle $L(\lambda, \alpha)$ is given by the Appell-Humbert cocycle

$$e_u(z) = \alpha(u) e^{\pi\lambda(z,u) + \frac{1}{2}\pi\lambda(u,u)}.$$

Let $\mathrm{Pic}(X)$ be the abelian group consisting of the isomorphism classes of line bundles on X and let $\mathrm{Pic}^0(X) \subset \mathrm{Pic}(X)$ be the kernel of the Chern class. We have an exact sequence :

$$0 \to \mathrm{Pic}^0(X) \to \mathrm{Pic}(X) \to \mathrm{NS}(X) \to 0.$$

Let us also write: $\hat{X} = \mathrm{Pic}^0(X)$; it is the group consisting of characters $\alpha : U \to S_1$ from U to the unit circle S_1. Let $V^*_{\mathbb{R}} = \mathrm{Hom}_{\mathbb{R}}(V, \mathbb{R})$. There is a homomorphism $V^*_{\mathbb{R}} \to \hat{X}$ sending $v^* \in V^*_{\mathbb{R}}$ to the line bundle $L(0, \alpha)$, where $\alpha : U \to S_1$ is the character

$$\alpha(u) = \exp(2i\pi\langle u, v^* \rangle).$$

This induces an isomorphism $V^*_{\mathbb{R}}/U^* \to \hat{X}$, where

$$U^* = \{u^* \in V^*_{\mathbb{R}} \text{ such that } \forall u \in U,\ \langle u, u^* \rangle \in \mathbb{Z}\}.$$

We can identify $\overline{V}_{\mathbb{C}}^*$ with the $\mathbb{R}$-dual $V_{\mathbb{R}}^*$ by the $\mathbb{R}$-linear bijection sending a semi-linear f to its imaginary part g (f can be recovered from g: use the formula $f(v) = -g(iv)+ig(v)$). This gives $\hat{X} = \overline{V}_{\mathbb{C}}^*/U^*$ a structure of complex torus; it is called *the dual complex torus of* X. With respect to this complex structure, the universal line bundle over $X \times \hat{X}$ given by Appell-Humbert formula is a holomorphic line bundle.

A Hermitian form on V induces a $\mathbb{C}$-linear map $V \to \overline{V}_{\mathbb{C}}^*$. If moreover its imaginary part takes integral values in U, the linear map $V \to \overline{V}_{\mathbb{C}}^*$ takes U into U^* and therefore induces a homomorphism $\lambda : X \to \hat{X}$ which is symmetric (i.e. such that $\hat{\lambda} = \lambda$ with respect to the obvious identification $X \simeq \hat{\hat{X}}$). In this way, we identify the Neron-Severi group $\mathrm{NS}(X)$ with the group of symmetric homomorphisms from X to $\hat{X}$.

Let (λ, α) be as in the theorem and let $\theta \in \mathrm{H}^0(X, L(\lambda, \alpha))$ be a global section of $L(\lambda, \alpha)$. Pulled back to V, θ becomes a holomorphic function on V which satisfies the equation

$$\theta(z+u) = e_u(z)\theta(z) = \alpha(u)e^{\pi\lambda(z,u)+\frac{1}{2}\pi\lambda(u,u)}\theta(z).$$

Such a function is called a *theta-function* with respect to the hermitian form λ and the multiplicator α. The Hermitian form λ needs to be positive definite for $L(\lambda, \alpha)$ to have a lot of sections, see [14, §3].

Theorem 1.1.4. *The line bundle $L(\lambda, \alpha)$ is ample if and only if the Hermitian form H is positive definite. In that case,*

$$\dim \mathrm{H}^0(X, L(\lambda, \alpha)) = \sqrt{\det(E)}.$$

Consider the case where H is degenerate. Let W be the kernel of H or of E, i.e.

$$W = \{x \in V | E(x, y) = 0, \forall y \in V\}.$$

Since E is integral on $U \times U$, $W \cap U$ is a lattice of W. In particular, $W/W \cap U$ is compact. For any $x \in X$, $u \in W \cap U$, we have

$$|\theta(x+u)| = |\theta(x)|$$

for all $d \in \mathbb{N}, \theta \in \mathrm{H}^0(X, L(\lambda, \alpha)^{\otimes d})$. By the maximum principle, it follows that θ is constant on the cosets of X modulo W and therefore $L(\lambda, \alpha)$ is not ample. A similar argument shows that if H is not positive definite, $L(H, \alpha)$ can not be ample, see [14, p.26].

If the Hermitian form H is positive definite, then the equality

$$\dim \mathrm{H}^0(X, L(\lambda, \alpha)) = \sqrt{\det(E)}$$

holds. In [14, p.27], Mumford shows how to construct a basis of the vector space $\mathrm{H}^0(X, L(\lambda, \alpha))$, well-defined up to a scalar, after choosing a sublattice $U' \subset U$ of rank n which is Lagrangian with respect to the symplectic form E and such that $U' = U \cap \mathbb{R}U'$. Based on the equality $\dim \mathrm{H}^0(X, L(\lambda, \alpha)^{\otimes d}) = d^n \sqrt{\det(E)}$, one can prove that $L(\lambda, \alpha)^{\otimes 3}$ gives rise to a projective embedding of X for any positive definite Hermitian form λ. See Theorem 2.2.3 for a more complete statement. □

Definition 1.1.5. (1) An abelian variety is a complex torus that can be embedded into a projective space.

(2) A polarization of an abelian variety $X = V/U$ is an alternating form $\lambda : \bigwedge^2 U \to \mathbb{Z}$ which is the Chern class of an ample line bundle.

With a suitable choice of a basis of U, λ can be represented by a matrix

$$E = \begin{pmatrix} 0 & D \\ -D & 0 \end{pmatrix},$$

where D is a diagonal matrix $D = (d_1, \ldots, d_n)$ for some non-negative integers $d_1, \ldots, d_n$ such that $d_1|d_2| \ldots |d_n$. The form E is non-degenerate if these integers are nonzero. We call $D = (d_1, \ldots, d_n)$ the *type of the polarization* λ. A polarization is called *principal* if its type is $(1, \ldots, 1)$.

Corollary 1.1.6 (Riemann). *A complex torus $X = V/U$ can be embedded as a closed complex submanifold into a projective space if and only if there exists a positive definite hermitian form λ on V such that the restriction of its imaginary part to U is a (symplectic) 2-form with integral values.*

Let us rewrite Riemann's theorem in term of matrices. We choose a $\mathbb{C}$-basis $e_1, \ldots, e_n$ for V and a $\mathbb{Z}$-basis $u_1, \ldots, u_{2n}$ of U. Let Π be the $n \times 2n$-matrix $\Pi = (\lambda_{ji})$ with $u_i = \sum_{j=1}^n \lambda_{ji} e_j$ for all $i = 1, \ldots, 2n$. Π is called *the period matrix*. Since $\lambda_1, \ldots, \lambda_{2n}$ form an $\mathbb{R}$-basis of V, the $2n \times 2n$-matrix $\begin{pmatrix} \Pi \\ \overline{\Pi} \end{pmatrix}$ is invertible. The alternating form $E : \bigwedge^2 U \to \mathbb{Z}$ is represented by an alternating matrix, also denoted by E, with respect to the $\mathbb{Z}$-basis $u_1, \ldots, u_{2n}$. The form $\lambda : V \times V \to \mathbb{C}$ given by $\lambda(x, y) = E(ix, y) + iE(x, y)$ is hermitian if and only if $\Pi E^{-1}\,{}^t\Pi = 0$. When this condition is satisfied, the Hermitian form λ is positive definite if and only if the symmetric matrix $i\Pi E^{-1}\,{}^t\overline{\Pi}$ is positive definite.

Corollary 1.1.7. *A complex torus $X = V/U$ defined by a period matrix Π is an abelian variety if and only if there is a nondegenerate alternating integral $2n \times 2n$ matrix E such that*

(1) $\Pi E^{-1}\,{}^t\Pi = 0$,

(2) $i\Pi E^{-1}\,{}^t\overline{\Pi} > 0$.

1.2 Quotients of the Siegel upper half space

Let X be an abelian variety of dimension n over $\mathbb{C}$ and let E be a polarization of X of type $D = (d_1, \ldots, d_n)$. There exists a basis $u_1, \ldots, u_n, v_1, \ldots, v_n$ of $\mathrm{H}_1(X, \mathbb{Z})$ with respect to which the matrix of E is of the form

$$E = \begin{pmatrix} 0 & D \\ -D & 0 \end{pmatrix}.$$

A datum $(X, E, (u_\bullet, v_\bullet))$ is called a *polarized abelian variety of type D with symplectic basis*. We are going to describe the moduli of polarized abelian varieties of type D with symplectic basis.

The Lie algebra V of X is an n-dimensional $\mathbb{C}$-vector space equipped with a lattice $U = \mathrm{H}_1(X, \mathbb{Z})$. Choose a $\mathbb{C}$-basis $e_1, \ldots, e_n$ of V. The vectors $e_1, \ldots, e_n, ie_1, \ldots, ie_n$ form an $\mathbb{R}$-basis of V. The isomorphism $\Pi_{\mathbb{R}} : U \otimes \mathbb{R} \to V$ is given by an invertible real $2n \times 2n$-matrix

$$\Pi_{\mathbb{R}} = \begin{pmatrix} \Pi_{11} & \Pi_{12} \\ \Pi_{21} & \Pi_{22} \end{pmatrix}.$$

The complex $n \times 2n$-matrix $\Pi = (\Pi_1, \Pi_2)$ is related to $\Pi_{\mathbb{R}}$ by the relations $\Pi_1 = \Pi_{11} + i\Pi_{21}$ and $\Pi_2 = \Pi_{12} + i\Pi_{22}$.

Lemma 1.2.1. *The set of polarized abelian varieties of type D with symplectic basis is canonically in bijection with the set of $GL_{\mathbb{C}}(V)$ orbits of isomorphisms of real vector spaces $\Pi_{\mathbb{R}} : U \otimes \mathbb{R} \to V$ such that for all $x, y \in V$, we have $E(\Pi_{\mathbb{R}}^{-1} ix, \Pi_{\mathbb{R}}^{-1} iy) = E(\Pi_{\mathbb{R}}^{-1} x, \Pi_{\mathbb{R}}^{-1} y)$ and such that the symmetric form $E(\Pi_{\mathbb{R}}^{-1} ix, \Pi_{\mathbb{R}}^{-1} y)$ is positive definite.*

There are at least two methods to describe this quotient. The first one is more concrete but the second one is more suitable for generalization.

In each $\mathrm{GL}_{\mathbb{C}}(V)$ orbit, there exists a unique $\Pi_{\mathbb{R}}$ such that $\Pi_{\mathbb{R}}^{-1} e_i = \frac{1}{d_i} v_i$ for $i = 1, \ldots, n$. Thus, the matrix $\Pi_{\mathbb{R}}$ has the form

$$\Pi_{\mathbb{R}} = \begin{pmatrix} \Pi_{11} & D \\ \Pi_{21} & 0 \end{pmatrix}$$

and Π has the form $\Pi = (Z, D)$, with

$$Z = \Pi_{11} + i\Pi_{21} \in M_n(\mathbb{C})$$

satisfying ${}^tZ = Z$ and $\mathrm{im}(Z) > 0$.

Proposition 1.2.2. *There is a canonical bijection from the set of polarized abelian varieties of type D with symplectic basis to the Siegel upper half-space*

$$\mathfrak{H}_n = \{Z \in M_n(\mathbb{C}) | {}^tZ = Z, \mathrm{im}(Z) > 0\}.$$

On the other hand, an isomorphism $\Pi_{\mathbb{R}} : U \otimes \mathbb{R} \to V$ defines a cocharacter $h : \mathbb{C}^\times \to \mathrm{GL}(U \otimes \mathbb{R})$ by transporting the complex structure of V to $U \otimes \mathbb{R}$. It follows from the relation $E(\Pi_{\mathbb{R}}^{-1} ix, \Pi_{\mathbb{R}}^{-1} iy) = E(\Pi_{\mathbb{R}}^{-1} x, \Pi_{\mathbb{R}}^{-1} y)$ that the restriction of h to the unit circle S_1 defines a homomorphism $h_1 : S_1 \to \mathrm{Sp}_{\mathbb{R}}(U, E)$. Moreover, the $\mathrm{GL}_{\mathbb{C}}(V)$-orbit of $\Pi_{\mathbb{R}} : U \otimes \mathbb{R} \to V$ is determined by the induced homomorphism $h_1 : S_1 \to \mathrm{Sp}_{\mathbb{R}}(U, E)$.

Proposition 1.2.3. *There is a canonical bijection from the set of polarized abelian varieties of type D with symplectic basis to the set of homomorphisms of real algebraic groups $h_1 : S_1 \to \mathrm{Sp}_{\mathbb{R}}(U, E)$ such that the following conditions are satisfied:*

(1) *the complexification $h_{1,\mathbb{C}} : \mathbb{G}_m \to \mathrm{Sp}(U \otimes \mathbb{C})$ gives rise to a decomposition as the direct sum of n-dimensional vector subspaces*

$$U \otimes \mathbb{C} = (U \otimes \mathbb{C})_+ \oplus (U \otimes \mathbb{C})_-$$

of weights $+1$ and -1;

(2) *the symmetric form $E(h_1(i)x, y)$ is positive definite.*

This set is a homogenous space under the action of $\mathrm{Sp}(U \otimes \mathbb{R})$ *acting by inner automorphisms.*

Let Sp_D be the $\mathbb{Z}$-algebraic group of automorphisms of the symplectic form E of type D. The discrete group $\mathrm{Sp}_D(\mathbb{Z})$ acts simply transitively on the set of symplectic bases of U.

Proposition 1.2.4. *There is a canonical bijection between the set of isomorphism classes of polarized abelian varieties of type D and the quotient* $\mathrm{Sp}_D(\mathbb{Z})\backslash\mathfrak{H}_n$.

According to H. Cartan, there is a way to give this quotient an analytic structure and then to prove that the quotient "is" indeed a quasi-projective normal variety over $\mathbb{C}$ (more precisely, that it can be endowed with a canonical embedding into a complex projective space with an image whose closure is a projective normal variety).

1.3 Torsion points and level structures

Let $X = V/U$ be an abelian variety of dimension n. For every integer N, the group of N-torsion points $X[N] = \{x \in X | Nx = 0\}$ can be identified with the finite group $N^{-1}U/U$, which is isomorphic to $(\mathbb{Z}/N\mathbb{Z})^{2n}$. Let E be a polarization of X of type $D = (d_1, \ldots, d_n)$ with $(d_n, N) = 1$. The alternating form $E : \bigwedge^2 U \to \mathbb{Z}$ can be extended to a non-degenerate symplectic form on $U \otimes \mathbb{Q}$. The Weil pairing

$$(\alpha, \beta) \mapsto \exp(2i\pi N E(\alpha, \beta))$$

defines a symplectic non-degenerate form

$$e_N : X[N] \times X[N] \to \mu_N,$$

where μ_N is the group of N-th roots of unity, provided that N is relatively prime to d_n. Let us *choose a primitive N-th root of unity*, so that the Weil pairing takes values in $\mathbb{Z}/N\mathbb{Z}$.

Definition 1.3.1. Let N be an integer relatively prime to d_n. A principal N-level structure of an abelian variety X with a polarization E is an isomorphism from the symplectic module $X[N]$ to the standard symplectic module $(\mathbb{Z}/N\mathbb{Z})^{2n}$ given by the matrix

$$J = \begin{pmatrix} 0 & I_n \\ -I_n & 0 \end{pmatrix},$$

where I_n is the identity $n \times n$-matrix.

Let $\Gamma(N)$ be the subgroup of $\mathrm{Sp}_D(\mathbb{Z})$ consisting of the automorphisms of (U, E) which induce the trivial action on U/NU.

Proposition 1.3.2. *There is a natural bijection between the set of isomorphism classes of polarized abelian varieties of type D equipped with principal N-level structures and the quotient* $\mathcal{A}^0_{n,N} = \Gamma_A(N)\backslash\mathfrak{H}_n$.

For $N \geq 3$, the group $\Gamma(N)$ does not contain torsion elements and acts freely on the Siegel half-space $\mathfrak{H}_n$. The quotient $\mathcal{A}^0_{n,N}$ is therefore a smooth complex analytic space.

2 The moduli space of polarized abelian schemes

2.1 Polarizations of abelian schemes

Definition 2.1.1. An abelian scheme over a scheme S is a smooth proper group scheme with connected geometric fibers. Being a group scheme, X is equipped with the following structures:

(1) a unit section $e_X : S \to X$;

(2) a multiplication morphism $X \times_S X \to X$;

(3) an inverse morphism $X \to X$,

such that the usual axioms for abstract groups hold.

Recall the following classical rigidity lemma.

Lemma 2.1.2. *Let X and X' be two abelian schemes over S. Let $\alpha : X \to X'$ be a morphism that sends the unit section of X to the unit section of X'. Then α is a homomorphism.*

Proof. We summarize the proof when S is a point. Consider the map $\beta : X \times X \to X'$ given by

$$\beta(x_1, x_2) = \alpha(x_1 x_2)\alpha(x_1)^{-1}\alpha(x_2)^{-1}.$$

We have $\beta(e_X, x) = e_{X'}$ for all $x \in X$. For any affine neighborhood U' of $e_{X'}$ in X', there exists an affine neighborhood U of e_X such that $\beta(U \times X) \subset U'$. For every $u \in U$, β maps the proper scheme $u \times X$ into the affine U'. It follows that the restriction of β to $u \times X$ is constant. Since $\beta(u, e_X) = e_{X'}$, $\beta(u, x) = e_{X'}$ for any $x \in X$. It follows that $\beta(u, x) = e_{X'}$ for any $u, x \in X$ since X is irreducible. □

Let us mention two useful consequences of the rigidity lemma. Firstly, the abelian scheme is necessarily commutative since the inverse morphism $X \to X$ is a homomorphism. Secondly, given the unit section, a smooth proper scheme can have *at most one* abelian scheme structure. It suffices to apply the rigidity lemma to the identity of X.

An *isogeny* $\alpha : X \to X'$ is a surjective homomorphism whose kernel $\ker(\alpha)$ is a finite group scheme over S. Let d be a positive integer. Let S be a scheme such that all its residue characteristics are relatively prime to d. Let $\alpha : X \to X'$ be an isogeny of degree d and let $K(\alpha)$ be the kernel of α. For every geometric point $\bar{s} \in S$, $K(\alpha)_{\bar{s}}$ is a discrete group isomorphic to $\mathbb{Z}/d_1\mathbb{Z} \times \cdots \times \mathbb{Z}/d_n\mathbb{Z}$ with $d_1 | \cdots | d_n$ and $d_1 \ldots d_n = d$. The function defined on the underlying topological space $|S|$ of S which maps a point $s \in |S|$ to the type of $K(\alpha)_{\bar{s}}$ for any geometric point $\bar{s}$ over s is a locally constant function. So it makes sense to talk about the type of an isogeny of degree prime to all residue characteristics.

Let X be an abelian scheme over S. Consider the functor $\mathrm{Pic}_{X/S}$ from the category of S-schemes to the category of abelian groups which assigns to every S-scheme T the group of isomorphism classes of (L, ι), where L is an invertible sheaf on $X \times_S T$ and ι is a trivialization $e_X^* L \simeq \mathcal{O}_T$ along the unit section. See [2, p.234] for the following theorem.

Theorem 2.1.3. *Let X be a projective abelian scheme over S. Then the functor* $\mathrm{Pic}_{X/S}$ *is representable by a smooth separated S-scheme which is locally of finite presentation over S.*

The smooth scheme $\mathrm{Pic}_{X/S}$ equipped with the unit section corresponding to the trivial line bundle $\mathcal{O}_X$ admits a neutral component $\mathrm{Pic}^0_{X/S}$ which is an abelian scheme over S.

Definition 2.1.4. Let X/S be a projective abelian scheme. The *dual abelian scheme* $\hat{X}/S$ is the neutral component $\mathrm{Pic}^0(X/S)$ of the Picard functor $\mathrm{Pic}_{X/S}$. The *Poincaré sheaf P* is the restriction of the universal invertible sheaf on $X \times_S \mathrm{Pic}_{X/S}$ to $X \times_S \hat{X}$.

For every abelian scheme X/S, its bidual abelian scheme (i.e. the dual abelian scheme of $\hat{X}/S$) is canonically identified to X/S. For every homomorphism $\alpha : X \to X'$, we have a homomorphism $\hat{\alpha} : \hat{X}' \to \hat{X}$. If α is an isogeny, the same is true for $\hat{\alpha}$. A homomorphism $\alpha : X \to \hat{X}$ is called *symmetric* if the equality $\alpha = \hat{\alpha}$ holds.

Lemma 2.1.5. *Let $\alpha : X \to Y$ be an isogeny and let $\hat{\alpha} : \hat{Y} \to \hat{X}$ be the dual isogeny. There is a canonical perfect pairing*

$$\ker(\alpha) \times \ker(\hat{\alpha}) \to \mathbb{G}_m.$$

Proof. Let $\hat{y} \in \ker(\hat{\alpha})$ and let $L_{\hat{y}}$ be the corresponding line bundle on Y with a trivialization along the unit section. Pulling it back to X, we get the trivial line bundle equipped with yet another trivialization on $\ker(\alpha)$. The difference between the two trivializations gives rise to a homomorphism $\ker(\alpha) \to \mathbb{G}_m$ which defines the desired pairing. It is not difficult to check that this pairing is perfect, see [14, p.143]. □

Let $L \in \mathrm{Pic}_{X/S}$ be an invertible sheaf over X with trivialized neutral fibre $L_e = 1$. For any point $x \in X$ over $s \in S$, let $T_x : X_s \to X_s$ be the translation by x. The invertible sheaf $T_x^* L \otimes L^{-1} \otimes L_x^{-1}$ has trivialized neutral fibre

$$(T_x^* L \otimes L^{-1} \otimes L_x^{-1})_e = L_x \otimes L_e^{-1} \otimes L_x^{-1} = 1,$$

so, L defines a morphism $\lambda_L : X \to \mathrm{Pic}_{X/S}$. Since the fibres of X are connected, λ_L factors through the dual abelian scheme $\hat{X}$ and gives rise to a morphism

$$\lambda_L : X \to \hat{X}.$$

Since λ_L sends the unit section of X to the unit section of $\hat{X}$, the morphism of schemes λ_L is necessarily a homomorphism of abelian schemes. Let us denote by $K(L)$ the kernel of λ_L.

Lemma 2.1.6. *For every line bundle L on X with a trivialization along the unit section, the homomorphism $\lambda_L : X \to \hat{X}$ is symmetric. If moreover, $L = \hat{x}^* P$ for some section $\hat{x} : S \to \hat{X}$, then $\lambda_L = 0$.*

Proof. By construction, the homomorphism $\lambda_L : X \to \hat{X}$ represents the line bundle $m^*L \otimes p_1^*L^{-1} \otimes p_2^*L^{-1}$ on $X \times X$, where m is the multiplication and p_1, p_2 are projections, equipped with the obvious trivialization along the unit section. As this line bundle is symmetric, the homomorphism λ_L is symmetric.

If $L = \mathcal{O}_X$ with the obvious trivialization along the unit section, it is immediate that $\lambda_L = 0$. Now for any $L = \hat{x}^*P$, L can be deformed continuously to the trivial line bundle and it follows that $\lambda_L = 0$. In order to make the argument rigorous, one can consider the universal family over $\hat{X}$ and apply the rigidity lemma. □

Definition 2.1.7. A line bundle L over an abelian scheme X equipped with a trivialization along the unit section is called non-degenerate if $\lambda_L : X \to \hat{X}$ is an isogeny.

In the case where the base S is $\mathrm{Spec}(\mathbb{C})$ and $X = V/U$, L is non-degenerate if and only if the associated Hermitian form on V is non-degenerate.

Let L be a non-degenerate line bundle on X with a trivialization along the unit section. The canonical pairing $K(L) \times K(L) \to \mathbb{G}_{m,S}$ is then symplectic. Assume S is connected with residue characteristics prime to the degree of λ_L. Then there exist $d_1 | \ldots | d_s$ such that for every geometric point $\bar{s} \in S$ the abelian group $K(L)_{\bar{s}}$ is isomorphic to $(\mathbb{Z}/d_1\mathbb{Z} \times \cdots \times \mathbb{Z}/d_n\mathbb{Z})^2$. We call $D = (d_1, \ldots, d_n)$ the type of the polarization λ

Definition 2.1.8. Let X/S be an abelian scheme. A *polarization* of X/S is a symmetric isogeny $\lambda : X \to \hat{X}$ which, locally for the étale topology of S, is of the form λ_L for some ample line bundle L of X/S.

In order to make this definition workable, we will need to recall basic facts about cohomology of line bundles on abelian varieties. See corollary 2.2.4 in the next paragraph.

2.2 Cohomology of line bundles on abelian varieties

We are going to recall some known facts about the cohomology of line bundles on abelian varieties. For the proofs, see [14, p.150]. Let X be an abelian variety over a field k. Let

$$\chi(L) = \sum_{i \in \mathbb{Z}} \dim_k \mathrm{H}^i(X, L)$$

be the Euler characteristic of L.

Theorem 2.2.1 (Riemann-Roch theorem). *For any line bundle L on X, if $L = \mathcal{O}_X(D)$ for a divisor D, we have*

$$\chi(L) = \frac{(D^g)}{g!},$$

where (D^g) is the g-fold self-intersection number of D.

Theorem 2.2.2 (Mumford's vanishing theorem). *Let L be a line bundle on X such that $K(L)$ is finite. There exists a unique integer $i = i(L)$ with $0 \leq i \leq n = \dim(X)$ such that*

$\mathrm{H}^j(X, L) = 0$ *for* $j \neq i$ *and* $\mathrm{H}^i(X, L) \neq 0$. *Moreover, L is ample if and only if* $i(L) = 0$. *For every* $m \geq 1$, $i(L^{\otimes m}) = i(L)$.

Assume $S = \mathrm{Spec}(\mathbb{C})$, $X = V/U$ with $V = \mathrm{Lie}(X)$ and U a lattice in V. Then the Chern class of L corresponds to a Hermitian form H and the integer $i(L)$ is the number of negative eigenvalues of H.

Theorem 2.2.3. *For any ample line bundle L on an abelian variety X, the line bundle* $L^{\otimes m}$ *is base-point free if* $m \geq 2$ *and it is very ample if* $m \geq 3$.

Since L is ample, $i(L) = 0$ and consequently $\dim_k \mathrm{H}^0(X, L) = \chi(L) > 0$. There exists an effective divisor D such that $L \simeq \mathcal{O}_X(D)$. Since $\lambda_L : X \to \hat{X}$ is a homomorphism, the divisor $T_x^*(D) + T_{-x}^*(D)$ is linearly equivalent to $2D$ and $T_x^*(D) + T_y^*(D) + T_{-x-y}^*(D)$ is linearly equivalent to $3D$. By moving $x, y \in X$ we get a lot of divisors linearly equivalent to $2D$ and to $3D$. The proof is based on this fact and on the formula for the dimension of $\mathrm{H}^0(X, L^{\otimes m})$. For a detailed proof, see [14, p.163]. □

Corollary 2.2.4. *Let* $X \to S$ *be an abelian scheme over a connected base and let L be an invertible sheaf on X such that* $K(L)$ *is a finite group scheme over S. If there exists a point* $s \in S$ *such that* L_s *is ample on* X_s, *then L is relatively ample for* X/S.

Proof. For t varying in S, the function $t \mapsto \dim \mathrm{H}^i(X_t, L_t)$ is upper semi-continuous. Hence

$$U_i := \{t \in S \mid \mathrm{H}^j(X, L) = 0 \text{ for all } j \neq i\}$$

is open. By Mumford's vanishing theorem, the collection of U_i's is a disjoint open partition of S. Since L_s is ample, $\mathrm{H}^0(X_s, L_s) \neq 0$ thus $s \in U_0$. As $U_0 \neq \emptyset$ and S is connected, we have $U_0 = S$. If L_t is ample (which is the case for any $t \in S$ as we have just seen), L is relatively ample on X over a neighborhood of t in S. □

2.3 An application of G.I.T

Let us fix two positive integers $n \geq 1$, N and a type $D = (d_1, \ldots, d_n)$ with $d_1 | \ldots | d_n$, where d_n is prime to N. Let $\mathcal{A}$ be the functor which assigns to a $\mathbb{Z}[(Nd_n)^{-1}]$-scheme S the set of isomorphism classes of polarized S-abelian schemes of type D: for any such S, $\mathcal{A}(S)$ is the set of isomorphism classes of triples (X, λ, η), where

(1) X is an abelian scheme over S ;

(2) $\lambda : X \to \hat{X}$ is a polarization of type D ;

(3) η is a symplectic similitude $(\mathbb{Z}/N\mathbb{Z})^{2n} \simeq X[N]$.

In the third condition, $(\mathbb{Z}/N\mathbb{Z})^{2n}$ and $X[N]$ are respectively endowed with the symplectic pairing (1.3.1) and with the *Weil pairing*, which is the symplectic pairing $X[N] \times_S X[N] \to \mu_{N,S}$ obtained from the pairing $X[N] \times_S \hat{X}[N] \to \mu_{N,S}$ of Lemma 2.1.5 (applied to the special case where α is multiplication by N) by composing it with the morphism $X[N] \to \hat{X}[N]$ induced by λ.

Theorem 2.3.1. *If N is large enough (with respect to D; in the special case of principal polarizations, where $D = (1,\dots 1)$, any $N \geq 3$ is large enough) the functor $\mathcal{A}$ defined above is representable by a smooth quasi-projective $\mathbb{Z}[(Nd_n)^{-1}]$-scheme.*

Proof. Let X be an abelian scheme over S and $\hat{X}$ its dual abelian scheme. Let P be the Poincaré line bundle over $X \times_S \hat{X}$ equipped with a trivialization over the neutral section $e_X \times_S \mathrm{id}_{\hat{X}} : \hat{X} \to X \times_S \hat{X}$ of X. Let $L^\Delta(\lambda)$ be the line bundle over X obtained by pulling back the Poincaré line bundle P

$$L^\Delta(\lambda) = (\mathrm{id}_X, \lambda)^* P$$

by the composite homomorphism

$$(\mathrm{id}_X, \lambda) = (\mathrm{id}_X \times \lambda) \circ \Delta : X \to X \times_S X \to X \times_S \hat{X},$$

where $\Delta : X \to X \times_S X$ is the diagonal. The line bundle $L^\Delta(\lambda)$ gives rise to a symmetric homomorphism $\lambda_{L^\Delta(\lambda)} : X \to \hat{X}$.

Lemma 2.3.2. *The equality $\lambda_{L^\Delta(\lambda)} = 2\lambda$ holds.*

Proof. Locally for the étale topology, we can assume $\lambda = \lambda_L$ for some line bundle over X which is relatively ample. Then

$$L^\Delta(\lambda) = \Delta^*(\mathrm{id}_X \times \lambda)^* P = \Delta^*(\mu^* L \otimes \mathrm{pr}_1 L^{-1} \otimes \mathrm{pr}_2 L^{-1}).$$

It follows that

$$L^\Delta(\lambda) = (2)^* L \otimes L^{-2}$$

where $(2) : X \to X$ is multiplication by 2. As for every $N \in \mathbb{N}$, $\lambda_{(N)^* L} = N^2 \lambda_L$, and in particular $\lambda_{(2)^* L} = 4\lambda_L$, we obtain the desired equality $\lambda_{L^\Delta(\lambda)} = 2\lambda$. □

Since locally over S, $\lambda = \lambda_L$ for a relatively ample line bundle L, the line bundle $L^\Delta(\lambda)$ is a relatively ample line bundle, and $L^\Delta(\lambda)^{\otimes 3}$ is very ample. It follows that its higher direct images by $\pi : X \to S$ vanish

$$\mathrm{R}^i \pi_* L^\Delta(\lambda)^{\otimes 3} = 0 \text{ for all } i \geq 1$$

and that $M = \pi_* L^\Delta(\lambda)$ is a vector bundle of rank

$$m + 1 := 6^n d$$

over S, where $d = d_1 \cdots d_n$.

Definition 2.3.3. A linear rigidification of a polarized abelian scheme (X, λ) is an isomorphism

$$\alpha : \mathbb{P}^m_S \to \mathbb{P}_S(M),$$

where $M = \pi_* L(\lambda)$. In other words, a linear rigidification of a polarized abelian scheme (X, λ) is a trivialization of the $\mathrm{PGL}(m+1)$-torsor associated to the vector bundle M of rank $m + 1$.

Let $\mathcal{H}$ be the functor that assigns to every scheme S the set of isomorphism classes of quadruples $(X, \lambda, \eta, \alpha)$, where (X, λ, η) is a polarized abelian scheme over S of type D with level structure η and α is a linear rigidification. Forgetting α, we get a functorial morphism

$$\mathcal{H} \to \mathcal{A}$$

which is a $\mathrm{PGL}(m+1)$-torsor.

The line bundle $L^{\Delta}(\lambda)^{\otimes 3}$ provides a projective embedding

$$X \hookrightarrow \mathbb{P}_S(M).$$

Using the linear rigidification α, we can embed X into the standard projective space

$$X \hookrightarrow \mathbb{P}^m_S.$$

For every $r \in \mathbb{N}$, the higher direct images vanish

$$\mathrm{R}^i\pi_* L^{\Delta}(\lambda)^{\otimes 3r} = 0 \text{ for all } i > 0$$

and $\pi_* L^{\Delta}(\lambda)^{\otimes 3r}$ is a vector bundle of rank $6^n dr^n$. Hence we have a functorial morphism

$$f : \mathcal{H}_n \to \mathrm{Hilb}^{Q(t),1}(\mathbb{P}_m)$$

(where $\mathrm{Hilb}^{Q(t),1}(\mathbb{P}_m)$ is the Hilbert scheme of 1-pointed subschemes of $\mathbb{P}^m$ with Hilbert polynomial $Q(t) = 6^n dt^n$) sending (X, λ, α) to the image of X in $\mathbb{P}^m$ pointed by the unit of X.

Proposition 2.3.4. *The morphism f identifies $\mathcal{H}$ with an open subfunctor of* $\mathrm{Hilb}^{Q(t),1}(\mathbb{P}^m)$ *consisting of pointed* smooth *subschemes of* $\mathbb{P}^m$.

Proof. Since a smooth projective pointed variety X has at most one abelian variety structure, the morphism f is injective. By Theorem 2.4.1 of the next paragraph, any smooth projective morphism $f : X \to S$ over a geometrically connected base S with a section $e : S \to X$ has an abelian scheme structure if and only if one geometric fiber X_s does. □

Since a polarized abelian variety with principal N-level structure has no non-trivial automorphisms (see [16]), $\mathrm{PGL}(m+1)$ acts freely on $\mathcal{H}$. We take $\mathcal{A}$ as the quotient of $\mathcal{H}$ by the free action of $\mathrm{PGL}(m+1)$. The construction of this quotient as a scheme requires nevertheless a quite technical analysis of stability. If N is large enough, then $X[N] \subset X \subset \mathbb{P}^m$ is not contained in any hyperplane; furthermore, no more than $N^{2n}/m+1$ of these N-torsion points can lie in the same hyperplane of $\mathbb{P}^m$. In that case, $(A, \lambda, \eta, \alpha)$ is a stable point. In the general case, we can increase the level structure and then perform a quotient by a finite group. See [15, p.138] for a complete discussion. □

2.4 Spreading the abelian scheme structure

Let us now quote a theorem of Grothendieck [15, theorem 6.14].

Theorem 2.4.1. *Let S be a connected noetherian scheme. Let $X \to S$ be a smooth projective morphism equipped with a section $e : S \to X$. Assume for one geometric point*

$s = \mathrm{Spec}(\kappa(s))$, X_s is an abelian variety over $\kappa(s)$ with neutral point $\epsilon(s)$. Then X is an abelian scheme over S with neutral section ϵ.

Let us consider first the infinitesimal version of this assertion.

Proposition 2.4.2. *Let $S = \mathrm{Spec}(A)$, where A is an Artin local ring. Let $\mathfrak{m}$ be the maximal ideal of A and let I be an ideal of A such that $\mathfrak{m}I = 0$. Let $S_0 = Spec(A/I)$. Let $f : X \to S$ be a proper smooth scheme with a section $e : S \to X$. Assume that $X_0 = X \times_S S_0$ is an abelian scheme with neutral section $e_0 = e|_{S_0}$. Then X is an abelian scheme with neutral section e.*

Proof. Let $k = A/\mathfrak{m}$ and $\overline{X} = X \otimes_A k$. Let $\mu_0 : X_0 \times_{S_0} X_0 \to X_0$ be the morphism $\mu_0(x, y) = x - y$ and let $\overline{\mu} : \overline{X} \times_k \overline{X} \to \overline{X}$ be the restriction of μ_0. The obstruction to extending μ_0 to a morphism $X \times_S X \to X$ is an element

$$\beta \in \mathrm{H}^1(\overline{X} \times \overline{X}, \overline{\mu}^* T_{\overline{X}} \otimes_k I)$$

where $T_{\overline{X}}$ is the tangent bundle of $\overline{X}$ which is a trivial vector bundle with fibre $\mathrm{Lie}(\overline{X})$. Thus, by the Künneth formula,

$$\mathrm{H}^1(\overline{X} \times \overline{X}, \overline{\mu}^* T_{\overline{X}} \otimes_k I) = (\mathrm{Lie}(\overline{X}) \otimes_k \mathrm{H}^1(\overline{X})) \oplus (\mathrm{H}^1(\overline{X}) \otimes_k \mathrm{Lie}(\overline{X})) \otimes_k I.$$

Consider $g_1, g_2 : X_0 \to X_0 \times_{S_0} X_0$ with $g_1(x) = (x, e)$ and $g_2(x) = (x, x)$. The endomorphisms of X_0, $\mu_0 \circ g_1 = \mathrm{id}_{X_0}$ and $\mu_0 \circ g_2 = (e \circ f)$ extend in an obvious way to X, so that the obstruction classes $\beta_1 = g_1^* \beta$ and $\beta_2 = g_2^* \beta$ must vanish. Since one can express β in terms of β_1 and β_2 by the Künneth formula, β vanishes too.

The set of all extensions μ of μ_0 is a principal homogenous space under

$$\mathrm{H}^0(\overline{X} \times_k \overline{X}, \overline{\mu}^* T_{\overline{X}} \otimes_k I).$$

Among these extensions, there exists a unique μ such that $\mu(e, e) = e$; this one provides a group scheme structure on X/S. □

We can extend the abelian scheme structure to an infinitesimal neighborhood of s. This structure can be algebraized and then descended to a Zariski neightborhood since the abelian scheme structure is unique if it exists. It remains to prove the following lemma due to Koizumi.

Lemma 2.4.3. *Let $S = \mathrm{Spec}(R)$, where R is a discrete valuation ring with generic point η. Let $f : X \to S$ be a proper and smooth morphism with a section $e : S \to X$. Assume that X_η is an abelian variety with neutral point $e(\eta)$. Then X is an abelian scheme with neutral section e.*

Proof. Suppose R is henselian. Since $X \to S$ is proper and smooth, the inertia group I acts trivially on $\mathrm{H}^i(X_{\overline{\eta}}, \mathbb{Q}_\ell)$. By the Néron-Ogg-Shafarevich criterion, there exists an abelian scheme A over S with $A_\eta = X_\eta$ and A is the Néron model of A_η. By the universal property of Néron's model there exists a morphism $\pi : X \to A$ extending the isomorphism $X_\eta \simeq A_\eta$. Let $\mathcal{L}$ be a relatively ample invertible sheaf on X/S. Choose a trivialization on the unit point of $X_\eta = A_\eta$. Then $\mathcal{L}_\eta$ with the trivialization on the unit section extends uniquely on A to a line bundle $\mathcal{L}'$, since $\mathrm{Pic}(A/S)$ satisfies the valuative criterion for properness. It follows that, over the closed point s of S, $\pi^*\mathcal{L}'_s$ and $\mathcal{L}_s$ have the same Chern class. If π has a fiber of positive dimension then the restriction to that fiber of $\pi^*\mathcal{L}'_s$ is trivial. On the other hand, the restriction of $\mathcal{L}_s$ to that fiber is still ample. This contradiction implies that all fibers of π have dimension zero. The finite birational morphism $\pi : X \to A$ is necessarily an isomorphism by Zariski's main theorem. □

2.5 Smoothness

In order to prove that $\mathcal{A}$ is smooth, we will need to review the Grothendieck-Messing theory of deformations of abelian schemes.

Let $S = \mathrm{Spec}(R)$ be a thickening of $\overline{S} = \mathrm{Spec}(R/I)$ with $I^2 = 0$, or more generally, locally nilpotent and equipped with a structure of divided powers. According to Grothendieck and Messing, we can attach to an abelian scheme $\overline{A}$ of dimension n over $\overline{S}$ a locally free $\mathcal{O}_S$-module of rank $2n$

$$\mathrm{H}^1_{\mathrm{cris}}(\overline{A}/\overline{S})_S$$

such that

$$\mathrm{H}^1_{\mathrm{cris}}(\overline{A}/\overline{S})_S \otimes_{\mathcal{O}_S} \mathcal{O}_{\overline{S}} = \mathrm{H}^1_{\mathrm{dR}}(\overline{A}/\overline{S}).$$

We can associate with every abelian scheme A/S such that $A \times_S \overline{S} = \overline{A}$ a sub-$\mathcal{O}_S$-module

$$\omega_{A/S} \subset \mathrm{H}^1_{\mathrm{dR}}(A/S) = \mathrm{H}^1_{\mathrm{cris}}(\overline{A}/\overline{S})_S$$

which is locally a direct factor of rank n and which satisfies

$$\omega_{A/S} \otimes_{\mathcal{O}_S} \mathcal{O}_{\overline{S}} = \omega_{\overline{A}/\overline{S}}.$$

Theorem 2.5.1 (Grothendieck-Messing). *The functor defined as above, from the category of abelian schemes A/S with $A \times_S \overline{S} = \overline{A}$ to the category of sub-$\mathcal{O}_S$-modules $\omega \subset \mathrm{H}^1(\overline{A}/\overline{S})_S$ which are locally a direct factor such that*

$$\omega \otimes_{\mathcal{O}_S} \mathcal{O}_{\overline{S}} = \omega_{\overline{A}/\overline{S}}$$

is an equivalence of categories.

See [11, p.151] for the proof of this theorem.

Let $S = \mathrm{Spec}(R)$ be a thickening of $\overline{S} = \mathrm{Spec}(R/I)$ with $I^2 = 0$. Let $\overline{A}$ be an abelian scheme over S and $\overline{\lambda}$ be a polarization of $\overline{A}$ of type $(d_1, \ldots, d_s)$ with integers d_i relatively prime to the residue characteristics of $\overline{S}$. The polarization $\overline{\lambda}$ induces an isogeny

$$\psi_{\overline{\lambda}} : \overline{A} \to \overline{A}^\vee,$$

where $\overline{A}^\vee$ is the dual abelian scheme of $\overline{A}/\overline{S}$. Since the degree of the isogeny is relatively prime to the residue characteristics, it induces an isomorphism

$$\mathrm{H}^1_{\mathrm{cris}}(\overline{A}^\vee/\overline{S})_S \to \mathrm{H}^1_{\mathrm{cris}}(\overline{A}/\overline{S})_S$$

or a bilinear form $\psi_{\overline{\lambda}}$ on $\mathrm{H}^1_{\mathrm{cris}}(\overline{A}/\overline{S})_S$ which is a symplectic form. The module of relative differentials $\omega_{\overline{A}/\overline{S}}$ is locally a direct factor of $\mathrm{H}^1_{\mathrm{cris}}(\overline{A}/\overline{S})_{\overline{S}}$ which is isotropic with respect to the symplectic form $\psi_{\overline{\lambda}}$. It is known that the Lagrangian grassmannian is smooth so that one can lift $\omega_{\overline{A}/\overline{S}}$ to a locally direct factor of $\mathrm{H}^1_{\mathrm{cris}}(\overline{A}/\overline{S})_S$ which is isotropic. By the Grothendieck-Messing theorem, we get a lifting of $\overline{A}$ to an abelian scheme A/S with a polarization λ that lifts $\overline{\lambda}$. □

2.6 Adelic description and Hecke correspondences

Let X and X' be abelian varieties over a base S. A homomorphism $\alpha : X \to X'$ is an isogeny if one of the following conditions is satisfied

- α is surjective and $\ker(\alpha)$ is a finite group scheme over S ;
- there exists $\alpha' : X' \to X$ such that $\alpha' \circ \alpha$ is multiplication by N on X and $\alpha \circ \alpha'$ is multiplication by N on X' for some positive integer N.

A *quasi-isogeny* is an equivalence class of pairs (α, N) formed by an isogeny $\alpha : X \to X'$ and a positive integer N, where $(\alpha, N) \sim (\alpha', N')$ if and only if $N'\alpha = N\alpha'$. Obviously, we think of the equivalence class (α, N) as $N^{-1}\alpha$.

Fix n, N, D as in 2.3. There is another description of the category $\mathcal{A}$ which is less intuitive but more convenient when we have to deal with level structures.

Let U be a free $\mathbb{Z}$-module of rank $2n$ and let E be an alternating form $U \times U \to M_U$ with value in some rank one free $\mathbb{Z}$-module M_U. Assume that the type of E is D. Let G be the group of symplectic similitudes of (U, M_U) which associates to any ring R the group $G(R)$ of pairs $(g, c) \in \mathrm{GL}(U \otimes R) \times R^\times$ such that

$$E(gx, gy) = cE(x, y)$$

for every $x, y \in U \otimes R$. Thus G is a group scheme defined over $\mathbb{Z}$ which is reductive over $\mathbb{Z}[1/d]$. For every prime $\ell \neq p$, let $K_{N,\ell}$ be the compact open subgroup of $G(\mathbb{Q}_\ell)$ defined as follows:

- if $\ell \nmid N$, then $K_\ell = G(\mathbb{Z}_\ell)$;
- if $\ell | N$, then K_ℓ is the kernel of the homomorphism $G(\mathbb{Z}_\ell) \to G(\mathbb{Z}_\ell/N\mathbb{Z}_\ell)$.

Fix a prime p not dividing either N or D. Let $\mathbb{Z}_{(p)}$ be the localization of $\mathbb{Z}$ obtained by inverting all the primes ℓ different from p.

For every scheme S whose residue characteristics are 0 or p, we consider the groupoid $\mathcal{A}'(S)$ defined as follows:

(1) objets of $\mathcal{A}'$ are triples $(X, \lambda, \tilde{\eta})$, where

- X is an abelian scheme over S;
- $\lambda : X \to \hat{X}$ is a $\mathbb{Z}_{(p)}$ multiple of a polarization of degree prime to p, such that for every prime ℓ and for every $s \in S$, the symplectic form induced by λ on $\mathrm{H}_1(X_s, \mathbb{Q}_\ell)$ is similar to $U \otimes \mathbb{Q}_\ell$;

- for every prime $\ell \neq p$, $\tilde{\eta}_\ell$ is a $K_{N,\ell}$-orbit of symplectic similitudes from $\mathrm{H}_1(X_s, \mathbb{Q}_\ell)$ to $U \otimes \mathbb{Q}_\ell$ which is invariant under $\pi_1(S, s)$. We assume that for almost all primes ℓ, this $K_{N,\ell}$-orbit corresponds to the self-dual lattice $\mathrm{H}_1(X_s, \mathbb{Z}_\ell)$.

(2) a homomorphism $\alpha \in \mathrm{Hom}_{\mathcal{A}'}((X, \lambda, \eta), (X', \lambda', \eta'))$ is a quasi-isogeny $\alpha : X \to X'$ of degree prime to p such that $\alpha^*(\lambda')$ and λ differ by a scalar in $\mathbb{Z}_{(p)}^\times$ and $\alpha^*(\eta') = \eta$.

Consider the functor $\mathcal{A} \to \mathcal{A}'$ which assigns to $(X, \lambda, \eta) \in \mathcal{A}(S)$ the triple $(X, \lambda, \tilde{\eta}) \in \mathcal{A}'(S)$, where the $\tilde{\eta}_\ell$ are defined as follows. Let s be a geometric point of S. Let ℓ be a prime not dividing N and D. Giving a symplectic similitude from $\mathrm{H}_1(X_s, \mathbb{Q}_\ell)$ to $U \otimes \mathbb{Q}_\ell$ up to the action of $K_{N,\ell}$ is equivalent to giving a self-dual lattice of $\mathrm{H}_1(X_s, \mathbb{Q}_\ell)$. The $K_{N,\ell}$-orbit is stable under $\pi_1(S, s)$ if and only the self-dual lattice is invariant under $\pi_1(S, s)$. We pick the obvious choice $\mathrm{H}_1(X_s, \mathbb{Z}_\ell)$ as the self-dual lattice of $\mathrm{H}_1(X_s, \mathbb{Q}_\ell)$ which is invariant under $\pi_1(S, s)$. If ℓ divides D, we want a $\pi_1(S, s)$-invariant lattice such that the restriction of the Weil symplectic pairing is of type D. Again, $\mathrm{H}_1(X_s, \mathbb{Z}_\ell)$ fulfills this property. If ℓ divides N, giving a symplectic similitude from $\mathrm{H}_1(X_s, \mathbb{Q}_\ell)$ to $U \otimes \mathbb{Q}_\ell$ up to the action of $K_{N,\ell}$ is equivalent to giving a self-dual lattice of $\mathrm{H}_1(X_s, \mathbb{Q}_\ell)$ and a rigidification of the pro-ℓ-part of N torsion points of X_s. But this is provided by the level structure η_ℓ in the moduli problem $\mathcal{A}$.

Proposition 2.6.1. *The above functor is an equivalence of categories.*

Proof. As defined, it is obviously faithful. It is full because a quasi-isogeny $\alpha : X \to X'$ which induces an isomorphism $\alpha^* : \mathrm{H}_1(X', \mathbb{Z}_\ell) \to \mathrm{H}_1(X, \mathbb{Z}_\ell)$ is necessarily an isomorphism of abelian schemes. By assumption α carries λ to a rational multiple of λ'. But both λ and λ' are polarizations of the same type, so α must carry λ to λ'. This proves that the functor is fully faithful.

The essential surjectivity is derived from the fact that we can modify an abelian scheme X equipped with a level structure $\tilde{\eta}$ by a quasi-isogeny $\alpha : X \to X'$ so that the composed isomorphism

$$U \otimes \mathbb{Q}_p \simeq \mathrm{H}_1(X, \mathbb{Q}_p) \simeq \mathrm{H}_1(X', \mathbb{Q}_p)$$

identifies $U \otimes \mathbb{Z}_p$ with $\mathrm{H}_1(X', \mathbb{Z}_p)$. There is a unique way to choose a rigidification η of $X'[N]$ compatible with $\tilde{\eta}_p$ for $p|N$. Since the symplectic form E on U is of type (D, D), the polarization λ on X' is also of this type. □

Let us now describe the points of $\mathcal{A}'$ with values in $\mathbb{C}$. Consider an objet of $(X, \lambda, \tilde{\eta}) \in \mathcal{A}'(\mathbb{C})$ equipped with a symplectic basis of $\mathrm{H}_1(X, \mathbb{Z})$. In this case, since λ is a $\mathbb{Z}_{(p)}$-multiple of a polarization of X, it is given by an element of

$$\mathfrak{h}_n^\pm = \{Z \in M_n(\mathbb{C}) | \, {}^tZ = Z, \pm\mathrm{im}(Z) > 0\}\,.$$

For all $\ell \neq p$, $\tilde{\eta}_\ell$ defines an element of $G(\mathbb{Q}_p)/K_p$. At p, the integral Tate module $\mathrm{H}_1(X, \mathbb{Z}_p)$ defines an element of $G(\mathbb{Q}_p)/G(\mathbb{Z}_p)$. It follows that

$$\mathcal{A}'(\mathbb{C}) = G(\mathbb{Q})\backslash[\mathfrak{h}_n^\pm \times G(\mathbb{A}_f)/K_N],$$

where $K_N = \prod_{\ell \neq p} K_{N,\ell} \times G(\mathbb{Z}_p)$.

One of the advantages of the prime description of the moduli problem is that we can replace the principal compact open subgroups K_N by any compact open subgroup $K = \prod K_\ell \subset G(\mathbb{A}_f)$ such that $K_p = G(\mathbb{Z}_p)$, thus obtaining a $\mathbb{Z}_{(p)}$-scheme $\mathcal{A}_K$. In the general case, the proof of the representability is reduced to the principal case.

Using this description, it is also easy to define the Hecke operators, as follows. Let $K = K^{(p)} \times G(\mathbb{Z}_p)$, where $K^{(p)}$ is a compact open subgroup of $G(\mathbb{A}_f^{(p)})$, and let $g \in G(\mathbb{A}_f^{(p)})$. We have a morphism

$$\begin{array}{ccc} \mathcal{A}_K & \longrightarrow & \mathcal{A}_{g^{-1}Kg} \\ (X, \lambda, \bar{\eta}K) & \mapsto & (X, \lambda, \bar{\eta}K \circ g = \bar{\eta} \circ g(g^{-1}Kg) \end{array}$$

(here we use the notation $\bar{\eta}K$ for the K-orbit $\bar{\eta}$ to emphasize the fact that it is an orbit). We then get a *Hecke correspondence*

$$\mathcal{A}_K \leftarrow \mathcal{A}_{K \cap gKg^{-1}} \rightarrow \mathcal{A}_{gKg^{-1}} \rightarrow \mathcal{A}_K,$$

where the right arrow is induced by g as above and the other ones are obvious morphisms (when $K' \subset K$, the orbit $\bar{\eta}K'$ defines an orbit $\bar{\eta}K$).

3 Shimura varieties of PEL type

3.1 Endomorphisms of abelian varieties

Let X be an abelian variety of dimension n over an algebraically closed field k. Let $\mathrm{End}(X)$ be the ring of endomorphisms of X and let $\mathrm{End}_{\mathbb{Q}}(X) = \mathrm{End}(X) \otimes \mathbb{Q}$. If $k = \mathbb{C}$ and $X = V/U$, then we have two faithful representations

$$\rho_a : \mathrm{End}(X) \rightarrow \mathrm{End}_{\mathbb{C}}(V) \text{ and } \rho_r : \mathrm{End}(X) \rightarrow \mathrm{End}_{\mathbb{Z}}(U).$$

It follows that $\mathrm{End}(X)$ is a torsion free abelian group of finite type. Over an arbitrary field k, we need to recall the notion of Tate module. Let ℓ be a prime different from the characteristic of k. Then for every m, the kernel $X[\ell^m]$ of the multiplication by ℓ^m in X is isomorphic to $(\mathbb{Z}/\ell^m\mathbb{Z})^{2n}$.

Definition 3.1.1. The Tate module $T_\ell(X)$ is the limit

$$T_\ell(X) = \varprojlim X[\ell^n]$$

of the inverse system where the transition maps are given by multiplication by $\ell : X[\ell^{n+1}] \rightarrow X[\ell^n]$. As an $\mathbb{Z}_\ell$-module, $T_\ell(X) \simeq \mathbb{Z}_\ell^{2n}$ (non canonically). The rational Tate module is $V_\ell = T_\ell \otimes_{\mathbb{Z}_\ell} \mathbb{Q}_\ell$.

We can identify the Tate module $T_\ell(X)$ with the first étale homology $\mathrm{H}_1(X, \mathbb{Z}_\ell)$, which by definition is the dual of $\mathrm{H}^1(X, \mathbb{Z}_\ell)$. Similarly, $V_\ell(X) = \mathrm{H}_1(X, \mathbb{Q}_\ell)$.

Theorem 3.1.2. *For any abelian varieties X, Y over k,* $\mathrm{Hom}(X, Y)$ *is a finitely generated abelian group, and the natural map*

$$\mathrm{Hom}(X, Y) \otimes \mathbb{Z}_\ell \rightarrow \mathrm{Hom}_{\mathbb{Z}_\ell}(T_\ell(X), T_\ell(Y))$$

is injective.

See [14, p.176] for the proof.

Definition 3.1.3. An abelian variety is called simple if it does not admit any strict abelian subvariety.

Proposition 3.1.4. *If X is a simple abelian variety,* $\mathrm{End}_{\mathbb{Q}}(X)$ *is a division algebra.*

Proof. Let $f : X \to X$ be a non-zero endomorphism of X. The identity component of its kernel is a strict abelian subvariety of X which must be zero. Thus the whole kernel of f is a finite group and the image of f is the whole of X for reasons of dimension. It follows that f is an isogeny and therefore invertible in $\mathrm{End}_{\mathbb{Q}}(X)$. Thus $\mathrm{End}_{\mathbb{Q}}(X)$ is a division algebra. □

Theorem 3.1.5 (Poincaré). *Every abelian variety X is isogenous to a product of simple abelian varieties.*

Proof. Let Y be an abelian subvariety of X. We want to prove the existence of a quasi-supplement of Y in X, in other words a subabelian variety Z of X such that the homomorphism $Y \times Z \to X$ is an isogeny. Let $\hat{X}$ be the dual abelian variety and let $\hat{\pi} : \hat{X} \to \hat{Y}$ be the dual homomorphism to the inclusion $Y \subset X$. Let L be an ample line bundle over X and $\lambda_L : X \to \hat{X}$ the isogeny attached to L. By restriction to Y, we get a homomorphism $\hat{\pi} \circ \lambda_L | Y : Y \to \hat{Y}$ which is surjective since $L|_Y$ is still an ample line bundle. Therefore the kernel Z of the homomorphism $\hat{\pi} \circ \lambda_L : X \to \hat{Y}$ is a quasi-supplement of Y in X. □

Corollary 3.1.6. $\mathrm{End}_{\mathbb{Q}}(X)$ *is a semi-simple algebra of finite dimension over* $\mathbb{Q}$.

Proof. If X is isogenous to $\prod_i X_i^{m_i}$, where the X_i are mutually non-isogenous abelian varieties and $m_i \in \mathbb{N}$. Then $\mathrm{End}_{\mathbb{Q}}(X) = \prod_i M_{m_i}(D_i)$, where $M_{m_i}(D_i)$ is the algebra of $m_i \times m_i$-matrices over the skew-field $D_i = \mathrm{End}_{\mathbb{Q}}(X_i)$. □

We have a function

$$\deg : \mathrm{End}(X) \to \mathbb{N}$$

defined by the following rule : $\deg(f)$ is the degree of the isogeny f if f is an isogeny and $\deg(f) = 0$ if f is not an isogeny. Using the formula $\deg(mf) = m^{2n}\deg(f)$ for all $f \in \mathrm{End}(X)$, $m \in \mathbb{Z}$ and $n = \dim(X)$, we can extend this function to $\mathrm{End}_{\mathbb{Q}}(X)$

$$\deg : \mathrm{End}_{\mathbb{Q}}(X) \to \mathbb{Q}_{+}.$$

For every prime $\ell \neq \mathrm{char}(k)$, we have a representation of the endomorphism algebra

$$\rho_\ell : \mathrm{End}_{\mathbb{Q}}(X) \to \mathrm{End}(V_\ell).$$

These representations for different ℓ are related by the degree function.

Theorem 3.1.7. *For every $f \in \mathrm{End}_{\mathbb{Q}}(X)$, we have*

$$\deg(f) = \det \rho_\ell(f) \text{ and } \deg(n.1_X - f) = P(n),$$

where $P(t) = \det(t - \rho_\ell(f))$ is the characteristic polynomial of $\rho_\ell(f)$. In particular, $\mathrm{tr}(\rho_\ell(f))$ is a rational number which is independent of ℓ.

Let $\lambda : X \to \hat{X}$ be a polarization of X. One attaches to λ an involution on the semi-simple $\mathbb{Q}$-algebra $\mathrm{End}_{\mathbb{Q}}(X)$. [1]

Definition 3.1.8. The Rosati involution on $\mathrm{End}_{\mathbb{Q}}(X)$ associated with λ is the involution defined by the following formula

$$f \mapsto f^* = \lambda^{-1}\hat{f}\lambda$$

for every $f \in \mathrm{End}_{\mathbb{Q}}(X)$.

The polarization $\lambda : X \to \hat{X}$ induces an alternating form $X[\ell^m] \times X[\ell^m] \to \mu_{\ell^m}$ for every m. By passing to the limit on m, we get a symplectic form

$$E : V_\ell(X) \times V_\ell(X) \to \mathbb{Q}_\ell(1).$$

By definition f^* is the adjoint of f for this symplectic form

$$E(fx, y) = E(x, f^*y).$$

Theorem 3.1.9. *The Rosati involution is positive. That is, for every* $f \in \mathrm{End}_{\mathbb{Q}}(X)$, $\mathrm{tr}(\rho_\lambda(ff^*))$ *is a positive rational number.*

Proof. Let $\lambda = \lambda_L$ for some ample line bundle L. One can prove the formula

$$\mathrm{tr}\rho_\ell(ff^*) = \frac{2n(L^{n-1}.f^*(L))}{(L^n)}.$$

Since L is ample, the cup-product $(L^{n-1}.f^*(L))$ (resp. L^n) is the number of intersections of an effective divisor $f^*(L)$ (resp. L) with $n-1$ generic hyperplanes of $|L|$. Since L is ample, these intersection numbers are positive integers. □

Let X be an abelian variety over $\mathbb{C}$ equipped with a polarization λ. The semi-simple $\mathbb{Q}$-algebra $B = \mathrm{End}_{\mathbb{Q}}(X)$ is equipped with

(1) a complex representation ρ_a and a rational representation ρ_r satisfying $\rho_r \otimes_{\mathbb{Q}} \mathbb{C} = \rho_a \oplus \overline{\rho}_a$, and

(2) an involution $b \mapsto b^*$ such that for all $b \in B - \{0\}$, we have $\mathrm{tr}\rho_r(bb^*) > 0$.

Suppose that B is a simple algebra with center F. Then F is a number field equipped with a positive involution $b \mapsto b^*$ restricted from B. There are two possibilities:

(1) The involution is trivial on F. Then F is a totally real number field (*involution of the first kind*). In this case, $B \otimes_{\mathbb{Q}} \mathbb{R}$ is a product of copies of $M_n(\mathbb{R})$ or a product of copies of $M_n(\mathbb{H})$, where $\mathbb{H}$ is the algebra of Hamiltonian quaternions, equipped with their respective positive involutions (case C and D).

[1] Our convention is that an involution of a non-commutative ring satisfies the relation $(xy)^* = y^*x^*$.

(2) The involution is non-trivial on F. Then its fixed points form a totally real number field F_0 and F is a totally imaginary quadratic extension of F_0 (*involution of the second kind*). In this case, $B \otimes_{\mathbb{Q}} \mathbb{R}$ is a product of copies of $M_n(\mathbb{C})$ equipped with its positive involution (case A).

3.2 Positive definite Hermitian forms

Let B be a semisimple algebra of finite dimension over $\mathbb{R}$ with an involution. A Hermitian form on a B-module V is a symmetric form $V \times V \to \mathbb{R}$ such that $(bv, w) = (v, b^*w)$. It is positive definite if $(v, v) > 0$ for all $v \in V$.

Lemma 3.2.1. *The following assertions are equivalent*

(1) *There exists a faithful B-module V such that* $\mathrm{tr}(xx^*, V) > 0$ *for all $x \in B - \{0\}$.*

(2) *The above is true for every faithful B-module V.*

(3) $\mathrm{tr}_{B/\mathbb{R}}(xx^*) > 0$ *for all nonzero $x \in B$.*

3.3 Skew-Hermitian modules

Summing up what has been said in the last two sections, the tensor product with $\mathbb{Q}$ of the algebra of endomorphisms of a polarized abelian variety is a finite-dimensional semi-simple $\mathbb{Q}$-algebra equipped with a positive involution. For every prime $\ell \neq \mathrm{char}(k)$, this algebra has a representation on the Tate module $V_\ell(X)$ which is equipped with a symplectic form. We are now going to look at this structure in a more axiomatic way.

Let k be a field. Let B be a finite-dimensional semisimple k-algebra equipped with an involution $*$. Let $\beta_1, \ldots, \beta_r$ be a basis of B as a k-vector space. For any finite-dimensional B-module V we can define a polynomial $\det_V \in k[x_1, \ldots, x_r]$ by the formula

$$\det_V = \det(x_1\beta_1 + \cdots + x_r\beta_r, V \otimes_k k[x_1, \ldots, x_r])$$

Lemma 3.3.1. *Two finite-dimensional B-modules V and U are isomorphic if and only if* $\det_V = \det_U$.

Proof. If k is an algebraically closed field, B is a product of matrix algebras over k. The lemma follows from the classification of modules over a matrix algebra. Now let k be an arbitrary field and $\bar{k}$ its algebraic closure. The group of automorphisms of a B-module is itself the multiplicative group of a semi-simple k-algebra, so it has trivial Galois cohomology. This allows us to descend from the algebraic closure $\bar{k}$ to k. □

Definition 3.3.2. A skew-Hermitian B-module is a B-module U endowed with a symplectic form

$$U \times U \to M_U$$

with values in a 1-dimensional k-vector space M_U such that $(bx, y) = (x, b^*y)$ for any $x, y \in V$ and $b \in B$.

An automorphism of a skew-Hermitian B-module U is a pair (g, c), where $g \in \mathrm{GL}_B(U)$ and $c \in \mathbb{G}_{m,k}$, such that $(gx, gy) = c(x, y)$ for any $x, y \in U$. The group of automorphisms of the skew-Hermitian B-module U is denoted by $G(U)$.

If k is an algebraically closed field, two skew-Hermitian modules V and U are isomorphic if and only if $\det_V = \det_U$. In general, the set of skew-Hermitian modules V with $\det_V = \det_U$ is classified by $\mathrm{H}^1(k, G(U))$.

Let $k = \mathbb{R}$, let B be a finite-dimensional semi-simple algebra over $\mathbb{R}$ equipped with an involution, and let U be a skew-Hermitian B-module. Let $h : \mathbb{C} \to \mathrm{End}_B(U_\mathbb{R})$ be such that $(h(z)v, w) = (v, h(\overline{z})w)$ and such that the symmetric bilinear form $(v, h(i)w)$ is positive definite.

Lemma 3.3.3. *Let $h, h' : \mathbb{C} \to \mathrm{End}_B(U_\mathbb{R})$ be two such homomorphisms. Suppose that the two $B \otimes_\mathbb{R} \mathbb{C}$-modules U induced by h and h' are isomorphic. Then h and h' are conjugate by an element of $G(\mathbb{R})$.*

Let B be a finite-dimensional simple $\mathbb{Q}$-algebra equipped with an involution and let $U_\mathbb{Q}$ be a skew-Hermitian module $U_\mathbb{Q} \times U_\mathbb{Q} \to M_{U_\mathbb{Q}}$. An integral structure is an order $\mathcal{O}_B$ of B and a free abelian group U equipped with multiplication by $\mathcal{O}_B$ and an alternating form $U \times U \to M_U$ of which the generic fibre is the skew Hermitian module $U_\mathbb{Q}$.

3.4 Shimura varieties of type PEL

Let us fix a prime p. We will describe the PEL moduli problem over a discrete valuation ring with residue characteristic p under the assumption that the PEL datum is unramified at p.

Definition 3.4.1. A rational PE-structure (polarization and endomorphism) is a collection of data as follows:

(1) B is a finite-dimensional simple $\mathbb{Q}$-algebra such that that $B_{\mathbb{Q}_p}$ is a product of matrix algebras over unramified extensions of $\mathbb{Q}_p$;

(2) $*$ is a positive involution of B;

(3) $U_\mathbb{Q}$ is a skew-Hermitian B-module;

(4) $h : \mathbb{C} \to \mathrm{End}_B(U_\mathbb{R})$ is a homomorphism such that $(h(z)v, w) = (v, h(\overline{z})w)$ and such that the symmetric bilinear form $(v, h(i)w)$ is positive definite.

The homomorphism h induces a decomposition $U_\mathbb{Q} \otimes_\mathbb{Q} \mathbb{C} = U_1 \oplus U_2$, where $h(z)$ acts on U_1 by z and on U_2 by $\overline{z}$. Let us choose a basis $\beta_1, \ldots, \beta_r$ of the $\mathbb{Q}$-vector space B. Let $x_1, \ldots, x_r$ be indeterminates. The determinant polynomial

$$\det_{U_1} = \det(x_1\beta_1 + \cdots + x_r\beta_r, U_1 \otimes \mathbb{C}[x_1, \ldots, x_r])$$

is a homogenous polynomial of degree $\dim_\mathbb{C} U_1$. The subfield of $\mathbb{C}$ generated by the coefficients of the polynomial $\det_{U_1}$ is a number field which is independent of the choice of the basis $\beta_1, \ldots, \beta_t$. The above number field E is called the *reflex field* of the PE-structure. Equivalently, E is the definition field of the *isomorphism class* of the $B_{\mathbb{C}}$-module U_1.

Definition 3.4.2. An integral PE structure consists of a rational PE structure equipped with the following extra data:

(5) $\mathcal{O}_B$ is an order of B which is stable under $*$ and maximal at p;

(6) U is an $\mathcal{O}_B$-integral structure of the skew-Hermitian module $U_{\mathbb{Q}}$.

Assuming that the basis $\beta_1, \ldots, \beta_r$ chosen above is a $\mathbb{Z}$-basis of $\mathcal{O}_B$, we see that the coefficients of the determinant polynomial $\det_{U_1}$ lie in $O = \mathcal{O}_E \otimes_{\mathbb{Z}} \mathbb{Z}_{(p)}$.

Fix an integer $N \geq 3$. Consider the moduli problem $\mathcal{B}$ of abelian schemes with a PE-structure and with principal N-level structures. The functor $\mathcal{B}$ assigns to any O-scheme S the category $\mathcal{B}(S)$ whose objects are

$$(A, \lambda, \iota, \eta),$$

where

(1) A is an abelian scheme over S;

(2) $\lambda : A \to \hat{A}$ is a polarization;

(3) $\iota : \mathcal{O}_B \to \mathrm{End}(A)$ is a homomorphism such that the Rosati involution induced by λ restricts to the involution $*$ of $\mathcal{O}_B$,

$$\det(\beta_1 X_1 + \cdots \beta_r x_r, \mathrm{Lie}(A)) = \det_{U_1}$$

and for every prime $\ell \neq p$ and every geometric point s of S, the Tate module $T_\ell(A_s)$, equipped with the action of $\mathcal{O}_B$ and with the alternating form induced by λ, is similar to $U \otimes \mathbb{Z}_\ell$;

(4) η is a similitude from $A[N]$ equipped with the symplectic form and the action of $\mathcal{O}_B$ to U/NU that can be lifted to an isomorphism $H_1(A_s, \mathbb{A}_f^p)$ with $U \otimes_{\mathbb{Z}} \mathbb{A}_f^p$, for every geometric point s of S.

Theorem 3.4.3. *The functor which assigns to each E-scheme S the set of isomorphism classes $\mathcal{B}(S)$ is smooth and representable by a quasi-projective scheme over $\mathcal{O}_E \otimes \mathbb{Z}_{(p)}$.*

Proof. For $\ell \neq p$, the isomorphism class of the skew-Hermitian module $T_\lambda(A_s)$ is locally constant with respect to s so that we can forget the condition on this isomorphism class in the representability problem.

By forgetting endomorphisms, we have a morphism $\mathcal{B} \to \mathcal{A}$. To have ι is equivalent to having actions of $\beta_1, \ldots, \beta_r$ satisfying certain conditions. Therefore, to prove that $\mathcal{B} \to \mathcal{A}$ is representable by a projective morphism it is enough to prove the following lemma.

Lemma 3.4.4. *Let A be a projective abelian scheme over a locally noetherian scheme S. Then the functor that assigns to every S-scheme T the set* $\mathrm{End}(A_T)$ *is representable by a disjoint union of projective schemes over S.*

Proof. The graph of an endomorphism b of A is a closed subscheme of $A \times_S A$ so that the functor of endomorphisms is a subfunctor of some Hilbert scheme. Let's check that this subfunctor is representable by a locally closed subscheme of the Hilbert scheme.

Let $Z \subset A \times_S A$ be a closed subscheme flat over a connected base S. Let's check that the condition

$s \in S$ such that Z_s is a graph

is an open condition. Suppose that $p_1 : Z_s \to A_s$ is an isomorphism over a point $s \in S$. By flatness, the relative dimension of Z over S is equal to that of A. For every $s \in S$ and every $a \in A$, the intersection $Z_s \cap \{a\} \times A_s$ is either of dimension bigger than 0 or consists of exactly one point since the intersection number is constant under deformation. This implies that the morphism $p_1 : Z \to A$ is a birational projective morphism. There is an open subset U of A over which $p_1 : Z \to A$ is an isomorphism. Since $\pi_A : A \to S$ is proper, $\pi_A(A - U)$ is closed. Its complement $S - \pi_A(A - U)$, which is open, is the set of $s \in S$ over which $p_1 : Z_s \to A_s$ is an isomorphism.

Let $Z \subset A \times_S A$ be the graph of a morphism $f : A \to A$. The morphism f is a homomorphism of abelian schemes if and only if f sends the unit to the unit, hence being a homomorphism of abelian schemes is a closed condition. So the functor which assigns to each S-scheme T the set $\mathrm{End}_T(A_T)$ is representable by a locally closed subscheme of a Hilbert scheme.

In order to prove that this subfunctor is represented by a closed subscheme of the Hilbert scheme, it is enough to check the valuative criterion of properness.

Let $S = \mathrm{Spec}(R)$ be the spectrum of a discrete valuation ring with generic point η. Let A be an S-abelian scheme and let $f_\eta : A_\eta \to A_\eta$ be an endomorphism. Then f_η can be extended in a unique way to an endomorphism $f : A \to A$ by the following extension lemma, due to Weil. □

Theorem 3.4.5 (Weil). *Let G be a smooth group scheme over S. Let X be a smooth scheme over S and let $U \subset X$ be an open subscheme whose complement $Y = X - U$ has codimension ≥ 2. Then a morphism $f : U \to G$ can be extended to X. In particular, if G is an abelian scheme, using its properness we can always extend even without the condition on the codimension of $X - U$.*

3.5 Adelic description

Let G be the $\mathbb{Q}$-reductive group defined as the automorphism group of the skew-Hermitian module $U_\mathbb{Q}$. For every $\mathbb{Q}$-algebra R, let

$$G(R) = \{(g, c) \in \mathrm{GL}_B(U)(R) \times R^\times | (gx, gy) = c(x, y)\}.$$

For all $\ell \neq p$ we have a compact open subgroup $K_\ell \subset G(\mathbb{Q}_\ell)$ which consists of $g \in G(\mathbb{Q}_\ell)$ such that $g(U \otimes \mathbb{Z}_\ell) = (U \otimes \mathbb{Z}_\ell)$ and which, when $\ell | N$, satisfy the extra condition that the action induced by g on $(U \otimes \mathbb{Z}_\ell)/N(U \otimes \mathbb{Z}_\ell)$ is trivial.

Lemma 3.5.1. *There exists a unique smooth group scheme $\mathcal{G}_{K_\ell}$ over $\mathbb{Z}_\ell$ such that $\mathcal{G}_{K_\ell} \otimes_{\mathbb{Z}_\ell} \mathbb{Q}_\ell = G \otimes_\mathbb{Q} \mathbb{Q}_\ell$ and $\mathcal{G}_{K_\ell}(\mathbb{Z}_\ell) = K_\ell$.*

Consider the functor $\mathcal{B}'$ which assigns to any E-scheme S the category $\mathcal{B}'(S)$ defined as follows. An object of this category is a quintuple

$$(A, \lambda, \iota, \tilde{\eta}),$$

where

(1) A is an S-abelian scheme over S,

(2) $\lambda : A \to \hat{A}$ is a $\mathbb{Z}_{(p)}$-multiple of a polarization,

(3) $\iota : \mathcal{O}_B \to \mathrm{End}(A)$ is a homomorphism such that the Rosati involution induced by λ restricts to the involution $*$ of $\mathcal{O}_B$ and such that

$$\det(\beta_1 X_1 + \cdots \beta_t X_t, \mathrm{Lie}(A)) = \det_{U_1}$$

(4) fixing a geometric point s of S, for every prime $\ell \neq p$, $\tilde{\eta}_\ell$ is a K_ℓ-orbit of isomorphisms from $V_\ell(A_s)$ to $U \otimes \mathbb{Q}_\ell$ compatible with symplectic forms and the action of $\mathcal{O}_B$ and stable under the action of $\pi_1(S, s)$.

A morphism from $(A, \lambda, \iota, \tilde{\eta})$ to $(A', \lambda, \iota, \tilde{\eta})$ is a quasi-isogeny $\alpha : A \to A'$ of degree prime to p carrying λ to a scalar (in $\mathbb{Q}^\times$) multiple of λ' and carrying $\tilde{\eta}$ to $\tilde{\eta}'$.

Proposition 3.5.2. *The obvious functor $\mathcal{B} \to \mathcal{B}'$ is an equivalence of categories.*

The proof is the same as in the Siegel case. □

3.6 Complex points

The isomorphism class of an object $(A, \lambda, \iota, \tilde{\eta}) \in \mathcal{B}'(\mathbb{C})$ gives rises to

(1) a skew-Hermitian B-module $\mathrm{H}_1(A, \mathbb{Q})$ and

(2) for every prime ℓ, a $\mathbb{Q}_\ell$-similitude $\mathrm{H}_1(A, \mathbb{Q}_\ell) \simeq U \otimes \mathbb{Q}_\ell$ as skew-Hermitian $B \otimes \mathbb{Q}_\ell$-modules, defined up to the action of K_ℓ.

For $\ell \neq p$, this is required in the moduli problem. For the prime p, for every $b \in B$, we have $\mathrm{tr}(b, \mathrm{H}_1(A, \mathbb{Q}_p)) = \mathrm{tr}(b, U \otimes \mathbb{Q})$ because both are equal to $\mathrm{tr}(b, \mathrm{H}_1(A, \mathbb{Q}_\ell))$ for any $\ell \neq p$. It follows that the skew-Hermitian modules $\mathrm{tr}(b, \mathrm{H}_1(A, \mathbb{Q}_p))$ and $\mathrm{tr}(b, \Lambda \otimes \mathbb{Q}_p)$ are isomorphic after base change to a finite extension of $\mathbb{Q}_p$ and therefore the isomorphism class of the skew-Hermitian module defines an element $\xi_p \in \mathrm{H}^1(\mathbb{Q}_p, G)$. Now, in the groupoïd $\mathcal{B}'(\mathbb{C})$ the arrows are given by prime to p isogenies, so $\mathrm{H}_1(A, \mathbb{Z}_p)$ is a well-defined self-dual lattice stable by multiplication by $\mathcal{O}_B$. It follows that the class $\xi_p \in \mathrm{H}^1(\mathbb{Q}_p, G)$ mentioned above comes from a class in $\mathrm{H}^1(\mathbb{Z}_p, G_{\mathbb{Z}_p})$, where $G_{\mathbb{Z}_p}$ is the reductive group scheme over $\mathbb{Z}_p$ which extends $G_{\mathbb{Q}_p}$. In the case where $G_{\mathbb{Z}_p}$ has connected fibres, this implies the vanishing of ξ_p. Kottwitz gave a further argument in the case where G is not connected.

The first datum gives rise to a class

$$\xi \in \mathrm{H}^1(\mathbb{Q}, G)$$

and the second datum implies that the image of ξ in each $\mathrm{H}^1(\mathbb{Q}_\ell, G)$ vanishes. We have

$$\xi \in \ker^1(\mathbb{Q}, G) = \ker(\mathrm{H}^1(\mathbb{Q}, G) \to \prod_\ell \mathrm{H}^1(\mathbb{Q}_\ell, G)).$$

According to a result of Borel and Serre, $\ker^1(\mathbb{Q}, G)$ is a finite set. For every $\xi \in \ker^1(\mathbb{Q}, G)$, fix a skew-Hermitian B-module $V^{(\xi)}$ whose class in $\ker^1(\mathbb{Q}, G)$ is ξ and fix a $\mathbb{Q}_\ell$-similitude of $V^{(\xi)} \otimes \mathbb{Q}_\ell$ with $U \otimes \mathbb{Q}_\ell$ as skew-Hermitian $B \otimes \mathbb{Q}_\ell$-module and also a similitude over $\mathbb{R}$.

Let $\mathcal{B}^{(\xi)}(\mathbb{C})$ be the subset of $\mathcal{B}^{(\xi)}(\mathbb{C})$ consisting of those $(A, \lambda, \iota, \tilde{\eta})$ such that $\mathrm{H}_1(A, \mathbb{Q})$ is isomorphic to $V^{(\xi)}$. Let $(A, \lambda, \iota, \tilde{\eta}) \in \mathcal{B}^{(\xi)}(\mathbb{C})$ and let β be an isomorphism of skew-Hermitian B-modules from $\mathrm{H}_1(A, \mathbb{Q})$ to $V^{(\xi)}$. The set of quintuples $(A, \lambda, \iota, \tilde{\eta}, \beta)$ can be described as follows.

(1) $\tilde{\eta}$ defines an element $\tilde{\eta} \in G(\mathbb{A}_f)/K$.

(2) The complex structure on $\mathrm{Lie}(A) = V \otimes_{\mathbb{Q}} \mathbb{R}$ defines a homomorphism $h : \mathbb{C} \to \mathrm{End}_B(V_\mathbb{R})$ such that $h(\overline{z})$ is the adjoint operator of $h(z)$ for the symplectic form on $V_\mathbb{R}$. Since $\pm\lambda$ is a polarization, $(v, h(i)w)$ is positive or negative definite. Moreover the isomorphism class of the $B \otimes \mathbb{C}$-module V is specified by the determinant condition on the tangent space. It follows that h lies in a $G(\mathbb{R})$-conjugacy class X_∞.

Therefore the set of quintuples is $X_\infty \times G(\mathbb{A}_f)/K$. Two different trivializations β and β' differ by an automorphism of the skew-Hermitian B-module $V^{(\xi)}$. This group is the inner form $G^{(\xi)}$ of G obtained by the image of $\xi \in \mathrm{H}^1(\mathbb{Q}, G)$ in $\mathrm{H}^1(\mathbb{Q}, G^{\mathrm{ad}})$. In conclusion we get

$$\mathcal{B}^{(\xi)}(\mathbb{C}) = G^{(\xi)}(\mathbb{Q})\backslash[X_\infty \times G(\mathbb{A}_f)/K]$$

and

$$\mathcal{B}(\mathbb{C}) = \bigsqcup_{\xi \in \ker^1(\mathbb{Q}, G)} \mathcal{B}^{(\xi)}(\mathbb{C}).$$

4 Shimura varieties

4.1 Review of Hodge structures

See [Deligne, Travaux de Griffiths]. Let Q be a subring of $\mathbb{R}$; we specifically have in mind the cases $Q = \mathbb{Z}, \mathbb{Q}$ or $\mathbb{R}$. A Q-Hodge structure will be called respectively an integral, rational or real Hodge structure.

Definition 4.1.1. A Q-Hodge structure is a projective Q-module V equipped with a bigrading of $V_\mathbb{C} = V \otimes_Q \mathbb{C}$

$$V_\mathbb{C} = \bigoplus_{p,q} H^{p,q}$$

such that $H^{p,q}$ and $H^{q,p}$ are complex conjugate, i.e. the semi-linear automorphism σ of $V_\mathbb{C} = V \otimes_Q \mathbb{C}$ given by $v \otimes z \mapsto v \otimes \overline{z}$ satisfies the relation $\sigma(H^{p,q}) = H^{q,p}$ for every $p, q \in \mathbb{Z}$.

The integers $h^{p,q} = \dim_{\mathbb{C}}(H^{p,q})$ are called Hodge numbers. We have $h^{p,q} = h^{q,p}$. If there exists an integer n such that $H^{p,q} = 0$ unless $p+q = n$ then the Hodge structure is said to be pure of weight n. When the Hodge structure is pure of weight n, the Hodge filtration $F^pV = \bigoplus_{r\geq p} V^{rs}$ determines the Hodge structure by the relation $V^{pq} = F^pV \cap \overline{F^qV}$.

Let $\mathbb{S} = \mathrm{Res}_{\mathbb{C}/\mathbb{R}}\mathbb{G}_m$ be the real algebraic torus defined as the Weil restriction from $\mathbb{C}$ to $\mathbb{R}$ of $\mathbb{G}_{m,\mathbb{C}}$. We have a norm homomorphism $\mathbb{C}^\times \to \mathbb{R}^\times$ whose kernel is the unit circle S^1. Similarly, we have an exact sequence of real tori

$$1 \to \mathbb{S}^1 \to \mathbb{S} \to \mathbb{G}_{m,\mathbb{R}} \to 1.$$

We have an inclusion $\mathbb{R}^\times \subset \mathbb{C}^\times$ whose cokernel can be represented by the homomorphism $\mathbb{C}^\times \to S^1$ given by $z \mapsto z/\overline{z}$. We have the corresponding exact sequence of real tori

$$1 \to \mathbb{G}_{m,\mathbb{R}} \to \mathbb{S} \to \mathbb{S}^1 \to 1.$$

The inclusion $w : \mathbb{G}_{m,\mathbb{R}} \to \mathbb{S}$ is called the *weight homomorphism*.

Lemma 4.1.2. *Let $G = GL(V)$ be the linear group defined over Q. A Hodge structure on V is equivalent to a homomorphism $h : \mathbb{S} \to G_{\mathbb{R}} = G \otimes_Q \mathbb{R}$. The Hodge structure is pure of weight n if the restriction of h to $\mathbb{G}_{m,\mathbb{R}} \subset \mathbb{S}$ factors through the center $\mathbb{G}_{m,\mathbb{R}} = Z(G_{\mathbb{R}})$ and the homomorphism $\mathbb{G}_{m,\mathbb{R}} \to Z(G_{\mathbb{R}})$ is given by $t \mapsto t^n$.*

Proof. A bi-grading $V_{\mathbb{C}} = \bigoplus_{p,q} V^{p,q}$ is the same as a homomorphism $h_{\mathbb{C}} : \mathbb{G}^2_{m,\mathbb{C}} \to G_{\mathbb{C}}$. The complex conjugation of $V_{\mathbb{C}}$ exchanges the factors $V^{p,q}$ and $V^{q,p}$ if and only if $h_{\mathbb{C}}$ descends to a homomorphism of real algebraic groups $h : \mathbb{S} \to G_{\mathbb{R}}$. □

Definition 4.1.3. A polarization of a Hodge structure $(V_{\mathbb{Q}}, V^{p,q})$ of weight n is a bilinear form Ψ_Q on V_Q such that the induced form Ψ on $V_{\mathbb{R}}$ is invariant under $h(S^1)$ and such that the form $\Psi(x, h(i)y)$ is symmetric and positive definite.

It follows from the identity $h(i)^2 = (-1)^n$ that the bilinear form $\Psi(x, y)$ is symmetric if n is even and alternating if n is odd :

$$\Psi(x, y) = (-1)^n\Psi(x, h(i)^2y) = (-1)^n\Psi(h(i)y, h(i)x) = (-1)^n\Psi(y, x).$$

Example. An abelian variety induces a typical Hodge structure. Let $X = V/U$ be an abelian variety. Let G be $GL(U \otimes \mathbb{Q})$ as an algebraic group defined over $\mathbb{Q}$. The complex structure V on the real vector space $U \otimes \mathbb{R} = V$ induces a homomorphism of real algebraic groups

$$\phi : \mathbb{S} \to G_{\mathbb{R}}$$

so that U is equipped with an integral Hodge structure of weight -1. A polarization of X is a symplectic form E on V, taking integral values on U such that $E(ix, iy) = E(x, y)$ and such that $E(x, iy)$ is a positive definite symmetric form.

Let V be a projective Q-module of finite rank. A Hodge structure on V induces Hodge structures on tensor products $V^{\otimes m} \otimes (V^*)^{\otimes n}$. Fix a finite set of tensors (s_i)

$$s_i \in V^{\otimes m_i} \otimes (V^*)^{\otimes n_i}.$$

Let $G \subset \mathrm{GL}(V)$ be the stabilizer of these tensors.

Lemma 4.1.4. *There is a bijection between the set of Hodge structures on V for which the tensors s_i are of type $(0,0)$ and the set of homomorphisms $\mathbb{S} \to G_{\mathbb{R}}$.*

Proof. A homomorphism $h : \mathbb{S} \to \mathrm{GL}(V)_{\mathbb{R}}$ factors through $G_{\mathbb{R}}$ if and only if the image $h(\mathbb{S})$ fixes all tensors s_i. This is equivalent to saying that these tensors are of type $(0,0)$ for the induced Hodge structures. □

There is a related notion of Mumford-Tate group.

Definition 4.1.5. Let G be an algebraic group over $\mathbb{Q}$. Let $\phi : S^1 \to G_{\mathbb{R}}$ be a homomorphism of real algebraic groups. The Mumford-Tate group of (G, ϕ) is the smallest algebraic subgroup $H = \mathrm{Hg}(\phi)$ of G defined over $\mathbb{Q}$ such that ϕ factors through $H_{\mathbb{R}}$.[2]

Let $\mathbb{Q}[G]$ be the ring of algebraic functions over G and $\mathbb{R}[G] = \mathbb{Q}[G] \otimes_{\mathbb{Q}} \mathbb{R}$. The group S^1 acts on $\mathbb{R}[G]$ through the homomorphism ϕ. Let $\mathbb{R}[G]^{\phi=1}$ be the subring of functions fixed by $\phi(S^1)$ and consider the subring

$$\mathbb{Q}[G] \cap \mathbb{R}[G]^{\phi=1}$$

of $\mathbb{Q}[G]$. For every $v \in \mathbb{Q}[G] \cap \mathbb{R}[G]^{\phi=1}$, let G_v be the stabilizer subgroup of G at v. Since G_v is defined over $\mathbb{Q}$ and ϕ factors through $G_{v,\mathbb{R}}$, we have the inclusion $H \subset G_v$. In particular, $v \in \mathbb{Q}[G]^H$. It follows that

$$\mathbb{Q}[G] \cap \mathbb{R}[G]^{\phi=1} = \mathbb{Q}[G]^H.$$

This property does not characterize H, however. In general, for any subgroup H of G, we have an obvious inclusion

$$H \subset H' = \bigcap_{v \in \mathbb{Q}[G]^H} G_v.$$

which is strict in general. If the equality $H = H'$ occurs, we say that H is an *observable* subgroup of G. To prove that this is indeed the case for the Mumford-Tate group of an abelian variety, we will need the following general lemma.

Lemma 4.1.6. *Let H be a reductive subgroup of a reductive group G. Then H is observable.*

Proof. Assume the base field $k = \mathbb{C}$. By a result of Chevalley (see [Borel]), for every subgroup H of G, there exists a representation $\rho : G \to \mathrm{GL}(V)$ and a vector $v \in V$ such that H is the stabilizer of the line kv. Since H is reductive, there exists a H-stable complement U of kv in V. Let $kv^* \subset V^*$ be the line orthogonal to U with some generator v^*. Then H is the stabilizer of the vector $v \otimes v^* \in V \otimes V^*$.

Let $G = \mathrm{GL}(U_{\mathbb{Q}})$ and let $\phi : S^1 \to G_{\mathbb{R}}$ be a homomorphism such that $\phi(i)$ induces a Cartan involution on $G_{\mathbb{R}}$. Let $\mathcal{C}$ be the smallest tensor subcategory of the category of Hodge structures that contains $(U_{\mathbb{Q}}, \phi)$ and is stable by subquotient. There is the forgetful functor $\mathrm{Fib}_{\mathcal{C}} : \mathcal{C} \to \mathrm{Vect}_{\mathbb{Q}}$. □

[2]Originally, Mumford called $\mathrm{Hg}(\phi)$ the Hodge group.

Lemma 4.1.7. *$H = Hg(\phi)$ is the automorphism group of the functor* Fib_C.

Proof. Let V be a representation of G defined over $\mathbb{Q}$ equipped with the Hodge structure defined by ϕ. Let U be a subvector space of V compatible with the Hodge structure. Then H must stabilize U. It follows that H acts naturally on Fib_C, i.e. we have a natural homomorphism $H \to \mathrm{Aut}^{\otimes}(\mathrm{Fib}_C)$. □

Proposition 4.1.8. *The Mumford-Tate group of a polarizable Hodge structure is a reductive group.*

Proof. Since we are working over fields of characteristic zero, H is reductive if and only if the category of representations of H is semi-simple. Using the Cartan involution, we can exhibit a positive definite bilinear form on V. This implies that every subquotient of $V^{\otimes m} \otimes (V^*)^{\otimes n}$ is a subobject. □

4.2 Variations of Hodge structures

Let S be a complex analytic variety. The letter Q denotes a ring contained in $\mathbb{R}$ which could be $\mathbb{Z}$, $\mathbb{Q}$ or $\mathbb{R}$.

Definition 4.2.1. A variation of Hodge structures (VHS) on S of weight n consists of the following data:

(1) a local system of projective Q-modules V;

(2) a decreasing filtration $F^p\mathcal{V}$ on the vector bundle $\mathcal{V} = V \otimes_Q \mathcal{O}_S$ such that Griffiths transversality is satisfied, i.e. for every integer p

$$\nabla(F^p\mathcal{V}) \subset F^{p-1}\mathcal{V} \otimes \Omega^1_S$$

where $\nabla : \mathcal{V} \to \mathcal{V} \otimes \Omega^1_S$ is the connection $v \otimes f \mapsto v \otimes df$ for which $V \otimes_Q \mathbb{C}$ is the local system of horizontal sections;

(3) for every $s \in S$, the filtration induces on V_s a pure Hodge structure of weight n.

There are obvious notions of the dual VHS and tensor product of VHS. The Leibniz formula $\nabla(v \otimes v') = \nabla(v) \otimes v' + v \otimes \nabla(v')$ assures that Griffiths transversality is satisfied for the tensor product.

Typical examples of polarized VHS are provided by cohomology of smooth projective morphisms. Let $f : X \to S$ be a smooth projective morphism over a complex analytic variety S. Then $H^n = \mathrm{R}^n f_* \mathbb{Q}$ is a local system of $\mathbb{Q}$-vector spaces. Since $\mathrm{H}^n \otimes_{\mathbb{Q}} \mathcal{O}_S$ is equal to the de Rham cohomology $\mathrm{H}^n_{dR} = \mathrm{R}^n f_* \Omega^{\bullet}_{X/S}$, where $\Omega^{\bullet}_{X/S}$ is the relative de Rham complex, and the Hodge spectral sequence degenerates at E^2, the abutment H^n_{dR} is equipped with a decreasing filtration by subvector bundles $F^p(\mathrm{H}^n_{dR})$ with

$$(F^p/F^{p+1})\mathrm{H}^n_{dR} = \mathrm{R}^q f_* \Omega^p_{X/S}$$

with $p+q=n$. The connection ∇ satisfies Griffiths transversality. By the Hodge decomposition, we have moreover a direct sum

$$\mathrm{H}^n_{dR}(X_s) = \bigoplus_{pq} H^{p,q}$$

with $H^{p,q} = \mathrm{H}^q(X_s, \Omega^p_{X_s})$ and $\overline{H^{p,q}} = H^{q,p}$ so that all the axioms of VHS are satisfied.

If we choose a projective embedding $X \to \mathbb{P}^d_S$, the line bundle $\mathcal{O}_{\mathbb{P}^d}(1)$ defines a class

$$c \in \mathrm{H}^0(S, \mathrm{R}^2 f_*\mathbb{Q}).$$

By the hard Lefschetz theorem, the cup product by c^{d-n} induces an isomorphism

$$\mathrm{R}^n f_*\mathbb{Q} \to \mathrm{R}^{2d-n} f_*\mathbb{Q} \text{ defined by } \alpha \mapsto c^{d-n} \wedge \alpha$$

so that by Poincaré duality we get a polarization on $\mathrm{R}^n f_*\mathbb{Q}$.

4.3 Reductive Shimura-Deligne data

The torus $\mathbb{S} = \mathrm{Res}_{\mathbb{C}/\mathbb{R}}\mathbb{G}_m$ plays a particular role in the formalism of Shimura varieties shaped by Deligne in [4], [5].

Definition 4.3.1. A Shimura-Deligne datum is a pair (G, X) consisting of a reductive group G over $\mathbb{Q}$ and a $G(\mathbb{R})$-conjugacy class X of homomorphisms $h : \mathbb{S} \to G_{\mathbb{R}}$ satisfying the following properties:

(SD1) For $h \in X$, only the characters $z/\overline{z}, 1, \overline{z}/z$ occur in the representation of $\mathbb{S}$ on $\mathrm{Lie}(G)$;

(SD2) $\mathrm{ad}h(i)$ is a Cartan involution of G^{ad}, i.e. the real Lie group $\{g \in G^{\mathrm{ad}}(\mathbb{C}) \mid \mathrm{ad}(h(i))\sigma(g) = g\}$ is compact (where σ denotes the complex conjugation and $G^{\mathrm{ad}} = G/Z(G)$ is the adjoint group of G).

The action $\mathbb{S}$, restricted to $\mathbb{G}_{m,\mathbb{R}}$ is trivial on $\mathrm{Lie}(G)$ so that $h : \mathbb{S} \to G_{\mathbb{R}}$ sends $\mathbb{G}_{m,\mathbb{R}}$ into the center $Z_{\mathbb{R}}$ of $G_{\mathbb{R}}$. The induced homomorphism $w = h|_{\mathbb{G}_{m,\mathbb{R}}} : \mathbb{G}_{m,\mathbb{R}} \to Z_{\mathbb{R}}$ is independent of the choice of $h \in X$. We call w the *weight homomorphism*.

After base change to $\mathbb{C}$, we have $\mathbb{S} \otimes_{\mathbb{R}} \mathbb{C} = \mathbb{G}_m \times \mathbb{G}_m$, where the factors are ordered in the way that $\mathbb{S}(\mathbb{R}) \to \mathbb{S}(\mathbb{C})$ is the map $z \mapsto (z, \overline{z})$. Let $\mu : \mathbb{G}_{m,\mathbb{C}} \to \mathbb{S}_{\mathbb{C}}$ the homomorphism defined by $z \to (z, 1)$. If $h : \mathbb{S} \to \mathrm{GL}(V)$ is a Hodge structure, then $\mu_h = h_{\mathbb{C}} \circ \mu : \mathbb{G}_m(\mathbb{C}) \to \mathrm{GL}(V_{\mathbb{C}})$ determines its Hodge filtration.

Siegel case. An abelian variety $A = V/U$ is equipped with a polarization E which is a non-degenerate symplectic form on $U_{\mathbb{Q}} = U \otimes \mathbb{Q}$. Let GSp be the group of symplectic similitudes

$$\mathrm{GSp}(U_{\mathbb{Q}}, E) = \{(g, c) \in \mathrm{GL}(U_{\mathbb{Q}}) \times \mathbb{G}_{m,\mathbb{Q}} | E(gx, gy) = cE(x, y)\}.$$

The scalar c is called the similitude factor. Base changed to $\mathbb{R}$, we get the group of symplectic similitudes of the real symplectic space $(U_{\mathbb{R}}, E)$. The complex vector space structure on $V = U_{\mathbb{R}}$ induces a homomorphism

$$h : \mathbb{S} \to \mathrm{GSp}(U_{\mathbb{R}}, E).$$

In this case, X is the set of complex structures on $U_{\mathbb{R}}$ such that $E(h(i)x, h(i)y) = E(x, y)$ and $E(x, h(i)y)$ is a positive definite symmetric form.

PEL case. Suppose B is a simple $\mathbb{Q}$-algebra with center F equipped with a positive involution $*$. Let $F_0 \subset F$ be the fixed field of $*$. Let G be the group of symplectic similitudes of a skew-Hermitian B-module V

$$G = \{(g, c) \in \mathrm{GL}_B(V) \times \mathbb{G}_{m,\mathbb{Q}} | (gx, gy) = c(x, y)\}.$$

Then the h of definition 3.4.1 induces a morphism $\mathbb{S} \to G_{\mathbb{R}}$.

Let G_1 be the subgroup of G defined by

$$G_1(R) = \{g \in \mathrm{GL}_B(V) \times \mathbb{G}_{m,\mathbb{Q}} | (gx, gy) = (x, y)\}$$

for any $\mathbb{Q}$-algebra R. We have an exact sequence

$$1 \to G_1 \to G \to \mathbb{G}_m \to 1.$$

The group G_1 is a scalar restriction of a group G_0 defined over F_0.

Since a simple $\mathbb{R}$-algebra with positive involution must be $M_n(\mathbb{C})$, $M_n(\mathbb{R})$ or $M_n(\mathbb{H})$ with their standard involutions, there will be three cases to be considered.

(1) Case (A) : If $[F : F_0] = 2$, then F_0 is a totally real field and F is a totally imaginary extension. Over $\mathbb{R}$, $B \otimes_{\mathbb{Q}} \mathbb{R}$ is the product of $[F_0 : \mathbb{Q}]$ copies of $M_n(\mathbb{C})$. $G_1 = \mathrm{Res}_{F_0/\mathbb{Q}} G_0$,where G_0 is an inner form of the quasi-split unitary group attached to the quadratic extension F/F_0.

(2) Case (C) : If $F = F_0$, then F is a totally real field. and $B \otimes \mathbb{R}$ is isomorphic to a product of $[F_0 : \mathbb{Q}]$ copies of $M_n(\mathbb{R})$ equipped with their positive involution. In this case, $G_1 = \mathrm{Res}_{F_0/\mathbb{Q}} G_0$, where G_0 is an inner form of a quasi-split symplectic group over F_0.

(3) Case (D) : $B \otimes \mathbb{R}$ is isomorphic to a product of $[F_0 : \mathbb{Q}]$ copies of $M_n(\mathbb{H})$ equipped with positive involutions. The simplest case is $B = \mathbb{H}$, with V a skew-Hermitian quaternionic vector space. In this case, $G_1 = \mathrm{Res}_{F_0/\mathbb{Q}} G_0$, where G_0 is an even orthogonal group.

Tori case. In the case where $G = T$ is a torus over $\mathbb{Q}$, both conditions (SD1) and (SD2) are obvious since the adjoint representation is trivial. The conjugacy class of $h : \mathbb{S} \to T_{\mathbb{R}}$ contains just one element since T is commutative.

Deligne proved the following statement in [5, prop. 1.1.14] which provides a justification for the not so natural notion of Shimura-Deligne datum.

Proposition 4.3.2. *Let (G, X) be a Shimura-Deligne datum. Then X has a unique structure of a complex manifold such that for every representation $\rho : G \to \mathrm{GL}(V)$, $(V, \rho \circ h)_{h \in X}$ is a variation of Hodge structures which is polarizable.*

Proof. Let $\rho : G \to \mathrm{GL}(V)$ be a faithful representation of G. Since $w_h \subset Z_G$, the weight filtration of V_h is independent of h. Since the weight filtration is fixed, the Hodge structure is determined by the Hodge filtration. It follows that the morphism to the Grassmannian

$$\omega : X \to \mathrm{Gr}(V_{\mathbb{C}})$$

– which sends h to the Hodge filtration attached to h – is injective. We need to prove that this morphism identifies X with a complex subvariety of $\mathrm{Gr}(V_{\mathbb{C}})$. It suffices to prove that

$$d\omega : T_h X \to T_{\omega(h)} \mathrm{Gr}(V_{\mathbb{C}})$$

identifies $T_h X$ with a complex vector subspace of $\mathrm{Gr}(V_{\mathbb{C}})$.

Let $\mathfrak{g}$ be the Lie algebra of G and $\mathrm{ad} : G \to \mathrm{GL}(\mathfrak{g})$ the adjoint representation. Let G_h be the centralizer of h, and $\mathfrak{g}_h$ its Lie algebra. We have $\mathfrak{g}_h = \mathfrak{g}_{\mathbb{R}} \cap \mathfrak{g}^{0,0}$ for the Hodge structure on $\mathfrak{g}$ induced by h. It follows that the tangent space to the real analytic variety X at h is

$$T_h X = \mathfrak{g}_{\mathbb{R}}/\mathfrak{g}_{\mathbb{C}}^{0,0} \cap \mathfrak{g}_{\mathbb{R}}.$$

Let W be a pure Hodge structure of weight 0. Consider the $\mathbb{R}$-linear morphism

$$W_{\mathbb{R}}/W_{\mathbb{R}} \cap W_{\mathbb{C}}^{0,0} \to W_{\mathbb{C}}/F^0 W_{\mathbb{C}},$$

which is injective. Since both vector spaces have the same dimension over $\mathbb{R}$, it is also surjective. It follows that $W_{\mathbb{R}}/W_{\mathbb{R}} \cap W_{\mathbb{C}}^{0,0}$ admits a canonical complex structure.

Since the above isomorphism is functorial on the category of pure Hodge structures of weight 0, we have a commutative diagram

$$\begin{array}{ccc} \mathfrak{g}_{\mathbb{R}}/\mathfrak{g}_{\mathbb{C}}^{0,0} \cap \mathfrak{g}_{\mathbb{R}} & \longrightarrow & \mathrm{End}(V_{\mathbb{R}})/\mathrm{End}(V_{\mathbb{R}}) \cap \mathrm{End}(V_{\mathbb{C}})^{0,0} \\ \downarrow & & \downarrow \\ \mathfrak{g}_{\mathbb{C}}/F^0\mathfrak{g}_{\mathbb{C}} & \longrightarrow & \mathrm{End}(V_{\mathbb{C}})/F^0\mathrm{End}(V_{\mathbb{C}}) \end{array}$$

which proves that the image of $T_h X = \mathfrak{g}_{\mathbb{R}}/\mathfrak{g}_{\mathbb{C}}^{0,0}$ in $T_{\omega(h)}\mathrm{Gr}(V_{\mathbb{C}}) = \mathrm{End}(V_{\mathbb{C}})/F^0\mathrm{End}(V_{\mathbb{C}})$ is a complex vector subspace.

The Griffiths transversality property of $V \otimes \mathcal{O}_X$ follows from the same diagram. There is a commutative triangle of vector bundles

$$\begin{array}{ccc} TX & \longrightarrow & \mathrm{End}(V \otimes \mathcal{O}_X) \\ & \searrow & \downarrow \\ & & \mathrm{End}(V \otimes \mathcal{O}_X)/F^0\mathrm{End}(V \otimes \mathcal{O}_X) \end{array}$$

where the horizontal arrow is the derivation in $V \otimes \mathcal{O}_X$. Griffiths transversality for $V \otimes \mathcal{O}_X$ is satisfied if and only if the image of the derivation is contained in $F^{-1}\mathrm{End}(V \otimes \mathcal{O}_X)$. But this follows from the fact that

$$\mathfrak{g}_{\mathbb{C}} = F^{-1}\mathfrak{g}_{\mathbb{C}}$$

and the map $TX \to \mathrm{End}(V \otimes \mathcal{O}_X)/F^0\mathrm{End}(V \otimes \mathcal{O}_X)$ factors through $(\mathfrak{g}_C \otimes \mathcal{O}_X)/F^0(\mathfrak{g}_C \otimes \mathcal{O}_X)$. □

4.4 The Dynkin classification

Let (G, X) be a Shimura-Deligne datum. Over $\mathbb{C}$, we have a conjugacy class of cocharacter

$$\mu^{ad} : \mathbb{G}_{m,\mathbb{C}} \to G^{ad}_{\mathbb{C}}.$$

The complex adjoint semi-simple group G^{ad} is isomorphic to a product of complex adjoint simple groups $G^{ad} = \prod_i G_i$. The simple complex adjoint groups are classified by their Dynkin diagrams. The axiom $SD1$ implies that μ^{ad} induces an action of $\mathbb{G}_{m,\mathbb{C}}$ on $\mathfrak{g}_i$ for which the set of weights is $\{-1, 0, 1\}$. Such cocharacters are called *minuscule*. Minuscule coweights belong to the set of fundamental coweights and therefore can be specified by special nodes in the Dynkin diagram. Every Dynkin diagram has at least one special node except the three diagrams F4, G2, E8. We can classify Shimura-Deligne data over the complex numbers with the help of Dynkin diagrams.

4.5 Semi-simple Shimura-Deligne data

Definition 4.5.1. A semi-simple Shimura-Deligne datum is a pair (G, X^+) consisting of a semi-simple algebraic group G over $\mathbb{Q}$ and a $G(\mathbb{R})^+$-conjugacy class of homomorphisms $h^1 : \mathbb{S}^1 \to G_{\mathbb{R}}$ satisfying the axioms (SD1) and (SD2). Here $G(\mathbb{R})^+$ denotes the neutral component of $G(\mathbb{R})$ for the real topology.

Let (G, X) be a reductive Shimura-Deligne datum. Let G^{ad} be the adjoint group of G. Every $h \in X$ induces a homomorphism $h^1 : \mathbb{S}^1 \to G^{\mathrm{ad}}$. The $G^{\mathrm{ad}}(\mathbb{R})^+$-conjugacy class X^+ of h^1 is isomorphic to the connected component of h in X.

The spaces X^+ are exactly the so-called Hermitian symmetric domains with the symmetry group $G(\mathbb{R})^+$.

Theorem 4.5.2 (Baily-Borel). *Let Γ be a torsion free arithmetic subgroup of $G(\mathbb{R})^+$. The quotient $\Gamma\backslash X^+$ has a canonical realization as a Zariski open subset of a complex projective algebraic variety. In particular, it has a canonical structure of a complex algebraic variety.*

These quotients $\Gamma\backslash X^+$, considered as complex algebraic varieties, are called *connected Shimura varieties*. The terminology is a little bit confusing, because they are not those Shimura varieties which are connected but rather connected components of Shimura varieties.

4.6 Shimura varieties

Let (G, X) be a Shimura-Deligne datum. For a compact open subgroup K of $G(\mathbb{A}_f)$, consider the double coset space

$$\mathrm{Sh}_K(G, X) = G(\mathbb{Q})\backslash[X \times G(\mathbb{A}_f)/K]$$

in which $G(\mathbb{Q})$ acts on X and $G(\mathbb{A}_f)$ on the left and K acts on $G(\mathbb{A}_f)$ on the right.

Lemma 4.6.1. *Let $G(\mathbb{Q})_+ = G(\mathbb{Q}) \cap G(\mathbb{R})_+$, where $G(\mathbb{R})_+$ is the inverse image of $G^{\mathrm{ad}}(\mathbb{R})$ under the morphism $G(\mathbb{R}) \to G^{\mathrm{ad}}(\mathbb{R})$. Let X_+ be a connected component of X. Then there is a homeomorphism*

$$G(\mathbb{Q})\backslash[X \times G(\mathbb{A}_f)/K] = \bigsqcup_{\xi \in \Xi} \Gamma_\xi \backslash X_+ ,$$

where ξ runs over a finite set Ξ of representatives of $G(\mathbb{Q})_+\backslash G(\mathbb{A}_f)/K$ and $\Gamma_\xi = \xi K \xi^{-1} \cap G(\mathbb{Q})$.

Proof. The map

$$\bigsqcup_{\xi \in \Xi} \Gamma_\xi \backslash X_+ \to G(\mathbb{Q})_+\backslash[X_+ \times G(\mathbb{A}_f)/K]$$

sending the class of $x \in X_+$ to the class of $(x, \xi) \in X_+ \times G(\mathbb{A}_f)$ is bijective by the very definition of the finite set Ξ and of the discrete groups Γ_ξ.

It follows from the theorem of real approximation that the map

$$G(\mathbb{Q})_+\backslash[X_+ \times G(\mathbb{A}_f)/K] \to G(\mathbb{Q})\backslash[X \times G(\mathbb{A}_f)/K]$$

is a bijection.

Lemma 4.6.2 (Real approximation). *For any connected group G over $\mathbb{Q}$, $G(\mathbb{Q})$ is dense in $G(\mathbb{R})$.*

See [18, p.415]. □

Remarks

(1) The group $G(\mathbb{A}_f)$ acts on the inverse limit

$$\varprojlim_K G(\mathbb{Q})\backslash[X \times G(\mathbb{A}_f]/K .$$

On Shimura varieties of finite level, there is an action of Hecke algebras by correspondences.

(2) In order to be arithmetically significant, Shimura varieties must have models over a number field. According to the theory of canonical models, there exists a number field called the reflex field E depending only on the Shimura-Deligne datum over which the Shimura variety has a model that can be characterized by certain properties.

(3) The connected components of Shimura varieties have canonical models over abelian extensions of the reflex field E; these extensions depend not only on the Shimura-Deligne datum but also on the level structure.

(4) Strictly speaking, the moduli space of abelian varieties with PEL structures is not a Shimura variety but a disjoint union of Shimura varieties. The union is taken over the set $\ker^1(\mathbb{Q}, G)$. For each class $\xi \in \ker^1(\mathbb{Q}, G)$, we have a $\mathbb{Q}$-group $G^{(\xi)}$ which is isomorphic to G over $\mathbb{Q}_p$ and over $\mathbb{R}$ but which might not be isomorphic to G over $\mathbb{Q}$.

(5) The Langlands correspondence has been proved in many particular cases by studying the commuting action of Hecke operators and of Galois groups of the reflex field on the cohomology of Shimura varieties.

5 CM tori and canonical models

5.1 The PEL moduli attached to a CM torus

Let F be a totally imaginary quadratic extension of a totally real number field F_0 of degree f_0 over $\mathbb{Q}$. We have $[F : \mathbb{Q}] = 2f_0$. Such a field F is called a CM field. Let τ_F denote the non-trivial element of $\mathrm{Gal}(F/F_0)$. This involution acts on the set $\mathrm{Hom}_{\mathbb{Q}}(F, \overline{\mathbb{Q}})$ of cardinality $2f_0$.

Definition 5.1.1. A CM-type of F is a subset $\Phi \subset \mathrm{Hom}_{\mathbb{Q}}(F, \overline{\mathbb{Q}})$ of cardinality f_0 such that

$$\Phi \cap \tau(\Phi) = \emptyset \text{ and } \Phi \cup \tau(\Phi) = \mathrm{Hom}_{\mathbb{Q}}(F, \overline{\mathbb{Q}}).$$

A CM type is a pair (F, Φ) consisting of a CM field F and a CM type Φ of F.

Let (F, Φ) be a CM type. The absolute Galois group $\mathrm{Gal}(\overline{\mathbb{Q}}/\mathbb{Q})$ acts on $\mathrm{Hom}_{\mathbb{Q}}(F, \overline{\mathbb{Q}})$. Let E be the fixed field of the open subgroup

$$\mathrm{Gal}(\overline{\mathbb{Q}}/E) = \{\sigma \in \mathrm{Gal}(\overline{\mathbb{Q}}/\mathbb{Q}) | \sigma(\Phi) = \Phi\}.$$

For every $b \in F$,

$$\sum_{\phi \in \Phi} \phi(b) \in E$$

and conversely E can be characterized as the subfield of $\overline{\mathbb{Q}}$ generated by the sums $\sum_{\phi\in\Phi} \phi(b)$ for $b \in F$.

For any number field K, we denote by $\mathcal{O}_K$ the maximal order of K. Let Δ be the finite set of primes where $\mathcal{O}_F$ is ramified over $\mathbb{Z}$. By construction, the scheme $Z_F = \mathrm{Spec}(\mathcal{O}_F[\ell^{-1}]_{\ell\in\Delta})$ is finite étale over $\mathrm{Spec}(\mathbb{Z}) - \Delta$. By construction the reflex field E is also unramified away from Δ; let $Z_E = \mathrm{Spec}(\mathcal{O}_E[\ell^{-1}]_{\ell\in\Delta})$. Then we have a canonical isomorphism

$$Z_F \times Z_E = (Z_{F_0} \times Z_E)_\Phi \sqcup (Z_{F_0} \times Z_E)_{\tau(\Phi)},$$

where $(Z_{F_0} \times Z_E)_\Phi$ and $(Z_{F_0} \times Z_E)_{\tau(\Phi)}$ are two copies of $(Z_{F_0} \times Z_E)$ with $Z_{F_0} = \mathrm{Spec}(\mathcal{O}_{F_0}[p^{-1}]_{p\in\Delta})$.

To complete the *PE-structure*, we take U to be the $\mathbb{Q}$-vector space F. The Hermitian form on U is given by

$$(b_1, b_2) = \mathrm{tr}_{F/\mathbb{Q}}(cb_1\tau(b_2))$$

for some element $c \in F$ such that $\tau(c) = -c$. The reductive group G associated to this PE-structure is a $\mathbb{Q}$-torus T equipped with a morphism $h : \mathbb{S} \to T$ which can be made explicit as follows.

Let $\widetilde{T} = \mathrm{Res}_{F/\mathbb{Q}}\mathbb{G}_m$. The CM-type Φ induces an isomorphism of $\mathbb{R}$-algebras and of tori

$$F \otimes_{\mathbb{Q}} \mathbb{C} = \prod_{\phi\in\Phi} \mathbb{C} \text{ and } \widetilde{T}(\mathbb{R}) = \prod_{\phi\in\Phi} \mathbb{C}^\times.$$

According to this identification, $\tilde{h} : \mathbb{S} \to \widetilde{T}_{\mathbb{R}}$ is the diagonal homomorphism

$$\mathbb{C}^\times \to \prod_{\phi\in\Phi} \mathbb{C}^\times.$$

The complex conjugation τ induces an involution τ on $\widetilde{T}$. The norm N_{F/F_0} given by $x \mapsto x\tau(x)$ induces a homomorphism $\mathrm{Res}_{F/\mathbb{Q}}\mathbb{G}_m \to \mathrm{Res}_{F_0/\mathbb{Q}}\mathbb{G}_m$.

The torus T is defined as the pullback of the diagonal subtorus $\mathbb{G}_m \subset \mathrm{Res}_{F_0/\mathbb{Q}}\mathbb{G}_m$. In particular

$$T(\mathbb{Q}) = \{x \in F^\times | x\tau(x) \in \mathbb{Q}^\times\}.$$

The morphism $\tilde{h} : \mathbb{S} \to \widetilde{T}_{\mathbb{R}}$ factors through T and defines a morphism $h : \mathbb{S} \to T$. As usual h defines a cocharacter

$$\mu : \mathbb{G}_{m,\mathbb{C}} \to T_{\mathbb{C}}$$

defined on points by

$$\mathbb{C}^\times \to \prod_{\phi\in\mathrm{Hom}(F,\overline{\mathbb{Q}})} \mathbb{C}^\times,$$

where the projection on the component $\phi \in \Phi$ is the identity and the projection on the component $\phi \in \tau(\Phi)$ is trivial. The reflex field E is the field of definition of μ.

Let $p \notin \Delta$ be an unramified prime of $\mathcal{O}_F$. Choose an open compact subgroup $K^p \subset T(\mathbb{A}_f^p)$ and take $K_p = T(\mathbb{Z}_p)$.

We consider the functor $\mathrm{Sh}(T, h_\Phi)$ which associates to a Z_E-scheme S the set of isomorphism classes of

$$(A, \lambda, \iota, \eta),$$

where

- A is an abelian scheme of relative dimension f_0 over S ;
- $\iota : \mathcal{O}_F \to \mathrm{End}(A)$ is an action of F on A such that for every $b \in F$ and every geometric point s of S, we have
$$\mathrm{tr}(b, \mathrm{Lie}(A_s)) = \sum_{\phi\in\Phi} \phi(a);$$
- λ is a polarization of A whose Rosati involution induces on F the complex conjugation τ;
- η is a level structure.

Proposition 5.1.2. $\mathrm{Sh}(T, h_\Phi)$ *is a finite étale scheme over* Z_E.

Proof. Since $\mathrm{Sh}(T, h_\Phi)$ is quasi-projective over Z_E, it suffices to check the valuative criterion for properness and the unique lifting property of étale morphisms.

Let $S = \mathrm{Spec}(R)$ be the spectrum of a discrete valuation ring with generic point $\mathrm{Spec}(K)$ and with closed point $\mathrm{Spec}(k)$. Pick a point $x_K \in \mathrm{Sh}(T, h_\Phi)(K)$ with

$$x_K = (A_K, \iota_K, \lambda_K, \eta_K).$$

The Galois group $\mathrm{Gal}(\overline{K}/K)$ acts on the $F \otimes \overline{\mathbb{Q}}_\ell$-module $\mathrm{H}^1(A \otimes_K \overline{K}, \overline{\mathbb{Q}}_\ell)$. It follows that $\mathrm{Gal}(\overline{K}/K)$ acts semisimply, and its inertia subgroup acts through a finite quotient. Using the Néron-Ogg-Shafarevich criterion for good reduction, we see that after replacing K by a finite extension K', and R by its normalization R' in K', A_K acquires good reduction, i.e. there exists an abelian scheme over R' whose generic fiber is $A_{K'}$. The endomorphisms extend by Weil's extension theorem. The polarization needs a little more care. The symmetric homomorphism $\lambda_K : A_K \to \hat{A}_K$ extends to a symmetric homomorphism $\lambda : A \to \hat{A}$. After finite étale base change of S, there exists an invertible sheaf L on A such that $\lambda = \lambda_L$. By assumption L_K is an ample invertible sheaf over A_K. λ is an isogeny, and L is non degenerate on the generic and on the special fibres. Mumford's vanishing theorem implies that $\mathrm{H}^0(X_K, L_K) \neq 0$. By the upper semi-continuity property, $\mathrm{H}^0(X_s, L_s) \neq 0$. But since L_s is non-degenerate, Mumford'vanishing theorem says that L_s is ample. This proves that $\mathrm{Sh}(T, h_\Phi)$ is proper.

Let $S = \mathrm{Spec}(R)$, where R is a local artinian $\mathcal{O}_E$-algebra with residue field $\overline{k}$, and let $\overline{S} = \mathrm{Spec}(\overline{R})$ with $\overline{R} = R/I$, $I^2 = 0$. Let $s = \mathrm{Spec}(\overline{k})$ be the closed point of S and $\overline{S}$. Let $\overline{x} \in \mathrm{Sh}(T, h_\Phi)(\overline{S})$ with $\overline{x} = (\overline{A}, \overline{\iota}, \overline{\lambda}, \overline{\eta})$. We have the exact sequence

$$0 \to \omega_{\overline{A}} \to \mathrm{H}^1_{dR}(\overline{A}) \to \mathrm{Lie}(\widehat{\overline{A}}) \to 0$$

with a compatible action of $\mathcal{O}_F \otimes_{\mathbb{Z}} \mathcal{O}_E$. As $\mathcal{O}_{Z_F \times Z_E}$-module, ω_{A_s} is supported on $(Z_{F_0} \times Z_E)_\Phi$ and $\mathrm{Lie}(\widehat{\overline{A}})$ is supported on $(Z_{F_0} \times Z_E)_{\tau(\Phi)}$ so that the above exact sequence splits. This extends to a canonical splitting of the cristalline cohomology $\mathrm{H}^1_{\mathrm{cris}}(\overline{A}/\overline{S})_S$. By the results of Grothendieck-Messing, this splitting induces a lift of the abelian scheme $\overline{A}/\overline{S}$ to an abelian scheme A/S. We are also able to lift the additional structures $\overline{\lambda}, \overline{\iota}, \overline{\eta}$ by the functoriality of Grothendieck-Messing theory. □

5.2 Description of its special fibre

We will keep the notation of the previous paragraph. Pick a place v of the reflex field E which does not lie over the finite set Δ of primes where $\mathcal{O}_F$ is ramified. $\mathcal{O}_E$ is unramified ovec $\mathbb{Z}$ at the place v. We want to describe the set $\mathrm{Sh}_K(T, h_\Phi)(\overline{\mathbb{F}}_p)$ equipped with the Frobenius operator Frob_v.

Theorem 5.2.1. *There is a natural bijection*

$$\mathrm{Sh}_K(T, h_\Phi)(\overline{\mathbb{F}}_p) = \bigsqcup_\alpha T(\mathbb{Q})\backslash Y^p \times Y_p,$$

where

(1) *α runs over the set of (compatible with the action of $\mathcal{O}_F$)-isogeny classes ;*

(2) *$Y^p = T(\mathbb{A}_f^p)/K^p$;*

(3) *$Y_p = T(\mathbb{Q}_p)/T(\mathbb{Z}_p)$;*

(4) *for every $\lambda \in T(\mathbb{A}_f^p)$ we have $\lambda(x^p, x_p) = (\lambda x^p, x_p)$;*

(5) *the Frobenius* Frob_v *acts by the formula*

$$(x^p, x_p) \mapsto (x^p, \mathrm{N}_{E_v/\mathbb{Q}_p}(\mu(p^{-1}))x_p).$$

Proof. Let $x_0 = (A_0, \lambda_0, \iota_0, \eta_0) \in \mathrm{Sh}_K(T, h_\Phi)(\overline{\mathbb{F}}_p)$. Let X be the set of pairs (x, ρ), where $x = (A, \lambda, \iota, \eta) \in \mathrm{Sh}_K(T, h_\Phi)(\overline{\mathbb{F}}_p)$ and

$$\rho : A_0 \to A$$

is a quasi-isogeny which is compatible with the action of $\mathcal{O}_F$ and transforms λ_0 into a rational multiple of λ.

We will need to prove the following two assertions:

(1) $X = Y^p \times Y_p$ with the prescribed action of Hecke operators and of Frobenius ;

(2) the group of quasi-isogenies of A_0 compatible with ι_0 and transforming λ_0 into a rational multiple is $T(\mathbb{Q})$.

Quasi-isogenies of degrees relatively prime to p. Let Y^p be the subset of X where the degree of the quasi-isogeny is relatively prime to p. Consider the prime description of the moduli problem. A point $(A, \lambda, \iota, \tilde{\eta})$ consists of

- an abelian variety A up to isogeny,
- a rational multiple λ of a polarization,
- the multiplication ι by $\mathcal{O}_F$ on A, and
- a class $\tilde{\eta}_\ell$ modulo the action of an open compact subgroup K_ℓ of isomorphisms from $\mathrm{H}_1(A, \mathbb{Q}_\ell)$ to U_ℓ which are compatible with ι and transform λ into a rational multiple of the symplectic form on U_ℓ.

By this description, an isogeny of degree prime to p compatible with ι and preserving the $\mathbb{Q}$-line of the polarization, is given by an element $g \in T(\mathbb{A}_f^p)$. The isogeny corresponding to g defines an isomorphism in the category $\mathcal{B}'$ if and only if $g\tilde{\eta} = \tilde{\eta}'$. Thus

$$Y^p = T(\mathbb{A}_f^p)/K^p$$

with the obvious action of Hecke operators and trivial action of Frob_v.

Quasi-isogenies whose degree is a power of p. Let Y_p the subset of X where the degree of the quasi-isogeny is a power of p. We will use the covariant Dieudonné theory to describe the set Y_p equipped with action of Frobenius operator.

Let $W(\overline{\mathbb{F}}_p)$ be the ring of Witt vectors with coefficients in $\overline{\mathbb{F}}_p$. Let L be the field of fractions of $W(\overline{\mathbb{F}}_p)$; we will write $\mathcal{O}_L$ instead of $W(\overline{\mathbb{F}}_p)$. The Frobenius automorphism $\sigma : x \mapsto x^p$ of $\overline{\mathbb{F}}_p$ induces by functoriality an automorphism σ on the Witt vectors. For every abelian variety A over $\overline{\mathbb{F}}_p$, $\mathrm{H}^1_{cris}(A/\mathcal{O}_L)$ is a free $\mathcal{O}_L$-module of rank $2n$ equipped with an operator φ which is σ-linear. Let $D(A) = \mathrm{H}_1^{cris}(A/\mathcal{O}_L)$ be the dual $\mathcal{O}_L$-module to $\mathrm{H}^1_{cris}(A/\mathcal{O}_L)$, on which φ acts in σ^{-1}-linearly. Furthermore, there is a canonical isomorphism

$$\mathrm{Lie}(A) = D(A)/\varphi D(A).$$

A quasi-isogeny $\rho : A_0 \to A$ induces an isomorphism $D(A_0) \otimes_{\overline{\mathcal{O}}_L} L \simeq D(A) \otimes_{\overline{\mathcal{O}}_L} L$ compatible with the multiplication by $\mathcal{O}_F$ and preserving the $\mathbb{Q}$-line of the polarizations. The following proposition is an immediate consequence of Dieudonné theory.

Proposition 5.2.2. *Let $H = D(A_0) \otimes_{\overline{\mathcal{O}}_L} L$. The above construction defines a bijection between Y_p and the set of lattices $D \subset H$ such that*

(1) $pD \subset \varphi D \subset D$,

(2) *D is stable under the action of $\mathcal{O}_B$ and*

$$\mathrm{tr}(b, D/\varphi D) = \sum_{\phi \in \Phi} \phi(b)$$

for every $b \in \mathcal{O}_B$,

(3) *D is self-dual up to a scalar in $\mathbb{Q}_p^\times$.*

Moreover, the Frobenius operator on Y_p, which transforms the quasi-isogeny $\rho : A_0 \to A$ into the quasi-isogeny $\varphi \circ \rho : A_0 \to A \to \sigma^* A$, acts on the above set of lattices by sending D to $\varphi^{-1} D$.

Since $\mathrm{Sh}(T, h_\Phi)$ is étale, there exists a unique lifting

$$\tilde{x} \in \mathrm{Sh}(T, h_\Phi)(\mathcal{O}_L)$$

of $x_0 = (A_0, \lambda_0, \iota_0, \eta_0) \in \mathrm{Sh}(T, h_\Phi)(\overline{\mathbb{F}}_p)$. By assumption,

$$D(A_0) = \mathrm{H}_1^{dR}(\tilde{A})$$

is a free $\mathcal{O}_F \otimes \mathcal{O}_L$-module of rank 1 equipped with a pairing given by an element $c \in (\mathcal{O}_{F_p}^\times)^{\tau=-1}$. The σ^{-1}-linear operator φ on $H = D(A_0) \otimes_{\overline{\mathcal{O}}_L} L$ is of the form

$$\varphi = t(1 \otimes \sigma^{-1})$$

for an element $t \in T(L)$.

Lemma 5.2.3. *The element t lies in the coset $\mu(p)T(\mathcal{O}_L)$.*

Proof. H is a free $\mathcal{O}_F \otimes L$-module of rank 1 and

$$\mathcal{O}_F \otimes L = \prod_{\psi \in \mathrm{Hom}(\mathcal{O}_F, \overline{\mathbb{F}}_p)} L$$

is a product of $2f_0$ copies of L. By ignoring the self-duality condition, t can be represented by an element

$$t = (t_\psi) \in \prod_{\psi \in \mathrm{Hom}(\mathcal{O}_F, \overline{\mathbb{F}}_p)} L^\times$$

It follows from the assumption

$$pD_0 \subset \varphi D_0 \subset D_0$$

that for all $\psi \in \mathrm{Hom}(\mathcal{O}_F, \overline{\mathbb{F}}_p)$ we have

$$0 \leq \mathrm{val}_p(t_\psi) \leq 1.$$

Bearing in mind the decomposition

$$Z_F \times Z_E = (Z_{F_0} \times Z_E)_\Phi \sqcup (Z_{F_0} \times Z_E)_{\tau(\Phi)},$$

induced by the CM type Φ, we see that the embedding of the residue field of v in $\overline{\mathbb{F}}_p$ induces a decomposition $\mathrm{Hom}(\mathcal{O}_F, \overline{\mathbb{F}}_p) = \Psi \sqcup \tau(\Psi)$ such that

$$\mathrm{tr}(b, D_0/\varphi D_0) = \sum_{\psi \in \Psi} \psi(b)$$

for all $b \in \mathcal{O}_F$. It follows that

$$\mathrm{val}_p(t_\psi) = \begin{cases} 0 & \text{if} \quad \psi \notin \Psi \\ 1 & \text{if} \quad \psi \in \Psi \end{cases}$$

By the definition of μ, it follows that $t \in \mu(p)T(\mathcal{O}_L)$. □

Description of Y_p continued. A lattice D stable under the action of $\mathcal{O}_F$ and self-dual up to a scalar can be uniquely written in the form

$$D = mD_0$$

for $m \in T(L)/T(\mathcal{O}_L)$. The condition $pD \subset \varphi D \subset D$ and the trace condition on the tangent space are equivalent to $m^{-1}t\sigma(m) \in \mu(p)T(\mathcal{O}_L)$ and thus m lies in the group of σ-fixed points in $T(L)/T(\mathcal{O})_L$, that is,

$$m \in [T(L)/T(\mathcal{O}_L)]^{\langle\sigma\rangle}.$$

Now one obtains a bijection between the cosets $m \in T(L)/T(\mathcal{O}_L)$ fixed by σ and the cosets $T(\mathbb{Q}_p)/T(\mathbb{Z}_p)$ by considering the exact sequence

$$1 \to T(\mathbb{Z}_p) \to T(\mathbb{Q}_p) \to [T(L)/T(\mathcal{O}_L]^{\langle\sigma\rangle} \to \mathrm{H}^1(\langle\sigma\rangle, T(\mathcal{O}_L)),$$

where the last cohomology group vanishes by Lang's theorem. It follows that

$$Y_p = T(\mathbb{Q}_p)/T(\mathbb{Z}_p)$$

and φ acts on it as $\mu(p)$.

On H, $\mathrm{Frob}_\nu(1 \otimes \sigma^r)$ acts as φ^{-r} so that

$$\begin{aligned}\mathrm{Frob}_\nu(1 \otimes \sigma^r) &= (\mu(p)(1 \otimes \sigma^{-1}))^{-r} \\ &= \mu(p^{-1})\sigma(\mu(p^{-1}))\ldots\sigma^{r-1}(\mu(p^{-1}))(1 \otimes \sigma^r).\end{aligned}$$

Thus the Frobenius Frob_ν acts on $Y^p \times Y_p$ by the formula

$$(x^p, x_p) \mapsto (x^p, \mathrm{N}_{E_\nu/\mathbb{Q}_p}(\mu(p^{-1}))x_p).$$

Self-isogenies. For every prime $\ell \neq p$, $\mathrm{H}^1(A_0, \mathbb{Q}_\ell)$ is a free $F \otimes_{\mathbb{Q}} \mathbb{Q}_\ell$-module of rank one. So

$$\mathrm{End}_{\mathbb{Q}}(A_0, \iota_0) \otimes_{\mathbb{Q}} \mathbb{Q}_\ell = F \otimes_{\mathbb{Q}} \mathbb{Q}_\ell.$$

It follows that $\mathrm{End}_{\mathbb{Q}}(A_0, \iota_0) = F$. The self-isogenies of A_0 form the group $F^\times$ and those that transform the polarization λ_0 to rational multiples of λ_0 form by definition the subgroup $T(\mathbb{Q}) \subset F^\times$. □

5.3 Shimura-Taniyama formula

Let (F, Φ) be a CM-type. Let $\mathcal{O}_F$ be an order of F which is maximal almost everywhere. Let p be a prime where $\mathcal{O}_F$ is unramified. We can either consider the moduli space of polarized abelian schemes endowed with CM-multiplication of CM-type Φ as in the previous paragraphs or consider the moduli space of abelian schemes endowed with CM-multiplication of CM-type Φ. Everything works in the same way for properness, étaleness, and the description of points, but we lose the obvious projective morphism to the Siegel moduli space. But since we know a posteriori that there is only a finite number of points, this loss is not serious.

Let (A, ι) be an abelian scheme of CM type (F, Φ) over a number field K which is unramified at p, equipped with a sufficiently large level structure. The field K must contain the reflex field E but might be bigger. Let $\mathfrak{q}$ be a place of K over p, and $\mathcal{O}_{K,\mathfrak{q}}$ be the localization of $\mathcal{O}_K$ at $\mathfrak{q}$, and let q be the cardinality of the residue field of $\mathfrak{q}$. By étaleness of the moduli space, A can be extended to an abelian scheme over $\mathrm{Spec}(\mathcal{O}_{K,\mathfrak{q}})$ equipped with a multiplication ι_ν by $\mathcal{O}_F$. As we have already seen, the CM type Φ and the choice of the place $\mathfrak{q}$ of K define a decomposition $\mathrm{Hom}(\mathcal{O}_F, \overline{\mathbb{F}}_p) = \Psi \sqcup \tau(\Psi)$ such that

$$\mathrm{tr}(b, \mathrm{Lie}\,(A_\nu)) = \sum_{\psi \in \Psi} \psi(b)$$

for all $b \in \mathcal{O}_F$.

Let $\pi_\mathfrak{q}$ be the relative Frobenius of A_ν. Since $\mathrm{End}_{\mathbb{Q}}(A_\nu, \iota_\nu) = F$, $\pi_\mathfrak{q}$ defines an element of F.

Theorem 5.3.1 (Shimura-Taniyama formula). *For every prime v of F, we have*

$$\frac{\mathrm{val}_v(\pi_{\mathfrak{q}})}{\mathrm{val}_v(q)} = \frac{|\Psi \cap H_v|}{|H_v|},$$

where $H_v \subset \mathrm{Hom}(\mathcal{O}_F, \overline{\mathbb{F}}_p)$ is the subset formed by the morphisms $\mathcal{O}_f \to \overline{\mathbb{F}}_p$ which factor through v.

Proof. As in the description of the Frobenius operator in Y_p, we have

$$\pi_{\mathfrak{q}} = \varphi^{-r},$$

where $q = p^r$. It is an elementary exercise to relate the Shimura-Taniyama formula to the group theoretical description of φ. □

5.4 Shimura varieties of tori

Let T be a torus defined over $\mathbb{Q}$ and $h : \mathbb{S} \to T_{\mathbb{R}}$ a homomorphism. Let $\mu : \mathbb{G}_{m,\mathbb{C}} \to T_{\mathbb{C}}\mathbb{C}$ be the associated cocharacter. Let E be the number field of definition of μ. Choose an open compact subgroup $K \subset T(\mathbb{A}_f)$. The Shimura variety attached to these data is

$$T(\mathbb{Q})\backslash T(\mathbb{A}_f)/K$$

since the conjugacy class X of h has just one element. This finite set is the set of $\mathbb{C}$-points of a finite étale scheme over $\mathrm{Spec}(E)$. We need to define how the absolute Galois group $\mathrm{Gal}(E)$ acts on this set.

The Galois group $\mathrm{Gal}(E)$ will act through its maximal abelian quotient $\mathrm{Gal}^{ab}(E)$. For almost all primes v of E, we will define how the Frobenius π_v at v acts.

A prime p is said to be unramified if T can be extended to a torus T over $\mathbb{Z}_p$ and if $K_p = T(\mathbb{Z}_p)$. Let v be a place of E over an unramified prime p. Then p is a uniformizing element of $\mathcal{O}_{E,v}$. The cocharacter $\mu : \mathbb{G}_m \to T$ is defined over $\mathcal{O}_{E,v}$, so that $\mu(p^{-1})$ is a well-defined element of $T(E_v)$. We require that the Frobenius π_v acts on $T(\mathbb{Q})\backslash T(\mathbb{A}_f)/K$ as the element

$$\mathrm{N}_{E_v/\mathbb{Q}_p}(\mu(p^{-1})) \in T(\mathbb{Q}_p).$$

By class field theory, this rule defines an action of $\mathrm{Gal}^{ab}(E)$ on the finite set $T(F)\backslash T(\mathbb{A}_f)/K$.

5.5 Canonical models

Let (G, h) be a Shimura-Deligne datum. Let $\mu : \mathbb{G}_{m,\mathbb{C}} \to G_{\mathbb{C}}$ be the attached cocharacter. Let E be the field of definition of the conjugacy class of μ, which is called the *reflex field* of (G, h).

Let (G_1, h_1) and (G_2, h_2) be two Shimura-Deligne data and let $\rho : G_1 \to G_2$ be an injective homomorphism of reductive $\mathbb{Q}$-groups which sends the conjugacy class h_1 to the conjugacy class h_2. Let E_1 and E_2 be the reflex fields of (G_1, h_1) and (G_2, h_2). Since the conjugacy class of $\mu_2 = \rho \circ \mu_1$ is defined over E_1, we have the inclusion $E_2 \subset E_1$.

Definition 5.5.1. A canonical model of $\mathrm{Sh}(G, h)$ is an algebraic variety defined over E such that for every Shimura-Deligne datum (G_1, h_1), where G_1 is a torus, and any injective homomorphism $(G_1, h_1) \to (G, h)$, the morphism

$$\mathrm{Sh}(G_1, h_1) \to \mathrm{Sh}(G, h)$$

is defined over E_1, where E_1 is the reflex field of (G_1, h_1) and the E_1-structure of $\mathrm{Sh}(G_1, h_1)$ was defined in the last paragraph.

Theorem 5.5.2 (Deligne). *There exists at most one canonical model up to unique isomorphism.*

Theorem 5.2.1 proves more or less that the moduli space gives rise to a canonical model for symplectic group. It follows that a PEL moduli space also gives rise to a canonical model. The same holds for Shimura varieties of Hodge type and abelian type. Some other crucial cases were obtained by Shih afterwards. In the general case, the existence of the canonical model is proved by Borovoi and Milne.

Theorem 5.5.3 (Borovoi, Milne). *The canonical model exists.*

5.6 Integral models

An integral model comes naturally with every PEL moduli problem. More generally, in the case of Shimura varieties of Hodge type (or even of "abelian type"), M. Kisin [6] proves the existence of a "canonical" integral model. In this case, the integral model is nothing but the normalization of the closure of the canonical model (over the reflex field) in the Siegel moduli space (base-changed to the maximal order of the reflex field). Kisin proves that this closure is smooth and satisfies an extension property expected for canonical integral models (see [6], (2.3.7) for a precise statement of this extension property, which characterizes this smooth model up to an unique isomorphism). Related results had previously been published by Vasiu in [20].

6 Points of Siegel varieties over a finite field

6.1 Abelian varieties over a finite field up to isogeny

Let $k = \mathbb{F}_q$ be a finite field of characteristic p with $q = p^s$ elements. Let A be a simple abelian variety defined over k and $\pi_A \in \mathrm{End}_k(A)$ its geometric Frobenius.

Theorem 6.1.1 (Weil). *The subalgebra $\mathbb{Q}(\pi_A) \subset \mathrm{End}_k(A)_{\mathbb{Q}}$ is a finite extension of $\mathbb{Q}$ such that for every inclusion $\phi : \mathbb{Q}(\pi_A) \hookrightarrow \mathbb{C}$, we have $|\phi(\pi_A)| = q^{1/2}$.*

Proof. Choose a polarization and let τ be the associated Rosati involution. We have

$$(\pi_A x, \pi_A y) = q(x, y)$$

so that $\tau(\pi_A)\pi_A = q$. For every complex embedding $\phi : \mathrm{End}(A) \to \mathbb{C}$, τ corresponds to the complex conjugation. It follows that $|\phi(\tau_A)| = q^{1/2}$. □

Definition 6.1.2. An algebraic number satisfying the conclusion of the above theorem is called a Weil q-number.

Theorem 6.1.3 (Tate). *The homomorphism*

$$\mathrm{End}_k(A) \to \mathrm{End}_{\pi_A}(V_\ell(A))$$

is an isomorphism.

The fact that there is a finite number of abelian varieties over a finite field with a given polarization type plays a crucial role in the proof of this theorem.

Theorem 6.1.4 (Honda-Tate). (1) *The category $M(k)$ of abelian varieties over k with $\mathrm{Hom}_{M(k)}(A, B) = \mathrm{Hom}(A, B) \otimes \mathbb{Q}$ is a semi-simple category.*

(2) *The map $A \mapsto \pi_A$ defines a bijection between the set of isogeny classes of simple abelian varieties over $\mathbb{F}_q$ and the set of Galois conjugacy classes of Weil q-numbers.*

Corollary 6.1.5. *Let A, B be abelian varieties over $\mathbb{F}_q$ of dimension n. They are isogenous if and only if the characteristic polynomials of π_A on $\mathrm{H}_1(\overline{A}, \mathbb{Q}_\ell)$ and π_B on $\mathrm{H}_1(\overline{B}, \mathbb{Q}_\ell)$ are the same.*

6.2 Conjugacy classes in reductive groups

Let k be a field and G be a reductive group over k. Let T be a maximal torus of G. The finite group $W = N(T)/T$ acts on T. Let

$$T/W := \mathrm{Spec}([k[T]^W])$$

where $k[T]$ is the ring of regular functions on T, i.e. $T = \mathrm{Spec}(k[T])$ and $k[T]^W$ is the ring of W-invariants regular functions on T. The following theorem is from [19].

Theorem 6.2.1 (Steinberg). *There exists a G-invariant morphism*

$$\chi : G \to T/W,$$

which induces a bijection between the set of semi-simple conjugacy classes of $G(k)$ and $(T/W)(k)$ if k is an algebraically closed field.

If $G = \mathrm{GL}(n)$, the map

$$[\chi](k) : \{ \text{ semisimple conjugacy classes of } G(k)\} \to (T/W)(k)$$

is still a bijection for any field of characteristic zero. For an arbitrary reductive group, this map is neither injective nor surjective.

For $a \in (T/W)(k)$, the obstruction to the existence of a (semi-simple) k-point in $\chi^{-1}(a)$ lies in some Galois cohomology group H^2. In some important cases this group vanishes.

Proposition 6.2.2 (Kottwitz). *If G is a quasi-split group with G^{der} simply connected, then $[\chi](k)$ is surjective.*

For now, we will assume that G is quasi-split and G^{der} is simply connected. In this case, the elements $a \in (T/W)(k)$ are called *stable conjugacy classes*. For every stable conjugacy class $a \in (T/W)(k)$, there might exist several semi-simple conjugacy classes of $G(k)$ contained in $\chi^{-1}(a)$.

Example. If $G = \mathrm{GL}(n)$, $(T_n/W_n)(k)$ is the set of monic polynomials of degree n

$$a = t^n + a_1 t^{n-1} + \cdots + a_0$$

with $a_0 \in k^\times$. If $G = \mathrm{GSp}(2n)$, $(T/W)(k)$ is the set of pairs (P, c), where P is a monic polynomial of degree $2n$ and $c \in k^\times$, satisfying

$$a(t) = c^{-n} t^{2n} a(c/t).$$

In particular, if $a = t^{2n} + a_1 t^{2n-1} + \cdots + a_{2n}$ then $a_{2n} = c^n$. The homomorphism $\mathrm{GSp}(2n) \to \mathrm{GL}(2n) \times \mathbb{G}_m$ induces a closed immersion

$$T/W \hookrightarrow (T_{2n}/W_{2n}) \times \mathbb{G}_m.$$

Semi-simple elements of $\mathrm{GSp}(2n)$ are stably conjugate if and only if they have the same characteristic polynomials and the same similitude factors.

Let $\gamma_0, \gamma \in G(k)$ be semisimple elements such that $\chi(\gamma_0) = \chi(\gamma) = a$. Since γ_0, γ are conjugate in $G(\bar{k})$, there exists $g \in G(\bar{k})$ such that $g\gamma_0 g^{-1} = \gamma$. It follows that for every $\varsigma \in \mathrm{Gal}(\bar{k}/k)$, $\varsigma(g)\gamma_0\varsigma(g)^{-1} = \gamma$ and thus

$$g^{-1}\varsigma(g) \in G_{\gamma_0}(\bar{k}).$$

The cocycle $\varsigma \mapsto g^{-1}\varsigma(g)$ defines a class

$$\mathrm{inv}(\gamma_0, \gamma) \in \mathrm{H}^1(k, G_{\gamma_0})$$

with trivial image in $\mathrm{H}^1(k, G)$. For $\gamma_0 \in \chi^{-1}(a)$ the set of semi-simple conjugacy classes stably conjugate to γ_0 is in bijection with

$$\ker(\mathrm{H}^1(k, G_{\gamma_0}) \to \mathrm{H}^1(k, G)).$$

It happens often that instead of an element $\gamma \in G(k)$ stably conjugate to γ_0, we have a G-torsor $\mathcal{E}$ over k with an automorphism γ such that $\chi(\gamma) = a$. We can attach to the pair $(\mathcal{E}, \gamma)$ a class in $\mathrm{H}^1(k, G_{\gamma_0})$ whose image in $\mathrm{H}^1(k, G)$ is the class of $\mathcal{E}$.

Consider the simplest case where γ_0 is semisimple and strongly regular. For $G = \mathrm{GSp}$, (g, c) is semisimple and strongly regular if and only if the characteristic polynomial of g is a separable polynomial. In this case, $T = G_{\gamma_0}$ is a maximal torus of G. Let $\hat{T}$ be the complex dual torus equipped with a finite action of $\Gamma = \mathrm{Gal}(\bar{k}/k)$.

Lemma 6.2.3 (Tate-Nakayama). *If k is a non-archimedean local field, then $\mathrm{H}^1(k, T)$ is the group of characters $\hat{T}^\Gamma \to \mathbb{C}^\times$ which have finite order.*

6.3 Kottwitz triples $(\gamma_0, \gamma, \delta)$

Let $\mathcal{A}$ be the moduli space of abelian schemes of dimension n with polarizations of type D and principal N-level structure. Let $U = \mathbb{Z}^{2n}$ be equipped with an alternating form of type D

$$U \times U \to M_U,$$

where M_U is a rank one free $\mathbb{Z}$-module. Let $G = \mathrm{GSp}(2n)$ be the group of automorphisms of the symplectic module U.

Let $k = \mathbb{F}_q$ be a finite field with $q = p^r$ elements. Let $(A, \lambda, \tilde{\eta}) \in \mathcal{A}'(\mathbb{F}_q)$. Let $\overline{A} = A \otimes_{\mathbb{F}_q} \overline{k}$ and $\pi_A \in \mathrm{End}(\overline{A})$ its relative Frobenius endomorphism. Let a be the characteristic polynomial of π_A on $\mathrm{H}_1(\overline{A}, \mathbb{Q}_\ell)$. This polynomial has rational coefficients and satisfies

$$a(t) = q^{-n} t^{2n} a(q/t),$$

so that (a, q) determines a stable conjugacy class a of $\mathrm{GSp}(\mathbb{Q})$. Weil's theorem implies that this is an elliptic class in $G(\mathbb{R})$. Since GSp is quasi-split and its derived group Sp is simply connected, there exists $\gamma_0 \in G(\mathbb{Q})$ lying in the stable conjugacy class a.

The partition of $\mathcal{A}'(\mathbb{F}_q)$ with respect to the stable conjugacy classes a in $G(\mathbb{Q})$ is the same as the partition by the isogeny classes of the underlying abelian variety A (forgetting the polarization). This follows from the fact that stable conjugacy classes in GSp are intersections of conjugacy classes of GL with GSp. In the general PEL case, we need a more involved description.

Now, let us partition such an isogeny class into classes of isogenies *respecting the polarization up to a multiple* and let us pick such a class. As in section 5, we choose a base point A_0, λ_0 in this class and define the set Y whose elements are quadruples $(A, \lambda, \tilde{\eta}, f : A \to A_0)$, where $(A, \lambda, \tilde{\eta}) \in \mathcal{A}'(\mathbb{F}_q)$ and f is a quasi-isogeny transforming λ to a $\mathbb{Q}^\times$-multiple of λ_0. The isogeny class is recovered as $I(\mathbb{Q})\backslash Y$, where $I(\mathbb{Q})$ is the group of quasi-isogenies of A_0 to itself sending λ_0 to a $\mathbb{Q}^\times$-multiple of itself.

Again, as in section 5, we can write $Y = Y^p \times Y_p$, where Y^p is the subset of Y consisting of the quadruples for which the degree of the quasi-isogeny f is prime to p (i.e., is an element of $\mathbb{Z}_{(p)}^\times$) and Y_p is the subset consisting of the quadruples for which the degree of f is a power of p. We will successively describe Y^p and Y^p.

Description of Y^p. For any prime $\ell \neq p$, $\rho_\ell(\pi_A)$ is an automorphism of the adelic Tate module $\mathrm{H}_1(\overline{A}, \mathbb{A}_f^p)$ preserving the symplectic form up to a similitude factor q

$$(\rho_\ell(\pi_A)x, \rho_\ell(\pi_A)y) = q(x, y).$$

The rational Tate module $\mathrm{H}_1(\overline{A}, \mathbb{A}_f^p)$ with the Weil pairing is similar to $U \otimes \mathbb{A}_f^p$ so that π_A defines a $G(\mathbb{A}_f^p)$-conjugacy class in $G(\mathbb{A}_f^p)$. We have

$$Y^p = \{\tilde{\eta} \in G(\mathbb{A}_f^p)/K^p | \tilde{\eta}^{-1}\gamma\tilde{\eta} \in K^p\}.$$

Note that for every prime $\ell \neq p$, γ_0 and γ_ℓ are stably conjugate. In the case where γ_0 is strongly regular semisimple, we have an invariant

$$\alpha_\ell : \hat{T}^{\Gamma_\ell} \to \mathbb{C}^\times$$

which is a character of finite order.

Description of Y_p. Recall that $\pi_A : \overline{A} \to \overline{A}$ is the composite of an isomorphism $u : \sigma^r(\overline{A}) \to \overline{A}$ and the r-th power of the Frobenius $\varphi^r : \overline{A} \to \sigma^r(\overline{A})$

$$\pi_A = u \circ \varphi^r.$$

On the covariant Dieudonné module $D = \mathrm{H}_1^{cris}(\overline{A}/\mathcal{O}_L)$, the operator φ acts σ^{-1}-linearly and u acts σ^r-linearly. We can extend these actions to $H = D \otimes_{\mathcal{O}_L} L$. Let $G(H)$ be the group of self-similitudes of H and we form the semi-direct product $G(H) \rtimes \langle\sigma\rangle$. The elements u, φ and π_A can be seen as commuting elements of this semi-direct product.

Since $u : \sigma^r(A) \to A$ is an isomorphism, u fixes the lattice $u(D) = D$. This implies that

$$H_r = \{x \in H \mid u(x) = x\}$$

is an L_r-vector space of dimension $2n$ over the field of fractions L_r of $W(\mathbb{F}_{p^r})$ and equipped with a symplectic form. Self-dual lattices in H fixed by u must come from self-dual lattices in H_r.

Since $\varphi \circ u = u \circ \varphi$, φ stabilizes H_r and its restriction to H_r induces a σ^{-1}-linear operator whose inverse will be denoted by $\delta \circ \sigma$ (with $\delta \in G(L)$). We have

$$Y_p = \{g \in G(L_r)/G(\mathcal{O}_{L_r}) \mid g^{-1}\delta\sigma(g) \in K_p\mu(p^{-1})K_p\}.$$

There exists an isomorphism between H and $U \otimes L$ that transports π_A to γ_0 and carries φ to an element $b\sigma \in T(L) \rtimes \langle\sigma\rangle$. By the definition of δ, the element $N\delta = \delta\sigma(\delta)\cdots\sigma^{r-1}(\delta)$ is stably conjugate to γ_0. Following Kottwitz, the σ-conjugacy class of b in $T(L)$ determines a character

$$\alpha_p : \hat{T}^{\Gamma_p} \to \mathbb{C}^\times.$$

The set of σ-conjugacy classes in $G(L)$ for any reductive group G is described in [8].

Invariant at ∞. Over $\mathbb{R}$, T is an elliptic maximal torus. The conjugacy class of the cocharacter μ induces a well-defined character

$$\alpha_\infty : \hat{T}^{\Gamma_\infty} \to \mathbb{C}^\times.$$

Let us state Kottwitz's theorem in a particular case which is more or less equivalent to theorem 5.2.1. The proof of the general case is much more involved.

Proposition 6.3.1. *Let* $(\gamma_0, \gamma, \delta)$ *be a triple with* γ_0 *semisimple strongly regular such that* γ *and* $N\delta$ *are stably conjugate to* γ_0. *Assume that the torus* $T = G_{\gamma_0}$ *is unramified at p. There exists a pair* $(A, \lambda) \in \mathcal{A}(\mathbb{F}_q)$ *for the triple* $(\gamma_0, \gamma, \delta)$ *if and only if*

$$\sum_v \alpha_v|_{\hat{T}^\Gamma} = 0.$$

In that case there are $\ker^1(\mathbb{Q}, T)$ *isogeny classes of* $(A, \lambda) \in \mathcal{A}(\mathbb{F}_q)$ *which are mapped to the triple* $(\gamma_0, \gamma, \delta)$.

Let γ_0 be as in the statement and $a \in \mathbb{Q}[t]$ its characteristic polynomial, which is a monic polynomial of degree $2n$ satisfying the equation

$$a(t) = q^{-n} t^{2n} a(q/t).$$

The algebra $F = \mathbb{Q}[t]/a$ is a product of CM-fields which are unramified at p. The moduli space of polarized abelian varieties with multiplication by $\mathcal{O}_F$ and with a given CM type is finite and étale at p. A point $A \in \mathcal{A}(\mathbb{F}_q)$ mapped to $(\gamma_0, \gamma, \delta)$ belongs to one of these Shimura varieties of dimension 0 by letting t act as the Frobenius endomorphism Frob_q.

We can lift A to a point $\tilde{A}$ with coefficients in $W(\mathbb{F}_q)$ by the étaleness. By choosing a complex embedding of $W(\mathbb{F}_q)$, we obtain a symplectic $\mathbb{Q}$-vector space by taking the first Betti homology $\mathrm{H}_1(\tilde{A} \otimes_{W(\mathbb{F}_q)} \mathbb{C}, \mathbb{Q})$, which is equipped with a non-degenerate symplectic form and multiplication by $\mathcal{O}_F$. This defines a conjugacy class in $G(\mathbb{Q})$ within the stable conjugacy class defined by the polynomial a. For every prime $\ell \neq p$, the ℓ-adic homology $\mathrm{H}_1(A \otimes_{\mathbb{F}_q} \overline{\mathbb{F}}_q, \mathbb{Q}_\ell)$ is a symplectic vector space equipped with an action of $t = \mathrm{Frob}_q$. This defines a conjugacy class γ_ℓ in $G(\mathbb{Q}_\ell)$. By the comparison theorem, we have a canonical isomorphism

$$\mathrm{H}_1(\tilde{A} \otimes_{W(\mathbb{F}_q)} \mathbb{C}, \mathbb{Q}) \otimes_{\mathbb{Q}} \mathbb{Q}_\ell = \mathrm{H}_1(A \otimes_{\mathbb{F}_q} \overline{\mathbb{F}}_q, \mathbb{Q}_\ell)$$

compatible with the action of t so that the invariant $\alpha_\ell = 0$ for $\ell \neq p$.

The cancellation between α_p ans α_∞ is essentially the equality $\varphi = \mu(p)(1 \otimes \sigma^{-1})$ occuring in the proof of Theorem 5.2.1. □

Kottwitz stated and proved a more general statement for all γ_0 and for all PEL Shimura varieties of type (A) and (C). In particular, he derived a formula for the number of points on $\mathcal{A}$

$$\mathcal{A}(\mathbb{F}_q) = \sum_{(\gamma_0,\gamma,\delta)} n(\gamma_0, \gamma, \delta) T(\gamma_0, \gamma, \delta),$$

where $n(\gamma_0, \gamma, \delta) = 0$ unless the Kottwitz vanishing condition is satisfied. In that case

$$n(\gamma_0, \gamma, \delta) = \ker^1(\mathbb{Q}, I)$$

and

$$T(\gamma_0, \gamma, \delta) = \mathrm{vol}(I(\mathbb{Q}\backslash I(\mathbb{A}_f)) O_\gamma(1_{K^p}) TO_\delta(1_{K_p \mu(p^{-1}) K_p}),$$

where I is an inner form of G_{γ_0}.

It is expected that this formula can be compared to the Arthur-Selberg trace formula, see [9].

Acknowledgements

We warmly thank J. Tilouine for inviting us to lecture on this subject in Villetaneuse in 2002. We are grateful to the participants of IHES summer school, particularly O. Gabber and S. Morel for many useful comments about the content of these lectures. We are especially thankful to Fu Lei for his careful reading of the manuscript.

References

[1] W. Baily and A. Borel, *Compactification of arithmetic quotients of bounded symmetric domains*, Ann. of Math 84(1966).

[2] S. Bosch, W. Lutkebohmert, M. Raynaud, *Neron models*, Ergeb. der Math. 21. Springer Verlag 1990.

[3] P. Deligne, *Travaux de Griffiths*, séminaire Bourbaki 376, in Springer Lecture Notes in Math. 180, Springer Verlag

[4] P. Deligne, *Travaux de Shimura*, séminaire Bourbaki 389, in Springer Lecture Notes in Math. 244, Springer Verlag (1971) 123–165.

[5] P. Deligne, *Interprétation modulaire et techniques de construction de modèles canoniques*. Proc. Symp. Pure Math. **33** (2) (1979) 247–289.

[6] M. Kisin, *Integral models for Shimura varieties of abelian type*, J.A.M.S. **23** (2010), no. 4, 967-1012

[7] R. Kottwitz, *Rational conjugacy classes in reductive groups*, Duke Math. J. **49** (1982), no. 4, 785-806.

[8] R. Kottwitz, *Isocrystals with additional structure*, Compositio Math. 56 (1985), no. 2, 201–220.

[9] R. Kottwitz, *Shimura varieties and λ-adic representations*, Automorphic forms, Shimura varieties, and L-functions, Vol. I (Ann Arbor, MI, 1988), 161-209, Perspect. Math., 10, Academic Press, Boston, MA, 1990.

[10] R. Kottwitz, *Points on some Shimura varieties over finite fields*, J. Amer. Math. Soc., **5** (1992), 373-444.

[11] W. Messing, *The crystals associated to Barsotti-Tate groups : with applications to abelian schemes*, Springer Lecture Notes in Math. 264, Springer Verlag.

[12] J. Milne, *Introduction to Shimura varieties* in Harmonic analysis, the trace formula, and Shimura varieties, 265–378, Amer. Math. Soc. 2005.

[13] B. Moonen, *Models of Shimura varieties in mixed characteristics*. in Galois representations in arithmetic algebraic geometry, 267–350, Cambridge Univ. Press, 1998.

[14] D. Mumford, *Abelian varieties*, Tata Institute of Fundamental Research Studies in Mathematics, No. 5, Oxford University Press, London 1970.

[15] D. Mumford, *Geometric invariant theory,* Ergebnisse der Mathematik und ihrer Grenzgebiete, Neue Folge, Band 34 Springer-Verlag 1965.

[16] J-P Serre, *Rigidité du foncteur de Jacobi d'échelon $n \geq 3$*, appendice à l' exposé 17 du séminaire Cartan 1960–61.

[17] D. Mumford, *Families of abelian varieties*, in Algebraic Groups and Discontinuous Subgroups, PSPM 9, pp. 347–351 Amer. Math. Soc.

[18] Platonov, Rapinchuk *Algebraic groups and number theory,* Pure and Applied Mathematics, 139. Academic Press, Inc., Boston, MA, 1994

[19] R. Steinberg *Conjugacy classes in algebraic groups* Springer Lecture Notes in Math. 366 Springer Verlag 1974.

[20] A. Vasiu *Integral canonical models of Shimura varieties of preabelian type*, fully corrected version, available at `people.math.binghamton.edu/adrian/#reductive`

UNITARY SHIMURA VARIETIES

MARC-HUBERT NICOLE

Aix Marseille Université, CNRS, Centrale Marseille
I2M, UMR 7373, 13453 Marseille, France
Email: marc-hubert.nicole@univ-amu.fr

Physical address:
Institut de Mathématiques de Marseille
Université d'Aix-Marseille, campus de Luminy, case 907
13288 Marseille cedex 9, FRANCE

Contents

1 Introduction

This chapter is an introduction to some of the trappings (and, inevitably, some of the traps) of compact unitary Shimura varieties. We focus on those Shimura varieties and tools which are instrumental in the production of the Galois representations in this book.

In some places, we shall repeat information already contained in other chapters to make this expository work more legible on its own. It is natural, before studying unitary Shimura varieties, to review e.g. modular and/or Shimura curves and to start perhaps with Siegel varieties as exposed in the chapter by Genestier-Ngô (a quick reading guide would be to read Sections 1 (on Siegel varieties), 2.6 (on the adelic description) and 3 (on PEL-type varieties) of Genestier-Ngô). To try making our description more concrete, we use examples of signature $(n, 1)$ for $n \geq 2$, which arise frequently in applications.

Reading guide: It might be wise to pick up the notation as necessary from Section 2 but skip the details about PEL data at first reading. When some passage or concept in the literature on Shimura varieties was reported to me as a possible source of confusion, I attempted a clarification, advertised with the symbol: $\lessdot$.

2 Unitary PEL-type Shimura data

The unitary Shimura varieties we are interested in can be studied via moduli theory. When studying their integral models and reduction modulo p in further chapters, this is immediately useful. Even over $\mathbb{C}$ this moduli interpretation is already worth mentioning, as the objects involved, polarized abelian varieties, can be described explicitly as quotients $\mathbb{C}^g/\Lambda$ by cocompact lattices Λ. This is explained in the case of Siegel varieties in the chapter by Genestier-Ngô. Here we introduce the additional structures which become necessary to describe the moduli problem in the unitary case, both in the isomorphism and isogeny categories of abelian varieties, relying heavily on Kottwitz's work [**17**].

2.1 The PEL data

Recall that PEL stands for Polarization, Endomorphisms and Level structures. Very roughly speaking, a polarization is a morphism $\lambda : A \longrightarrow A^\vee$ between an abelian variety A and its dual $A^\vee$, induced by a line bundle on A; and a level structure, at least over $\mathbb{C}$, is a collection of points of finite order. Polarizations and level structures already play a prominent rôle for Siegel varieties and they have been defined in Genestier-Ngô's chapter, so we focus on the endomorphism datum. To that end, we introduce some arithmetic data to enrich the moduli problem of polarized abelian varieties. Let A be an abelian variety over $\mathbb{C}$, and let $\lambda : A \longrightarrow A^\vee$ be a polarization. Recall that $\mathrm{End}(A) \otimes \mathbb{Q}$ is a finite dimensional semisimple algebra with a positive (Rosati) involution. The PEL data introduce structures which reflect in particular the properties of $\mathrm{End}(A) \otimes \mathbb{Q}$ with the goal of studying the corresponding family of abelian varieties. For a concise and insightful introduction to PEL data over $\mathbb{C}$ in the compact case, see also the expository paper of Shimura himself: [**26**, Sections 4–5].

Definition 2.1. A unitary PEL datum is a quintuple $(F, D, *, V, \langle , \rangle)$.

We explain all the ingredients in the PEL datum. Let F be a CM field[1] that is, a totally imaginary quadratic extension of a totally real field F^+. For example, it is a good idea to keep in mind throughout the text the example of an imaginary quadratic extension of $\mathbb{Q}$; or even better: to assume that the CM field F contains any imaginary quadratic field.[2]

Let D be a simple division $\mathbb{Q}$-algebra of dimension d^2 over its center F, and let $*$ be a positive involution of the second kind, which means that $*$ restricted to $\mathrm{Gal}(F/F^+)$ is the non-trivial element, complex conjugation.[3] As noted above, the positivity condition on the involution is imposed from the positivity of the Rosati involution on endomorphism algebras of abelian varieties. In Albert's classification (see [**26**, p.317]), our algebra is of type IV. Over $\mathbb{R}$, recall that the $\mathbb{R}$-algebra becomes familiar: $D \otimes \mathbb{R}$ is the product of $[F^+ : \mathbb{Q}]$ copies of $\mathbb{M}_d(\mathbb{C})$, the complex matrix algebra with the usual involution induced by complex conjugation and transposition.

Remark 2.2. Note that in the general PEL set-up, there is no need to restrict ourselves to division algebras, as is explained in other chapters. There is no loss in doing so for the purpose of applications to Galois representations attached to *self-dual* automorphic representations, see below.

To give the PEL construction extra flexibility, we introduce a (left) D-module V (see [**16**, Section 1] and [**13**, Section 5.2, p.15–17] for the case $V = D$ and $F^+ = \mathbb{Q}$ and extra details). Indeed, let V be a D-module of finite dimension, equipped with a non-degenerate $\mathbb{Q}$-valued alternating form $\langle , \rangle$ i.e.,

$$\langle dv, w\rangle = \langle v, d^* w\rangle \quad \forall v, w \in V \text{ and } d \in D.$$

As D is a division algebra, it follows automatically that $V \cong D^r$ for some $r \in \mathbb{N}$.

Remark 2.3. If $r = 1$, the Shimura variety constructed from the PEL data is automatically compact.

Up to choosing a basis of V over D, $V \cong D^r$ implies that $\mathrm{End}_D(V) \cong M_r(D^{opp})$, where D^{opp} is the opposite algebra.[4]. This is an algebra with an involution[5] that is defined as the adjoint involution of $\langle , \rangle$, and, by the Skolem-Noether theorem, it can be written down explicitly as $c \mapsto s\, {}^t c^* s^{-1}$ for some matrix $s \in \mathbb{M}_r(D^{opp})$ such that $-s = {}^t s^*$, depending on $\langle , \rangle$ and the choice of basis of V over D. Furthermore, we are given a complex structure on $(V, \langle , \rangle) \otimes \mathbb{R}$. This complex structure is encoded as a choice of a morphism

$$h : \mathbb{C} \longrightarrow \mathrm{End}_D(V_{\mathbb{R}}),$$

[1] CM = **C**omplex **M**ultiplication.

[2] If not, the shape of the Galois representations arising in the cohomology of the corresponding Shimura varieties is more complicated, see Harris's note in this book.

[3] The other possibility is when the center of D is a totally real field, and this is called an involution of the first kind.

[4] $x \cdot^{opp} y := y \cdot x$

[5] Beware that in general $\mathrm{End}_D(V)$ is not necessarily an algebra with a *positive* involution.

which may be seen as part of the PEL data, and which satisfies the following two properties: After choosing a square root[6] of -1, we require that $\langle \bullet, h(i)\bullet \rangle_{\mathbb{R}}$ is symmetric and positive definite, cf. the definition of a Riemann form. Moreover, $h(z)$ is an adjoint operator to $h(\overline{z})$ with respect to $\langle , \rangle$ i.e., $\langle h(z)\bullet, \bullet \rangle = \langle \bullet, h(\overline{z})\bullet \rangle$.[7] The homomorphism $h : \mathbb{C} \longrightarrow \mathrm{End}_D(V_{\mathbb{R}})$ induces a decomposition

$$V \otimes_{\mathbb{Q}} \mathbb{C} = V_1 \oplus V_2,$$

where $h(z)$ acts via z on V_1 (resp. $\overline{z}$ on V_2). Chose $\beta_1, \ldots, \beta_r$, a $\mathbb{Q}$-basis of D. The determinant polynomial $\det_{V_1}$ is defined as:

$$\det\left(x_1\beta_1 + \cdots + x_r\beta_r, V_1 \otimes \mathbb{C}[x_1, \ldots, x_r]\right),$$

This is a homogeneous polynomial of degree $\dim_{\mathbb{C}}(V_1)$. Its coefficients generate a number subfield of $\mathbb{C}$ which is independent of the choice of basis. This number field is called the reflex field E, and it is also the field of definition of the isomorphism class of the $D_{\mathbb{C}}$-module V_1.

2.2 The Shimura datum

Let (G, X) be a Shimura datum, that is, G is a reductive group defined over $\mathbb{Q}$, and X is a $G(\mathbb{R})$-conjugacy class of morphisms

$$h : \mathbb{S} \longrightarrow G_{\mathbb{R}},$$

where $\mathbb{S} = \mathrm{Res}_{\mathbb{C}/\mathbb{R}}\mathbb{G}_m$, satisfying the three axioms of Deligne (see Chap. 4 of Genestier-Ngô's chapter).[8] In particular, X is necessarily a conjugacy class of a *minuscule* cocharacter usually denoted $\mu := \mu_h : \mathbb{G}_{m,\mathbb{C}} \longrightarrow G_{\mathbb{C}}$, for $\mu(z) := h_{\mathbb{C}}(z, 1)$.

Indeed, X may be identified with $G(\mathbb{R})/K_\infty$ for K_∞ the centralizer of a given h in X, and Deligne's axioms ensure that the connected components of $G(\mathbb{R})/K_\infty$ are Hermitian symmetric domains. One of the guiding threads for Deligne's axiomatization of Shimura varieties is to try viewing them as moduli spaces of suitable motives. More concretely, from Hodge theory it is known that all Hermitian symmetric domains are moduli spaces of Hodge structures, see [**10**, Cor. 1.1.17]. Of course, abelian varieties are a very special kind of Hodge structures. We come back to this point below with explicit details in the proof of Prop. 3.6.

As recalled below in Section 3.2, to a pair (G, X) we can associate a projective system $\{Sh_K(G, X)(\mathbb{C})\}_K$, indexed by K any compact, open subgroup of $G(\mathbb{A}^\infty)$. This is called the Shimura variety associated to the Shimura datum (G, X). The compactness of the subgroups

---

[6] Since we insist the alternating form has values in $\mathbb{Q}$ and not the twist $\mathbb{Q}(1) := \mathbb{Q} \otimes \mathbb{Z}(1)$, where $\mathbb{Z}(1)$ is the kernel of the exponential map $\exp(x) : \mathbb{C} \longrightarrow \mathbb{C}^\times$, we need to choose a square root of $\sqrt{-1}$, in effect determining a (non-canonical) isomorphism $\frac{1}{2\pi\sqrt{-1}} : \mathbb{Q}(1) \xrightarrow{\cong} \mathbb{Q}$.

[7] This is usually noted: $h(\overline{z}) = h(z)^*$, using the notation $*$ also for the involution on $\mathrm{End}_D(V)$.

[8] Note that this map $h : \mathbb{S} \longrightarrow G_{\mathbb{R}}$ is the inverse of the map $h : \mathbb{C} \longrightarrow \mathrm{End}_D(V_{\mathbb{R}})$ defined in the previous section.

$K \subset G(\mathbb{A}^\infty)$ we consider implies that the groups $\Gamma_K := G(\mathbb{Q}) \cap K$ are discrete in $G(\mathbb{R})$. Combined with openness, it translates in classical terms in Γ_K being congruence subgroups, a special subclass of arithmetic subgroups.

In this section, we content ourselves with explaining how to get the group G in the unitary PEL-type set-up, as "unitary" is often used in a loose sense. We provide below an explicit description of the Hermitian symmetric domains arising in $G(\mathbb{R})/K_\infty$ in Section 3.1.

We define concretely the unitary group and the group of unitary similitudes. Both groups are connected reductive groups over $\mathbb{Q}$.

Definition 2.4. Let R be a $\mathbb{Q}$-algebra.

(1) Define

$$U(R) := \mathrm{Aut}_{D\otimes_{\mathbb{Q}}R}(V \otimes_{\mathbb{Q}} R, \langle , \rangle),$$

i.e., $g \in \mathrm{U}(R)$ is an element of $\mathrm{GL}_{D\otimes R}(V \otimes R)$ such that

$$\langle gx, gy\rangle = \langle x, y\rangle \quad \forall x, y \in V.$$

(2) Define

$$\mathrm{GU}(R) := \mathrm{Aut}_{D\otimes_{\mathbb{Q}}R}(V \otimes_{\mathbb{Q}} R, R^\times \cdot \langle , \rangle),$$

i.e., $g \in \mathrm{GU}(R)$ is an element of $\mathrm{GL}_{D\otimes R}(V \otimes R)$ such that there exists an $r \in R^\times$ such that

$$\langle gx, gy\rangle = r \cdot \langle x, y\rangle \quad \forall x, y \in V.$$

An alternative and more standard definition (which is equivalent as long as $V \neq 0$) of GU goes as follows:

$$\mathrm{GU}(R) \subset \mathrm{Aut}_{D\otimes\mathbb{Q}}(V \otimes \mathbb{Q}, R^\times \cdot \langle , \rangle) \times \mathbb{G}_m(R),$$

i.e., $(g, r) \in \mathrm{GU}(R)$ is an element of $\mathrm{GL}_{D\otimes R}(V \otimes R) \times \mathbb{G}_m(R)$ such that

$$\langle gx, gy\rangle = r \cdot \langle x, y\rangle \quad \forall x, y \in V.$$

As g uniquely determines r (as long as V is non-trivial), these two groups are canonically isomorphic.

To be precise, the $\mathbb{Q}$-group U defined above is an inner form of a standard unitary group (see Section 3.1 below). E.g., in the case V is a single copy of D, the factorization over $\mathbb{R}$ of $V_{\mathbb{C}}$ according to embeddings $\tau \in I := \{\tau_i | \tau_i : F \hookrightarrow \mathbb{C}, i = 1, \dots, [F^+ : \mathbb{Q}]\}$[9] gives rise to the isomorphism:

$$G(\mathbb{R}) \cong \prod_{\tau \in I} \mathrm{GU}(p_\tau, q_\tau)(\mathbb{R}),$$

for signatures (p_τ, q_τ), $p_\tau + q_\tau = d$, and where (p_τ, q_τ) is the signature indexed by τ.

[9] i.e., one embedding per pair of conjugate complex embeddings.

2.3 Moduli interpretations

There are (at least) two moduli interpretations in terms of abelian varieties for PEL-type Shimura varieties. The first one, perhaps more natural, considers abelian schemes up to isomorphism; the second considers them up to isogeny, usually controlled, for example a prime-to-p isogeny. The isogeny category allows the considerations of other kinds of level structures than the principal level structures. We give a simplified account of both moduli problems. For details on compatibilities, the determinant condition and the precise definition of the level structures, we refer to Genestier-Ngô.

To allow a quick presentation, let G be an integral model of the group GU, so that $G(\widehat{\mathbb{Z}})$ is defined; cf. Lem. 5.3.1 of Genestier-Ngô states that smooth integral models of G_ℓ are unique over $\mathbb{Z}_\ell$ compatibly with the level structure.

For short, we abuse terminology and also call an abelian scheme over a complex analytic base S the analytic variant of a proper smooth group object with connected fibers i.e., a family $\mathcal{A} \xrightarrow{\pi} S$ where $\mathcal{A}$ is compact group object, and π is a C^∞-submersion.

2.3.1 The isomorphism category

Let K be an open compact subgroup of $G(\widehat{\mathbb{Z}})$.[10] The moduli problem $\mathcal{M}_{K,\cong}$ associates to any complex analytic space S over $\mathbb{C}$ the quadruple

$$\underline{A} = (A, \iota, \lambda, \overline{\eta})/_{\cong},$$

where:

- A is an abelian scheme over S;
- the map $\iota : D \longrightarrow \mathrm{End}_S(A) \otimes \mathbb{Q}$ is a $\mathbb{Q}$-algebra homomorphism;
- the map $\lambda : A \longrightarrow A^\vee$ is a polarization of A;
- the Lie algebra $\mathrm{Lie}(A)$ satisfies the determinant condition as a $D \otimes \mathcal{O}_S$-module;
- $\overline{\eta}$ is an (integral) K-level structure.

Moreover, the data of ι and λ are intertwined: the polarization λ is $(D, *)$-linear, and it induces the Rosati involution on $\mathrm{End}^0(A)$, which coincides with the positive involution (of the second kind) on the division algebra D.

The notion of equivalence $\cong$ of quadruples $\underline{A} \cong \underline{A}'$ is given by isomorphisms $A \cong A'$ of abelian schemes over S which are compatible with the polarization, the endomorphism and level structures.

2.3.2 The isogeny category

Let K be an open compact subgroup of $G(\mathbb{A}^\infty)$. The moduli problem $\mathcal{M}_{K,\sim}$ associates to any complex analytic space S over $\mathbb{C}$ the quadruple

$$\underline{A} = (A, \iota, \lambda, \overline{\eta})/_{\sim},$$

[10] Here we do not allow $G(\mathbb{A}^\infty)$.

where:

- A is an abelian scheme over S;
- the map $\iota : D \longrightarrow \mathrm{End}_S(A) \otimes \mathbb{Q}$ is a $\mathbb{Q}$-algebra homomorphism;
- the map $\lambda : A \longrightarrow A^\vee$ is a $\mathbb{Q}$-polarization of A;
- the Lie algebra Lie(A) satisfies the determinant condition as a $D \otimes \mathcal{O}_S$-module;
- $\overline{\eta}$ is a K-level structure.

Moreover, the data of ι and λ are intertwined as in the isomorphism category: we recall that the $(D, *)$-linear polarization λ induces the Rosati involution on $\mathrm{End}^0(A)$, which coincides with the involution (of the second kind) on the division algebra D.

Remark 2.5. When $S = \mathbb{C}$, the determinant condition amounts to saying that Lie(A) is isomorphic to V_1 as $D_\mathbb{C}$-modules. Further, over $\mathbb{C}$, we may impose the stronger requirement that

$$(V, \mathbb{Q}^\times \cdot \langle , \rangle) \cong \left(H_1(A, \mathbb{Q}), \mathbb{Q}^\times \cdot \langle , \rangle_\lambda\right), \tag{1}$$

where $\langle , \rangle_\lambda$ is the $*$-Hermitian structure induced by the polarization λ. Also, the determinant condition, in the unitary case, restricts the signature considered, as opposed to considering all signatures at once.

The notion of equivalence $\sim$ of quadruples $\underline{A} \cong \underline{A}'$ is given by $\mathbb{Q}$-isogenies $A \sim A'$ of abelian schemes over S which are compatible with the polarization (up to $\mathbb{Q}^{\times,+}$), the endomorphism and level structures. [11] The following comparison theorem is proved in the chapter of Genestier-Ngô:

Theorem 2.6. *Let $K \subset G(\widehat{\mathbb{Z}})$. There is a canonical isomorphism:*

$$\mathcal{M}_{K,\cong} \xrightarrow{\cong} \mathcal{M}_{K,\sim}.$$

Remark 2.7. The argument sketched in Genestier-Ngô for representability of the PEL moduli in the algebraic category is a dévissage to the case of Siegel moduli schemes, where it is a major result of Mumford's Geometric Invariant Theory.

2.3.3 *Digression on polarizations*

$\lessdot$ A possible source of confusion stems from polarization data, and this is reflected at the level of the groups we consider.

For example, why consider GU instead of U? The difference is not so clear in the Siegel case. Indeed, the problem is that if the group U is not e.g., split over $\mathbb{Z}$, there might not exist any self-dual lattice with additional properties that is, no *principal* polarization. Hence the need to relax the polarization condition to allow non-principal degrees. For example,

[11] Elsewhere in the book, $\mathbb{Q}^\times$-isogenies are replaced e.g., by $\mathbb{Z}_{(p)}^\times$-isogenies, etc.

we may decide to consider not only an element, but rather a $\mathbb{Q}$-line (or a polarization up to a positive unit, a module with a notion of positivity, etc.) in the polarization module $\mathrm{Hom}(A, A^\vee)^{sym}$ composed of symmetric[12] isogenies between A and $A^\vee$. In Subsection 2.3, we explain the moduli problem with $\mathbb{Q}$-polarizations with associated group GU. A closely related moduli problem involving instead F^+-lines in a suitable sense (by considering F^+-valued Hermitian pairings instead of $\mathbb{Q}$-valued pairings) may be defined by considering a slightly larger (but still connected) group. Beware that in general the moduli functor is actually never representable because of the presence of infinitely many units (hence, automorphisms), so one considers instead the associated coarse moduli scheme (that is, with functorial isomorphisms over k-points, for any algebraically closed field k). Let

$$\mathrm{GU}_{F^+}(R) := \mathrm{Aut}_{D\otimes\mathbb{Q}}(V \otimes_{\mathbb{Q}} R, (F^+ \otimes R)^\times \cdot \langle,\rangle),$$

Recall that we have the short exact sequence, with the map $g \mapsto r$ as above:

$$1 \longrightarrow U \longrightarrow \mathrm{GU} \overset{g\mapsto r}{\longrightarrow} \mathbb{G}_m \longrightarrow 1.$$

The definition of GU_{F^+} is obtained by replacing $\mathbb{G}_m$ by $\mathrm{Res}_{F^+/\mathbb{Q}}\mathbb{G}_m$:

$$1 \longrightarrow U \longrightarrow \mathrm{GU}_{F^+} \longrightarrow \mathrm{Res}_{F^+/\mathbb{Q}}\mathbb{G}_m \longrightarrow 1.$$

The Hermitian symmetric domain associated to GU_{F^+} is the same as for GU, see below.

Remark 2.8. The center Z of GU_{F^+} is cohomologically trivial, as $Z \cong \mathrm{Res}_{F/\mathbb{Q}}\mathbb{G}_m$, by Hilbert's Satz 90.

See [**14**, Section 7.1.3] for more details, but beware that our notations are different (Hida's group GU is our GU_{F^+}).

3 Compact complex varieties

3.1 The explicit Hermitian symmetric domain $\mathcal{D}_{p,q}$

We give a few different realizations of the Hermitian symmetric domain arising in the general unitary case, and specialize in examples to signatures $(n, 1)$ and (n, n). More details are contained in [**14**, 5.2.1], cf. [**27**].

Let us recall a general description valid for any Shimura datum. Given the datum of a reductive group G over $\mathbb{Q}$, it is often practical to write $X = G(\mathbb{R})/K_\infty$, where $G(\mathbb{R})$ are the $\mathbb{R}$-points of G, and $K_\infty \subset G(\mathbb{R})$ is the centralizer of h.
∢ This, in general, is not a connected Hermitian symmetric domain, as $\pi_0(X) \cong G(\mathbb{R})/G(\mathbb{R})_+$, and $G(\mathbb{R})$ does not necessarily coincide with $G(\mathbb{R})_+$, where $G(\mathbb{R})_+$ is the inverse image in $G(\mathbb{R})$ of the identity component of $G^{ad}(\mathbb{R})$.

For the unitary Shimura varieties we are interested in, the Hermitian symmetric domain is a product of spaces indexed by (p, q) of all p-by-q complex matrices u such that $1_p - {}^t\overline{u}u$ is positive hermitian. If $p = 0$ or $q = 0$, the convention is that the space consists of only

[12] An isogeny is symmetric if $\lambda = \lambda^\vee$.

one point. We assume $p \geq q$; note that in Section 3.3.1, the cases (p, q) and (q, p) are not equivalent.

The unbounded realizations of symmetric Hermitian domains are useful for *non-compact* Shimura varieties (compactifications, expansion at the cusps, etc.).

Let $I_{p,q} := \begin{pmatrix} 1_p & 0 \\ 0 & -1_q \end{pmatrix}$. The standard real unitary group of signature (p, q) is:

$$\mathrm{U}_{p,q}(\mathbb{R}) := \left\{ g \in \mathrm{GL}_{p+q}(\mathbb{C}) | {}^t\overline{g} I_{p,q} g = I_{p,q} \right\}.$$

The Hermitian symmetric domain is defined as follows:

Definition 3.1 (Bounded realization).

$$\mathcal{D}_{p,q} := \left\{ u \in \mathbb{M}_{p,q}(\mathbb{C}) | \, {}^t \begin{pmatrix} \overline{u} \\ 1 \end{pmatrix} \begin{pmatrix} 1_p & 0 \\ 0 & -1_q \end{pmatrix} \begin{pmatrix} u \\ 1 \end{pmatrix} = {}^t\overline{u}u - 1_q < 0 \right\}.$$

Example 3.2. *Let $p = q = 1$. Then $\mathcal{D}_{1,1} = \{u \in \mathbb{C} | \, 1 > |u|\} \subset \mathbb{C}$ is the unit disc.*

The group $\mathrm{U}_{p,q}(\mathbb{R})$ acts on $\mathcal{D}_{p,q}$ by identifying the latter with the quotient of the space of matrices $\left\{ X \in \mathbb{M}_{d,d}(\mathbb{C}) | X^* I_{p,q} X = \begin{pmatrix} A & 0 \\ 0 & -B \end{pmatrix}, A, B > 0 \right\}$ by $\mathrm{GL}_p(\mathbb{C}) \times \mathrm{GL}_q(\mathbb{C})$, see proof of Cor. 3.9 for an indication and [**14**, Prop. 5.7] for fuller details.

An equivalent definition of the real unitary group $\mathrm{U}_{p,q}(\mathbb{R})$ may be given by using the matrix:

$$J_{p,q} := \begin{pmatrix} & & 1_q \\ & S & \\ -1_q & & \end{pmatrix},$$

a $(p + q)$-by-$(p + q)$ $*$-Hermitian matrix such that $-iS > 0$ i.e., positive definite, where S is a certain $(p - q)$-by-$(p - q)$ matrix; cf. [**27**, Chap. I, Sec. 6], cf. with [**14**, p.240, last paragraph]. Then

$$\mathrm{U}_{p,q}(\mathbb{R}) = \left\{ g \in \mathrm{GL}_{p+q}(\mathbb{C}) | \, {}^t\overline{g} J_{p,q} g = J_{p,q} \right\}.$$

Definition 3.3. The (p, q)-upper half space $\mathcal{H}_{p,q}$ is defined as pairs of matrices $(Z, W) \in \mathbb{M}_{q,q}(\mathbb{C}) \times \mathbb{M}_{(p-q)\times q}(\mathbb{C})$ such that

$$-i \begin{pmatrix} \overline{Z} & \overline{W} & 1 \end{pmatrix} J_{p,q} \begin{pmatrix} Z \\ W \\ 1 \end{pmatrix} = -i \left({}^t\overline{Z} - Z + {}^t\overline{W} S W \right) < 0.$$

Example 3.4. *Let $p = q = 1$. Then $\mathcal{H}_{1,1}$ is the upper half plane.*

We refer to [**14**, Prop. 5.9] for the details of the group action of $\mathrm{U}_{p,q}(\mathbb{R})$ on $\mathcal{H}_{p,q}$, which a bit more notationally involved than for the bounded realization.

Fact 3.5. (1) *The two groups denoted by* $\mathrm{U}_{p,q}(\mathbb{R})$ *are isomorphic.*

(2) *The Hermitian symmetric domain* $\mathcal{D}_{p,q}$ *(bounded realization) is isomorphic to* $\mathcal{H}_{p,q}$ *(unbounded realization).*

3.1.1 Example: $\mathrm{U}(n,1)$

The Hermitian symmetric domain $\mathcal{D}_{n,1}$ can be described in many different manners. The first two descriptions generalize directly to arbitrary signatures.

1. The following description of $\mathcal{D}_{n,1}$ makes clear where the complex structure comes from. We backtrack for a moment to signature $(2,1)$ for simplicity of presentation. $\mathcal{D}_{2,1}$ can be viewed as a subvariety of a Grassmannian variety. Fix an embedding of an imaginary quadratic field $F \hookrightarrow \mathbb{C}$. Then the real vector space $V(\mathbb{R})$ has dimension 3. Fix a basis of $V(\mathbb{R})$, and consider the matrix $J = \begin{pmatrix} 1 & & \\ & 1 & \\ & & -1 \end{pmatrix}$ in that basis.

   The space $Gr_1(V(\mathbb{R}))$ is the Grassmannian variety of complex lines in $V(\mathbb{R})$, which clearly has a complex structure. We embed $\mathcal{D}_{2,1}$ as an open complex submanifold of $Gr_1(V(\mathbb{R}))$ as the space of complex lines on which J is negative definite.

2. We recast the previous concrete description slightly more abstractly in terms of the PEL data. Let $(V, \langle,\rangle)$ be a $D \otimes \mathbb{R}$-module equipped with an alternating form $\langle,\rangle$, such that $\mathrm{Aut}_D(V, \langle,\rangle) \cong \mathrm{U}(n,1)(\mathbb{R})$, where $(n,1)$ is the signature of $(V, \langle,\rangle)$. In this example, we drop the $\mathbb{R}$ from the notation, and we write $V = V_{\mathbb{R}}$. Let $V = V_+ \perp V_-$ be a decomposition of V such that the Hermitian form associated to $\langle,\rangle$ is positive definite on V_+ and negative definite on V_-. Let I be the $\mathbb{C}$-linear endomorphism of V coinciding with $\pm i$ on $V_\pm$. Then $I \in \mathcal{D}_{n,1}$, and the stabilizer of I in $G(\mathbb{R})$ is:

   $$\mathrm{U}(V_+) \times \mathrm{U}(V_-) \cong \mathrm{U}(n) \times \mathrm{U}(1).$$

   To sum up, we get that $\mathcal{D}_{n,1} = \mathrm{U}(n,1)/\big(\mathrm{U}(n) \times \mathrm{U}(1)\big)$.

3. The complex unit ball in $\mathbb{C}^n$. It follows by choosing a basis $z_1, \ldots, z_{n+1}$ generalizing the basis in (2) above, getting:

   $$\left\{(z_1, z_2, \ldots, z_{n+1}) \in \mathbb{C}^{n+1} \mid |z_1|^2 + |z_2|^2 + \cdots + |z_n|^2 - |z_{n+1}|^2 < 0\right\} / \mathbb{C}^\times,$$

   and by dividing by z_{n+1}^2, we get the unit ball in $\mathbb{C}^n$. This is the most concrete geometric description, and of course it uses crucially that the signature is $(n,1)$. We leave it as an exercise to the reader to make the action of $\mathrm{U}(n,1)(\mathbb{R})$ explicit on the unit ball.

3.1.2 Example: $\mathrm{U}(n,n)$

Let $p = q = n$. In this case,

$$\mathcal{H}_{p,q} \equiv \mathcal{H}_{n,n} = \left\{Z \in \mathbb{M}_{n,n}(\mathbb{C}) \mid -i({}^t\overline{Z} - Z) < 0\right\}.$$

Since $\mathbb{M}_{n,n}(\mathbb{C})$ is the same $\mathbb{C}$-vector space as two copies of the spaces of n-by-n Hermitian matrices (as any matrix is the sum $M + iM'$ of a Hermitian matrix M and a skew-Hermitian matrix iM', for M' also a Hermitian matrix), the last condition is equivalent to the imaginary part of Z being positive definite i.e., $\mathrm{Im}(Z) > 0$.

3.2 Complex points as double coset spaces

We describe the complex points of Shimura varieties explicitly in terms of the group G and the Hermitian symmetric domain X arising in the Shimura datum (G, X). This is how Shimura varieties are first introduced in Deligne's papers [**9**], [**10**]. Let $X = G(\mathbb{R})/K_\infty$ be a Hermitian symmetric domain as in the previous section. Write $G(\mathbb{Q})$ (resp. $G(\mathbb{A}^\infty)$) for the $\mathbb{Q}$-points (resp. $\mathbb{A}^\infty$-points) of G.[13] Let K be an open, compact subgroup of $G(\mathbb{A}^\infty)$. $\sphericalangle$ To avoid confusion, we spell out in complete detail how we obtain the double coset space (mind the parentheses!):

$$G(\mathbb{Q})\backslash\big(X \times G(\mathbb{A}^\infty)/K\big).$$

The group K acts on $G(\mathbb{A}^\infty)$ (and not on X) by multiplication on the right, and $G(\mathbb{Q})$ acts diagonally on the left on $X \times G(\mathbb{A}^\infty)$, where the action on $G(\mathbb{A}^\infty)$ is by multiplication on the left, and the action on X is via $G(\mathbb{Q}) \hookrightarrow G(\mathbb{R})$. Note that in this convention, we take the quotient on the same side as the action.

3.2.1 Connected components

Let X^+ be a connected component of X. Since $G(\mathbb{Q}) \subset G(\mathbb{R})$ is dense, $G(\mathbb{Q})$ acts transitively on the set of connected components, and the stabilizer of X^+ in $G(\mathbb{Q})$ is

$$G(\mathbb{Q})_+ := G(\mathbb{R})_+ \cap G(\mathbb{Q}).$$

We therefore obtain another double coset space description of $Sh_K(G, X)$ as:

$$G(\mathbb{Q})_+\backslash X^+ \times G(\mathbb{A}^\infty)/K.$$

This dissection is useful to clarify the link with the classical (i.e., non-adelic) description of varieties that we already mentioned above. Let $\mathcal{R}$ be a system of representatives for the double cosets $G(\mathbb{Q})_+\backslash G(\mathbb{A}^\infty)/K$, which is finite by strong approximation.[14] Then the Shimura variety is a disjoint sum of quotients of Hermitian symmetric domains:

$$Sh_K(G, X)(\mathbb{C}) \cong \bigsqcup_{g \in \mathcal{R}} \Gamma_{gKg^{-1}}\backslash X^+,$$

where $\Gamma_{gKg^{-1}}$ is the stabilizer of $g \cdot K$ in $G(\mathbb{Q})_+$ i.e., $\Gamma_{gKg^{-1}} = G(\mathbb{Q})_+ \cap gKg^{-1}$.

[13] We use lowerscript (resp. superscript) for ∞ as for p: $K_p \subset G(\mathbb{Q}_p)$, while K^p excludes the part at p. Another common notation for $\mathbb{A}^\infty$ is $\mathbb{A}_f$, the finite part of the adeles $\mathbb{A}$. Reality check: $\mathbb{A}_\infty = \mathbb{R}$ for the adeles $\mathbb{A}$ over $\mathbb{Q}$.

[14] Strong approximation also allows to compute the set of connected components of the quotients i.e., the Shimura varieties $Sh_K(G, X)$ themselves, cf. [**10**].

3.3 Representability over $\mathbb{C}$ for unitary PEL data

Our first goal is to give evidence over $\mathbb{C}$ for the moduli theoretic interpretation of the double coset spaces: $G(\mathbb{Q})\backslash X \times G(\mathbb{A}^\infty)/K$, when $G = \mathrm{GU}$ and X is a union of products of $\mathcal{D}_{p,q}$'s, for $p + q = d$. In particular, points of the quotients of the domain $\mathcal{D}_{p,q}$ are parametrized by abelian varieties over $\mathbb{C}$ with additional structures. We focus on providing some details about the representability in the complex analytic category, which is more accessible thanks to the concrete nature of abelian varieties over $\mathbb{C}$.

Proposition 3.6. *Let K be sufficiently small.*[15] *The moduli problem in the complex analytic isogeny category 2.3.2 is representable by a smooth complex analytic space $\widetilde{Sh}_K$, a finite disjoint union of double coset spaces:*

$$G^\xi(\mathbb{Q})\backslash X \times G(\mathbb{A}^\infty)/K.$$

We wish to show that the analytic moduli space above and the analytification of the (complex fiber of the) algebraic moduli scheme (supposing its existence as a black box) are isomorphic as complex analytic spaces.

Here are the three main steps of our strategy.

Step 1. Instead of only quoting Mumford's G.I.T. **[25]** and leaving its intricate details to the reader, we work rather in the complex analytic category, where taking quotients is much less involved. This shows a representability result over $\mathbb{C}$ without assuming G.I.T. We also give a closely related, hands-on interpretation of the points of the Hermitian symmetric domain $\mathcal{D}_{p,q}$.

Step 2. Nonetheless, we also assume thanks to Mumford **[25]** that the moduli problem is representable in the algebraic category by a suitable scheme denoted $\widetilde{Sh}_K^{alg}$ over the reflex field E, see Genestier-Ngô's chapter.

Step 3. Next, we check carefully that the resulting complex analytic moduli space is isomorphic to the analytification of the algebraic moduli scheme $\left(\widetilde{Sh_K^{alg}} \otimes_E \mathbb{C}\right)^{an}$ arising from Mumford's G.I.T. ⋖ This last point is not trivial at all, and proving it in detail typically requires a careful description of the spaces involved.[16]

3.3.1 Points on $\mathcal{D}_{p,q}$ as polarized abelian varieties

For concreteness, we imitate, for F an imaginary quadratic field, the succinct analysis provided by Genestier-Ngô in their Section 1.2 for the Siegel variety. To simplify, we consider only the case of signature (p, q). Recall that

$$\mathcal{D}_{p,q} := \left\{ u \in \mathbb{M}_{p,q}(\mathbb{C}) | \, {}^t\begin{pmatrix} \overline{u} \\ 1 \end{pmatrix} \begin{pmatrix} 1_p & 0 \\ 0 & -1_q \end{pmatrix} \begin{pmatrix} u \\ 1 \end{pmatrix} = {}^t\overline{u}u - 1_q < 0 \right\}.$$

---

[15] Sufficiently small means that the groups $\Gamma_{gKg^{-1}}$ arising in the quotients are torsionfree for all $g \in G(\mathbb{A}^\infty)$.

[16] For example, a similar issue arises for toroidal compactifications of Shimura varieties: the algebraic and analytic toroidal compactifications of Shimura varieties are canonically isomorphic, see **[18]** for the PEL case.

Let $A = V_{\mathbb{C}}/\Lambda$ be an abelian variety of dimension $d = p + q$ over $\mathbb{C}$. The Lie algebra $V_{\mathbb{C}}$ of A is a d-dimensional $\mathbb{C}$-vector space equipped with the lattice $\Lambda \cong H_1(A, \mathbb{Z})$. Let $\langle , \rangle$ be an alternating form corresponding to a polarization λ (of fixed, non-trivial type) of A such that the associated non-degenerate Hermitian form has signature (p, q). The type is given by the elementary divisors of the polarization matrix, see Section 1.1 of Genestier-Ngô. A principal polarization is of type $(1, \ldots, 1)$.

Choose a hermitian basis $e_1, \ldots, e_p, e_{p+1}, \ldots, e_{p+q}$ of $V_{\mathbb{R}}$ such that the Hermitian matrix is $I_{p,q} = \begin{pmatrix} 1_p & 0 \\ 0 & -1_q \end{pmatrix}$, with corresponding $\mathbb{R}$-basis $\{e_\bullet, ie_\bullet\}$ of $V_{\mathbb{C}}$. The isomorphism $\pi_{\mathbb{R}} : \Lambda \otimes \mathbb{R} \longrightarrow V_{\mathbb{C}}$ defines a cocharacter $h : \mathbb{C}^\times \longrightarrow \mathrm{U}_{\mathbb{R}}(V_{\mathbb{C}}, \langle , \rangle)$.

We say that an abelian variety has signature (p, q) if $\mathrm{Lie}(A) \cong V_1$ as Hermitian modules, and therefore decomposes in p- and q-dimensional parts. In analogy with Prop. 1.2.3 of Genestier-Ngô, we have:

Proposition 3.7. *There is a canonical bijection between the set of polarized abelian varieties of fixed type and signature (p, q) with an hermitian basis and the set of homomorphisms*

$$h_1 : S^1 \longrightarrow \mathrm{U}_{\mathbb{R}}(\Lambda_{\mathbb{R}}, \langle , \rangle)$$

such that $h_{1,\mathbb{C}} : \mathbb{G}_m \longrightarrow \mathrm{U}_{\mathbb{R}}(\Lambda_{\mathbb{C}}, \langle , \rangle)$ give rise to a decomposition into a direct sum

$$\Lambda \otimes \mathbb{C} = (\Lambda \otimes \mathbb{C})_p \oplus (\Lambda \otimes \mathbb{C})_q,$$

of dimension p (resp. q), where the decomposition is induced by the Hermitian form associated to $\langle , \rangle$ being positive (resp. negative) definite.

Corollary 3.8. *There is a canonical bijection between the set of polarized abelian varieties of fixed type and signature (p, q) with a hermitian basis and the quotient* $\mathrm{U}(p, q)/\mathrm{U}(p) \times \mathrm{U}(q)$.

Proof. Combine Prop. 3.7 and the argument of Subsection 3.1.1, part 2. □

Corollary 3.9. *There is a canonical bijection from the set of complex polarized abelian varieties of fixed type and signature (p, q) with a hermitian basis and*

$$\mathcal{D}_{p,q} = \left\{ u \in \mathbb{M}_{p,q}(\mathbb{C}) \mid 1_q - {}^t\bar{u}u > 0 \right\}.$$

Proof. (sketch) The bijection goes through identifying both sets with a subset of matrices Z in $\mathbb{M}_{p+q}(\mathbb{C})$ via the map $u \mapsto \begin{pmatrix} 1_p & u \\ {}^t\bar{u} & 1_q \end{pmatrix}$. Note that having signature (p, q) implies the positive definiteness: $1_q - {}^t\bar{u}u > 0$, or equivalently $1_p - u{}^t\bar{u} > 0$. See [**14**, Prop. 5.7] for more details. □

Let F be a general CM field, and suppose that the associated Hermitian symmetric domain remains $\mathcal{D}_{p,q}$ i.e., the Hermitian form is assumed to be definite everywhere except at one pair of complex conjugate embeddings. Then Prop. 3.7 and 3.8 generalize mutatis mutandis.

3.3.2 Proof of the isomorphism of complex varieties

• Step 1. We start by taking stock of the fact that abelian varieties are a special kind of Hodge structures, and that families of abelian varieties correspond to variations of Hodge structures, following Deligne [**10**]. Adding PEL structures will provide the desired representability result in the complex category, cf. [**9**, 4.15].

Recall that there is an equivalence of categories between polarized abelian varieties over $\mathbb{C}$ and polarized $\mathbb{Z}$-Hodge structures of type $\{(-1,0),(0,-1)\}$ without torsion, see [**9**, Thm 4.7] and Section 4.1 of Genestier-Ngô. Further, this extends to families.

Recall that we have a short exact sequence:

$$0 \longrightarrow R^1\pi_*\mathbb{Z} \longrightarrow \mathrm{Lie}(A) \longrightarrow A \longrightarrow 0.$$

This motivates the following definition: A $\mathbb{Z}$-variation of Hodge structures of type $\{(-1,0),(0,-1)\}$ is a local system $\mathfrak{F} = R^1\pi_*\mathbb{Z}$ of rank $2g$ over S with its Hodge filtration i.e., $Fil := Fil(\mathfrak{F} \otimes \mathcal{O}_S) \subset \mathfrak{F} \otimes_{\mathbb{Z}} \mathcal{O}_S$,[17] which induces fiberwise $\mathbb{Z}$-Hodge structures. Note that by the Betti-to-deRham comparison theorem, $\mathfrak{F} \otimes \mathcal{O}_S = \mathcal{H}^1_{dR}(\mathcal{A}/S) := R^1\pi_*(\Omega^\bullet_{\mathcal{A}/S})$, and $Fil(\mathfrak{F} \otimes \mathcal{O}_S) = \omega_{\mathcal{A}/S}$, so we indeed get the usual Hodge filtration

$$0 \longrightarrow \omega_{\mathcal{A}/S} \longrightarrow R^1\pi_*(\Omega^\bullet_{\mathcal{A}/S}) \longrightarrow \mathrm{Lie}(A) \longrightarrow 0.$$

A polarization of a variation of $\mathbb{Z}$-Hodge structures of weight n is a morphism

$$\phi : \mathfrak{F} \otimes_{\mathbb{Z}} \mathfrak{F} \longrightarrow \mathbb{Z}(-n),$$

inducing polarizations fiberwise. The construction of the polarization ϕ_λ of a variation of Hodge structures associated to an abelian scheme with polarization λ is the direct globalization of the construction for abelian varieties.

Proposition 3.10. *Let S be an analytic space over $\mathbb{C}$. The category of polarized abelian schemes over S is equivalent to the category of polarized variations of $\mathbb{Z}$-Hodge structures of type $\{(-1,0),(0,-1)\}$ without torsion.*

Proof. The first functor associates to (A,λ) a polarized abelian scheme over S, the triple $(R^1\pi_*\mathbb{Z}, Fil, \phi_\lambda)$. Its inverse functor associates to $(\mathfrak{F}, Fil, \phi)$ the polarized abelian scheme $Fil\backslash\mathfrak{F} \otimes \mathcal{O}_S/\mathfrak{F}$. □

Corollary 3.11. *Let S be an analytic space over $\mathbb{C}$. The category of polarized abelian schemes over S up to isogeny is equivalent to the category of polarized variations of $\mathbb{Q}$-Hodge structures of type $\{(-1,0),(0,-1)\}$.*

Adding endomorphisms and level structures, we get a reformulation of our original moduli functor in the isogeny category of Section 2.3.2. Fix a Hermitian space V. Consider the functor Φ associating to any smooth complex analytic space S the set of isomorphism classes of quintuples:

$$(\mathfrak{F}, Fil, \phi, \iota, \overline{\eta}),$$

[17] We ignore Griffiths's transversality condition, as it is automatically verified for abelian schemes.

where $(\mathfrak{F}, Fil, \phi)$ is a polarized variation of $\mathbb{Q}$-Hodge structures of type $\{(-1,0),(0,-1)\}$ over S; $\iota : D \otimes \mathbb{Q} \longrightarrow \mathrm{End}_{\mathcal{O}_S}(\mathfrak{F} \otimes \mathcal{O}_S)$ is an $\mathbb{Q}$-algebra homomorphism; $\overline{\eta}$ is a K-level structure, with the usual compatibilities.

Theorem 3.12. *The functor Φ is representable by $\widetilde{Sh_K}$, a finite disjoint union of double coset spaces:*

$$G^{\xi}(\mathbb{Q})\backslash X \times G(\mathbb{A}^{\infty})/K.$$

Note that the disjoint union is indexed by the isomorphism classes of Hermitian modules V over $\mathbb{Q}$ as in condition (1) of Remark 2.5, and each G^{ξ} is locally isomorphic to G.

Proof. (sketch) Consider the subfunctor Φ_V where we impose condition (1) on a given V as in Remark 2.5. It is enough to show that Φ_V is representable by:

$$Sh_K := G^{\xi_V}(\mathbb{Q})\backslash X \times G(\mathbb{A}^{\infty})/K.$$

where G^{ξ_V} is the unitary group GU associated to V. Thanks to the reinterpretation of the moduli problem over $\mathbb{C}$ in terms of linear algebraic data and additional PEL structures, the representability can be shown after embedding the data in a suitable Grassmannian complex analytic variety and quotienting to take into account the level structure; note that taking quotients in the analytic category is much simpler than in the category of schemes, by [**5**]. To simplify further, we may assume, up to shrinking K again, that $\Gamma_{gKg^{-1}}$ acts on X^+ freely properly discontinuously. We then show that the $\mathbb{C}$-points $Sh_K(G^{\xi_V}, X)(\mathbb{C})$ of the solution of the moduli problem Φ_V, are naturally in bijection with $G^{\xi_V}(\mathbb{Q})\backslash X \times G(\mathbb{A}^{\infty})/K$, and that $Sh_K(G^{\xi_V}, X)$ is the analytic moduli space of Φ_V (see [**22**, Prop. 2.8, 2.12, cf. Prop. 3.10]). □

• Step 2: see Genestier-Ngô or [**14**] for brief accounts of Mumford's GIT [**25**].
• Step 3. Recall that we need to establish an isomorphism of complex analytic varieties: 1) the bijection at the level of points first; and then 2) the local isomorphism, thanks to an appeal to Grothendieck-Messing deformation theory.
1) We start by recalling the bijection at the level of $\mathbb{C}$-points:

Proposition 3.13. *The natural map*

$$\left(\widetilde{Sh_K^{alg}} \otimes_E \mathbb{C}\right)^{an} \longrightarrow \bigsqcup_{\xi \in \ker^1(\mathbb{Q}, G)} G^{\xi}(\mathbb{Q})\backslash\left(X \times G(\mathbb{A}^{\infty})/K\right),$$

is a bijection.

Proof. This is the essential content of [**18**, Lem. 2.3.2 and discussion before], cf. Section 3.6 of Genestier-Ngô and [**17**, Section 8, pp. 399–400] whose notation we borrowed, or [**9**, Scholie 4.11]. □

Recall that for even signature, the so-called Hasse principle holds i.e., $\ker^1(\mathbb{Q}, G) \subset H^1(\mathbb{Q}, G)$ is trivial, the condition (1) of Remark 2.5 holds automatically and there is nothing to do. For odd signature, the Hasse principle fails in a mild way: the solution of the moduli problem is equal to finitely many disjoint copies $|\ker^1(\mathbb{Q}, G)|$ of the same unitary Shimura variety, where

$$\xi \in \ker^1(\mathbb{Q}, G) := \ker\Big(H^1(\mathbb{Q}, G) \longrightarrow \prod_v H^1(\mathbb{Q}_v, G)\Big),$$

and condition (1) ensures that we end up with a single copy.

2) We quote Harish-Chandra's embedding theorem:

Theorem 3.14 (Harish-Chandra). *Let X be a Hermitian symmetric space of non-compact type. Then there exists a bounded symmetric domain $Y \subset X^\vee$ in the compact dual $X^\vee$ and a holomorphic diffeomorphism $X \longrightarrow Y$.*

Example 3.15. *The Harish-Chandra embedding generalizes the classical map from the unit disc to the Riemann sphere, cf. [15, p.394].*

We skip further details about the compact dual of a Hermitian symmetric space (a reference for this is: [**15**, Chap. VIII]). In short, the compact dual is $X^\vee = G_{\mathbb{C}}/P_\mu$, for the parabolic subgroup P_μ associated to the minuscule cocharacter μ of the Shimura datum, cf. [**18**, Lem. 2.1.2 (3)] in the PEL context. Instead, we give the main idea by using an argument of Deligne see e.g., [**8**, Proof of Prop. 6.1], cf. Prop. 4.3.2 in Genestier-Ngô. Recall that X is the $G(\mathbb{R})$-conjugacy class of a given map $h : \mathbb{C}^\times \longrightarrow G(\mathbb{R})$. Let $x \in X$ define a Hodge structure of type $\{(-1, 0), (0, -1)\}$ corresponding to a complex abelian variety A. Let $H = H_1(A, \mathbb{Q})$. Let $Fil(x) := H^{(0,-1)} \subset H \otimes \mathbb{C}$. Since $X = G(\mathbb{R})/K_\infty$, at $x = h$, the tangent space to X is $\mathrm{Lie}(G_{\mathbb{R}})/\mathrm{Lie}(K_\infty)$. Consider the Grassmannian variety:

$$Grass_d(H \otimes \mathbb{C}) := \{W \subset H \otimes \mathbb{C} | \dim(W) = d\},$$

where $d = \dim(A)$.

The map:

$$\phi : X \longrightarrow Grass_d(H \otimes \mathbb{C}), \quad x \mapsto Fil(x),$$

is injective because the Hodge filtration determines the Hodge decomposition. We obtain a map $d\phi$ on tangent spaces via the maps

$$G(\mathbb{R}) \hookrightarrow G(\mathbb{C}) \hookrightarrow \mathrm{GL}(H \times \mathbb{C}),$$

that is,

$$\mathrm{Lie}(G_{\mathbb{R}})/\mathrm{Lie}(K_\infty) \cong \mathrm{Lie}(G_{\mathbb{C}})/Fil(\mathrm{Lie}(G_{\mathbb{C}}) \hookrightarrow \mathrm{End}(H \otimes \mathbb{C})/Fil(\mathrm{End}(H \otimes \mathbb{C})),$$

where the filtrations on $\mathrm{Lie}(G_{\mathbb{C}})$ and $\mathrm{End}(H \otimes \mathbb{C})$ are defined via h. This shows that the tangent space of X at x embeds into the tangent space of $Grass_d(H \otimes \mathbb{C})$ at $\phi(x)$. Finally, we apply deformation theory, which says in a nutshell that deforming the Hodge filtration is the same as deforming the abelian variety itself. Everything mentioned above behaves well with respect to PEL structures. This finishes our rough sketch of proof.

Let K be sufficiently small. By Chow's theorem, the complex analytic (or holomorphic) variety $Sh_K(G, X)(\mathbb{C}) = G(\mathbb{Q})\backslash X \times G(\mathbb{A}^\infty)/K$, when projective i.e., isomorphic to a closed subset of some projective space $\mathbb{P}^N(\mathbb{C})$, arises from a smooth projective algebraic variety defined over $\mathbb{C}$. In general, even when G is not necessarily anisotropic, theorems of Baily and Borel ensure that $Sh_K(G, X)(\mathbb{C})$ arises anyway from a quasi-projective algebraic variety.

3.3.3 The tower $Sh(G, X)$

In general, note that

$$Sh(G, X)(\mathbb{C}) \neq G(\mathbb{Q})\backslash X \times G(\mathbb{A}^\infty),$$

where $Sh(G, X) := \varprojlim Sh_K(G, X)$.

Let Z be the center of G. Let $Z(\mathbb{Q})^-$ be the closure of the center $Z(\mathbb{Q})$ in $Z(\mathbb{A}^\infty)$. Then $Z(\mathbb{Q}) \cdot K = Z(\mathbb{Q})^- \cdot K$ in $G(\mathbb{A}^\infty)$ and:

$$Sh_K(G, X) = G(\mathbb{Q})\backslash X \times (G(\mathbb{A}^\infty)/K)$$
$$\cong \big(G(\mathbb{Q})/Z(\mathbb{Q})\big)\backslash\big(X \times (G(\mathbb{A}^\infty/Z(\mathbb{Q}) \cdot K)\big) \cong G(\mathbb{Q})/Z(\mathbb{Q})\backslash X \times (G(\mathbb{A}^\infty)/Z(\mathbb{Q})^- \cdot K).$$

Proposition 3.16

$$\varprojlim Sh_K(G, X) = \frac{G(\mathbb{Q})}{Z(\mathbb{Q})}\backslash X \times (G(\mathbb{A}^\infty)/Z(\mathbb{Q})^-).$$

Proof. Cf. Deligne's [**10**, 2.1.10] and Milne [**23**]. □

On the other hand, we have the following observation: if $Z(\mathbb{Q})$ is discrete in $Z(\mathbb{A}^\infty)$, then $Z(\mathbb{Q}) = Z(\mathbb{Q})^-$, hence

$$Sh(G, X)(\mathbb{C}) = G(\mathbb{Q})\backslash X \times G(\mathbb{A}^\infty).$$

4 Hecke matters

4.1 Hecke correspondences

Suppose that K, K' are small enough subgroups of $G(\mathbb{A}^\infty)$, so that the Shimura varieties over $\mathbb{C}$ associated to K, K' are smooth.

Let $K \subset K' \subset G(\mathbb{A}^\infty)$.

Proposition 4.1. *The natural map*

$$\pi_{K'/K} : Sh_K \longrightarrow Sh_{K'}$$

is finite étale, and Galois if K is normal in K'.

In general, relax the condition $K \subset K'$ to the following: there exists an element $g \in G(\mathbb{A}^\infty)$ such that $g^{-1}Kg \subset K'$ is a subgroup, necessarily of finite index $[gK'g^{-1} : K] < \infty$.

Consider the map:

$$\pi_{KgK'} : Sh_K \longrightarrow Sh_{K'}, \quad [x, h] \mapsto [x, hg].$$

Its degree is $[gK'g^{-1} : K]$.

Proposition 4.2. *The application $\pi_{KgK'}$ is a covering map. Further, this map depends only on the double class KgK'.*

Proof. We prove that the map only depends on the double class. Replace g by $\gamma g \gamma'$, for $\gamma \in K', \gamma' \in K$ in the definition of $\pi_{KgK'}$ and check that the resulting maps are the same. We verify that $K \subset (\gamma g \gamma')^{-1} \cdot K' \cdot \gamma g \gamma' = \gamma'^{-1} g^{-1} \gamma^{-1} \cdot K' \cdot \gamma g \gamma' = \gamma'^{-1} g^{-1} \cdot K' \cdot g \gamma'$, hence $\gamma' K \gamma'^{-1} = K \subset g^{-1} K' g$, and the degree of course remains the same. □

Definition 4.3. The set of morphisms between Sh_K and $Sh_{K'}$, denoted

$$\mathrm{Mor}(Sh_K, Sh_{K'}),$$

is defined to be: $\left\{ K'gK \in K' \backslash G(\mathbb{A}^\infty)/K \text{ where } K \subset g^{-1}K'g \text{ an open subgroup} \right\}$.

Definition 4.4. The set of Hecke correspondences on $Sh_{K'}$ is defined as the set of pairs

$$(K, (\pi'_1, \pi'_2)),$$

where π'_1, π'_2 are two morphisms in $\mathrm{Mor}(Sh_K, Sh_{K'})$.

The picture to keep in mind for Hecke correspondences is this:

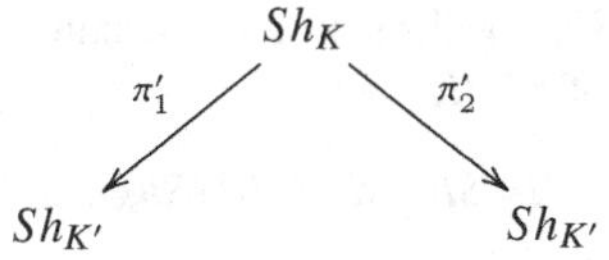

Roughly speaking, we pullback via $(\pi'_1)^*$ from the copy of $Sh_{K'}$ on the left and via $(\pi'_2)_*$ we pushdown to the copy of $Sh_{K'}$ on the right.

Let $K^\sharp$ be another subgroup of $G(\mathbb{A}^\infty)$ of finite index in $g^{-1}K'g$. We say that two pairs of morphisms (π'_1, π'_2), $(\pi^\sharp_1, \pi^\sharp_2)$ are equivalent if there exists an invertible morphism

$$Sh_{K'} \xrightarrow{\alpha} Sh_{K^\sharp},$$

such that the following diagram commutes:

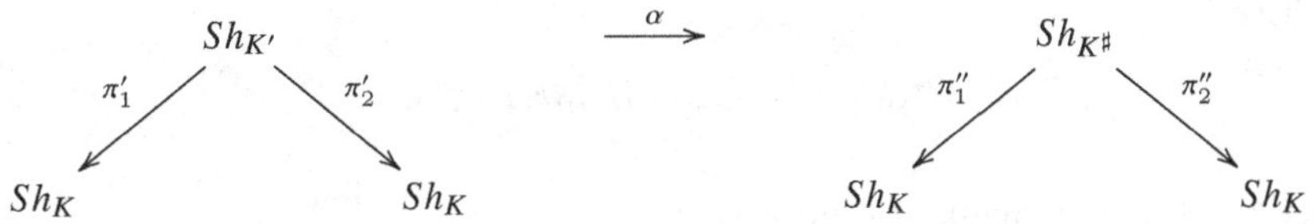

More generally, let $K, K' \subset G(\mathbb{A}^\infty)$ be two open compact subgroups. Write $K^g := g^{-1}Kg$. Consider the following two natural projection maps:

$$\begin{array}{lccc} \pi_{1,K'}: & Sh_{K^g \cap K'} & \longrightarrow & Sh_{K'} \\ & [x,h] \mod K^g \cap K' & \mapsto & [x,h] \mod K' \\ \pi_{K,g^{-1}}: & Sh_{K^g \cap K'} & \longrightarrow & Sh_K \\ & [x,h] \mod K^g \cap K' & \mapsto & [x, hg^{-1}] \mod K \end{array}$$

N.B. A theorem of Borel implies that the projection maps defined above are algebraic, and moreover, they are defined over the reflex field E.

Proposition 4.5. *The image of* $\pi_{K,g^{-1}} \times \pi_{1,K'}$ *in* $Sh_K \times Sh_{K'}$ *is an algebraic correspondence depending only on the double coset* KgK'.

Proof. This follows from the fact that $Sh_K = Sh/K$. □

4.2 Trace map and Hecke operators on cohomology

In this section, we recall some of the properties of the system of complex vector spaces provided by Betti cohomology: $H^*(Sh, \mathbb{C})$. More generally, we consider locally constant sheaves $\mathcal{L}$ on Sh. This is a system of locally constant sheaves $\mathcal{L}$ on Sh_K for all small enough open, compact subgroups K compatibly with pullbacks of the maps $Sh \longrightarrow Sh_K$. By definition,

$$H^*(Sh, \mathcal{L}) := \varinjlim H^*(Sh_K, \mathcal{L}).$$

The following properties hold:

(1) the spaces $H^*(Sh_K, \mathcal{L})$ is finite dimensional.

(2) Let $\pi_{K'/K} : Sh_K \longrightarrow Sh_{K'}$ be the map above making $Sh_{K'}$ into the quotient of Sh_K under the right action of K'/K. Then

$$H^*(Sh_K, \mathcal{L}) = H^*(Sh_{K'}, \mathcal{L})^{K'/K},$$

the $\mathbb{C}$-subspace of K'/K-fixed vectors.

Going back to Sh, these two properties show that $H^*(Sh, \mathcal{L})$ is a smooth, admissible representation of $G(\mathbb{A}^\infty)$, and that:

$$H^*(Sh_K, \mathcal{L}) = H^*(Sh, \mathcal{L})^K,$$

the $\mathbb{C}$-subspace of K-fixed vectors.

Moreover, for any lisse ℓ-adic sheaf $\mathcal{L}$, we have $G(\mathbb{A}^\infty)$-equivariant comparison isomorphisms between étale ℓ-adic cohomology and Betti cohomology, that is, for any $\iota : \overline{\mathbb{Q}}_\ell \longrightarrow \mathbb{C}$, we have:

$$H^*(Sh, \mathcal{L}) \otimes_\iota \mathbb{C} \xrightarrow{\cong} H^*(Sh, \mathcal{L}) \otimes_\iota \mathbb{C}.$$

Fact 4.6 (The trace map). *The trace map:*

$$\mathrm{Tr}_{K'/K} : H^*(Sh_K, \mathcal{L}_K) \longrightarrow H^*(Sh_{K'}, \mathcal{L}_{K'})$$

for $K \subset K'$, *satisfies the following two equalities:*

$$\mathrm{Tr}_{K''/K'} \circ \mathrm{Tr}_{K'/K} = \mathrm{Tr}_{K''/K} \quad \text{for } K \subset K' \subset K'', \tag{2}$$

and, for $[K' : K]$ *the degree of the covering map* $Sh_K \longrightarrow Sh_{K'}$

$$\mathrm{Tr}_{K'/K} \circ \pi^*_{K'/K}(\bullet) = \bullet \times [K' : K]. \tag{3}$$

Definition 4.7. Let K, K' be arbitrary small open compact subgroups of $G(\mathbb{A}^\infty)$. Define the Hecke operator:

$$T_{KgK'} : H^*(Sh_K, \mathcal{L}_K) \longrightarrow H^*(Sh_{K'}, \mathcal{L}_{K'})$$

via

$$T_{KgK'} = |\det(g)|_{\mathbb{A}} \cdot \mathrm{Tr}_{K'/K^{g^{-1}} \cap K'} \circ [g] \circ \pi_{K/K \cap K'^g},$$

where $[g] : H^*(Sh_{K \cap K'^g}, \mathcal{L}_{K \cap K'^g}) \longrightarrow H^*(Sh_{K^{g^{-1}} \cap K'}, \mathcal{L}_{K^{g^{-1}} \cap K'})$ is induced by

$$g : Sh_{K \cap K'^g} \xrightarrow{\cong} Sh_{K^{g^{-1}} \cap K'}.$$

To sum up: the double coset ring $\mathbb{Z}[K\backslash G(\mathbb{A}^\infty)/K]$ acts on the cohomology group $H^*(Sh_K, \mathcal{L}_K)$. The image of $\mathbb{Z}[K\backslash G(\mathbb{A}^\infty)/K]$ is usually called a Hecke algebra.

4.3 Automorphic representations as Hecke modules

We recall the classical link, due to Borel and Bernstein, between smooth representations π of $G(\mathbb{A}^\infty)$ such that π^K generates π as a $G(\mathbb{A}^\infty)$-representation (i.e., $\pi^K \neq 0$ if π is irreducible) and the Hecke algebra $\mathcal{H}_K(G(\mathbb{A}^\infty), \mathbb{Z})$.

Definition 4.8. The Hecke algebra $\mathcal{H}_K(G(\mathbb{A}^\infty), \mathbb{Z})$ is the algebra of Hecke correspondences, that is:

$$\mathcal{H}_K(G(\mathbb{A}^\infty), \mathbb{Z}) := \mathrm{End}_{\mathbb{Z}[G(\mathbb{A}^\infty)]}(\mathbb{Z}[Sh_K]).$$

Note that in our concrete geometric set-up described above, an element f of the Hecke algebra can be identified with a function

$$f : K\backslash G(\mathbb{A}^\infty)/K \longrightarrow \mathbb{Z},$$

which is locally constant and has compact support. In short: characteristic functions form a basis of the free abelian group $\mathcal{H}_K(G(\mathbb{A}^\infty), \mathbb{Z})$.

Recall that $K \subset G(\mathbb{A}^\infty)$ is a compact open subgroup. Let π be a smooth complex representation of $G(\mathbb{A}^\infty)$. Then we can equip π with the structure of a Hecke module that is, the K-fixed part $\pi^K = \mathrm{Hom}_{\mathbb{Z}[G(\mathbb{A}^\infty)]}(\mathbb{Z}[Sh_K], \pi)$ is a $\mathcal{H}_K(G(\mathbb{A}^\infty), \mathbb{C})$-module.

Theorem 4.9 ([1]). *There is an equivalence of categories between the category of smooth complex representations π of $G(\mathbb{A}^\infty)$ such that π^K generates π and non-degenerate $\mathcal{H}_K(G(\mathbb{A}^\infty), \mathbb{C})$-modules. In particular, irreducible representations such that $\pi^K \neq 0$ correspond to simple $\mathcal{H}_K(G(\mathbb{A}^\infty), \mathbb{C})$-modules.*

5 Automorphic bundles and local systems

5.1 The Matsushima-Murakami formula

We now assume that $Sh = Sh(G, X)(\mathbb{C})$ is compact. The Matsushima-Murakami formula describes $H^*(Sh, \mathcal{L})$ as the subgroups K vary in terms of irreducible representations of

$G(\mathbb{A}^\infty)$ itself and $(\mathfrak{g}, K)$-modules. The Matsushima-Murakami formula (see [**21**], [**20**] and [**3**, Chap VII]) decomposes the Betti cohomology of Sh as a direct sum of certain $\mathbb{C}$-vector spaces appearing with suitable multiplicity, indexed by the cuspidal automorphic representations π of $G(\mathbb{A})$. We provide an adelic formulation of the formula suited to our presentation of Shimura varieties as double coset spaces. Let $\rho : G(\mathbb{R}) \longrightarrow \mathrm{Aut}(W_\rho)$ be a finite dimensional representation on a $\mathbb{C}$-vector space W_ρ. Let $\mathcal{L}_\rho$ be a $G(\mathbb{Q})$-equivariant constant sheaf $\mathcal{L}_\rho$ on $X \times (G(\mathbb{A}^\infty)/K)$.[18] We note also $\mathcal{L}_\rho$ the locally constant sheaf on Sh_K obtained by descent:

$$\mathcal{L}_\rho := G(\mathbb{Q})\backslash\big(X \times (G(\mathbb{A}^\infty)/K) \times W_\rho\big).$$

As before,

$$H^*(Sh, \mathcal{L}_\rho) := \varinjlim H^*(Sh_K, \mathcal{L}_\rho),$$

a smooth, admissible representation of $G(\mathbb{A}^\infty)$.

Theorem 5.1 (Matsushima-Murakami). *Suppose that $Sh = Sh(G, X)(\mathbb{C})$ is compact. Let $\mathcal{L}_\rho$ be a locally constant sheaf on Sh defined via an irreducible finite dimensional representation $\rho : G(\mathbb{R}) \longrightarrow \mathrm{Aut}(W_\rho)$. Then we have an equality:*

$$H^*(Sh, \mathcal{L}_\rho) = \bigoplus_\pi m(\pi)\pi^\infty \otimes H^*(\mathfrak{g}, K_\infty; \pi_\infty \otimes \rho), \tag{4}$$

as admissible representations of $G(\mathbb{A}^\infty)$, where $\pi = \pi^\infty \otimes \pi_\infty$ goes through all automorphic representations of $G(\mathbb{A})$, and $m(\pi)$ is the multiplicity of π in

$$L^2\big(G(\mathbb{Q})\backslash G(\mathbb{A})\big)[\omega_\infty],$$

where $\omega_\infty : Z(\mathbb{R}) \longrightarrow \mathbb{C}^\times$ is the inverse of the central character of ρ.

The desired $\mathbb{C}$-vector spaces decomposing the Betti cohomology of Sh are the $(\mathfrak{g}, K)$-cohomology spaces $H^*(\mathfrak{g}, K_\infty, \pi_\infty)$ (cf. [**3**, Chap. I]), where $\mathfrak{g}$ is the complexified Lie algebra $\mathrm{Lie}(G)$ of $G(\mathbb{R})$ and $K_\infty \subset G(\mathbb{R})$.

When restricting to finite level, we have the corresponding equality:

$$H^*(Sh_K, \mathcal{L}_\rho) = \bigoplus_\pi m(\pi)(\pi^\infty)^K \otimes H^*(\mathfrak{g}, K_\infty; \pi_\infty \otimes \rho), \tag{5}$$

which is compatible with the Hecke action of $G(\mathbb{A}^\infty)$.[19] Vogan and Zuckerman determined in [**28**] the representations π_∞ of $G(\mathbb{R})$ such that $H^*(\mathfrak{g}, K_\infty, \pi_\infty) \neq 0$.

A remarkable consequence of the Matsushima-Murakami formula for compact Shimura varieties is a generalization of the Eichler-Shimura isomorphism for modular curves, cf. [**14**, Thm 5.13, p. 246] for a statement in classical (i.e., non-adelic) language.

---

[18] Note that $K \subset G(\mathbb{A}^\infty)$, as we do not use the notation K^∞.

[19] A result of Harish-Chandra states that the set of isomorphism classes of irreducible representations π of $G(\mathbb{A}^\infty)$ such that $\pi^K \neq 0$ and there exists an automorphic representation π of $G(\mathbb{A})$ such that $\pi = \pi_\infty \otimes \pi$, is finite.

5.2 From local systems to Galois representations

In this section we replace $\mathbb{C}$ by a number field L and recast the local systems in this set-up. Let $\rho : G \longrightarrow \mathrm{Aut}(W_\rho)$ be a finite dimensional algebraic representation on a L-vector space W_ρ, assuming that ρ is irreducible over $\overline{L}$. Then ρ gives rise to a local system $\mathcal{L}_\rho$ of L-vector spaces on $Sh_K(\mathbb{C})$. For every finite place λ of L, the system $\mathcal{L}_\rho$ on $Sh_K(\mathbb{C})$ comes from a lisse L_λ-sheaf on the Shimura variety Sh_K over its reflex field E, see [**19**, p.38]. Let $\mathcal{H}_L$ be the Hecke algebra $\mathcal{H}_L := \mathcal{H}_K(G(\mathbb{A}^\infty), L)$ with coefficients in L. $\mathcal{H}_L$ acts on $H^*(Sh_K(\mathbb{C}), \mathcal{L}_\rho)$. Fix an admissible representation π^∞ on $G(\mathbb{A}^\infty)$ on an L-vector space, assuming it is irreducible over $\overline{L}$. From Matsushima's formula, we get that the L-vector space

$$W^*(\pi^\infty) = \mathrm{Hom}_{\mathcal{H}_L}\Big((\pi^\infty)^K, H^*(Sh_K(\mathbb{C}), \mathcal{L}_\rho)\Big),$$

can be described, after tensoring with $\mathbb{C}$ via an embedding of $L \hookrightarrow \mathbb{C}$, as:

$$W^*(\pi^\infty) \otimes_L \mathbb{C} \cong \oplus_{\pi_\infty} m(\pi) H^*(\mathfrak{g}, K_\infty; \pi_\infty \otimes (\rho \otimes_L \mathbb{C})).$$

The L_λ-vector space $W^*(\pi^\infty) \otimes L_\lambda$ carries a representation of $\mathrm{Gal}(\overline{E}/E)$, since the actions of $\mathcal{H}_L$ and $\mathrm{Gal}(\overline{E}/E)$ on $H^*(Sh_K(\mathbb{C}), \mathcal{L}_\rho) \otimes L_\lambda$ commute. Further details can be found in subsequent chapters about the resulting Galois representations.

Acknowledgements

I thank Michael Harris for the invitation to contribute to this book series, and for his support through the FSMP. I thank Laurent Fargues for repeatedly sharing his insights about Shimura varieties and in particular his suggestion to enrich the discussion of representability in the complex category. I also thank: Ashay Burungale, Ana Caraiani, Christopher Daw, Gabriel Dospinescu, Arno Kret, Kai-Wen Lan and Benoît Stroh for various suggestions and/or discussions. I am quite grateful to Andreas Mihatsch for contributing many comments about the whole text in its preliminary version; additional thanks are also due to Arno Kret. I also thank the anonymous referee for a careful reading of the text.

My work on this chapter was started at IISc Bangalore, during a short stay in July 2014; I thank warmly the faculty members of its mathematics department for their hospitality. The chapter was entirely written at MSRI (Berkeley) during the Fall 2014 trimester program "New Geometric Methods in Number Theory and Automorphic Forms"; I thank MSRI for its support (NSF grant No. 0932078000) and superlative working conditions. I would like to thank Aix-Marseille University for granting me an ad hoc "délégation" in Fall 2014, thanks to complementary funding from FSMP (Fondation sciences mathématiques de Paris).

6 Appendix

We collect here a non-trivial criterion for compactness:

Fact 6.1 ([2], [24]). *$Sh_K(\mathbb{C})$ is compact if and only if the group $G_{/\mathbb{Q}}$ in the Shimura datum (G, X) is anisotropic modulo center over $\mathbb{Q}$ i.e., G^{ad} does not contain any $\mathbb{Q}$-split torus. In* *particular, a unitary Shimura variety is compact if the group G has signature* $(0, d)$ *at least at one archimedean place.*

Recall that G^{ad} is the adjoint group of G.

Bibliography

[1] Bernstein, I. N., Zelevinsky, A. V., Induced representations of reductive p-adic groups. I. *Ann. Sci. cole Norm. Sup.* (4) **10**, 1977, no. 4, 441–472.

[2] Borel, A., Harish-Chandra, Arithmetic subgroups of algebraic groups. *Ann. of Math.* (2), **75**, 1962, 485–535.

[3] Borel, A., Wallach, N., *Continuous cohomology, discrete subgroups, and representations of reductive groups*. Second edition. Mathematical Surveys and Monographs, **67**. American Mathematical Society, Providence, RI, 2000. xviii+260 pp.

[4] *On the stabilization of the trace formula*. Edited by L. Clozel, M. Harris, J.-P. Labesse and B.-C. Ngô. Stabilization of the Trace Formula, Shimura Varieties, and Arithmetic Applications, **1**. International Press, Somerville, MA, 2011. xiv+527 pp.

[5] Cartan, H., Quotient d'un espace analytique par un groupe d'automorphismes. *A symposium in honor of S. Lefschetz, Algebraic geometry and topology*. pp. 90–102. Princeton University Press, Princeton, N. J. ,1957.

[6] Chenevier G., Lannes, J., Formes automorphes et voisins de Kneser des réseaux de Niemeier, 388p., arxiv arXiv:1409.7616v1.

[7] Cornut, C., course notes on Shimura varieties, IMJ, Paris, 2011.

[8] Deligne, P., Hodge cycles on abelian varieties (notes by J.S. Milne), in *Hodge cycles, motives, and Shimura varieties*. Lecture Notes in Mathematics, **900**. Springer-Verlag, Berlin-New York, 1982. ii+414 pp.

[9] Deligne, P., Travaux de Shimura. *Séminaire Bourbaki*, 23ème année (1970/71), Exp. No. **389**, pp. 123–165. Lecture Notes in Math., Vol. **244**, Springer, Berlin, 1971.

[10] Deligne, P., *Variétés de Shimura: interprétation modulaire, et techniques de construction de modèles canoniques*. Automorphic forms, representations and L-functions (Proc. Sympos. Pure Math., Oregon State Univ., Corvallis, Ore., 1977), Part 2, pp. 247–289, Proc. Sympos. Pure Math., **XXXIII**, Amer. Math. Soc., Providence, R.I., 1979.

[11] Gordon, B., Canonical models of Picard modular surfaces. *The zeta functions of Picard modular surfaces*, 1–29, Univ. Montréal, Montreal, Qc, 1992.

[12] Goresky, M., MacPherson, R., Lefschetz numbers of Hecke correspondences. *The zeta functions of Picard modular surfaces*, 465–478, Univ. Montréal, Montréal, Qc, 1992.

[13] Haines, T., Introduction to Shimura varieties with bad reduction of parahoric type. *Harmonic analysis, the trace formula, and Shimura varieties*, 583–642, Clay Math. Proc., **4**, Amer. Math. Soc., Providence, RI, 2005.

[14] Hida, H., *p-adic automorphic forms on Shimura varieties*. Springer Monographs in Mathematics. Springer-Verlag, New York, 2004. xii+390 pp.

[15] Helgason, S., *Differential geometry, Lie groups, and symmetric spaces*. Graduate Studies in Mathematics, **34**. American Mathematical Society, Providence, RI, 2001. xxvi+641 pp.

[16] Kottwitz, R., On the λ-adic representations associated to some simple Shimura varieties. *Invent. Math.* **108**, 1992, no. 3, 653–665.

[17] Kottwitz, R., Points on some Shimura varieties over finite fields, *J. Amer. Math. Soc.* **5**, 1992, no. 2, 373–444.

[18] Lan, K.-W., Comparison between analytic and algebraic constructions of toroidal compactifications of PEL-type Shimura varieties. *J. Reine Angew. Math.* **664** (2012), 163–228.

[19] Langlands, R., Modular forms and ℓ-adic representations. *Modular functions of one variable, II* (Proc. Internat. Summer School, Univ. Antwerp, Antwerp, 1972), pp. 361–500. Lecture Notes in Math., Vol. **349**, Springer, Berlin, 1973.

[20] Matsushima, Y., A formula for the Betti numbers of compact locally symmetric Riemannian manifolds. *J. Differential Geometry* **1**, 1967, 99–109.

[21] Matsushima, Y., Murakami, S., On vector bundle valued harmonic forms and automorphic forms on symmetric riemannian manifolds. *Ann. of Math.* (2) **78** 1963, 365–416.

[22] Milne, J. S., Shimura varieties and motives. *Motives* (Seattle, WA, 1991), 447–523, Proc. Sympos. Pure Math., **55**, Part 2, Amer. Math. Soc., Providence, RI, 1994.

[23] Milne, J. S., Introduction to Shimura varieties. *Harmonic analysis, the trace formula, and Shimura varieties*, 265–378, Clay Math. Proc., **4**, Amer. Math. Soc., Providence, RI, 2005.

[24] Mostow, G., Tamagawa, T., On the compactness of arithmetically defined homogeneous spaces. *Ann. of Math.* (2), **76**, 1962 446–463.

[25] Mumford, D., Fogarty, J., Kirwan, F., *Geometric invariant theory.* Third edition. Ergebnisse der Mathematik und ihrer Grenzgebiete (2), **34**. Springer-Verlag, Berlin, 1994. xiv+292 pp.

[26] Shimura, G., Moduli of abelian varieties and number theory. *Algebraic Groups and Discontinuous Subgroups*, Proc. Sympos. Pure Math. **IX**, Boulder, Colo., 1965, pp. 312–332 Amer. Math. Soc., Providence, R.I.

[27] Shimura, G., *Euler products and Eisenstein series.* CBMS Regional Conference Series in Mathematics, **93**. Published for the Conference Board of the Mathematical Sciences, Washington, DC; by the American Mathematical Society, Providence, RI, 1997. xx+259 pp.

[28] Vogan, D., Zuckerman G, Unitary representations with nonzero cohomology. *Compositio Math.*, **53** (1), 51–90, 1984.

INTEGRAL MODELS OF SHIMURA VARIETIES OF PEL TYPE

SANDRA ROZENSZTAJN

UMPA, ÉNS de Lyon, UMR 5669 du CNRS, 46, allée d'Italie, 69364 Lyon Cedex 07, France
E-mail address: sandra.rozensztajn@ens-lyon.fr

1 PEL data for integral models

1.1 PEL Shimura data

We start with a set $\mathcal{D}$ of Shimura data of PEL type, as explained in [Del71, § 4.9] or [GN, § 4.3]. That is, $\mathcal{D}$ consists of:

(1) a finite semisimple $\mathbf{Q}$-algebra B, endowed with a positive involution $*$.

(2) a finite B-module V, endowed with a non-degenerate bilinear alternating pairing $\langle\cdot,\cdot\rangle$, such that $\langle bx, y\rangle = \langle x, b^* y\rangle$ for all $b \in B$ and $(x, y) \in V$.

(3) an $\mathbf{R}$-morphism $h : \mathbf{C} \to \mathrm{End}_B(V)_{\mathbf{R}}$ such that complex conjugation on $\mathbf{C}$ corresponds by h to the adjunction in $\mathrm{End}_B(V)_{\mathbf{R}}$ with respect to the pairing $\langle\cdot,\cdot\rangle$, and such that $(u, v) \mapsto \langle u, h(i)v\rangle$ is a symmetric definite positive pairing over $V_{\mathbf{R}}$.

Let G be the reductive group over $\mathbf{Q}$ defined by:

$$G(R) = \{g \in \mathrm{GL}(V \otimes_{\mathbf{Q}} R), \exists \mu \in R^{\times}, \forall x, y \in V \otimes R,$$
$$\langle gx, gy\rangle = \mu\langle x, y\rangle \text{ and the action of } g \text{ is } B\text{-linear}\}$$

We can attach to h a morphism $\mu_h : \mathbf{C}^{\times} \to G_{\mathbf{C}}$ that induces a decomposition $V_{\mathbf{C}} = V_0 \oplus V_1$, where $\mu_h(z)$ acts by z on V_1 and by 1 on V_0. The reflex field E of the Shimura data is then the subfield of $\bar{\mathbf{Q}}$ generated by the traces of the elements of B acting on V_0.

Let $\mathcal{X}$ be the $G(\mathbf{R})$-conjugacy class of $h^{-1} : \mathbf{C}^{\times} \to G_{\mathbf{R}}$. Then for each compact open subgroup K of $G(\mathbf{A}_f)$ that is small enough, consider the analytic space $G(\mathbf{Q})\backslash\mathcal{X} \times (G(\mathbf{A}_f)/K)$. We recall now some of the results of [Del71] (see also a detailled explanation in [GN]). There is an algebraic variety $\mathrm{Sh}(G, \mathcal{X})_K$ such that $\mathrm{Sh}(G, \mathcal{X})_K(\mathbf{C}) = G(\mathbf{Q})\backslash\mathcal{X} \times (G(\mathbf{A}_f)/K)$. The variety $\mathrm{Sh}(G, \mathcal{X})_K$ actually has a model over the reflex field E, and under additional conditions, this model is actually unique. We call it the canonical model of the Shimura variety $\mathrm{Sh}(G, \mathcal{X})_K$ ([Del71, Définition 3.13]). In the situation we are studying, where the Shimura data is of PEL type, the canonical model can be constructed as a union of connected components of a moduli space parameterizing abelian varieties with given polarization, endomorphisms and level structure, that can be expressed in terms of the Shimura data $\mathcal{D}$.

1.2 New data

We now want to construct an integral model at p of the Shimura variety $\mathrm{Sh}(G, \mathcal{X})_K$, that is a smooth model over the ring $\mathcal{O}_E \otimes_{\mathbf{Z}} \mathbf{Z}_{(p)}$. We need some extra data and assumptions in order to ensure that this is possible, and in order to define this integral model as a moduli space of abelian varieties.

We add to the Shimura data $\mathcal{D}$ the following element: let $\mathcal{O}_B$ be a $\mathbf{Z}_{(p)}$-order in B that is stable under the involution $*$ of B and becomes maximal after tensorization with $\mathbf{Z}_p$.

Moreover, we require additional conditions:

(1) B is unramified at p, that is $B_{\mathbf{Q}_p} = B \otimes_{\mathbf{Q}} \mathbf{Q}_p$ is isomorphic to a product of matrix algebras over unramified extensions of $\mathbf{Q}_p$.

(2) there exists a $\mathbf{Z}_p$-lattice Λ in $V_{\mathbf{Q}_p}$ that is stable under $\mathcal{O}_B$, and such that the pairing $\langle \cdot, \cdot \rangle$ induces a perfect duality of Λ with itself.

Note that condition (1) implies that $\mathcal{O}_B \otimes_{\mathbf{Z}} \mathbf{Z}_p$ is isomorphic to a product of matrix algebras over rings of integers of unramified extensions of $\mathbf{Q}_p$.

Example 1.1. Let B be an imaginary quadratic extension of $\mathbf{Q}$. Then condition (1) simply means that p does not ramify in B. We can choose for $\mathcal{O}_B$ the $\mathbf{Z}_{(p)}$-order generated by the ring of integers of B.

1.3 The reductive group

Consider the reductive group G attached to the Shimura data $\mathcal{D}$. Because of the additional conditions, $G_{\mathbf{Q}_p}$ is unramified, as we can define a smooth reductive model $\mathcal{G}$ of $G_{\mathbf{Q}_p}$ over $\mathbf{Z}_p$. Indeed, fix a lattice Λ as in §1.2, (2). Let C_0 be the subgroup of $G(\mathbf{Q}_p)$ that stabilizes the lattice Λ. Then C_0 is the hyperspecial subgroup of the $\mathbf{Z}_p$-points of $\mathcal{G}$.

Remark 1.2. A consequence of [Kot92, Lemma 7.2] is that, when the Shimura data is of type A or C (see §1.4), the lattice Λ as in §1.2 (2) is essentially unique, as two such lattices differ by the action of an element of $G(\mathbf{Q}_p)$. As a consequence, the subgroup C_0 of $G(\mathbf{Q}_p)$ is uniquely defined up to conjugation in these cases.

1.4 Classification

For later use, we recall the classification of Shimura data. Let $C = \mathrm{End}_B(V)$. It is endowed with an involution $*$ coming from the involution of B, and by construction $G(\mathbf{Q}) = \{x \in C, xx^* \in \mathbf{Q}^\times\}$. Let $C_{\mathbf{R}} = C \otimes_{\mathbf{Q}} \mathbf{R}$. Then we have three families of Shimura data:

Type A: $C_{\mathbf{R}}$ is a product of copies of $M_n(\mathbf{C})$ for some n.
Type C: $C_{\mathbf{R}}$ is a product of copies of $M_{2n}(\mathbf{R})$ for some n.
Type D: $C_{\mathbf{R}}$ is a product of copies of $M_n(\mathbf{H})$ for some n.

Unitary groups correspond to type A. More precisely, if the data is of type A, then the subgroup G_1 of G such that $G_1(\mathbf{Q}) = \{x \in C, x^*x = 1\}$ is the restriction of scalars to $\mathbf{Q}$ of some inner form of a unitary group. The group of symplectic similitudes GSp corrsponds to type C, and orthogonal groups in even dimension to type D. As type D can be more complicated and we are mostly interested in the case of unitary groups, we will sometimes state results only for types A and C.

2 Preliminaries

In this Section we fix $\mathcal{D}$ a set of Shimura data as in §1.

2.1 Polarized abelian schemes with an action of $\mathcal{O}_B$

Let S be an $\mathcal{O}_E \otimes_{\mathbf{Z}} \mathbf{Z}_{(p)}$-scheme.

Definition 2.1. Let R be a subring of $\mathbf{Q}$. An R-isogeny between two abelian schemes A and A' over S is an isomorphism in the category where the objects are abelian schemes and the set of morphisms from A to A' is $\mathrm{Hom}(A, A') \otimes_{\mathbf{Z}} R$. An R-polarization of A is a polarization of A that is also an R-isogeny from A to the dual abelian scheme A^t.

Definition 2.2. We say that (A, λ, ι) is a $\mathbf{Z}_{(p)}$-polarized abelian scheme with an action of $\mathcal{O}_B$ if:

(1) A is an abelian scheme over S.

(2) λ is a $\mathbf{Z}_{(p)}$-polarization.

(3) ι is an injective ring homorphism $\mathcal{O}_B \to \mathrm{End}(A) \otimes_{\mathbf{Z}} \mathbf{Z}_{(p)}$ which respects involutions on both sides: the involution $*$ on the left side, and the Rosati involution $\dagger$ coming from λ on the right side.

2.2 The determinant condition of Kottwitz

We now have to find a way to explain how $\mathcal{O}_B$ acts on the abelian scheme. More precisely we want to be able to express the fact that $\mathcal{O}_B$ acts on $\mathrm{Lie}\, A$ the same way it acts on V_0.

2.2.1 The determinant condition for projective modules

We fix once and for all a generating family $\alpha_1, \ldots \alpha_t$ of $\mathcal{O}_B$ as a $\mathbf{Z}_{(p)}$-module.

Let R be an algebra over $\mathcal{O}_E \otimes_{\mathbf{Z}} \mathbf{Z}_{(p)}$, and M be a finitely generated projective R-module. Suppose that $\mathcal{O}_B$ acts on M by R-linear endomorphisms. We then say that M is a R-module with an action of $\mathcal{O}_B$. A special case of such a module with an action of $\mathcal{O}_B$ is given by V_0 (see definition in §1.1).

We consider the action of $\mathcal{O}_B[X_1, \ldots X_t]$ on $M \otimes_R R[X_1, \ldots X_t]$. We denote by $\det_M \in R[X_1, \ldots X_t]$ the determinant of the element $X_1\alpha_1 + \ldots X_t\alpha_t$ for this action. Here the ring R is understood.

It is clear that $\det_M$ is functorial in R : that is, if $f : R \to R'$ is a homomorphism of $\mathcal{O}_E \otimes_{\mathbf{Z}} \mathbf{Z}_{(p)}$-algebras, and $M' = M \otimes_R R'$, then $\mathcal{O}_B$ acts on M' by R'-linear endomorphisms and $\det_{M'} = f(\det_M)$.

We have the following result:

Lemma 2.3. $\det_{V_0} \in \left(\mathcal{O}_E \otimes_{\mathbf{Z}} \mathbf{Z}_{(p)}\right)[X_1, \ldots X_t]$.

Proof. By definition of the reflex field E, all the elements $\det(b; V_0)$ lie in E, so the coefficients of $\det_{V_0}$ are in E. Let F be a number field such that the action of B on V_0 is defined on F. Then the image of $\mathcal{O}_B$ in the algebra of $g \times g$ matrices over F is an $\mathcal{O}_F \otimes \mathbf{Z}_{(p)}$-order. Hence $\det(b, V_0)$ is in $\mathcal{O}_F \otimes \mathbf{Z}_{(p)}$ for all $b \in \mathcal{O}_B$. So the coefficients of $\det_{V_0}$ are integral over $\mathbf{Z}_{(p)}$. This proves the Lemma. □

Lemma 2.4. *Let k be a field, and V and W be two finite-dimensional k-vector spaces with an action of $\mathcal{O}_B$. Then V and W are isomorphic if and only if* $\det_V = \det_W$.

Proof. Let us denote $\mathcal{O}_B \otimes_{\mathbf{Z}} k$ by A. Then A is a finite dimensional semisimple algebra over k. Indeed, if $\operatorname{char}(k) = 0$ then $A = B \otimes_{\mathbf{Q}} k$, and if $\operatorname{char}(k) = p$ then A is a product of matrix algebras over extensions of $\mathbb{F}_p$, as $\mathcal{O}_B \otimes \mathbf{Z}_p$ is a maximal order of $B \otimes \mathbf{Q}_p$, which is itself a product of matrix algebras over unramified extensions of $\mathbf{Q}_p$. Moreover, V and W are isomorphic as k-vector spaces with an action of $\mathcal{O}_B$ if and only if they are isomorphic as A-modules.

We write $A = A_1 \times \cdots \times A_n$ where the A_i are simple k-algebras. We consider V and W as A-modules. Then we have decompositions $V = V_1 \times \cdots \times V_n$ and $W = W_1 \times \cdots \times W_n$ where V_i and W_i are A_i-modules. As A_i is simple it has only one isomorphism class of irreducible representation. Hence V and W are isomorphic if and only if $\dim V_i = \dim W_i$ for all i, and it is clear that this information can be recovered from $\det_V$ and $\det_W$. □

Definition 2.5. If R is an $(\mathcal{O}_E \otimes_{\mathbf{Z}} \mathbf{Z}_{(p)})$-algebra, and M an R-module with an action of $\mathcal{O}_B$, then we say that M satisfies the determinant condition if $\det_M$ equals the image of $\det_{V_0}$ in $R[X_1, \ldots X_t]$.

We then show how the isomorphism class varies under specialization:

Lemma 2.6. *Suppose R is an $\mathcal{O}_E \otimes_{\mathbf{Z}} \mathbf{Z}_{(p)}$-algebra that is a local ring with residue field k, and let M be a finitely generated free R-module with an action of $\mathcal{O}_B$. Then M satisfies the determinant condition if and only if $M \otimes_R k$ does.*

Proof. We need only prove that M satisfies the determinant condition when $M \otimes_R k$ does. Note that if $R \subset R'$, then the R'-module $R' \otimes_R M$ satisfies the determinant condition if and only if M satisfies the condition. Note also that as $\mathcal{O}_B$ is finitely generated over $\mathbf{Z}_p$, there exists a noetherian subring $R_0 \subset R$, and a free R_0-module M_0 with an action of $\mathcal{O}_B$ such that $M = M_0 \otimes_{R_0} R$. So we can assume that R is noetherian if needed.

Fix K a finite extension of $\mathbf{Q}_p$ containing E such that there exists a K-vector space W with an action of B satisfying the determinant condition. As $\mathcal{O}_B \otimes_{\mathbf{Z}} \mathbf{Z}_p$ is compact, W contains an $\mathcal{O}_K$-lattice $\mathcal{L}$ with an action of $\mathcal{O}_B$, and $\mathcal{L}$ satisfies the determinant condition. Suppose first that R is in fact an $\mathcal{O}_K$-algebra. Then by Lemma 2.4 $M \otimes_R k$ is isomorphic to $\mathcal{L} \otimes_{\mathcal{O}_K} k$ as a k-module with action of $\mathcal{O}_B$, as $\det_{M \otimes_R k} = \det_{\mathcal{L} \otimes_{\mathcal{O}_K} k}$. Let us show now that $\mathcal{L} \otimes_{\mathcal{O}_K} R$ is a projective $\mathcal{O}_B \otimes_{\mathbf{Z}} R$-module. It is enough to show it when $R = \mathcal{O}_K$. But $\mathcal{O}_B \otimes_{\mathbf{Z}} \mathcal{O}_K$ is a product of matrix algebras over extensions of $\mathbf{Z}_p$. So the result holds by Morita equivalence, as $\mathcal{L}$ is torsion-free. So the $\mathcal{O}_B \otimes_{\mathbf{Z}} R$-linear isomorphism

$\mathcal{L} \otimes_{\mathcal{O}_K} k \to M \otimes_R k$ lifts to an $\mathcal{O}_B \otimes_{\mathbf{Z}} R$-linear morphism $\mathcal{L} \otimes_{\mathcal{O}_K} R \to M$. Now forget the action of $\mathcal{O}_B$. Nakayama's Lemma implies that this morphism is an isomorphism. As $\mathcal{L} \otimes_{\mathcal{O}_K} R$ satisfies the determinant condition, so does M.

Fix now F a finite extension of E such that the action of B on V_0 can be realized on F, that is, there exists an F-vector space $\mathcal{L}$ with an action of $\mathcal{O}_B$ satisfying the determinant condition. Suppose now that R is an F-algebra. Then by the same reasoning as before, we show that M satisfies the determinant condition if $M \otimes_R k$ does, using this time the fact that $\mathcal{L}$ is a projective $\mathcal{O}_B \otimes_{\mathbf{Z}} F$-module.

Assume now that we have an R-algebra R' which is either an $\mathcal{O}_K$-algebra or an F-algebra, which is such that $R \subset R'$, and such that for any maximal ideal $\mathfrak{m}'$ of R', we have $\mathfrak{m} \subset (\mathfrak{m}' \cap R)$. Then if the determinant condition holds for $M \otimes_R k$, it holds for $M \otimes_R (R'/\mathfrak{m}')$ for all $\mathfrak{m}'$ as $k \subset R'/\mathfrak{m}'$, hence it holds for $M \otimes_R R'_{\mathfrak{m}'}$ by the study of the two previous cases, and so it holds also for M as $R \subset R'$, and $R' \subset \prod_{\mathfrak{m}'} R'_{\mathfrak{m}'}$ (by [Mat89, Theorem 4.6]).

So we need only find such an R'. If p is invertible in R, we set $R' = R \otimes_E F$. If p is not invertible in R, we set $R' = \hat{R} \otimes_{\mathbf{Z}_p} \mathcal{O}_K$, where $\hat{R}$ is the p-adic completion of R (in this situation we need to assume that R is noetherian). □

We conclude that $\det_M$ depends only on specializations:

Corollary 2.7. *Let R be an $\mathcal{O}_E \otimes_{\mathbf{Z}} \mathbf{Z}_{(p)}$-algebra such that* spec R *is connected, and let M be a finitely generated projective R-module with an action of $\mathcal{O}_B$. Then M satisfies the determinant condition if and only if there exists a maximal ideal $\mathfrak{m}$ of R, with residue field k, such that $M \otimes_R k$ satisfies the determinant condition.*

2.2.2 *The determinant condition for abelian schemes with an action of $\mathcal{O}_B$*

Let (A, λ, ι) be a polarized abelian scheme with an action of $\mathcal{O}_B$ over the base scheme S. Then $\mathcal{O}_B$ acts on Lie A, which is a locally free $\mathcal{O}_S$-module. For each open affine subset U of S, we can define $\det_{\mathrm{Lie}\,A}(U) \in \Gamma(U, \mathcal{O}_S)[X_1, \dots X_t]$ as in §2.2.1. By functoriality of the definition of det, these sections are compatible, hence glue to define a global section $\det_{\mathrm{Lie}\,A} \in \Gamma(S, \mathcal{O}_S)[X_1, \dots X_t]$. As $\Gamma(S, \mathcal{O}_S)$ is naturally an $\mathcal{O}_E \otimes_{\mathbf{Z}} \mathbf{Z}_{(p)}$-algebra it makes sense to compare $\det_{\mathrm{Lie}\,A}$ to the image of $\det_{V_0}$ in $\Gamma(S, \mathcal{O}_S)[X_1, \dots X_t]$. Following [Kot92], we set the following definition:

Definition 2.8. The triple (A, λ, ι) satisfies the determinant condition of Kottwitz if $\det_{\mathrm{Lie}\,A}$ is the image of $\det_{V_0}$.

One consequence of the definition is the following: the dimension of Lie A and hence that of A is equal to that of V_0.

2.2.3 *Some geometric properties of the determinant condition*

We state what geometric consequences we can deduce from Proposition 2.6 and Corollary 2.7:

Proposition 2.9. *Let S be an $\mathcal{O}_E \otimes_{\mathbf{Z}} \mathbf{Z}_{(p)}$-scheme, and S_0 a closed subscheme of S with nilpotent definition ideal. Let (A, λ, ι) be an abelian scheme over S with an action of $\mathcal{O}_B$. Suppose that the base change of (A, λ, ι) to S_0 satisfies the determinant condition. Then so does (A, λ, ι).*

Proposition 2.10. *Let S be an $\mathcal{O}_E \otimes_{\mathbf{Z}} \mathbf{Z}_{(p)}$-scheme, and (A, λ, ι) an abelian scheme over S with an action of $\mathcal{O}_B$. Then there is a closed subscheme T of S that is a union of connected components, such that for any closed point x of S, $(A_x, \lambda_x, \iota_x)$ satisfies the determinant condition if and only if x is a point of T.*

2.2.4 *The example of unitary groups*

In the case of unitary groups over $\mathbf{Q}$, we give a condition on Lie A that is equivalent to the determinant condition of Kottwitz and that is simpler to state.

Let B be a quadratic imaginary extension of $\mathbf{Q}$, and let τ be in B such that $\mathcal{O}_B = \mathbf{Z}[\tau]$. Fix a prime p that is unramified in B.

Let R be an $\mathcal{O}_{B,(p)}$-algebra, and M a locally free R-module with an action of $\mathcal{O}_B$. Then we have a decomposition $M = M^+ \oplus M^-$ where M^+ and M^- are also locally free. Here M^+ is defined as the submodule of M where the action of τ from the action of $\mathcal{O}_B$ and the action of the image of τ in R coincide, and M^- is the submodule where these action differ by conjugation in $\mathcal{O}_B$. When R is connected we can then define the type of M as the pair of integers $(\mathrm{rk}\, M^+, \mathrm{rk}\, M^-)$.

Then, if S is an $\mathcal{O}_E \otimes_{\mathbf{Z}} \mathbf{Z}_{(p)}$-scheme, an abelian scheme A with an action of $\mathcal{O}_B$ over the base S satisfies the determinant condition if and only if Lie A has the same type as the B-module V_0.

2.3 Level structures

2.3.1 *Tate modules*

Denote by $\mathbf{A}_f^p$ the ring of finite adeles of $\mathbf{Q}$ away from p, and by $\hat{\mathbf{Z}}^{(p)}$ the ring $\prod_{\ell \neq p} \mathbf{Z}_\ell$. We denote $V \otimes_{\mathbf{Q}} \mathbf{A}_f^p$ by $V^{(p)}$.

Let (A, λ, ι) be a polarized abelian scheme with an action of $\mathcal{O}_B$, defined on an $\mathcal{O}_E \otimes_{\mathbf{Z}} \mathbf{Z}_{(p)}$-schemes. Let s be any geometric point of S, and consider the Tate modules: $T(A_s) = \varprojlim_N A_s[N]$, $T^{(p)}(A_s) = \varprojlim_{N \text{ prime to } p} A_s[N] = T(A_s) \otimes_{\hat{\mathbf{Z}}} \hat{\mathbf{Z}}^{(p)}$ and $V^{(p)}(A_s) = H_1(A_s, \mathbf{A}_f^p) = T^{(p)}(A_s) \otimes_{\hat{\mathbf{Z}}^{(p)}} \mathbf{A}_f^p$.

They are endowed with a non-degenerate bilinear form, coming from the polarization λ, and an action of $\mathcal{O}_B$, coming from the action of $\mathcal{O}_B$ on A itself.

Let f be a separable isogeny from A to A' with kernel C. Then f induces a morphism $T(f) : T(A_s) \to T(A'_s)$ which is injective with cokernel isomorphic to C_s. If f is of prime-to-p degree then f induces $T^{(p)}(f) : T^{(p)}(A_s) \to T^{(p)}(A'_s)$ which is injective with cokernel C_s. We also have an isomorphism $V^{(p)}(f) : V^{(p)}(A_s) \to V^{(p)}(A'_s)$.

Suppose f is a separable R-isogeny for some subring R of $\mathbf{Q}$. Then $V(f)$ is still well-defined but f does not necessarily map $T(A_s)$ into $T(A'_s)$. In fact f maps $T(A_s)$ into $T(A'_s)$ if and only if f is an isogeny in the usual sense.

2.3.2 *Level subgroups*

Definition 2.11. A level subgroup of G is a compact open subgroup of $G(\mathbf{A}_f^p)$.

In particular, for us a level structure is always "away from p".

Example 2.12. Fix a lattice $L \subset V$ that is self-dual for $\langle \cdot, \cdot \rangle$. Fix an integer N prime to p. We define the principal level subgroup of level N of $G(\mathbf{A}_f^p)$ as the compact open subgroup $\{g \in G(\mathbf{A}_f^p), (g-1)(L \otimes_{\mathbf{Z}} \hat{\mathbf{Z}}^{(p)}) \subset N(L \otimes_{\mathbf{Z}} \hat{\mathbf{Z}}^{(p)})\}$.

2.3.3 Definition of the level structures

Let S be an $\mathcal{O}_E \otimes_{\mathbf{Z}} \mathbf{Z}_{(p)}$-scheme, and s a geometric point of S. Let (A, λ, ι) be a polarized abelian scheme with an action of $\mathcal{O}_B$. We say that a map η^p from $V^{(p)}$ to $V^{(p)}(A_s)$ respects the structures on both sides if it respects the bilinear forms up to a scalar in $(\mathbf{A}_f^p)^\times$, and if it is compatible with the $\mathcal{O}_B$-action on both sides.

Let g be in $G(\mathbf{A}_f^p)$. If η^p respects the structures on both sides then so does $\eta^p \circ g$. Hence $G(\mathbf{A}_f^p)$ acts on the set of such maps.

Definition 2.13. Let K^p be a level subgroup. A level structure of level K^p on (A, λ, ι) is a choice of a geometric point s for each connected component of S, and for each s a choice of a K^p-orbit $\bar{\eta}^p$ of morphisms $\eta^p : V^{(p)} \to V^{(p)}(A_s)$ respecting the structures on both sides and such that the orbit is fixed under the action of $\pi_1(s, S)$.

Remark 2.14. The last condition ensures that a level structure is in fact independant of the choice of s. Moreover, a level structure exists at some point s if and only if for any geometric point s' in the same connected component as s there exists a level structure at s'.

3 The integral model as a moduli scheme

3.1 Definition of the moduli problem

Let us fix a set of Shimura data $\mathcal{D}$ as in §1. We also fix a compact open subgroup K^p of $G(\mathbf{A}_f^p)$. We will define a moduli problem classifying abelian schemes with an action of $\mathcal{O}_B$ and K^p-level structure.

Definition 3.1. Let $\mathcal{F}_{K^p}$ be the following category fibered in groupoids over the category $(Sch/\mathcal{O}_E \otimes_{\mathbf{Z}} \mathbf{Z}_{(p)})$ of $\mathcal{O}_E \otimes_{\mathbf{Z}} \mathbf{Z}_{(p)}$-schemes:

- The objects over a scheme S are quadruples $\underline{A} = (A, \lambda, \iota; \bar{\eta}^p)$, where (A, λ, ι) is a $\mathbf{Z}_{(p)}$-polarized projective abelian scheme over S with an action of $\mathcal{O}_B$ which respects the determinant condition of Kottwitz (Definition 2.8 of §2.2), and $\bar{\eta}^p$ is a level structure of level K^p over each connected component of S.

- The morphisms from $\underline{A} = (A, \lambda, \iota; \bar{\eta}^p)$ to $\underline{A'} = (A', \lambda', \iota'; \bar{\eta}'^p)$ over S are given by a $\mathbf{Z}_{(p)}$-isogeny $f : A \to A'$ compatible with the action of $\mathcal{O}_B$ and the level structures, that is:

  (1) there exists a locally constant function r on S with values in $\mathbf{Z}_{(p)}^\times$ such that $\lambda = r(f^t \circ \lambda' \circ f)$.

  (2) f induces a morphism from End $(A) \otimes_{\mathbf{Z}} \mathbf{Z}_{(p)}$ to End $(A') \otimes_{\mathbf{Z}} \mathbf{Z}_{(p)}$, that we still denote by f; then for all $b \in \mathcal{O}_B$, $f \circ \iota(b) = \iota'(b)$.

(3) $\bar{\eta}'^p = V^{(p)}(f) \circ \bar{\eta}^p$, where we denote by $V^{(p)}(f)$ the morphism induced by f from $V^{(p)}(A_s)$ to $V^{(p)}(A'_s)$ for any s.

3.2 Known results about the moduli problem

We summarize here what is known about the moduli problem. We will explain these results in more detail in the rest of the Chapter.

3.2.1 Representability

Theorem 3.2. *$\mathcal{F}_{K^p}$ is a smooth Deligne-Mumford stack and it is representable by a quasi-projective scheme when K^p is small enough.*

The part of the Theorem concerning representability will be proved in § 4, where we will also explain what "small enough" means. The part about smoothness will be proved in § 5.

When $\mathcal{F}_{K^p}$ is representable by a scheme, we denote this scheme by S_{K^p}. In some cases we know a little more about S_{K^p} ([Kot92, end of § 5]:

Proposition 3.3. *Suppose that* $\mathrm{End}_B(V)$ *is a division algebra and let K^p be small enough that $\mathcal{F}_{K^p}$ is representable by a scheme S_{K^p}. Then S_{K^p} is projective over* $\mathrm{spec}\, \mathcal{O}_E \otimes_{\mathbf{Z}} \mathbf{Z}_{(p)}$.

3.2.2 Hecke operators

Theorem 3.4. *The family of schemes S_{K^p}, for K^p the small enough compact open subgroups of $G(\mathbf{A}_f^p)$, form a tower of schemes with finite smooth transition morphisms. The group $G(\mathbf{A}_f^p)$ acts on the tower via Hecke operators.*

This is the object of §6.

3.2.3 Generic fiber

The generic fiber of the schemes we have constructed are isomorphic (except for some special cases) to the Shimura varieties S_K in characteristic zero that were defined in the first Chapter, when $K = K^p C_0$, where C_0 is the hyperspecial subgroup of $G(\mathbf{Q}_p)$ defined in §1.3. Moreover this is compatible with the action of Hecke operators. More precisely we will see in §7:

Theorem 3.5. *When the Shimura data is of type A or C we have the following isomorphism for each compact open subgroup K^p of $G(\mathbf{A}_f^p)$:*

$$S_{K^p} \otimes_{\mathcal{O}_E \otimes_{\mathbf{Z}} \mathbf{Z}_{(p)}} E \xrightarrow{\sim} S_{K^p C_0}$$

Moreover, the induced isomorphism between the towers $(S_{K^p} \otimes E)_{K^p}$ and $(S_{K^p C_0})_{K^p}$ is compatible with the action of $G(\mathbf{A}_f^p)$ on both sides.

4 Representability of the moduli problem

4.1 Statement of the Theorem

The goal of this Section is the proof of the following Theorem (which is a more precise version of Theorem 3.2):

Theorem 4.1. *For all level subgroups K^p, $\mathcal{F}_{K^p}$ is a Deligne-Mumford stack. Moreover if K^p is small enough so that it is contained in a principal level subgroup of level $N \geq 3$, the functor $\mathcal{F}_{K^p}$ is representable by a quasi-projective scheme over $\mathcal{O}_E \otimes_{\mathbf{Z}} \mathbf{Z}_{(p)}$.*

We will prove this Theorem by comparing our moduli problem to the case of the Siegel modular varieties, that is the case where the endomorphism ring is trivial, which is already known by the results of [MFK94]. This is the proof outlined in [Kot92]. Another strategy to study the representability of the moduli problem would be via Artin's criterion. This has the advantage of being more direct and not to rely on the difficult results of [MFK94], but it has the drawback that it only shows that the moduli problem is representable by an algebraic space when the level is small enough. To prove that it is in fact representable by a quasi-projective scheme one then has to use additional arguments. A detailed proof using this strategy can be found in [Lan13].

4.2 The Siegel case

We first study the so-called Siegel case. In this case the scheme we obtain is the Siegel moduli space of abelian varieties.

4.2.1 Siegel Shimura data

We define the Siegel Shimura data $\mathcal{D}$ as follows: we take $B = \mathbf{Q}$, $\mathcal{O}_B = \mathbf{Z}_{(p)}$, and $V = \mathbf{Q}^{2g}$ endowed with the standard symplectic form. In this case the group G is the similitude symplectic group GSp_{2g}, and there exists only one conjugacy class of maps h satisfying the conditions of §1.1 (3). The reflex field is $\mathbf{Q}$. As $B = \mathbf{Q}$, this means that we can forget about the action of $\mathcal{O}_B$ in the definition of the moduli problem.

4.2.2 The result of [MFK94]

We describe the result of [MFK94] concerning the case of the Siegel Shimura data. Fix an integer $g \geq 1$ and an integer $N \geq 3$.

Definition 4.2. Let $\mathcal{A}(N)$ be the category fibered in groupoids on $\mathbf{Z}[1/N]$-schemes such that: For any $\mathbf{Z}[1/N]$-scheme S, the set of objects of $\mathcal{A}_S$ is the set of triples $\underline{A} = (A, \lambda; \alpha)$ where A is a projective abelian scheme of dimension g, λ is a principal polarization, and α is a symplectic similitude (with multiplicator in $(\mathbf{Z}/N\mathbf{Z})^\times$) between $(\mathbf{Z}/N\mathbf{Z})^{2g}_S$ and $A[N]_S$. Here we consider $(\mathbf{Z}/N\mathbf{Z})^{2g}$ to be endowed with the standard symplectic form. If $\underline{A}$, $\underline{A}'$ are objects of $\mathcal{A}_S$, then the morphisms from $\underline{A}$ to $\underline{A}'$ are the isomorphisms $f : A \to A'$ such that $\lambda = f^t \circ \lambda' \circ f$ and $\alpha' = f \circ \alpha$.

The main result is the following ([MFK94], Theorem 7.9, see also [MB85], Theorem 3.2 of Chapter VII):

Proposition 4.3. *The category fibered in groupoids $\mathcal{A}(N)$ is representable by a (smooth) quasi-projective scheme.*

4.2.3 Reformulation of the moduli problem

In order to compare more easily our situation to that studied in [MFK94] we give another formulation of the moduli problem. As $\mathbf{Q}^{2g} = V$ is endowed with the standard symplectic form, $\mathbf{Z}^{2g}$ is a self-dual lattice. We write L for $\mathbf{Z}^{2g} \otimes_{\mathbf{Z}} \hat{\mathbf{Z}}^{(p)}$. Let $K(N)$ be the principal level subgroup of level N for some N prime to p, as defined in Example 2.12, with respect to the lattice L. Let $\mathcal{F}_{K(N)}$ be the category fibered in groupoids as in Definition 3.1, with the Siegel Shimura data.

Proposition 4.4. *In the case of the Siegel Shimura data, when N is prime to p, $\mathcal{F}_{K(N)}$ and the restriction of $\mathcal{A}(N)$ to the subcategory of $\mathbf{Z}_{(p)}$-schemes are isomorphic.*

Proof. In the proof, we abbreviate $\mathcal{A}(N)$ by $\mathcal{A}$ and $\mathcal{F}_{K(N)}$ by $\mathcal{F}$. Let S be a $\mathbf{Z}_{(p)}$-scheme, and let $\mathcal{A}_S$ and $\mathcal{F}_S$ be the categories of objects of $\mathcal{A}$ and $\mathcal{F}$ over S. We must construct an equivalence of categories between $\mathcal{A}_S$ and $\mathcal{F}_S$. We can assume that S is connected, and fix a geometric point s of S.

Let $(A, \lambda; \alpha)$ be an object of $\mathcal{A}_S$. We want to construct an object of $\mathcal{F}_S$. The morphism $\alpha : (\mathbf{Z}/N\mathbf{Z})^{2g}_S \to A[N]_S$ gives a symplectic similitude $\alpha_s : (\mathbf{Z}/N\mathbf{Z})^{2g} \to A[N]_s$ that is invariant under the action of $\pi_1(s, S)$. This morphism α_s then extends to a symplectic similitude $\eta^p_0 : L \to T^{(p)}(A_s)$, which also induces a symplectic similitude $\eta^p : V^{(p)} \to V^{(p)}(A_s)$. The map η^p is not uniquely defined, but its $K(N)$-conjugation class is uniquely determined by α_s, and so is invariant under the action of $\pi_1(s, S)$. Hence the orbit $\bar{\eta}^p$ of η^p defines a level structure $\bar{\eta}^p$ extending α. Hence to each $(A, \lambda; \alpha)$ we can attach $(A, \lambda; \bar{\eta}^p)$ which is an object of $\mathcal{F}_S$. Morphisms in $\mathcal{A}_S$ also define morphisms in $\mathcal{F}_S$, hence our construction is functorial.

Let us show now that this functor is faithful. Let $\underline{A} = (A, \lambda; \alpha)$ and $\underline{A'} = (A', \lambda'; \alpha')$ be objects in $\mathcal{A}_S$. Let $f : A \to A'$ be a $\mathbf{Z}_{(p)}$-isogeny which is a morphism in the category $\mathcal{F}_S$. Then f is an isomorphism, as by construction of the level structures f induces an isomorphism from $T^{(p)}(A)$ to $T^{(p)}(A')$. Moreover we necessarily have $\lambda' = f^t \circ \lambda \circ f$, as λ and λ' are principal polarizations. Hence f is a morphism in the category $\mathcal{A}_S$.

Finally let us show that this functor is essentially surjective. We need to see that any object in $\mathcal{F}_S$ is isomorphic to an object coming from $\mathcal{A}_S$. Let $\underline{A} = (A, \lambda; \bar{\eta}^p)$ be an object of $\mathcal{F}_S$. We need to find $\underline{A'} = (A', \lambda', \bar{\eta}^{p\prime})$ and a morphism $\underline{A} \to \underline{A'}$ in $\mathcal{F}_S$ such that λ' is a principal polarization and $\bar{\eta}'^p : V^{(p)} \to V^{(p)}(A'_s)$ induces a symplectic similitude between $(\mathbf{Z}/N\mathbf{Z})^{2g}_S$ and $A[N]_S$.

Observe first that we need only find an object $\underline{A'}$ in $\mathcal{F}_S$ with a map $\underline{A'} \to \underline{A}$ (or, equivalently, with a map $\underline{A} \to \underline{A'}$) such that η'^p induces an isomorphism between L and $T^{(p)}(A'_s)$. Indeed such a level structure then induces a $\pi_1(s, S)$-invariant symplectic similitude $\alpha_s : (\mathbf{Z}/N\mathbf{Z})^{2g} \to A[N]_s$ that gives us the isomorphism $\alpha : (\mathbf{Z}/N\mathbf{Z})^{2g}_S \to A[N]_S$. Moreover, if we have such an $\underline{A'}$ we can change the polarization λ' to make it principal. We know that the bilinear forms on L and $T^{(p)}(A'_s)$ differ by a scalar a in $(\mathbf{A}^p_f)^\times$. By multiplying λ' by some prime-to-p integer n, we multiply the bilinear form on $T^{(p)}(A'_s)$ by n, and so we change a into na. So we can assume that a is in $(\mathbf{A}^p_f)^\times \cap \hat{\mathbf{Z}}^{(p)}$. Let ℓ be a prime not dividing p. We know that the isogeny λ' is divisible by ℓ if and only if the pairing on $T_\ell A_s$ coming from λ' is divisible by ℓ. Hence we can divide λ' by some integer n prime to p so that the pairing on L and the pairing on $T^{(p)}(A'_s)$ coming from the new λ' differ by an element of $\hat{\mathbf{Z}}^{(p)\times}$. But then the new polarization λ' induces an isomorphism $T^{(p)}(A'_s) \to T^{(p)}(A'^t_s)$ and so is principal, as we already know that its degree is prime to p.

Now we try to find $\underline{A'} = (A', \lambda', \bar{\eta}^{p\prime})$ and a morphism $\underline{A} \to \underline{A'}$ in $\mathcal{F}_S$ such that $\bar{\eta}'^p : V^{(p)} \to V^{(p)}(A'_s)$ induces a symplectic similitude between L and $T^{(p)}(A'_s)$. We note first that $T^{(p)}(A_s)$ and $\eta^p(L)$ are commensurable lattices in $V^{(p)}(A_s)$.

Fix a submodule $M \subset T^{(p)}(A_s)$ with finite index which is invariant under the action of $\pi_1(s, S)$. Then, from the interpretation of the Tate module as a π_1, we see that there exists A' and a prime-to-p isogeny $f : A' \to A$ such that $f(T^{(p)}(A'_s)) = M$. We can define a polarization on A' by $\lambda' = f^t \circ \lambda \circ f$. If moreover $M \subset \eta^p(L)$, then we can define a level structure on A' by the condition that $\eta^p = f \circ \eta'^p$. Then $\underline{A'} = (A', \lambda'; \bar{\eta}'p)$ is an object of $\mathcal{F}_S$ and f induces a map $\underline{A'} \to \underline{A}$ in $\mathcal{F}_S$.

We apply this to $M = T^{(p)}(A_s) \cap \eta^p(L)$. The construction gives us an $\underline{A'}$ with a map to $\underline{A}$ which satisfies moreover that $T^{(p)}(A'_s) \subset \eta'^p(L)$. So now we need only treat the case where $T^{(p)}(A_s) \subset \eta^p(L)$.

Fix a submodule $M \subset V^{(p)}(A_s)$ that is invariant under the action of $\pi_1(s, S)$, and suppose that $T^{(p)}(A_s) \subset M$ with finite index. Then there exists an étale subgroup scheme $C \subset A$ with C_s isomorphic to $M/T^{(p)}(A_s)$, such that the isogeny $f : A \to A' = A/C$ induces an isomorphism from M to $T^{(p)}(A'_s)$. There is some prime-to-p integer n such that (nf^{-1}) is an isogeny from A' to A, then we can endow the abelian scheme A' with a polarization $\lambda' = (nf^{-1})^t \circ \lambda \circ (nf^{-1})$. We can define a level structure on (A', λ') by $\bar{\eta}'p = f \circ \bar{\eta}^p$. Assume now that $T^{(p)}(A_s) \subset \eta^p(L) = M$. Then f defines a morphism in $\mathcal{F}_S$ from $(A, \lambda; \bar{\eta}^p)$ to $(A', \lambda'; \bar{\eta}'p)$. Moreover $\bar{\eta}'p$ has the property that $\bar{\eta}'p(L) = T^{(p)}(A'_s)$. So this treats the case where $T^{(p)}(A_s) \subset \eta^p(L)$. □

4.3 From the Siegel case to the PEL case

Fix a Shimura data $\mathcal{D}$ as in §1. We denote by $\mathrm{GSp}(V)$ the symplectic similitude group attached to the vector space V and the given alternating pairing on it, forgetting the action of B. Then the reductive group G from the data $\mathcal{D}$ is naturally a subgroup of $\mathrm{GSp}(V)$.

We fix a level subgroup $K^p \subset \mathrm{GSp}(\mathbf{A}_f^p)$, and set $K_G^p = K^p \cap G(\mathbf{A}_f^p)$ which is a level subgroup of $G(\mathbf{A}_f^p)$. We can define two moduli problems as in Definition 3.1. One is attached Shimura data $\mathcal{D}$, and the level subgroup K_G^p, we denote it by $\mathcal{F}$. The other is attached to the Siegel Shimura data, and the level subgroup K^p, we denote it by $\mathcal{S}$.

We have a natural transformation from $\mathcal{F}$ to $\mathcal{S}$, which is defined by sending the quadruple $(A, \lambda, \iota; \bar{\eta})$ over an S-scheme to $(A, \lambda; \bar{\bar{\eta}})$, where $\bar{\bar{\eta}}$ is the K^p-orbit generated by $\bar{\eta}$.

We have to prove that the functor $\mathcal{F}$ is relatively representable over $\mathcal{S}$, and that $\mathcal{F}$ is projective over $\mathcal{S}$. More precisely we will show:

Proposition 4.5. *Let K^p be such that $\mathcal{S}$ is representable by a scheme. Then $\mathcal{F}$ is relatively representable over $\mathcal{S}$ by a scheme that is projective over $\mathcal{S}$.*

By §4.2, $\mathcal{S}$ is representable by a scheme for example when K^p is a principal level subgroup of level $N \geq 3$ prime to p.

4.3.1 Construction of a scheme

In this Paragraph we fix an $\mathcal{O}_E \otimes_{\mathbf{Z}} \mathbf{Z}_{(p)}$-scheme S, an abelian scheme A over S, a $\mathbf{Z}_{(p)}$-polarization λ of A, and a K^p-level structure $\bar{\eta}^p$ on (A, λ).

We will need the following result:

Lemma 4.6. *Let S be a locally noetherian scheme, and A a projective abelian scheme over S. Then the functor from S-schemes to the category of sets that attaches $End(A_T)$ to T is representable by a union of projective schemes over S. We will denote by $\mathcal{E}$ the scheme representing the functor $T \mapsto End(A_T) \otimes_{\mathbf{Z}} \mathbf{Z}_{(p)}$. It is also a union of projective schemes over S.*

Proof. This follows from the theory of Hilbert schemes, as an endomorphism of A_T is a special case of a subscheme of $A \times_T A$. A detailed proof of this Lemma can be found in [Hid04, Section 6.1]. □

In our special case, the abelian scheme A is endowed with a prime-to-p polarization λ. So $\mathcal{E}$ naturally comes with an involution r, which is the Rosati involution. Let $m = 2n$ and $a_1, \dots a_m$ be a set of generators of $\mathcal{O}_B$ as a $\mathbf{Z}_{(p)}$-algebra with $a_{n+i} = a_i^*$. We define a closed subscheme Z of $\mathcal{E}^m$ as follows: let T be an S-scheme, and $(x_1, \dots x_m) \in \mathcal{E}^m(T)$. Then $(x_1, \dots x_m)$ is in Z if and only if any $\mathbf{Z}_{(p)}$-polynomial relation verified by $(a_1, \dots a_m)$ is also verified by $(x_1, \dots x_m)$ and by $(r(x_1), \dots r(x_m))$, where $r(x_i) = x_{n+i}$. The abelian scheme A_Z is then endowed with an algebra homomorphism $\mathcal{O}_B \to \mathrm{End}(A_Z) \otimes_{\mathbf{Z}} \mathbf{Z}_{(p)}$, which is compatible with the Rosati involution. That is, A_Z is a polarized abelian scheme with an action of $\mathcal{O}_B$ as in §2.1.

We know by Proposition 2.10 that the locus where the $\mathcal{O}_B$-action on A_Z satisfies the determinant condition is a union of connected components of Z.

Moreover, A_Z is also endowed with a K^p-level structure, by base change from the level structure on A. We want to understand the locus where this level structure comes from a K^p_G-level structure. Let Z' be a connected component of Z, and fix a geometric point z of Z'. Then the K^p-level structure corresponds to a $\pi_1(z, Z')$-invariant K^p-orbit $\bar{\eta}^p$ of symplectic similitudes $V^{(p)} \to V^{(p)}(A_z)$. Consider the condition: there exists an element f in $\bar{\eta}^p$ which is $\mathcal{O}_B$-equivariant, and such that its K^p_G-orbit is $\pi_1(z, Z')$- invariant. This condition does not depend on the choice of z on Z' by Remark 2.14. This condition is satisfied if and only if $\bar{\eta}^p$ comes from a K^p_G-level structure on $A_{Z'}$, and then is comes from a unique such structure, as two elements f as in the condition differ by an element of K^p_G. So the locus where the K^p-level structure comes from a K^p_G-level structure is a union of connected components of Z.

We denote by $X_{\mathcal{D}}$ the union of the connected components of Z where the determinant condition holds and the K^p-level structure comes from a K^p_G-level structure. Let $A_{X_{\mathcal{D}}}$ be the abelian scheme over $X_{\mathcal{D}}$ coming from A. As follows from the construction of $X_{\mathcal{D}}$, we have:

Lemma 4.7. *The abelian scheme $A_{X_{\mathcal{D}}}$ is naturally endowed with a structure of a polarized abelian scheme with K^p_G-level structure.*

4.3.2 Comparing $\mathcal{F}$ to $\mathcal{S}$

We now show the relative representability of $\mathcal{F}$ over $\mathcal{S}$ when $\mathcal{S}$ is representable by a scheme. We fix a scheme S, and a morphism $S \to \mathcal{S}$, and consider the functor $\mathcal{F}' = \mathcal{F} \times_{\mathcal{S}} S$. We have to show that $\mathcal{F}'$ is representable by a scheme.

The given morphism $S \to \mathcal{S}$ amounts to an equivalence class of triples $(\mathcal{A}, \lambda; \bar{\eta})$ where $\mathcal{A}$ is an abelian variety over S, endowed with a prime-to-p polarization λ, and a level structure $\bar{\eta}$ of level K^p. We choose a representative of this equivalence class. We can then construct a scheme $X_{\mathcal{D}}$ over S as in §4.3.1.

We then define a natural transformation $\mathcal{F}' \to X_{\mathcal{D}}$. Let T be an S-scheme. An element of $\mathcal{F}'(T)$ is an equivalence class of quadruples $(A, \lambda, \iota; \bar{\eta})$, such that its image by the forgetful functor $\mathcal{F} \to \mathcal{S}$ is in the same equivalence class as $(\mathcal{A}, \lambda; \bar{\eta})_T$. That is, there is a prime-to-p isogeny $f : A \to \mathcal{A}_T$, compatible with the polarizations and the level structures. Then f induces an isomorphism between $\mathrm{End}(A) \otimes_{\mathbf{Z}} \mathbf{Z}_{(p)}$ and $\mathrm{End}(\mathcal{A}_T) \otimes_{\mathbf{Z}} \mathbf{Z}_{(p)}$. We use this isomorphism to define a morphism $\iota : \mathcal{O}_B \to \mathrm{End}(\mathcal{A}_T) \otimes_{\mathbf{Z}} \mathbf{Z}_{(p)}$. Hence we get a point in $X_{\mathcal{D}}(T)$.

We have to show that this construction is well-defined, that is, it does not depend on the choice of $(A, \lambda, \iota; \bar{\eta})$ in the equivalence class. But this comes from the fact that any element of $(\mathcal{A}, \lambda; \bar{\eta})_T$ has no non-trivial automorphism, as we have chosen the level such that $\mathcal{S}$ is representable by a scheme.

Lemma 4.8. *This natural transformation is an isomorphism.*

Proof. We only have to find a natural transformation $X_{\mathcal{D}} \to \mathcal{F}'$ that is a quasi-inverse to the transformation we have just defined. But this is Lemma 4.7. □

Hence $\mathcal{F}'$ is representable by the scheme $X_{\mathcal{D}}$. The connected components of $X_{\mathcal{D}}$ are projective over the scheme representing $\mathcal{S}$, which is itself quasi-projective over $\mathcal{O}_E \otimes_{\mathbf{Z}} \mathbf{Z}_{(p)}$. To finish the proof of Proposition 4.5, we only have to show that $X_{\mathcal{D}}$ has only a finite number of connected components. But this comes from the fact that $\mathcal{F}$ is locally of finite presentation over $\mathcal{O}_E \otimes_{\mathbf{Z}} \mathbf{Z}_{(p)}$, as can be seen using the criterion of Proposition 4.15 of [LMB00].

4.4 Reduction to the case of principal level structures

We now finish the proof of Theorem 4.1 for the set of Shimura data $\mathcal{D}$. Fix a lattice $L \subset V$ as in Example 2.12, and denote by $K_G(N) \subset G(\mathbf{A}_f^p)$ the principal level subgroup of level N relative to this choice of L. If $N \geq 3$ and N is prime to p, then Proposition 4.5 implies that $\mathcal{F}_{K_G(N)}$ is representable by a scheme.

Let K^p be a level subgroup of $G(\mathbf{A}_f^p)$, and let K'^p level subgroup that is contained in K^p. Then we have a functor $\mathcal{F}_{K'^p} \to \mathcal{F}_{K^p}$ that sends the object $(A, \lambda, \iota; \bar{\eta}^p)/S$ to the object $(A, \lambda, \iota; \bar{\bar{\eta}}^p)/S$ where $\bar{\bar{\eta}}^p$ is the K^p-orbit generated by $\bar{\eta}^p$ (see also §6).

Assume now that K'^p is $K_G(N)$ for some $N \geq 3$ and prime to p, so that $\mathcal{F}_{K'^p}$ is a scheme, and in particular a Deligne-Mumford stack. This functor makes the scheme $\mathcal{F}_{K'^p}$ an étale presentation of the stack $\mathcal{F}_{K^p}$. Hence $\mathcal{F}_{K^p}$ is a Deligne-Mumford stack by [LMB00, Proposition 4.3.1].

If $K^p \subset K_G(N)$ for an $N \geq 3$, then $\mathcal{F}_{K^p}$ is representable by an algebraic space. This follows from Lemma 4.9 below and [LMB00, Corollary 8.1.1]: a Deligne-Mumford stack where the objects have only the trivial automorphism is representable by an algebraic space. But then we have a finite morphism $\mathcal{F}_{K^p} \to \mathcal{F}_{K_G(N)}$, which is hence schematic, so $\mathcal{F}_{K^p}$ is representable by a scheme as $\mathcal{F}_{K_G(N)}$ is.

We used the following rigidity lemma:

Lemma 4.9. *Let K^p be a level subgroup contained in a principal level subgroup $K_G(N)$ with $N \geq 3$. Then for any scheme S over $\mathcal{O}_E \otimes_{\mathbf{Z}} \mathbf{Z}_{(p)}$ and any object $\underline{A}$ of $\mathcal{F}_{K^p,S}$, $\underline{A}$ has only the trivial automorphism.*

Proof. It follows from the fact that an automorphism of a polarized abelian variety over an algebraically closed field that acts as the identity on the N-torsion subgroup for some $N \geq 3$ is the identity automorphism (see [Ser], or [Mil86] in the book [CS86] for a proof). □

5 Smoothness

Theorem 5.1. *The stack $\mathcal{F}_{K^p}$ is a smooth Deligne-Mumford stack. When K^p is small enough so that S_{K^p} is a scheme, then it is a smooth scheme.*

We need only prove this when K^p is small enough so that S_{K^p} is a scheme, as the transition morphisms between the S_{K^p} with varying level subgroups are étale. As S_{K^p} is locally of finite presentation, we only have to prove that S_{K^p} is formally smooth. We make use of the infinitesimal lifting criterion for smoothness, noting that as $\mathcal{O}_E \otimes \mathbf{Z}_{(p)}$ is noetherian, we need only test artinian algebras. That is, it suffices to show:

Proposition 5.2. *Let R be an artinian $\mathcal{O}_E \otimes_{\mathbf{Z}} \mathbf{Z}_{(p)}$-algebra. Let $S_0 = \operatorname{spec} R_0$ and $S = \operatorname{spec} R$ such that $R_0 = R/I$ with $I^2 = 0$. If $(A_0, \lambda_0, \iota_0; \bar{\eta}_0)$ on S_0 satisfies the determinant condition of Kottwitz, then it lifts to a $(A, \lambda, \iota; \bar{\eta})$ on S that also satisfies the determinant condition.*

5.1 First reductions

Let us first take care of the level structure:

Lemma 5.3. *If $(A_0, \lambda_0, \iota_0)$ lifts to (A, λ, ι) then any level structure $\bar{\eta}_0$ on $(A_0, \lambda_0, \iota_0)$ lifts to a level structure $\bar{\eta}$ on (A, λ, ι).*

Lifting $\bar{\eta}$ amounts to lifting some sections of $A_0[N]$ to sections of $A[N]$, for a family of integers N prime to p. As $A[N]$ is étale over S, this is automatic.

We now take the determinant condition out of the picture: if $(A_0, \lambda_0, \iota_0)$ on S_0 satisfying the determinant condition of Kottwitz lifts to (A, λ, ι) on S, then the lift automatically satisfies the determinant condition by Proposition 2.9.

Moreover we also know the following result, which is a consequence of the "rigidity lemma" ([MFK94, Theorem 6.1]):

Lemma 5.4. *Let A and B be two abelian schemes over S, then the restriction $\operatorname{Hom}(A, B) \to \operatorname{Hom}(A_0, B_0)$ is injective.*

From this we can deduce that if λ_0 and ι_0 both extend to a lifting A of A_0, then the compatibility condition between involutions is automatically satisfied.

5.2 The theory of Grothendieck-Messing

Consider the situation as in the hypothesis of Proposition 5.2. As R is artinian, on each component of S, p is either invertible or nilpotent. The generic fiber of the moduli space is smooth, as it is the canonical model of a Shimura variety (see §7), so the existence of a lift in the case where p is invertible is already known. So we can assume that p is nilpotent on S. Hence we can use the theory of Grothendieck-Messing to study the problem of lifting A_0. Let us recall the part of the theory relevant to the situation. The complete constructions and proofs can be found in [Mes72].

There is a functor from the category of abelian schemes over S_0 to the category of locally free sheaves on S associating to an abelian scheme A_0/S_0 the evaluation of the Dieudonné crystal $\mathbb{D}(A_0)$ on the inclusion $S_0 \to S$, that we will denote by $\mathbb{D}(A_0)_S$. For any abelian variety A/S lifting A_0, $\mathbb{D}(A_0)_S$ is canonically isomorphic to $\mathcal{H}^1_{DR}(A/S)$.

In the case where A_0 is a $\mathbf{Z}_{(p)}$-polarized abelian scheme with an action of $\mathcal{O}_B$, $\mathbb{D}(A_0)_S$ also has an action of $\mathcal{O}_B$. Moreover the polarization induces a morphism $\mathbb{D}(A_0)_S \to \mathbb{D}(A_0^t)_S = \mathbb{D}(A_0)_S^*$, which is an isomorphism because the polarization is separable, and which is compatible with the action of $\mathcal{O}_B$ on both sides. Hence the polarization induces a non-degenerate alternating pairing on $\mathbb{D}(A_0)_S$ that is skew-hermitian with respect to $\mathcal{O}_B$.

A submodule of $\mathbb{D}(A_0)_S$ is said to be admissible if it is locally a direct factor, and reduces to $(\mathrm{Lie}\, A_0)^*$ on S_0.

Theorem 5.5 (Grothendieck-Messing). *There is an equivalence of categories between the category of abelian schemes over S and the category of pairs (A_0, F), where A_0 is an abelian scheme over S_0 and F an admissible submodule of $\mathbb{D}(A_0)_S$, given by $A \mapsto (A_{|S_0}, (\mathrm{Lie}\, A)^*)$.*

In order for the lifting A of A_0 to be polarized with an action of $\mathcal{O}_B$, it is enough that $(\mathrm{Lie}\, A)^*$ is an $\mathcal{O}_B$-stable totally isotropic submodule of $\mathbb{D}(A_0)_S$.

We are then reduced to the following linear algebra problem: Let M be a projective module of rank $2g$ over R with an action of $\mathcal{O}_B$ and a non-degenerate alternating pairing that is skew-hermitian with respect to $\mathcal{O}_B$. Let $M_0 = M \otimes_R R_0$, and let $N_0 \subset M_0$ be a locally direct factor submodule of M_0 of rank g stable under the action of $\mathcal{O}_B$ and totally isotropic for the alternating pairing. Then we need to find a lifting of N_0 to a submodule N of M that has the same properties.

The way to find such a submodule depends of the type of the group G. Details can be found in [LR87] and [Zin82]. We will only treat a simple example: the case of unitary groups over $\mathbf{Q}$.

5.3 An example: unitary groups over Q

Let B be an imaginary quadratic extension of $\mathbf{Q}$, with involution the complex conjugation, and suppose that the prime p is split in B. Then $A = \mathcal{O}_B \otimes_{\mathbf{Z}} \mathbf{Z}_p$ is $\mathbf{Z}_p \times \mathbf{Z}_p$ and the involution exchanges the factors. Let $e_1 = (1, 0)$ and $e_2 = (0, 1)$. Then $e_i M$ is totally isotropic for $i = 1, 2$ as $e_1^* = e_2$. Moreover a submodule Q of an A-module is A-stable if and only if $Q = e_1 Q \oplus e_2 Q$.

We can further simplify the problem: Let A_0 be the universal abelian scheme over S_{K^p}. We know that $\mathcal{H}^1_{DR}(A_0/S)$ and $(\mathrm{Lie}\, A_0)^*$ are locally free modules on S_{K^p}. As smoothness is a local question on S_{K^p}, we can assume that the exact sequence $0 \to (\mathrm{Lie}\, A_0)^* \to \mathcal{H}^1_{DR}(A_0/S) \to \mathrm{Lie}(A_0^t) \to 0$ is split and that these modules are in fact free, This amounts to assuming that we have a decomposition $M_0 = N_0 \oplus P_0$, with N_0 and P_0 free. This implies that M is a free R-module with basis any lifting of a basis of M_0. Let us denote $e_i N_0$ by $N_{0,i}$ for $i = 1, 2$. Then the $N_{0,i}$ are projective. We can also assume that they are free, by the same reasoning as before.

Let us choose a basis of M_0 consisting of the union of a basis of $N_{0,i}$, $i = 1, 2$, and a basis of P_0. We can lift the basis of $N_{0,i}$ to a family in $e_i M$, which gives us free liftings $N_i \subset e_i M$ of $N_{0,i}$. They are totally isotropic, but not necessarily orthogonal. As the bilinear form is non-degenerate, we can modify the lifting of the basis of $N_{0,2}$ such that N_2 is orthogonal to N_1, and still $N_1 \subset e_1 M$. Then $N = N_1 \oplus N_2$ is the lifting of N_0 we were looking for. Indeed, N is A-stable, totally isotropic, projective (even free), and M/N is projective, as it is isomorphic to the submodule P of M generated by any lifting of the chosen basis of P_0.

6 Hecke operators

We explain here the relation between the Shimura varieties when the level varies and the action of the Hecke operators.

6.1 The tower of Shimura varieties

Let K^p and K'^p be compact open subgroups of $G(\mathbf{A}_f^p)$, such that $K^p \subset K'^p$. Then we have a natural morphism from $\mathcal{F}_{K^p}$ to $\mathcal{F}_{K'^p}$ which sends a quadruple $(A, \lambda, \iota; \bar{\eta})$ over the base S to the quadruple $(A, \lambda, \iota; \bar{\eta}')$, where $\bar{\eta}'$ is the K'^p-orbit generated by η. Hence we have a morphism of moduli schemes $S_{K^p} \to S_{K'^p}$. As in the characteristic zero case, we then have a whole tower of integral models $(S_{K^p})_{K^p}$.

If K^p is a normal subgroup of K'^p, then $S_{K^p} \to S_{K'^p}$ is an étale Galois covering of Galois group K'^p/K^p. More generally, for all $K^p \subset K'^p$ compact open subgroups of $G(\mathbf{A}_f^p)$, the morphism $S_{K^p} \to S_{K'^p}$ is finite étale and surjective.

The tower is smooth in the following sense: each of the schemes is smooth for K^p small enough (as explained in §5), and the maps in the tower are also smooth.

6.2 Action of the Hecke operators

We also have Hecke operators: the group $G(\mathbf{A}_f^p)$ acts on the tower via its action on the level structure. That is: for each $g \in G(\mathbf{A}_f^p)$, g maps $\mathcal{F}_{K^p}$ to $\mathcal{F}_{g^{-1}K^p g}$ by sending $(A, \lambda, \iota; \bar{\eta})$ to $(A, \lambda, \iota; \bar{\eta} \circ g)$.

7 Relation to the generic fiber

We will now see how the scheme S_{K^p} relates to the Shimura variety $\mathrm{Sh}(G, X)_K(\mathbf{C}) = G(\mathbf{Q})\backslash\mathcal{X} \times G(\mathbf{A}_f)/K$ and to its canonical model.

7.1 Modular definition of the canonical model

We first recall the construction of the canonical model. Let K be a compact open subgroup of $G(\mathbf{A}_f)$, and let $\mathrm{Sh}(G,X)_K(\mathbf{C}) = G(\mathbf{Q})\backslash \mathcal{X} \times G(\mathbf{A}_f)/K$.

We can obtain a canonical model of this Shimura variety via a moduli space, as follows:

Definition 7.1. Let $\mathcal{Q}$ be the following category fibered in groupoids over the category (Sch/E) of E-schemes:

- The objects over a scheme S are quadruples $\underline{A} = (A, \lambda, \iota; \bar{\eta})$, where (A, λ, ι) is a polarized projective abelian scheme over S with an action of $\mathcal{O}_B$ which respects the determinant condition of Kottwitz (Definition 2.8 of §2.2), and $\bar{\eta}$ is a level structure of level K over each connected component of S, that is, a K-orbit of isomorphisms between $V \otimes \mathbf{A}_f$ and $H_1(A_s, \mathbf{A}_f)$, for s a geometric point of S, that is invariant under the action of $\pi_1(s, S)$ (so that the definition does not depend on the choice of s).
- The morphisms from $\underline{A}$ to $\underline{A'}$ over S are given by a $\mathbf{Q}$-isogeny $f : A \to A'$ compatible with the action of $\mathcal{O}_B$ and the level structures, that is:
    (1) there exists a locally constant function r on S with values in $\mathbf{Q}^\times$ such that $\lambda = r(f^t \circ \lambda' \circ f)$.
    (2) f induces a morphism from $\mathrm{End}\,(A) \otimes_{\mathbf{Z}} \mathbf{Z}_{(p)}$ to $\mathrm{End}\,(A') \otimes_{\mathbf{Z}} \mathbf{Z}_{(p)}$, that we still denote by f; then for all $b \in \mathcal{O}_B$, $f \circ \iota(b) = \iota'(b)$.
    (3) $\bar{\eta}' = V(f) \circ \bar{\eta}$, where we denote by $V(f)$ the morphism induced from $V(A_s)$ to $V(A'_s)$.

The functor $\mathcal{Q}$ is representable by a scheme S_K when K is small enough. Then S_K is a disjoint union of canonical models over E of the Shimura variety $\mathrm{Sh}(G,X)_K$. More precisely, as is explained in [Kot92, § 8]:

Proposition 7.2.

$$S_K = \sqcup_{\ker^1(\mathbf{Q},G)} Sh(G', \mathcal{X})_K$$

where $\ker^1(\mathbf{Q}, G)$ *is the set of locally trivial elements of* $H^1(\mathbf{Q}, G)$ *and parametrizes the inner forms* G' *of* G *that are locally isomorphic to* G *at every place.*

The failure of the Hasse principle is essentially harmless, as follows from the study of $\ker^1(\mathbf{Q}, G)$ in [Kot92, § 7]. Recall that if the data is of type A, then G is the restriction of scalars to $\mathbf{Q}$ of a unitary group. We denote by n the dimension of the hermitian space giving rise to this group.

Proposition 7.3. *When G is of type C, or of type A with even n,* $\ker^1(\mathbf{Q}, G)$ *is trivial. When G is of type A with odd n, all the groups G' are isomorphic to G.*

In particular, under the hypotheses of the proposition, S_K is in fact isomorphic to a finite union of copies of the canonical model of the Shimura variety $\mathrm{Sh}(G, \mathcal{X})_K$.

7.2 Relationship to the integral model

Let C_0 be the hyperspecial maximal compact open subgroup of $G(\mathbf{A}_f)$ at p defined in §1.3, that stabilizes a lattice $\Lambda \subset V_{\mathbf{Q}_p}$. Note that when the Shimura data is of type A or C, then by Remark 1.2 there is no ambiguity about C_0 up to conjugation by an element of $G(\mathbf{Q}_p)$. If K^p is a compact open subgroup of $G(\mathbf{A}_f^p)$, then $K = K^p C_0$ is a compact open subgroup of $G(\mathbf{A}_f)$.

Theorem 7.4. *We have then the following isomorphism when the Shimura data is of type A or C:*

$$S_{K^p} \otimes E \xrightarrow{\sim} S_{K^p C_0}$$

Moreover, the induced isomorphism between the towers $(S_{K^p} \otimes E)_{K^p}$ and $(S_{K^p C_0})_{K^p}$ is compatible with the action of $G(\mathbf{A}_f^p)$ on both sides.

It follows from this result that the generic fiber of the integral model S_{K^p} is a union of copies of the canonical model of the Shimura variety $\mathrm{Sh}(G, \mathcal{X})_K$.

Let $\mathcal{F}_{K^p}$ the category we introduced in Definition 3.1 in order to define the moduli problem for the integral model of the Shimura variety, relative to the level subgroup K^p. In what follows we will abbreviate $\mathcal{F}_{K^p}$ by $\mathcal{F}$. We write $\mathcal{F}_{|E}$ for the restriction $\mathcal{F}$ to the set of E-schemes. Hence if $\mathcal{F}$ is representable by the $\mathcal{O}_E \otimes_{\mathbf{Z}} \mathbf{Z}_{(p)}$-scheme S_{K^p} then $\mathcal{F}_{|E}$ is representable by $S_{K^p} \otimes E$.

Theorem 7.4 is a consequence of the following Proposition (which is similar to Proposition 4.4):

Proposition 7.5. *When the Shimura data is of type A or C, the categories fibered in groupoids $\mathcal{Q}$ and $\mathcal{F}_{|E}$ are isomorphic. The isomorphism is compatible with the action of the Hecke operators on the towers on both sides when K^p varies.*

Proof. Let S be an E-scheme. Let us explain how to define an equivalence of categories from $\mathcal{F}_{|E,S}$ to $\mathcal{Q}_S$. Let $\underline{A} = (A, \lambda, \iota; \bar{\eta}^p)$ be an object of $\mathcal{F}_{|E,S}$. The problem is in the definition of $\bar{\eta}'$: we already have a K^p-orbit of isomorphisms $\bar{\eta}$ between $V \otimes \mathbf{A}_f^p$ and $H_1(A_s, \mathbf{A}_f^p)$ and we have to extend it to the whole of $\mathbf{A}_f$. That is, we have to find a C_0-orbit of isomorphisms between Λ and $H_1(A_s, \mathbf{Z}_p)$.

Observe that V and $H_1(A_s, \mathbf{Q})$ are isomorphic B-modules, as they become so after tensorization by $\mathbf{Q}_\ell$ for any $\ell \neq p$ (this follows from the existence of the level structure outside p). Then $V \otimes \mathbf{Q}_p$ and $H_1(A_s, \mathbf{Q}_p)$ are isomorphic as B-modules. Moreover both have self-dual $\mathcal{O}_B$-lattices. Now we use the condition on the Shimura data: as it is of type A or C, we know by [Kot92, Lemma 7.2] that the lattices Λ and $H_1(A_s, \mathbf{Z}_p)$ are isomorphic as hermitian modules with an action of $\mathcal{O}_B$. Moreover the C_0-orbit of the isomorphism is then well-defined independently of choices. Hence we can uniquely extend the level structure $\bar{\eta}^p$ to $\bar{\eta}$. □

References

[CS86] Gary Cornell and Joseph H. Silverman, editors. *Arithmetic geometry*. Springer-Verlag, New York, 1986. Papers from the conference held at the University of Connecticut, Storrs, Connecticut, July 30–August 10, 1984.

[Del71] Pierre Deligne. Travaux de Shimura. In *Séminaire Bourbaki, 23ème année (1970/71), Exp. No. 389*, pages 123–165. Lecture Notes in Math., Vol. 244. Springer, Berlin, 1971.

[GN] Alain Genestier and Bao Chau Ngo. Lectures on Shimura varieties. First chapter of this book.

[Hid04] Haruzo Hida. *p-adic automorphic forms on Shimura varieties*. Springer Monographs in Mathematics. Springer-Verlag, New York, 2004.

[Kot92] Robert E. Kottwitz. Points on some Shimura varieties over finite fields. *J. Amer. Math. Soc.*, 5(2):373–444, 1992.

[Lan13] Kai-Wen Lan. *Arithmetic compactifications of PEL-type Shimura varieties*, volume 36 of *London Mathematical Society Monographs Series*. Princeton University Press, Princeton, NJ, 2013.

[LMB00] Gérard Laumon and Laurent Moret-Bailly. *Champs algébriques*, volume 39 of *Ergebnisse der Mathematik und ihrer Grenzgebiete. 3. Folge. A Series of Modern Surveys in Mathematics [Results in Mathematics and Related Areas. 3rd Series. A Series of Modern Surveys in Mathematics]*. Springer-Verlag, Berlin, 2000.

[LR87] R. P. Langlands and M. Rapoport. Shimuravarietäten und Gerben. *J. Reine Angew. Math.*, 378:113–220, 1987.

[Mat89] Hideyuki Matsumura. *Commutative ring theory*, volume 8 of *Cambridge Studies in Advanced Mathematics*. Cambridge University Press, Cambridge, second edition, 1989. Translated from the Japanese by M. Reid.

[MB85] Laurent Moret-Bailly. Pinceaux de variétés abéliennes. *Astérisque*, (129):266, 1985.

[Mes72] William Messing. *The crystals associated to Barsotti-Tate groups: with applications to abelian schemes*. Lecture Notes in Mathematics, Vol. 264. Springer-Verlag, Berlin, 1972.

[MFK94] D. Mumford, J. Fogarty, and F. Kirwan. *Geometric invariant theory*, volume 34 of *Ergebnisse der Mathematik und ihrer Grenzgebiete (2) [Results in Mathematics and Related Areas (2)]*. Springer-Verlag, Berlin, third edition, 1994.

[Mil86] J. S. Milne. Abelian varieties. In *Arithmetic geometry (Storrs, Conn., 1984)*, pages 103–150. Springer, New York, 1986.

[Ser] Jean-Pierre Serre. Rigidité du foncteur de Jacobi d'échelon $n \geq 3$. Appendice à l'exposé 17 du séminaire Cartan 1960–1961.

[Zin82] Thomas Zink. Über die schlechte Reduktion einiger Shimuramannigfaltigkeiten. *Compositio Math.*, 45(1):15–107, 1982.

INTRODUCTION TO THE LANGLANDS–KOTTWITZ METHOD

YIHANG ZHU

1 Introduction

This chapter is a survey of Kottwitz's two papers [Kot92] and [Kot90]. The first step towards understanding the (étale) cohomology of Shimura varieties is to study the trace of a Hecke operator away from p composed with a power of the Frobenius at p, where p is a prime of good reduction. In the paper [Kot92], Kottwitz obtains a formula for this trace, for PEL Shimura varieties (which are assumed to be compact and not of type D). Kottwitz's formula involves orbital integrals away from p and twisted orbital integrals at p. We shall refer to it as the *unstabilized Langlands–Kottwitz formula*. The shape of this formula is reminiscent of another formula, which stems from automorphic representation theory rather than Shimura varieties, namely the elliptic part of the Arthur–Selberg trace formula. In fact, the main idea of the so-called Langlands–Kottwitz method is that the Langlands–Kottwitz formula could be compared with the Arthur–Selberg trace formula, and the comparison should eventually reflect the reciprocity between Galois representations and automorphic representations.

However, apart from very special situations, the unstabilized Langlands–Kottwitz formula is not directly comparable with the Arthur–Selberg trace formula. One has to *stabilize* both the formulas in order to compare them. Recall from [Har11] that the elliptic part of the Arthur–Selberg trace formula can be stabilized, in the sense that it can be rewritten in terms of stable orbital integrals on the endoscopic groups. In a similar sense, the unstabilized Langlands–Kottwitz formula could also be stabilized, and this is the main result of the paper [Kot90]. The stabilization procedure for the Langlands–Kottwitz formula in [Kot90] is analogous to the stabilization procedure for the elliptic part of the Arthur–Selberg trace formula, to which Kottwitz also made major contributions (see [Kot84] [Kot86], and also the exposition in [Har11]). At the same time there are also important differences between the two stabilization procedures, which mainly happen at the places p and ∞.

After the stabilization, the Langlands–Kottwitz formula is expressed in terms of the elliptic parts of the stable Arthur–Selberg trace formulas for the endoscopic groups. See Theorem 6.1.2 for the precise statement. In the special case of unitary Shimura varieties, the stabilized formula is stated in [CHL11, (4.3.5)], which is among the main inputs to the chapter [CHL11]. The present chapter, to some extent, could be viewed as an extended explanation of the formula [CHL11, (4.3.5)].

In our exposition of the unstabilized Langlands–Kottwitz formula, we shall specialize to the case of unitary PEL Shimura varieties. More precisely, we only consider the unitary similitude group of a global Hermitian space over a field, as opposed to over a more general

Last updated: December 5, 2018.

division algebra. This is already enough for the chapter [CHL11], and our main purpose for restricting to this special case is to gain some concreteness and to simplify certain proofs. We hope that nothing essential in the general PEL case is lost. Our exposition of the stabilization procedure is, however, for a general Shimura datum, under some simplifying assumptions of a group theoretic nature.

In the paper [Kot90], Kottwitz assumes several hypotheses in local harmonic analysis, including the Fundamental Lemma for Endoscopy, the Fundamental Lemma for Unstable Base Change, and the Langlands–Shelstad Transfer Conjecture. These hypotheses are all proved theorems now, and we shall point out the references when they are invoked in the argument.

Here is the organization of this chapter. In § 2 we describe the problem. In § 3–§ 5 we discuss the unstabilized Langlands–Kottwitz formula in [Kot92]. In § 6 we discuss the stabilization procedure in [Kot90]. In § 7 we make some brief remarks on developments following Kottwitz's work.

Acknowledgement. I would like to thank the editors of this volume for inviting me to contribute, and the anonymous referee for many helpful remarks, suggestions, and corrections. I would like to thank Mark Kisin for introducing me to Shimura varieties. I also thank Tasho Kaletha, Kai-Wen Lan, Sophie Morel, Sug Woo Shin, and Wei Zhang for explaining to me many subtle details about Shimura varieties and endoscopy over the years. Last but not least, I would like to thank the authors and editors of the first volume of this book series, which served as a great guide for me, when I was entering the field of automorphic forms and trace formula as a student.

2 Description of the problem

2.1 The unitary PEL and Shimura data

We follow the setting of [CHL11] but work slightly more generally. Let F be a totally real field and let $\mathcal{K}/F$ be a totally imaginary quadratic extension.[1] Denote by $\mathbf{c}$ the nontrivial element in $\mathrm{Gal}(\mathcal{K}/F)$. Let $(V, [\cdot,\cdot])$ be a non-degenerate Hermitian space for $\mathcal{K}/F$, i.e., V is a finite dimensional $\mathcal{K}$-vector space and $[\cdot,\cdot]$ is a $\mathbf{c}$-sesquilinear form satisfying $[x, y] = \mathbf{c}([y, x])$ for all $x, y \in V$. We assume that $(V, [\cdot,\cdot])$ is *anisotropic*, which means $[x, x] \neq 0$ for all $x \in V - \{0\}$. For instance, the Hermitian spaces V_1 and V_2 considered in [CHL11, § 3] are both anisotropic.

Fix an element $\epsilon \in \mathcal{K}$ such that $\mathbf{c}(\epsilon) = -\epsilon$. We define

$$\langle\cdot,\cdot\rangle := \mathrm{Tr}_{\mathcal{K}/\mathbb{Q}}(\epsilon[\cdot,\cdot]). \tag{2.1.1}$$

Then $\langle\cdot,\cdot\rangle$ is a $\mathbb{Q}$-bilinear $\mathbb{Q}$-valued non-degenerate alternating form on V. It satisfies

$$\langle ax, y\rangle = \langle x, \mathbf{c}(a)y\rangle,\ \forall x, y \in V, a \in \mathcal{K}.$$

It can be easily checked that $[\cdot,\cdot]$ is in turn uniquely determined by $\langle\cdot,\cdot\rangle$, via (2.1.1).

[1]Our $\mathcal{K}$ plays the same role as the simple $\mathbb{Q}$-algebra B in [Kot92]. In [Kot92], the center of B is denoted by F, which for us is of course still $B = \mathcal{K}$. Note our different usage of the notation F.

Finally we choose $h : \mathbb{C} \to \mathrm{End}_{\mathbb{R},\mathcal{K}}(V \otimes_{\mathbb{Q}} \mathbb{R})$ to be an $\mathbb{R}$-algebra morphism such that $\langle h(z)x, y\rangle = \langle x, h(\bar{z})y\rangle$ for all $x, y \in V \otimes_{\mathbb{Q}} \mathbb{R}$ and such that the symmetric bilinear form $(x, y) \mapsto \langle x, h(i)y\rangle$ on $V \otimes_{\mathbb{Q}} \mathbb{R}$ is positive definite. Then in the terminology of [GN, Definition 3.4.1], the datum

$$(B := \mathcal{K}, * := \mathbf{c}, V, \langle\cdot,\cdot\rangle, h)$$

is a *rational PE-structure*. Our datum is in fact a special case of the unitary PEL data (in a broader sense) considered in [Nic], and of course they are all special cases of the PEL data considered in [GN], following [Kot92].

We write C for the algebra $\mathrm{End}_{\mathcal{K}}(V)$ and write $*$ for the involution on C given by the alternating form $\langle\cdot,\cdot\rangle$, characterized by

$$\langle cx, y\rangle = \langle x, c^* y\rangle, \ \forall x, y \in V, c \in C.$$

The reductive group G over $\mathbb{Q}$ associated to the above data as in [GN, § 3.5] and [Roz, § 1.1] is the same as the unitary similitude group $GU(V, [\cdot,\cdot])$ defined in [CHL11, § 1.1]. By definition, for any $\mathbb{Q}$-algebra R, we have

$$\begin{aligned} G(R) &= \{g \in \mathrm{GL}_{\mathcal{K}}(V \otimes_{\mathbb{Q}} R) | \exists \nu(g) \in R^{\times}\ \forall x, y \in V,\ [gx, gy] = \nu(g)[x, y]\} \\ &= \{g \in \mathrm{GL}_{\mathcal{K}}(V \otimes_{\mathbb{Q}} R) | \exists \nu(g) \in R^{\times}\ \forall x, y \in V,\ \langle gx, gy\rangle = \nu(g)\langle x, y\rangle\} \\ &= \{g \in (C \otimes_{\mathbb{Q}} R)^{\times} | \exists \nu(g) \in R^{\times}, g^* g = \nu(g)\}. \end{aligned}$$

Moreover, the factors of similitude $\nu(g)$ in the above three descriptions are the same, and we have a $\mathbb{Q}$-morphism $\nu : G \to \mathbb{G}_m$.

Let E be the reflex field of the PE-structure (see [GN, § 3.4]).[2] The $\mathbb{R}$-algebra morphism h induces a homomorphism between $\mathbb{R}$-algebraic groups

$$\mathbb{S} = \mathrm{Res}_{\mathbb{C}/\mathbb{R}}\, \mathbb{G}_m \longrightarrow G_{\mathbb{R}},$$

which we still denote by h. Let X be the $G(\mathbb{R})$-conjugacy class of

$$h^{-1} : \mathbb{S} \longrightarrow G_{\mathbb{R}}.$$

Then (G, X) is a Shimura datum whose reflex field is E.[3]

Remark 2.1.1. We record two important facts about the reductive group G, which are easy to check and simplify the theory greatly.

(1) The derived subgroup G^{der} is simply connected.

(2) The center Z_G satisfies the condition that its maximal $\mathbb{R}$-split subtorus is $\mathbb{Q}$-split. (This subtorus is $\mathbb{G}_m$.)

---

[2]This notation is not to be confused with the imaginary quadratic field E in [CHL11].

[3]To be more precise, when $G(\mathbb{R})$ is compact modulo center, which is possible in our setting, the pair (G, X) would violate the axiom [Del79, (2.1.1.3)] for a Shimura datum. However for our purpose this does not matter too much. One may actually use the PEL datum and the associated moduli problem to replace the theory of Shimura varieties in such a case.

The following lemma also plays an important role in [Kot92]:

Lemma 2.1.2 ([Kot92, Lemma 7.1]). *Let Q be any algebraically closed field of characteristic zero. Then two elements in $G(Q)$ are conjugate if and only if they have the same image under ν and their images in $\mathrm{GL}_{\mathcal{K}}(V \otimes_{\mathbb{Q}} Q) \cong \mathrm{GL}_{\dim_{\mathcal{K}} V}(Q) \times \mathrm{GL}_{\dim_{\mathcal{K}} V}(Q)$ are conjugate.* □

We next fix *p-integral data* as in [Roz, § 1.2]. Let p be a rational prime that is unramified in $\mathcal{K}$. We need to choose a $\mathbb{Z}_{(p)}$-order $\mathcal{O}_B$ in B and a $\mathbb{Z}_p$-lattice Λ in $V_{\mathbb{Q}_p}$ such that Λ is stable under $\mathcal{O}_B$ and self-dual under $\langle\cdot,\cdot\rangle$. We take $\mathcal{O}_B := \mathcal{O}_{\mathcal{K}} \otimes \mathbb{Z}_{(p)}$, and fix the choice of Λ.

Of course the existence of Λ is a non-trivial assumption on the prime p, and in particular implies that $G_{\mathbb{Q}_p}$ is unramified. More precisely, as in [Roz], the choice of Λ determines a reductive model $\mathcal{G}$ over $\mathbb{Z}_p$ of $G_{\mathbb{Q}_p}$ and a hyperspecial subgroup $K_p := \mathcal{G}(\mathbb{Z}_p)$ of $G(\mathbb{Q}_p)$.

2.2 The integral models

Fix a prime $\mathfrak{p}$ of E above p. Given the above data, we obtain a smooth quasi-projective scheme S_{K^p} over $\mathcal{O}_{E,(\mathfrak{p})}$ for each small enough compact open subgroup K^p of $G(\mathbb{A}_f^p)$. (In fact, the choice of Λ is not needed for the construction of S_{K^p}.) Roughly speaking, S_{K^p} parametrizes $\mathbb{Z}_{(p)}$-polarized projective abelian schemes with an action of $\mathcal{O}_B$ respecting the determinant condition, equipped with K^p-level structure, up to prime-to-p isogeny. This is the topic of [Roz]. Moreover, our assumption that $(V, [\cdot,\cdot])$ is anisotropic implies that G is anisotropic modulo center, which further implies that S_{K^p} is projective over $\mathcal{O}_{E,(\mathfrak{p})}$. See the Introduction of this volume for a general discussion of the issue of projectivity of the integral models. In the following, all compact open subgroups K^p of $G(\mathbb{A}_f^p)$ that appear are understood to be small enough.

When K^p varies, we obtain a natural tower $\varprojlim_{K^p} S_{K^p,E}$ with finite étale transition maps. The group $G(\mathbb{A}_f^p)$ acts on the tower by acting on the level structures in the moduli problem. Thus for any $g \in G(\mathbb{A}_f^p)$ and any compact open subgroups $K_1^p, K_2^p \subset G(\mathbb{A}_f^p)$ satisfying $g^{-1}K_1^p g \subset K_2^p$, we have a finite étale morphism

$$[\cdot g] = [\cdot g]_{K_1^p,K_2^p} : S_{K_1^p} \longrightarrow S_{K_2^p}.$$

These morphisms are called the *Hecke operators*. In particular the transition maps in the tower $\varprojlim_{K^p} S_{K^p,E}$ are the Hecke operators $[\cdot 1]_{K_1^p,K_2^p}$.

For any reductive group H over $\mathbb{Q}$, denote by $\ker^1(\mathbb{Q}, H)$ the pointed set

$$\ker(\mathbf{H}^1(\mathbb{Q}, H) \to \mathbf{H}^1(\mathbb{A}, H)).$$

Recall from [Roz, § 7] that the generic fiber of the tower $\varprojlim_{K^p} S_{K^p,E}$, as a tower of schemes over E, is identified with the disjoint union of $\left|\ker^1(\mathbb{Q}, G)\right|$ copies of the tower $\varprojlim_{K^p} \mathrm{Sh}_{K^pK_p}(G, X)$ of canonical models for the Shimura datum $(G, X) = (G, h^{-1})$.[4] The

[4] When $G(\mathbb{R})$ is compact modulo center, the pair (G, X) is not a Shimura datum, and we understand $\mathrm{Sh}_{K^pK_p}(G, X)$ simply as the (zero-dimensional) locally symmetric space $G(\mathbb{Q})\backslash(X \times G(\mathbb{A}_f)/K^pK_p)$. See footnote [3].

reader is referred to [GN] or [Mil05] for the theory of canonical models. This identification is equivariant with respect to the $G(\mathbb{A}_f^p)$-actions on both sides. On $\mathbb{C}$-points, the Hecke operator

$$[\cdot g] : \mathrm{Sh}_{K_1^p K_p}(\mathbb{C}) \longrightarrow \mathrm{Sh}_{K_2^p K_p}(\mathbb{C})$$

where g, K_1^p, K_2^p are as above, is the map

$$G(\mathbb{Q})\backslash(X \times G(\mathbb{A}_f)/K_1^p K_p) \longrightarrow G(\mathbb{Q})\backslash(X \times G(\mathbb{A}_f)/K_2^p K_p)$$
$$(x, h) \mapsto (x, hg).$$

See [Nic] for more discussion on the $\mathbb{C}$-points of unitary Shimura varieties and Hecke operators between them.

Remark 2.2.1. There are different conventions in the literature on the notion of canonical models, and they are related by replacing h by h^{-1}. Our convention here, which is the same as that in [CHL11], follows [Kot92]. Note that there is a mistake in [Kot90] which switches h and h^{-1}, see the remark at the end of [Kot92], cf. also [CHL11, § 3].

2.3 λ-adic automorphic sheaves

Fix a finite dimensional representation ξ of $G(\mathbb{C})$. It descends to an algebraic representation over a number field $L(\xi)$. Let λ be a place of $L(\xi)$ coprime to p. Let $\ell := \lambda|_{\mathbb{Q}}$. Then ξ induces a λ-adic representation of $G(\mathbb{Q}_\ell)$. Using the action of $G(\mathbb{Q}_\ell) \subset G(\mathbb{A}_f^p)$ on the tower $\varprojlim_{K^p} S_{K^p}$, we obtain a *$\lambda$-adic automorphic sheaf on the tower*. More precisely, we obtain the following datum:

(1) For each K^p (small enough), we have a λ-adic sheaf $\mathscr{F}_{K^p}$ on S_{K^p}.

(2) For each Hecke operator $[\cdot g] : S_{K_1^p} \to S_{K_2^p}$, where K_1^p, K_2^p are as in (1) and $g \in G(\mathbb{A}_f^p)$ is such that $g^{-1}K_1^p g \subset K_2^p$, we have a canonical identification $\eta : [\cdot g]^* \mathscr{F}_{K_2^p} \xrightarrow{\sim} \mathscr{F}_{K_1^p}$, satisfying certain compatibility conditions.

For more details of the construction see [Kot92, § 6], cf. also [Pin92a, § 5.1], [HT01, § III.2], [Car86, § 2.1], in which the constructions can be easily adapted to the $G(\mathbb{A}_f^p)$-tower of integral models here.

Remark 2.3.1. In order to construct $\mathscr{F}_{K^p}$ for a fixed K^p, it is crucial that the sub-tower $\varprojlim_{K'^p, K'^p \subset K^p} S_{K'^p}$ over S_{K^p} has Galois group K^p, or equivalently, that for any open normal subgroup K'^p of K^p the Galois covering $S_{K'^p} \to S_{K^p}$ has Galois group K^p/K'^p. This is indeed guaranteed by property (2) in Remark 2.1.1, when K^p is small enough. This point is not emphasized explicitly in [Kot92], but in fact all the PEL Shimura data of types A and C treated in [Kot92] satisfy property (2) in Remark 2.1.1. (This holds even more generally for all Hodge type Shimura data, but may fail for abelian-type Shimura data.)

2.4 The goal

Let $\mathbb{F}_q = \mathbb{F}_{p^r}$ be the residue field of E at $\mathfrak{p}$. Fix a small enough compact open subgroup $K^p \subset G(\mathbb{A}_f^p)$. Let the symbol k denote either $\overline{\mathbb{F}}_q$ or $\overline{E}$. Consider the λ-adic cohomology groups

$$\mathbf{H}^*_{k,K^p} := \mathbf{H}^*(S_{K^p,k}, \mathscr{F}_{K^p}).$$

Of course $\mathbf{H}^*_{\overline{E},K^p}$ is the direct sum of $\left|\ker^1(\mathbb{Q}, G)\right|$ copies of the analogous cohomology group of $\mathrm{Sh}_{K^pK_p}(G, X)_{\overline{E}}$. The $G(\mathbb{A}_f^p)$ action on the tower $\varprojlim_{K'^p, K'^p \subset K^p} S_{K'^p}$ endows $\mathbf{H}^*_{k,K^p}$ with the structure of a module over the Hecke algebra

$$\mathcal{H}_{K^p} := \mathcal{H}(G(\mathbb{A}_f^p)//K^p, L(\xi))$$

consisting of $L(\xi)$-valued K^p-biinvariant locally constant compactly supported functions on $G(\mathbb{A}_f^p)$. See § 2.5 below for a more precise description of the Hecke action.[5] Moreover $\mathrm{Gal}(\overline{E}/E)$ (respectively $\mathrm{Gal}(\overline{\mathbb{F}}_q/\mathbb{F}_q)$) acts on $\mathbf{H}^*_{\overline{E},K^p}$ (respectively $\mathbf{H}^*_{\overline{\mathbb{F}}_q,K^p}$) by transport of structure, and this action commutes with the action of $\mathcal{H}_{K^p}$.

Since S_{K^p} is proper smooth over $\mathcal{O}_{E,(\mathfrak{p})}$, we know that the $\mathrm{Gal}(\overline{E}/E)$-action on $\mathbf{H}^*_{\overline{E},K^p}$ is unramified at $\mathfrak{p}$, and we have a canonical isomorphism

$$\mathbf{H}^*_{\overline{E},K^p} \cong \mathbf{H}^*_{\overline{\mathbb{F}}_q,K^p}$$

as $(\mathcal{H}_{K^p} \times \mathrm{Gal}(\overline{\mathbb{F}}_q/\mathbb{F}_q))$-modules.

The main goal of this chapter is to understand the Euler characteristic

$$\begin{aligned} T(j, f^{p,\infty}) : &= \sum_i (-1)^i \operatorname{Tr}(\Phi_q^j \times f^{p,\infty}, \mathbf{H}^i_{\overline{E},K^p}) \\ &= \sum_i (-1)^i \operatorname{Tr}(\Phi_q^j \times f^{p,\infty}, \mathbf{H}^i_{\overline{\mathbb{F}}_q,K^p}) \in L(\xi)_\lambda, \end{aligned}$$

where

- Φ_q is the **geometric** q-Frobenius in $\mathrm{Gal}(\overline{\mathbb{F}}_q/\mathbb{F}_q)$
- $f^{p,\infty} \in \mathcal{H}_{K^p}$
- $j \in \mathbb{N}$ is large enough.

Without loss of generality, we may assume $f^{p,\infty}$ is the characteristic function of $K^p g^{-1} K^p$ for some $g \in G(\mathbb{A}_f^p)$, as such functions span $\mathcal{H}_{K^p}$. We write $T(j, g)$ for $T(j, f^{p,\infty})$ in this case.

[5] Strictly speaking one also needs to choose a Haar measure on $G(\mathbb{A}_f^p)$ to pin down the action of $\mathcal{H}_{K^p}$. We always choose the one giving volume 1 to K^p.

2.5 Applying the Lefschetz–Verdier trace formula

We shall apply the Lefschetz–Verdier trace formula (or rather its simplified version conjectured by Deligne) to compute $T(j,g)$. Denote

$$K_g^p := K^p \cap gK^pg^{-1}.$$

Consider the cohomological correspondence

$$u(j,g) : [\cdot 1]^*(\Phi^j)^*\mathscr{F}_{K^p} \cong [\cdot 1]^*\mathscr{F}_{K^p} \xrightarrow[\eta]{\sim} \mathscr{F}_{K_g^p} \xrightarrow[\eta^{-1}]{\sim} [\cdot g]^*\mathscr{F}_{K^p}$$

supported on the geometric correspondence

$$\begin{array}{ccccc} & & S_{K_g^p,\overline{\mathbb{F}}_q} & & \\ & {\scriptstyle a:=[\cdot g]} \swarrow & & \searrow {\scriptstyle c:=\Phi^j\circ[\cdot 1]_{K_g^p,K^p}} & \\ S_{K^p,\overline{\mathbb{F}}_q} & & & & S_{K^p,\overline{\mathbb{F}}_q} \end{array} \tag{2.5.1}$$

Here Φ denotes the q-Frobenius endomorphism of the variety $S_{K^p,\overline{\mathbb{F}}_q}$, and we have used the natural isomorphism $\mathscr{F}_{K^p} \cong (\Phi^j)^*\mathscr{F}_{K^p}$ in the above definition of $u(j,g)$. We have an induced endomorphism $u(j,g)_*$ on $\mathbf{H}^*_{\overline{\mathbb{F}}_q,K^p}$. See [Pin92b, § 1] for more details on this formalism.[6]

By definition, when $j = 0$, the endomorphism $u(j,g)_*$ gives the action of $f^{p,\infty} = 1_{K^pg^{-1}K^p} \in \mathcal{H}_{K^p}$ on $\mathbf{H}^*_{\overline{\mathbb{F}}_q,K^p}$.

Lemma 2.5.1. *$T(j,g)$ is equal to the Euler characteristic*

$$\operatorname{Tr} u(j,g)_* := \sum_i (-1)^i \operatorname{Tr}\left(u(j,g)_*, \mathbf{H}^*_{\overline{\mathbb{F}}_q,K^p}\right)$$

of $u(j,g)_*$.

Proof. This follows from the definition of the $\mathcal{H}_{K^p}$-module structure on $\mathbf{H}^*(\overline{\mathbb{F}}_q, K^p)$ mentioned above, and the compatibility between the geometric Frobenius $\Phi_q \in \operatorname{Gal}(\overline{\mathbb{F}}_q/\mathbb{F}_q)$ and the Frobenius endomorphism Φ of $S_{K^p,\overline{\mathbb{F}}_q}$ acting on λ-adic cohomology. □

To compute $\operatorname{Tr} u(j,g)_*$, let $\operatorname{Fix}(j,g)$ denote the (geometric) fixed points of (2.5.1), namely

$$\operatorname{Fix}(j,g) = \left\{x' \in S_{K_g^p}(\overline{\mathbb{F}}_q) : a(x') = c(x')\right\}.$$

By general consideration (see [Pin92b, Lemma 7.1.2]) we know that $\operatorname{Fix}(j,g)$ is finite if j is large enough. Recall that the general Lefschetz–Verdier trace formula asserts:

$$\operatorname{Tr} u(j,g)_* = \sum_{x'\in \operatorname{Fix}(j,g)} \operatorname{LT}_{x'},$$

[6] Note that in [Pin92b] a would be denoted by a right arrow and c would be denoted by a left arrow.

where $\mathrm{LT}_{x'}$ are abstractly defined *local terms* and are in general difficult to compute. According to a conjecture of Deligne, when j is large enough, we have a similar formula:

$$\mathrm{Tr}\, u(j,g)_* = \sum_{x' \in \mathrm{Fix}(j,g)} \mathrm{LT}_{x'}^{naive}, \tag{2.5.2}$$

where $\mathrm{LT}_{x'}^{naive}$ are the so-called *naive local terms*, which are much easier to compute. Deligne's conjecture has been proved by Pink [Pin92b] in special cases (which already suffices for us) and by Fujiwara [Fuj97] in general (cf. also [Var05]). It is this version (2.5.2) of the Lefschetz–Verdier trace formula that we shall use.

In our case, for $x' \in \mathrm{Fix}(j,g)$ and $x := a(x')$, the naive local term $\mathrm{LT}_{x'}^{naive}$ is given by the trace on the stalk $\mathscr{F}_{K^p,x}$ of the endomorphism

$$\mathscr{F}_{K^p,x} \cong (c^* \mathscr{F}_{K^p})_{x'} \xrightarrow{u(j,g)} (a^* \mathscr{F}_{K^p})_{x'} \cong \mathscr{F}_{K^p,x}.$$

See [Kot92, § 16] and [Pin92b, § 1.5]. This turns out to be quite easy to understand once we have a good parametrization of $\mathrm{Fix}(j,g)$. Hence the main problem is to understand $\mathrm{Fix}(j,g)$, or to "count its points".

Example 2.5.2. When $g = 1$, we have $\mathrm{Fix}(j,g) = S_{K^p}(\mathbb{F}_{q^j})$.

2.6 The strategy for understanding Fix(j,g)

Recall from [Roz, § 3.1] or [Kot92, § 5] that the $\overline{\mathbb{F}}_q$-points on S_{K^p} correspond to isomorphism classes in the $\mathbb{Z}_{(p)}$-linear category $\mathscr{C}_{\overline{\mathbb{F}}_q}$, where

- the objects in $\mathscr{C}_{\overline{\mathbb{F}}_q}$ are quadruples
$$(\bar{A}, \lambda, \iota, \bar{\eta})$$
where $(\bar{A}, \lambda, \iota)$ is a $\mathbb{Z}_{(p)}$-polarized abelian variety over $\overline{\mathbb{F}}_q$ with an action of $\mathcal{O}_B = \mathcal{O}_{\mathcal{K}} \otimes \mathbb{Z}_{(p)}$ satisfying the compatibility between $\mathbf{c}$ and the Rosati involution for λ, and satisfying the determinant condition with respect to h, and $\bar{\eta}$ is a K^p-level structure.
- the morphisms from $(\bar{A}, \lambda, \iota, \bar{\eta})$ to $(\bar{A}_1, \lambda_1, \iota_1, \bar{\eta}_1)$ are the elements of $\mathrm{Hom}(A, B) \otimes_{\mathbb{Z}} \mathbb{Z}_{(p)}$ that preserve the additional structures (where we only require the polarizations to be preserved up to a scalar).

To compare with [Roz, § 3.1], the groupoid $\mathcal{F}_{K^p}(\overline{\mathbb{F}}_q)$ in that chapter is obtained by considering all objects in $\mathscr{C}_{\overline{\mathbb{F}}_q}$ and only isomorphisms between them.

Write q^j as p^n. For a moment we suppose $g = 1$. Then by Example 2.5.2, we know that an object $(\bar{A}, \lambda, \iota, \bar{\eta}) \in \mathscr{C}_{\overline{\mathbb{F}}_q}$ represents a point in $\mathrm{Fix}(j,g)$ if and only if $(\bar{A}, \lambda, \iota, \bar{\eta})$ is "defined over $\mathbb{F}_{p^n}$" in a suitable sense. In this case, by forgetting $\bar{\eta}$ we obtain a triple $(\bar{A}, \lambda, \iota)$, which should be "defined over $\mathbb{F}_{p^n}$" in a suitable sense.

Now let g be general. If $(\bar{A}, \lambda, \iota, \bar{\eta})$ represents a point in $\mathrm{Fix}(j,g)$, we can still equip $(\bar{A}, \lambda, \iota)$ with a "virtual $\mathbb{F}_{p^n}$-structure". (The precise definition is to be given later.) We will define a $\mathbb{Q}$-linear category $\mathscr{V}_{n,\mathcal{K},pol}$ of such triples $(\bar{A}, \lambda, \iota)$ equipped with a virtual $\mathbb{F}_{p^n}$-structure. Thus we get a well-defined map $\mathbf{f}$ from $\mathrm{Fix}(j,g)$ to the set $\mathscr{I}$ of isomorphism classes in $\mathscr{V}_{n,\mathcal{K},pol}$. In order to understand $\mathrm{Fix}(j,g)$, it suffices to understand the subset $\mathrm{im}(\mathbf{f})$ of $\mathscr{I}$, and to study the fibers of $\mathbf{f}$.

We carry out the strategy outlined above in § 3–§ 5, arriving at the unstabilized Langlands–Kottwitz formula for $T(j,g)$ in Theorem 5.3.1.

3 Honda-Tate theory for virtual abelian varieties

3.1 Virtual abelian varieties with a polarizable condition

Definition 3.1.1. Let $n \in \mathbb{N}$ and let σ be the arithmetic p-Frobenius. Let $\mathscr{A}_{\overline{\mathbb{F}}_p}$ be the category of abelian varieties over $\overline{\mathbb{F}}_p$. Define the $\mathbb{Z}_{(p)}$-linear category $\mathscr{V}_n^p$ of *virtual abelian varieties over* $\mathbb{F}_{p^n}$ *up to prime-to-p isogeny* as follows:

- the objects are pairs $A = (\bar{A}, u)$, where $\bar{A}$ is an object of $\mathscr{A}_{\overline{\mathbb{F}}_p} \otimes \mathbb{Z}_{(p)}$ and u is an isomorphism $\sigma^n(\bar{A}) \to \bar{A}$ in $\mathscr{A}_{\overline{\mathbb{F}}_p} \otimes \mathbb{Z}_{(p)}$. Here $\sigma^n(\bar{A})$ denotes the base change of $\bar{A}$ along the map $\sigma^n : \overline{\mathbb{F}}_p \to \overline{\mathbb{F}}_p$.
- the morphisms from (A_1, u_1) to (A_2, u_2) are morphisms $f : A_1 \to A_2$ in $\mathscr{A}_{\overline{\mathbb{F}}_p} \otimes \mathbb{Z}_{(p)}$ such that $f u_1 = u_2 \sigma^n(f)$.

For $A = (\bar{A}, u) \in \mathscr{V}_n^p$, define its *Frobenius endomorphism*

$$\pi_A \in \mathrm{End}_{\mathscr{A}_{\overline{\mathbb{F}}_p} \otimes \mathbb{Z}_{(p)}}(\bar{A}) = \mathrm{End}(\bar{A}) \otimes \mathbb{Z}_{(p)}$$

to be the composition of the p^n-Frobenius homomorphism $\bar{A} \to \sigma^n(\bar{A})$ with u. In particular, $\mathrm{End}_{\mathscr{V}_n^p}(A)$ is the centralizer of π_A in $\mathrm{End}(\bar{A}) \otimes \mathbb{Z}_{(p)}$.

Definition 3.1.2. Let $\mathscr{V}_n$ be the isogeny category $\mathscr{V}_n^p \otimes_{\mathbb{Z}_{(p)}} \mathbb{Q}$. For $A, B \in \mathscr{V}_n$, we write $\mathrm{End}(A)_{\mathbb{Q}}$ and $\mathrm{Hom}(A,B)_{\mathbb{Q}}$ for the $\mathbb{Q}$-vector spaces $\mathrm{End}_{\mathscr{V}_n}(A)$ and $\mathrm{Hom}_{\mathscr{V}_n}(A,B)$ respectively. Denote by $\mathbb{Q}[\pi_A]$ the $\mathbb{Q}$-subalgebra of $\mathrm{End}(A)_{\mathbb{Q}}$ generated by π_A.

Definition 3.1.3. Let $c \in \mathbb{Q}$ and $A \in \mathscr{V}_n$. By a *c-polarization of* A, we mean a $\mathbb{Q}$-polarization λ of $\bar{A}$ such that $\pi_A^* \lambda = c\lambda$. This only exists when $c > 0$. If there exists a c-polarization of A we say A is *c-polarizable*. Let $\mathscr{V}_{n,c}$ be the full subcategory of $\mathscr{V}_n$ consisting of c-polarizable objects.

Remark 3.1.4. Let A be an abelian variety over $\mathbb{F}_{p^n}$. Then A defines an object of $\mathscr{V}_n$, by taking $\bar{A}$ to be $A_{\overline{\mathbb{F}}_p}$ and u to be the canonical isomorphism. For such A, a p^n-polarization is the same as a polarization $A \to A^\vee$ defined over $\mathbb{F}_{p^n}$.

Definition 3.1.5. Let $c \in \mathbb{Q}_{>0}$. By a *c-number* we mean an element $\pi \in \overline{\mathbb{Q}}$ whose images under all embeddings $\overline{\mathbb{Q}} \to \overline{\mathbb{Q}}_p$ have non-negative valuations and whose images under all embeddings $\overline{\mathbb{Q}} \to \mathbb{C}$ have absolute value $c^{1/2}$.

Example 3.1.6. Suppose $c = p^n$. Recall that a *Weil* p^n*-number* is by definition an **algebraic integer** all of whose complex embeddings have absolute value $p^{n/2}$. Hence a p^n-number π

in the sense of Definition 3.1.5 is a Weil p^n-number if and only if π is a unit at all places of $\mathbb{Q}(\pi)$ coprime to p. (To check this equivalence, use the product formula.) For example, let $p = 7$. Then $7(1 + 2i)/(1 - 2i)$ is a 7^2-number but not a Weil 7^2-number.

In the following theorem we summarize results about $\mathscr{V}_{n,c}$ proved in [Kot92, § 10]. The theorem can be viewed as a generalization of Honda-Tate theory, and its proof also heavily relies on the latter.

Theorem 3.1.7 (Kottwitz). *For $c \in \mathbb{Q}$, the category $\mathscr{V}_{n,c}$ is nonempty only if $c > 0$ and $v_p(c) = n$. Assume this is the case. Then $\mathscr{V}_{n,c}$ is a semi-simple abelian category with all objects having finite length. For any simple object A of $\mathscr{V}_{n,c}$, the $\mathbb{Q}$-algebra $\mathrm{End}(A)_{\mathbb{Q}}$ is a division algebra with center $\mathbb{Q}[\pi_A]$ (which is a field). The Hasse invariant of $\mathrm{End}(A)_{\mathbb{Q}}$ at all places of $\mathbb{Q}[\pi_A]$ are given by explicit formulas, see [Kot92, Lemma 10.11]. The image of π_A under any embedding $\mathbb{Q}[\pi_A] \to \overline{\mathbb{Q}}$ is a c-number. Moreover $A \mapsto \pi_A$ induces a bijection from the set of isomorphism classes of simple objects in $\mathscr{V}_{n,c}$ to the set of Galois orbits of c-numbers.* □

3.2 The endomorphism structure

Definition 3.2.1. Let $\mathscr{V}_{n,c,\mathcal{K}}$ be the category whose objects are pairs (A, ι) with $A \in \mathscr{V}_{n,c}$ and ι a $\mathbb{Q}$-algebra map $\mathcal{K} \to \mathrm{End}(A)_{\mathbb{Q}}$, and whose morphisms are morphisms in $\mathscr{V}_{n,c}$ that are compatible with the $\mathcal{K}$-actions. In the following we also use the simpler notations $\mathrm{End}_{\mathcal{K}}(A, \iota)_{\mathbb{Q}}$ or $\mathrm{End}_{\mathcal{K}}(A)_{\mathbb{Q}}$ to denote $\mathrm{End}_{\mathscr{V}_{n,c,\mathcal{K}}}(A, \iota)$.

As shown in [Kot92, Lemmas 3.2, 3.3], we can formally deduce the structure of $\mathscr{V}_{n,c,\mathcal{K}}$ from that of $\mathscr{V}_{n,c}$, and obtain the following corollary.

Corollary 3.2.2 ([Kot92, Lemma 10.13]). *Let $c \in \mathbb{Q}$ such that $c > 0$ and $v_p(c) = n$. Then $\mathscr{V}_{n,c,\mathcal{K}}$ is a semi-simple abelian category with all objects having finite length. Fix an embedding $\mathcal{K} \to \overline{\mathbb{Q}}$. The simple objects of $\mathscr{V}_{n,c,\mathcal{K}}$ are classified by c-numbers in $\overline{\mathbb{Q}}$ up to conjugacy over $\mathcal{K}$.*

Sketch of proof. We only give the definition of the simple object X_π in $\mathscr{V}_{n,c,\mathcal{K}}$ (up to isomorphism) corresponding to a c-number π, in terms of the classification in Theorem 3.1.7. Let A_π be the simple object in $\mathscr{V}_{n,c}$ corresponding to π as in Theorem 3.1.7. Choose a $\mathbb{Q}$-vector space isomorphism $\mathcal{K} \cong \mathbb{Q}^d$. Then the action of $\mathcal{K}$ on itself by multiplication gives an embedding $\mathcal{K} \hookrightarrow M_d(\mathbb{Q})$. Consider $A_\pi \otimes_{\mathbb{Q}} \mathcal{K} := A_\pi^{\oplus d} \in \mathscr{V}_{n,c}$. Define ι to be the composition

$$\mathcal{K} \hookrightarrow M_d(\mathbb{Q}) \xrightarrow{\mathbb{Q} \to \mathrm{End}(A_\pi)_{\mathbb{Q}}} M_d(\mathrm{End}(A_\pi)_{\mathbb{Q}}) = \mathrm{End}(A_\pi \otimes_{\mathbb{Q}} \mathcal{K})_{\mathbb{Q}}.$$

Then X_π is given by $(A_\pi \otimes_{\mathbb{Q}} \mathcal{K}, \iota)$. □

Remark 3.2.3. With the notation in the above proof, we have

$$\mathrm{End}_{\mathcal{K}}(X_\pi)_{\mathbb{Q}} = \mathcal{K} \otimes_{\mathbb{Q}} \mathrm{End}(A_\pi)_{\mathbb{Q}}.$$

4 Kottwitz triples

4.1 The notion of a Kottwitz triple and its Kottwitz invariant

Denote $\Gamma := \mathrm{Gal}(\overline{\mathbb{Q}}/\mathbb{Q})$, and $\Gamma_v := \mathrm{Gal}(\overline{\mathbb{Q}}_v/\mathbb{Q}_v)$ for all places v of $\mathbb{Q}$. For each v we choose an embedding $\overline{\mathbb{Q}} \to \overline{\mathbb{Q}}_v$, and correspondingly we have an embedding $\Gamma_v \to \Gamma$. For each $n \in \mathbb{N}$, denote $W_n := W(\mathbb{F}_{p^n})$ (the ring of Witt vectors) and $L_n := W_n[1/p] = \mathrm{Frac}(W_n)$. Denote $W := W(\overline{\mathbb{F}}_p)$ and $L := W[1/p] = \mathrm{Frac}(W)$.

Definition 4.1.1. Let $n \in \mathbb{N}$. An $\mathbb{F}_{p^n}$-*Kottwitz triple* is a triple $(\gamma_0, \gamma, \delta)$, where γ_0 is a stable conjugacy class of semi-simple $\mathbb{R}$-elliptic elements in $G(\mathbb{Q})$, γ is a conjugacy class in $G(\mathbb{A}_f^p)$, and δ is a σ-conjugacy class in $G(L_n)$, satisfying the following conditions:

(1) γ_0 is stably conjugate to γ.

(2) γ_0 (viewed as an element of $G(\mathbb{Q}_p)$) is an n-th norm of δ, i.e., γ_0 is conjugate to $\delta\sigma(\delta)\cdots\sigma^{n-1}(\delta)$ in $G(\overline{\mathbb{Q}}_p)$. Here recall that the n-th norm of δ is well-defined as a stable conjugacy class in $G(\mathbb{Q}_p)$. See [Kot82] for more details.

(3) Let $B(G_{\mathbb{Q}_p})$ be the set of σ-conjugacy classes in $G(L)$. We require that the image of δ under the Kottwitz map $B(G_{\mathbb{Q}_p}) \to X^*(Z(\widehat{G})^{\Gamma_p})$ is equal to $-\mu_1$, where μ_1 is the natural image in $X^*(Z(\widehat{G})^{\Gamma_p})$ of the Hodge cocharacter $\mu_h : \mathbb{G}_m \to G_{\mathbb{C}}$ associated to $h : \mathbb{S} \to G_{\mathbb{R}}$.[7] See [Kot85] for fundamental facts about $B(G_{\mathbb{Q}_p})$ and the definition of the Kottwitz map, and see [Kot90, § 2] for how μ_h has a natural image in $X^*(Z(\widehat{G})^{\Gamma_p})$.

Remark 4.1.2. Note that γ_0 (up to stable conjugacy) is uniquely determined by γ.

Remark 4.1.3. In general, if G is a reductive group $\mathbb{Q}$ that has a Shimura datum, then $G_{\mathbb{R}}$ has elliptic maximal tori. For such G, the two notions of ellipticity over $\mathbb{R}$ for semi-simple elements of $G(\mathbb{R})$ (p. 392 of [Kot86]) are equivalent. For a semi-simple element $\gamma_0 \in G(\mathbb{Q})$, we say it is $\mathbb{R}$-*elliptic* if $\gamma_0 \in G(\mathbb{R})$ is elliptic over $\mathbb{R}$.

Definition 4.1.4 (cf. [Kot86], [Har11, § 4]). Let $\gamma_0 \in G(\mathbb{Q})$ be a semi-simple elliptic element. Let $I_0 := G_{\gamma_0}$. Define $\mathfrak{K}(\gamma_0/\mathbb{Q}) = \mathfrak{K}(I_0/\mathbb{Q})$ to be the finite abelian group consisting of elements in $\pi_0((Z(\widehat{I_0})/Z(\widehat{G}))^{\Gamma}) = (Z(\widehat{I_0})/Z(\widehat{G}))^{\Gamma}$ whose images in $\mathbf{H}^1(\Gamma, Z(\widehat{G}))$ are locally trivial.

In [Kot90, § 2], Kottwitz attaches an element $\alpha(\gamma_0, \gamma, \delta) \in \mathfrak{K}(\gamma_0/\mathbb{Q})^D$ (where the superscript D denotes the Pontryagin dual) to each $\mathbb{F}_{p^n}$-Kottwitz triple $(\gamma_0, \gamma, \delta)$. The construction also depends on the Shimura datum. We call $\alpha(\gamma_0, \gamma, \delta)$ the *Kottwitz invariant of* $(\gamma_0, \gamma, \delta)$. We will sketch its construction below. Of course, now we view γ_0 as an element

[7] Recall that h is involved in the PEL datum, and the Shimura datum in question is (G, h^{-1}). On p. 419 of [Kot92] Kottwitz formulates this condition by requiring δ to have image equal to μ_1, which should be a mistake. In fact, any δ that contributes non-trivially to the right hand side of [Kot92, (19.5)] must have image $-\mu_1$.

of $G(\mathbb{Q})$ on the nose as opposed to a stable conjugacy class, so that the group $I_0 := G_{\gamma_0}$ is defined. When γ_0 varies in the stable conjugacy class, the groups $\mathfrak{K}(\gamma_0/\mathbb{Q})$ are canonically identified with each other and the elements $\alpha(\gamma_0, \gamma, \delta)$ are compatible with respect to the these identifications.

We now sketch the construction of $\alpha(\gamma_0, \gamma, \delta)$. As recalled in [Har11, § 3], at each finite place $v \neq p$ the difference between the v-component of γ and the element $\gamma_0 \in G(\mathbb{Q}_v)$ is measured by a class α_v in

$$\ker(\mathbf{H}^1(\mathbb{Q}_v, I_0) \to \mathbf{H}^1(\mathbb{Q}_v, G)).$$

Recall that Kottwitz's "abelianization map" gives a canonical bijection (see e.g. [Har11, § 4]):

$$\mathbf{H}^1(\mathbb{Q}_v, H) \xrightarrow{\sim} \pi_0(Z(\widehat{H})^{\Gamma_v})^D,$$

for $H = I_0$ or $H = G$. Hence α_v can be viewed as a character

$$\alpha_v : Z(\widehat{I_0})^{\Gamma_v} \longrightarrow \mathbb{C}^\times$$

whose restriction to $(Z(\widehat{I_0})^{\Gamma_v})^0 Z(\widehat{G})^{\Gamma_v}$ is trivial.

At $v = p$, by a theorem of Steinberg we can find $c \in G(L)$ such that $c\gamma_0 c^{-1} = \delta\sigma(\delta)\cdots\sigma^{n-1}(\delta)$. Define $b := c^{-1}\delta\sigma(c)$. Then b is a well-defined element in $B(I_{0,\mathbb{Q}_p})$. Define α_p to be the image of b under the Kottwitz map (see [Kot85])

$$B(I_{0,\mathbb{Q}_p}) \longrightarrow X^*(Z(\widehat{I_0})^{\Gamma_p}).$$

At $v = \infty$, Kottwitz uses the homomorphism $h : \mathbb{S} \to G_{\mathbb{R}}$ to construct an element

$$\alpha_\infty \in X^*(Z(\widehat{I_0})^{\Gamma_\infty}), \tag{4.1.1}$$

which depends only on the $G(\mathbb{R})$-conjugacy class of h. Its image in $X^*(Z(\widehat{G})^{\Gamma_\infty})$ is equal to μ_1, see [Kot90, p. 167].

Thus at every place v of $\mathbb{Q}$, we have constructed a character

$$\alpha_v : Z(\widehat{I_0})^{\Gamma_v} \longrightarrow \mathbb{C}^\times. \tag{4.1.2}$$

By condition (3) in Definition 4.1.1, the α_v's can be extended to characters

$$\beta_v : Z(\widehat{I_0})^{\Gamma_v} Z(\widehat{G}) \longrightarrow \mathbb{C}^\times,$$

such that

$$\beta_v|_{Z(\widehat{G})} = \begin{cases} \mu_1, & v = \infty \\ -\mu_1, & v = p \\ 1, & \text{else.} \end{cases}$$

Recall that our goal is to define a character

$$\alpha(\gamma_0, \gamma, \delta) : \mathfrak{K}(\gamma_0/\mathbb{Q}) \to \mathbb{C}^\times.$$

Let $\kappa \in \mathfrak{K}(\gamma_0/\mathbb{Q})$. Using the Chebotarev density theorem, it is easy to see that

$$\mathfrak{K}(\gamma_0/\mathbb{Q}) = \left[\bigcap_v Z(\widehat{I_0})^{\Gamma_v} Z(\widehat{G})\right] / Z(\widehat{G}).$$

Thus κ has lifts $\kappa_v \in Z(\widehat{I_0})^{\Gamma_v} Z(\widehat{G})$ for all v. We define

$$\alpha(\gamma_0, \gamma, \delta)(\kappa) := \prod_v \beta_v(\kappa_v).$$

This definition is independent of the choices of κ_v.

4.2 Associating Kottwitz triples to virtual abelian varieties with additional structures

Let A be an abelian variety over $\mathbb{F}_{p^n}$. Let $\bar{A} := A_{\overline{\mathbb{F}}_p}$. We have the following well-known constructions:

(1) We have the $\mathbb{A}_f^p$-Tate module $\mathbf{H}_1(\bar{A}, \mathbb{A}_f^p)$ of $\bar{A}$, which is a finite free $\mathbb{A}_f^p$-module with a distinguished endomorphism, namely the action of π_A.

(2) We have the cristalline homology $\bar{\Lambda} = \mathbf{H}_1(\bar{A}/W)$, which is by definition the W-linear dual of the usual cristalline cohomology $\mathbf{H}^1_{\mathrm{cris}}(\bar{A}/W)$. Thus $\bar{\Lambda}$ is a finite free W-module, and the usual Frobenius on $\mathbf{H}^1_{\mathrm{cris}}(\bar{A}/W)$ induces a σ-linear automorphism Φ of $\bar{\Lambda}[1/p]$, still called the Frobenius. We have $\Phi(\bar{\Lambda}) \supset \bar{\Lambda}$. Moreover, the W_n-linear dual $\Lambda = \mathbf{H}_1(A/W_n)$ of $\mathbf{H}^1_{\mathrm{cris}}(A/W_n)$ provides a canonical W_n-structure for $\bar{\Lambda}$. In other words we have a canonical isomorphism $\Lambda \otimes_{W_n} W \xrightarrow{\sim} \bar{\Lambda}$. The Frobenius Φ on $\bar{\Lambda}[1/p]$ stabilizes $\Lambda[1/p]$. We refer the reader to [Hai05, § 14] for a good introduction to these constructions, including a discussion on the duality.

We would like to generalize the above constructions to all $A \in \mathscr{V}_n$. The generalization of (1) is verbatim. On p. 403 of [Kot92] Kottwitz shows how to generalize (2), which is in fact already needed in the proof of Theorem 3.1.7. More precisely:

Lemma 4.2.1. *For all $A = (\bar{A}, u) \in \mathscr{V}_n$, let $\bar{\Lambda} = \mathbf{H}_1(\bar{A}/W)$ be the W-linear dual of $\mathbf{H}^1_{\mathrm{cris}}(\bar{A}/W)$. There is a canonical W_n-structure for $\bar{\Lambda}$, denoted by $\Lambda = \mathbf{H}_1(A/W_n)$, which is functorial with respect to the morphisms in $\mathscr{V}_n$. The Frobenius Φ on $\bar{\Lambda}[1/p]$ stabilizes $\Lambda[1/p]$. Moreover, the operator Φ^{-n} on Λ coincides with the endomorphism of Λ induced functorially by $\pi_A \in \mathrm{End}(A)_{\mathbb{Q}}$.*

Sketch of proof. The key point is that $\bar{\Lambda}$ and $\bar{\Lambda}' := \mathbf{H}_1(\sigma^n(\bar{A})/W)$ are related by a canonical isomorphism

$$\bar{\Lambda} \otimes_{W, \sigma^n} W \cong \bar{\Lambda}'.$$

Hence u induces a σ^n-linear automorphism u of $\bar{\Lambda}$. We define

$$\Lambda := \left\{x \in \bar{\Lambda} | ux = x\right\}.$$

The rest of the properties are easy to check. □

Fix $c \in \mathbb{Q}_{>0}$ with $v_p(c) = n$. Consider $A \in \mathscr{V}_{n,c}$ and a c-polarization λ of it. Then λ induces an $\mathbb{A}_f^p(1)$-valued alternating form on $\mathbf{H}_1(\bar{A}, \mathbb{A}_f^p)$, denoted by $(\cdot,\cdot)_\lambda$, satisfying

$$(\pi_A v, \pi_A w)_\lambda = c(v, w)_\lambda, \ \forall v, w \in \mathbf{H}_1(\bar{A}, \mathbb{A}_f^p). \tag{4.2.1}$$

Similarly, we have an alternating form $(\cdot,\cdot)_\lambda$ on the L_n-vector space $\mathbf{H}_1(A/W_n)[1/p]$, well-defined up to a scalar in $W_n^\times$, satisfying

$$(\Phi v, \Phi w)_\lambda = c'\sigma(v, w), \ \forall v, w \in \mathbf{H}_1(A/W_n)[1/p],$$

for some $c' \in L_n^\times$ whose norm to $\mathbb{Q}_p^\times$ is c^{-1}. See [Kot92, p. 404] for details. Now suppose in addition that we have a $\mathbb{Q}$-algebra map $\iota : \mathcal{K} \to \operatorname{End}(A)_\mathbb{Q}$ which is *compatible with* λ in the sense that ι is equivariant with respect to the automorphism $\mathbf{c}$ on $\mathcal{K}$ and the Rosati involution on $\operatorname{End}(A)_\mathbb{Q}$ defined by λ. Then by functoriality we have actions of $\mathcal{K}$ on $\mathbf{H}_1(\bar{A}, \mathbb{A}_f^p)$ and on $\mathbf{H}_1(A/W_n)[1/p]$, commuting with π_A and Φ respectively.

We now refine the category $\mathscr{V}_{n,c,\mathcal{K}}$ in § 3.2 by including a c-polarization as part of the structure.

Definition 4.2.2. Define the category $\mathscr{V}_{n,c,\mathcal{K},pol}$ of *c-polarized virtual abelian varieties over* $\mathbb{F}_{p^n}$ as follows. The objects are triples (A, ι, λ), where $(A, \iota) \in \mathscr{V}_{n,c,\mathcal{K}}$ and λ is a c-polarization of A which is compatible with ι in the sense mentioned above. The morphisms are the morphisms of $\mathscr{V}_{n,c,\mathcal{K}}$ that preserve the polarizations up to a scalar. Define $\mathscr{V}_{n,c}^G$ to be the full subcategory of $\mathscr{V}_{n,c,\mathcal{K},pol}$ consisting of objects (A, ι, λ) satisfying the following two conditions:

(1) Fix an arbitrary $\mathbb{A}_f^p$-linear isomorphism $\mathbb{A}_f^p \cong \mathbb{A}_f^p(1)$. There is an $\mathbb{A}_f^p$-linear isomorphism

$$(\mathbf{H}_1(\bar{A}, \mathbb{A}_f^p), (\cdot,\cdot)_\lambda) \to (V, \langle\cdot,\cdot\rangle) \otimes_\mathbb{Q} \mathbb{A}_f^p$$

which is $\mathcal{K}$-equivariant and preserves the alternating forms up to a scalar.

(2) There is an L_n-linear isomorphism

$$(\mathbf{H}_1(A/W_n)[1/p], (\cdot,\cdot)_\lambda) \to (V, \langle\cdot,\cdot\rangle) \otimes_\mathbb{Q} L_n$$

which is $\mathcal{K}$-equivariant and preserves the alternating forms up to a scalar.

Definition 4.2.3. Let $(A, \iota, \lambda) \in \mathscr{V}_{n,c}^G$. Choose isomorphisms as in (1) and (2) of Definition 4.2.2. Using these isomorphisms, we translate π_A^{-1} to an element γ of $\mathrm{GL}(V)(\mathbb{A}_f^p)$, and translate Φ to a σ-linear automorphism δ' of $V \otimes_\mathbb{Q} L_n$. Then it is easy to see that $\gamma \in G(\mathbb{A}_f^p)$ and $\delta' = \delta \circ \sigma$ for some $\delta \in G(L_n)$. Moreover the conjugacy class of γ in $G(\mathbb{A}_f^p)$ and the σ-conjugacy class of δ in $G(L_n)$ are independent of the choices. We say that (γ, δ) is associated to (A, ι, λ).

Remark 4.2.4. From (4.2.1) we know that γ in the above definition has factor of similitude $\nu(\gamma) = c^{-1}$.

Definition 4.2.5. Let $\mathscr{V}_{n,c}^{(G,h^{-1})}$ be the full subcategory of $\mathscr{V}_{n,c}^{G}$ consisting of objects whose associated δ as in Definition 4.2.3 satisfies condition (3) in Definition 4.1.1.

Proposition 4.2.6 ([Kot92, § 14]). *Let $(A, \iota, \lambda) \in \mathscr{V}_{n,c}^{(G,h^{-1})}$. Let (γ, δ) be associated to it as in Definition 4.2.3. Then the stable conjugacy class of γ contains a $\mathbb{Q}$-point γ_0, which in particular has factor of similitude $\nu(\gamma_0) = c^{-1}$ (see Remark 4.2.4). Moreover, $(\gamma_0, \gamma, \delta)$ is an $\mathbb{F}_{p^n}$-Kottwitz triple.*

Sketch of proof. For any $\mathbb{Q}$-algebra R, let $I(R)$ be the automorphism group of (A, ι, λ) in the R-linear category $\mathscr{V}_{n,c}^{(G,h^{-1})} \otimes_{\mathbb{Q}} R$. Then the functor $R \mapsto I(R)$ is representable by a reductive group I over $\mathbb{Q}$. To show the existence of γ_0, the key claim is that any maximal torus T of I defined over $\mathbb{Q}$ admits an embedding to G defined over $\mathbb{Q}$. Once the claim is proved, noting that π_A is a $\mathbb{Q}$-point of the center of I, we define γ_0 to be the image of π_A^{-1} under a $\mathbb{Q}$-embedding $T \to G$.

The proof of the claim contains two steps. Let N be the centralizer of $T(\mathbb{Q})$ in $\mathrm{End}_{\mathcal{K}}(A)_{\mathbb{Q}}$. Then the Rosati involution $*$ on $\mathrm{End}(A)_{\mathbb{Q}}$ induced by λ stabilizes N and we have $\dim_{\mathcal{K}} N = \dim_{\mathcal{K}} V$. In fact it can be shown that $(N, *|_N)$ is a CM algebra (i.e. a product of CM fields). The algebraic group T is recovered from N by

$$T(R) = \{g \in (N \otimes_{\mathbb{Q}} R)^{\times} | gg^* \in R^{\times}\}$$

for any $\mathbb{Q}$-algebra R. The first step in the proof of the claim is to show the existence of a $\mathcal{K}$-algebra embedding $N \to C = \mathrm{End}_{\mathcal{K}}(V)$. This step is trivial in our case, as C is a matrix algebra over $\mathcal{K}$ of the right dimension. The second step is to show that one can modify the embedding $N \to C$ such that it respects the involutions on N and C. By cohomological considerations and the theory of transferring tori, we reduce to checking that certain local obstructions vanish. At finite places $v \neq p$, the local obstructions vanish because of condition (1) in Definition 4.2.2. At p and ∞, the obstructions vanish more or less automatically.

Once the existence of γ_0 is proved, it is not too difficult to check that $(\gamma_0, \gamma, \delta)$ is an $\mathbb{F}_{p^n}$-Kottwitz triple with the help of Lemma 2.1.2. For example, condition (3) in Definition 4.1.1 holds by definition, and condition (2) in that definition boils down to the relation $\Phi^{-n} = \pi_A$ in Lemma 4.2.1. □

Definition 4.2.7. In the situation of Proposition 4.2.6, we say that $(\gamma_0, \gamma, \delta)$ is the $\mathbb{F}_{p^n}$-Kottwitz triple associated to $(A, \iota, \lambda) \in \mathscr{V}_{n,c}^{(G,h^{-1})}$.

The following result is arguably the technical core of [Kot92].

Theorem 4.2.8 ([Kot92, Lemma 18.1]). *Let $(\gamma_0, \gamma, \delta)$ be an $\mathbb{F}_{p^n}$-Kottwitz triple. Then $(\gamma_0, \gamma, \delta)$ is associated to an object (A, ι, λ) in $\mathscr{V}_{n,c}^{(G,h^{-1})}$ if and only if the following conditions hold:*

(1) *The factor of similitude $\nu(\gamma_0)$ of γ_0 is equal to c^{-1}.*

(2) *The Kottwitz invariant $\alpha(\gamma_0, \gamma, \delta)$ is trivial.*

(3) *There is a W_n-lattice Λ in $V \otimes_{\mathbb{Q}} L_n$ such that $(\delta\sigma)\Lambda \supset \Lambda$.*

4.3 Sketch of the "only if" part in Theorem 4.2.8

We only sketch the proof of (2), which is the most difficult part. Firstly Kottwitz observes that by special properties of G, the group $\mathfrak{K}(\gamma_0/\mathbb{Q})^D$ always maps injectively into $X^*(Z(\widehat{I_0})^\Gamma)$, where $I_0 := G_{\gamma_0}$. For $(A, \iota, \lambda) \in \mathscr{V}_{n,c}^{(G,h^{-1})}$, denote by $\alpha(A, \iota, \lambda)$ the image in $X^*(Z(\widehat{I_0})^\Gamma)$ of the Kottwitz invariant of the Kottwitz triple associated to (A, ι, λ). It suffices to show that $\alpha(A, \iota, \lambda)$ is trivial.

In fact Kottwitz proves a generalization (see Proposition 4.3.2 below) of the statement $\alpha(A, \iota, \lambda) = 0$, and this generalization is also needed in the "if" part of Theorem 4.2.8. To state it we need an auxiliary set.

Definition 4.3.1. Let $\mathscr{V}^{aux}$ be the set of quadruples $(\gamma_0; A, \iota, \lambda)$, where (A, ι, λ) is an object of $\mathscr{V}_{n,c,\mathcal{K},pol}$ and $\gamma_0 \in G(\mathbb{Q})$, satisfying the following conditions:

- γ_0 is a semi-simple $\mathbb{R}$-elliptic element with $\nu(\gamma_0) = c^{-1}$.
- Let Y be an indeterminate over $\mathcal{K}$. Define a $\mathcal{K}[Y]$-module structure on V (resp. $\mathbf{H}_1(\bar{A}, \mathbb{A}_f^p)$) by letting Y act as γ_0^{-1} (resp. π_A). We require that there exists a $\mathcal{K}[Y] \otimes_{\mathbb{Q}} \mathbb{A}_f^p$-module isomorphism $\phi : V \otimes_{\mathbb{Q}} \mathbb{A}_f^p \xrightarrow{\sim} \mathbf{H}_1(\bar{A}, \mathbb{A}_f^p)$. Here we do **not** require ϕ to preserve the alternating forms $\langle\cdot,\cdot\rangle$ on $V \otimes_{\mathbb{Q}} \mathbb{A}_f^p$ and $(\cdot,\cdot)_\lambda$ on $\mathbf{H}_1(\bar{A}, \mathbb{A}_f^p)$ up to a scalar.

Let $(A, \iota, \lambda) \in \mathscr{V}_{n,c}^{(G,h^{-1})}$ and let $(\gamma_0, \gamma, \delta)$ be the associated $\mathbb{F}_{p^n}$-Kottwitz triple. Then $(\gamma_0; A, \iota, \lambda)$ is an element in $\mathscr{V}^{aux}$. However in the definition of $\mathscr{V}^{aux}$ above we only require $(A, \iota, \lambda) \in \mathscr{V}_{n,c,\mathcal{K},pol}$. Kottwitz shows that the construction

$$\mathscr{V}_{n,c}^{(G,h^{-1})} \ni (A, \iota, \lambda) \mapsto \alpha(A, \iota, \lambda) \tag{4.3.1}$$

can be extended to a map

$$\mathscr{V}^{aux} \ni (\gamma_0; A, \iota, \lambda) \mapsto \alpha(\gamma_0; A, \iota, \lambda) \in X^*(Z(\widehat{I_0})^\Gamma),\ I_0 := G_{\gamma_0}. \tag{4.3.2}$$

Kottwitz then proves:

Proposition 4.3.2 ([Kot92, Lemma 15.2]). *The element $\alpha(\gamma_0; A, \iota, \lambda)$ is trivial for all $(\gamma_0; A, \iota, \lambda) \in \mathscr{V}^{aux}$.*

4.3.1 Sketch of the construction (4.3.2)

Consider a finite place $v \neq p$. Let ϕ be as in Definition 4.3.1. The difference between the two alternating forms on $V \otimes_{\mathbb{Q}} \mathbb{Q}_v$, given by $\langle\cdot,\cdot\rangle$ and $\phi^{-1}((\cdot,\cdot)_\lambda)$, is measured by a class in $\mathbf{H}^1(\mathbb{Q}_v, I_0)$. In fact, the $\mathbb{Q}$-algebraic group I_0 is the automorphism group of $(V, \langle\cdot,\cdot\rangle)$, as a $\mathcal{K}[Y]$-module equipped with an alternating form up to a scalar. The $\mathcal{K}[Y] \otimes_{\mathbb{Q}} \mathbb{Q}_v$-module $\mathbf{H}_1(\bar{A}, \mathbb{Q}_v)$ is by assumption isomorphic to $V \otimes_{\mathbb{Q}} \mathbb{Q}_v$, and therefore

$$(\mathbf{H}_1(\bar{A}, \mathbb{Q}_v), (\cdot,\cdot)_\lambda) \otimes_{\mathbb{Q}_v} \overline{\mathbb{Q}}_v$$

is isomorphic to

$$(V, \langle\cdot,\cdot\rangle) \otimes_{\mathbb{Q}} \overline{\mathbb{Q}}_v,$$

as $\mathcal{K}[Y] \otimes_{\mathbb{Q}} \overline{\mathbb{Q}}_v$-modules equipped with alternating forms up to a scalar. Thus their difference over $\mathbb{Q}_v$ is measured by a class in $\mathbf{H}^1(\mathbb{Q}_v, I_0)$.

Using this class in $\mathbf{H}^1(\mathbb{Q}_v, I_0)$ we construct, as in the end of § 4.1, a character

$$\alpha_v : Z(\widehat{I_0})^{\Gamma_v} \to \mathbb{C}^\times.$$

We also have a character

$$\alpha_\infty : Z(\widehat{I_0})^{\Gamma_\infty} \to \mathbb{C}^\times$$

as in (4.1.1). At p, by Steinberg's theorem there exists a $\mathcal{K}(Y) \otimes_{\mathbb{Q}} L$-module isomorphism

$$\mathbf{H}_1(\bar{A}/W)[1/p] \xrightarrow{\sim} V \otimes_{\mathbb{Q}} L$$

that preserves the alternating forms, where Y acts on the left hand side by π_A and on the right hand side by γ_0^{-1}. Then Φ acting on the left hand side translates to some $b\sigma$ acting on the right hand side, for $b \in I_0(L)$ well-defined up to σ-conjugacy. We define $\alpha_p \in X^*(Z(\widehat{I_0})^{\Gamma_p})$ to be the image of b under the Kottwitz map for $I_{0,\mathbb{Q}_p}$. Define

$$\alpha(\gamma_0; A, \iota, \lambda) := \prod_v (\alpha_v)|_{Z(\widehat{I_0})^\Gamma}. \tag{4.3.3}$$

The above is a sketch of the definition of (4.3.2). We leave it as an exercise for the reader to check that (4.3.2) indeed extends (4.3.1).

4.3.2 *Sketch of the proof of Proposition 4.3.2*

Our specific PEL datum with $B = \mathcal{K}$ simplifies the proof. Let I be the reductive group over $\mathbb{Q}$ of automorphisms of (A, ι, λ) in $\mathscr{V}_{n,c,\mathcal{K},pol}$, as in the proof of Proposition 4.2.6. Let $M := \mathrm{End}_{\mathcal{K}}(A)_{\mathbb{Q}}$. Then M is equipped with the Rosati involution $*$ and

$$I(R) = \{g \in (M \otimes R)^\times | g^* g \in R^\times\}$$

for any $\mathbb{Q}$-algebra R. Let M_0 be the centralizer of γ_0 in $C = \mathrm{End}_{\mathcal{K}}(V)$. Then similarly M_0 is stable under the involution $*$ on C and we have

$$I_0(R) = \{g \in (M_0 \otimes R)^\times | g^* g \in R^\times\}.$$

Consider the I_0-torsor of $\mathcal{K}$-linear $*$-isomorphisms from M_0 to M that take $\gamma_0 \in M_0$ to $\pi_A^{-1} \in M$. The second condition in Definition 4.3.1 guarantees that this torsor is nonempty, and it follows that there is an inner twisting $i : I \to I_0$ (as algebraic groups over $\mathbb{Q}$) canonical up to $I_0(\overline{\mathbb{Q}})$-conjugation. By the general theory of transferring tori, we find a maximal torus T of I that transfers to I_0 along i, i.e., there exists $g_0 \in I_0(\overline{\mathbb{Q}})$ such that $(\mathrm{Int}(g_0) \circ i)|_T : T \to I_0$ is defined over $\mathbb{Q}$. Let N be the centralizer of T in M and let N' be the centralizer of $(\mathrm{Int}(g_0) \circ i)(T)$ in M_0. Then $N \cong N'$, and it is a CM algebra containing $\mathcal{K}$ with $\mathcal{K}$-dimension equal to $\dim_{\mathcal{K}} V$.

Essentially by replacing $\mathcal{K}$ with N' in the whole discussion, we reduce to the special case where $\dim_{\mathcal{K}} V = 1$. In this case we know that $I = I_0 = G = T$ is a torus, given by

$$T(R) = \{g \in (\mathcal{K} \otimes R)^{\times} | g \cdot \mathrm{c}(g) \in R^{\times}\}.$$

Moreover $\dim A = [\mathcal{K} : \mathbb{Q}]/2$. The element $\alpha(\gamma_0; A, \iota, \lambda)$, which we would like to show is trivial, now lies in $X^*(\widehat{T}^{\Gamma}) = X_*(T)_{\Gamma}$. The vanishing of $\alpha(\gamma_0; A, \iota, \lambda)$ in this case is shown in [Kot92, Lemma 13.2]. We sketch its proof below.

Note that if we replace λ by some λ' such that $(\gamma_0; A, \iota, \lambda')$ is still in $\mathscr{V}^{aux}$, then $\alpha(\gamma_0; A, \iota, \lambda')$ differs from $\alpha(\gamma_0; A, \iota, \lambda)$ by the image in $X_*(T)_{\Gamma}$ of a global class in $\mathbf{H}^1(\mathbb{Q}, T)$. By Tate-Nakayama, any such image is trivial. Hence $\alpha(\gamma_0; A, \iota, \lambda)$ is independent of λ. Moreover $\alpha(\gamma_0; A, \iota, \lambda)$ is independent of the virtual $\mathbb{F}_{p^n}$-structure on the abelian variety $\bar{A}$ over $\overline{\mathbb{F}}_p$, as long as the virtual structure is compatible with the $\mathcal{K}$-action on $\bar{A}$. Hence we may write $\alpha(\gamma_0; \bar{A}, \iota)$ for $\alpha(\gamma_0; A, \iota, \lambda)$. By Tate's CM lifting theorem, the pair $(\bar{A}, \iota)$ (up to $\mathbb{Q}$-isogeny) lifts to a CM abelian variety $(\mathcal{A}, \iota)$ with CM by $\mathcal{K}$ over characteristic zero.

We view $\bar{A}$ simply as the reduction of $\mathcal{A}$, ignoring the "up to $\mathbb{Q}$-isogeny". After suitably choosing a $\mathbb{Q}$-symmetrization on $\mathcal{A}$ and a non-degenerate alternating form on V (which need not be $\langle\cdot,\cdot\rangle$), we obtain an element $b \in B(T_{\mathbb{Q}_p})$ that describes the isocrystal $\mathbf{H}_1(\bar{A}/W)[1/p]$. Moreover these choices can be made in such a convenient way that the vanishing of $\alpha(\gamma_0; \bar{A}, \iota)$ is easily reduced to the following claim.

Claim ([Kot92, Lemma 13.1]). The image of b under the Kottwitz map for $T_{\mathbb{Q}_p}$

$$B(T_{\mathbb{Q}_p}) \to X^*(\widehat{T}^{\Gamma_p})$$

(which is a bijection) is equal to the image of $\mu_{(\mathcal{A},\iota)}$ under the natural projection

$$X_*(T) \to X_*(T)_{\Gamma_p} = X^*(\widehat{T}^{\Gamma_p}).$$

Here $\mu_{(\mathcal{A},\iota)} \in X_*(T)$ is the cocharacter corresponding to the CM type of $(\mathcal{A}, \iota)$.

The proof of [Kot92, Lemma 13.1] involves p-adic Hodge theory and the Shimura–Taniyama reciprocity, see [Kot92, §§ 12, 13]. We note that Reimann and Zink proved a stronger form of [Kot92, Lemma 13.1] in [RZ88] (for $p \neq 2$). We refer the reader to the proof of [Hai02, Theorem 7.2] for an explanation of how the result of Reimann–Zink can be interpreted so as to imply [Kot92, Lemma 13.1]. □

4.4 Sketch of the "if" part in Theorem 4.2.8

Let $(\gamma_0, \gamma, \delta)$ be such an $\mathbb{F}_{p^n}$-Kottwitz triple. As usual we write I_0 for G_{γ_0}. The first step is to construct an object X' in $\mathscr{V}_{n,c,\mathcal{K},pol}$, such that $(\gamma_0; X') \in \mathscr{V}^{aux}$ (see Definition 4.3.1). Let f be the characteristic polynomial of $\gamma_0^{-1} \in \mathrm{GL}_{\mathcal{K}}(V)$ over $\mathcal{K}$. Let $f = \prod_i f_i$ be the irreducible factorization over $\mathcal{K}$. It can be shown that all the roots of f_i are c-numbers, so each f_i gives rise to a simple object X_i in $\mathscr{V}_{n,c,\mathcal{K}}$ via Corollary 3.2.2. Then X' can be constructed by forming the direct sum of the X_i's with suitable multiplicities and choosing a suitable c-polarization.

Having constructed X', we recall from (4.3.3) that the invariant $\alpha(\gamma_0; X') \in X^*(Z(\widehat{I_0})^{\Gamma})$ is constructed from local invariants

$$\alpha_v(\gamma_0; X') \in X^*(Z(\widehat{I_0})^{\Gamma_v}).$$

Also recall from (4.1.2) that the invariant $\alpha(\gamma_0, \gamma, \delta)$ is constructed from local invariants

$$\alpha_v(\gamma_0, \gamma, \delta) \in X^*(Z(\widehat{I_0})^{\Gamma_v}).$$

Let

$$\beta_v := \alpha_v(\gamma_0, \gamma, \delta) - \alpha_v(\gamma_0; X').$$

By assumption the invariant $\alpha(\gamma_0, \gamma, \delta)$ is trivial, and by Proposition 4.3.2 the invariant $\alpha(\gamma_0; X')$ is trivial. Therefore

$$\prod_v \beta_v = 0 \in \mathfrak{K}(I_0/\mathbb{Q})^D.$$

Let I be the reductive group over $\mathbb{Q}$ of automorphisms of X' in $\mathscr{V}_{n,c,\mathcal{K},pol}$, as in the proof of Proposition 4.2.6. We have seen in § 4.3 that there is an inner twisting $I \to I_0$ which is canonical up to $I_0(\overline{\mathbb{Q}})$-conjugation. Hence we have canonical identifications

$$X^*(Z(\widehat{I_0})^{\Gamma_v}) \cong X^*(Z(\widehat{I})^{\Gamma_v})$$

$$\mathfrak{K}(I_0/\mathbb{Q}) \cong \mathfrak{K}(I/\mathbb{Q}).$$

The elements β_v can be viewed as elements of $X^*(Z(\widehat{I})^{\Gamma_v})$ whose product is zero in $\mathfrak{K}(I/\mathbb{Q})^D$. It follows that the β_v's are localizations of a common global class $\beta \in \mathbf{H}^1(\mathbb{Q}, I)$. We can then *twist X' by β*, to obtain an object $X = (X')^{\beta} \in \mathscr{V}_{n,c,\mathcal{K},pol}$. Here X is characterized by the property that it is isomorphic to X' in $\mathscr{V}_{n,c,\mathcal{K},pol} \otimes \overline{\mathbb{Q}}$, and such that the functor that assigns to each $\mathbb{Q}$-algebra R the set of isomorphisms $X' \to X$ in $\mathscr{V}_{n,c,\mathcal{K},pol} \otimes R$ is represented by the I-torsor corresponding to β. It can be checked that $X \in \mathscr{V}_{n,c}^{(G,h^{-1})}$ and that $(\gamma_0, \gamma, \delta)$ is its associated Kottwitz triple.

Q.E.D. for Theorem 4.2.8.

5 The unstabilized Langlands–Kottwitz formula

5.1 Definition of orbital integrals

Let $(\gamma_0, \gamma, \delta)$ be an $\mathbb{F}_{p^n}$-Kottwitz triple. Assume $\alpha(\gamma_0, \gamma, \delta)$ is trivial. Let $I_0 := G_{\gamma_0}$. It is shown in [Kot90, § 3] that there exists an inner form I of I_0, such that

- $I(\mathbb{R})$ is compact modulo Z_G.
- $I_{\mathbb{Q}_v}$ is isomorphic to G_{γ_v} (as inner forms of $I_{0,\mathbb{Q}_v}$), for all finite places $v \neq p$.
- $I_{\mathbb{Q}_p}$ is isomorphic to the σ-centralizer G_δ^σ of δ (as inner forms of $I_{0,\mathbb{Q}_p}$).

We refer the reader to [Kot90, § 3] for the more precise statement.

Recall that $\mathbb{F}_q = \mathbb{F}_{p^r}$ is the residue field of E at $\mathfrak{p}$. Assume $r|n$. The cocharacter μ_h : $\mathbb{G}_m \to G_{\mathbb{C}}$ descends to a conjugacy class of cocharacters $\mu_{\overline{\mathbb{Q}}}$ over $\overline{\mathbb{Q}}$, and this conjugacy class is defined over E. Since $G_{\mathbb{Q}_p}$ is unramified (with reductive model $\mathcal{G}$ over $\mathbb{Z}_p$), the $G(\overline{\mathbb{Q}}_p)$-conjugacy class of $\mu_{\overline{\mathbb{Q}}}$ contains a cocharacter $\mu : \mathbb{G}_m \to G_{L_r}$ which is defined over

$L_r = E_\mathfrak{p}$ and factors through a maximal W_r-split torus of $\mathcal{G}_{W_r}$. We fix such a cocharacter μ and define

$$\phi_n : G(L_n) \to \mathbb{Z}$$

to be the characteristic function of $\mathcal{G}(W_n)\mu(p^{-1})\mathcal{G}(W_n)$. Let

$$f^{p,\infty} : G(\mathbb{A}_f^p) \to \mathbb{Z}$$

be the characteristic function of $K^p g^{-1} K^p$, as in the end of § 2.4.

Fix Haar measures on $I_0(\mathbb{A}_f^p)$ and $I_0(\mathbb{Q}_p)$, and *transfer* them to $I(\mathbb{A}_f^p) \cong G_\gamma(\mathbb{A}_f^p)$ and $I(\mathbb{Q}_p) \cong G_\delta^\sigma(\mathbb{Q}_p)$ respectively. We refer the reader to [Kot88, p. 631] for the notion of transferring Haar measures between inner forms. Fix the Haar measure on $G(\mathbb{A}_f^p)$ that gives volume 1 to K^p. (Compare footnote [5].) Fix the Haar measure on $G(L_n)$ that gives volume 1 to $\mathcal{G}(W_n)$. Define the *orbital integral*

$$O_\gamma(f^{p,\infty}) := \int_{G_\gamma(\mathbb{A}_f^p)\backslash G(\mathbb{A}_f^p)} f^{p,\infty}(x^{-1}\gamma x)dx$$

and the *twisted orbital integral*

$$TO_\delta(\phi_n) := \int_{G_\delta^\sigma(\mathbb{Q}_p)\backslash G(L_n)} \phi_n(y^{-1}\delta\sigma(y))dy.$$

Here

$$G_\delta^\sigma(\mathbb{Q}_p) := \left\{y \in G(L_n) | y^{-1}\delta\sigma(y) = \delta\right\}.$$

Define

$$O(\gamma_0, \gamma, \delta) := \mathrm{vol}(I(\mathbb{Q})\backslash I(\mathbb{A}_f))O_\gamma(f^{p,\infty})TO_\delta(\phi_n) \tag{5.1.1}$$

$$c(\gamma_0, \gamma, \delta) := \mathrm{vol}(I(\mathbb{Q})\backslash I(\mathbb{A}_f)) \left|\ker(\ker^1(\mathbb{Q}, I_0) \to \ker^1(\mathbb{Q}, G))\right|. \tag{5.1.2}$$

Then $O(\gamma_0, \gamma, \delta)$ is independent of the choices of Haar measures on $I_0(\mathbb{A}_f^p)$ and $I_0(\mathbb{Q}_p)$, whereas $c(\gamma_0, \gamma, \delta)$ depends on the product measure on $I_0(\mathbb{A}_f)$.

Lemma 5.1.1 ([Kot92, p. 441]). *Let $(\gamma_0, \gamma, \delta)$ be an $\mathbb{F}_{p^n}$-Kottwitz triple. Assume $\alpha(\gamma_0, \gamma, \delta)$ is trivial, $r|n$, and $O(\gamma_0, \gamma, \delta) \neq 0$. Then $(\gamma_0, \gamma, \delta)$ satisfies the conditions (1) (3) in Theorem 4.2.8 for some $c \in \mathbb{Q}_{>0}$ with $v_p(c) = n$.* □

5.2 Counting Fix(j, g)

Let $n := rj$, so that $p^n = q^j$. Consider a point $x' \in \mathrm{Fix}(j, g)$. Then x' is represented by an object $(\bar{A}, \lambda, \iota, \bar{\eta}) \in \mathscr{C}_{\bar{\mathbb{F}}_q}$ (see § 2.6) such that there is an isomorphism

$$u : (\bar{A}, \lambda, \iota, \bar{\eta} \cdot g) \xrightarrow{\sim} \sigma^n(\bar{A}, \lambda, \iota, \bar{\eta}) := (\sigma^n(\bar{A}), \sigma^n(\lambda), \sigma^n(\iota), \sigma^n(\bar{\eta}))$$

in the category $\mathscr{C}_{\overline{\mathbb{F}}_q}$. Then $(\bar{A}, u, \iota, \lambda)$ defines an object X in $\mathscr{V}_{n,c,\mathcal{K},pol}$ for some $c \in \mathbb{Q}$. The isomorphism class of X only depends on x'. Moreover it can be checked that $X \in \mathscr{V}_{n,c}^{(G,h^{-1})}$, see [Kot92, pp. 430–431].

Let $\mathscr{I}_{n,c}$ denote the set of isomorphism classes in $\mathscr{V}_{n,c}^{(G,h^{-1})}$, and let $\mathscr{I}_n := \bigsqcup_{c \in \mathbb{Q}} \mathscr{I}_{n,c}$. Thus we have a well-defined map

$$\mathrm{f} : \mathrm{Fix}(j, g) \to \mathscr{I}_n.$$

We omit the proofs of the following three results.

Lemma 5.2.1 ([Kot92, § 16]). *Let $x' \in \mathrm{Fix}(j, g)$. Let $(\gamma_0, \gamma, \delta)$ be the Kottwitz triple associated to $\mathrm{f}(x')$ as in Definition 4.2.7. Then $\alpha(\gamma_0, \gamma, \delta)$ is trivial by Theorem 4.2.8, and so $O(\gamma_0, \gamma, \delta)$ in (5.1.1) is defined. The fiber of* f *containing x' has cardinality $O(\gamma_0, \gamma, \delta)$. Moreover, for any point x'_1 in this fiber, the naive local term $\mathrm{LT}^{naive}_{x'_1}$ at x'_1 is equal to $\mathrm{Tr}_{\xi}(\gamma_0)$.* □

Lemma 5.2.2 ([Kot92, § 17 and p. 441]). *Let $(\gamma_0, \gamma, \delta)$ be an $\mathbb{F}_{p^n}$-Kottwitz triple. Then the set of $y \in \mathscr{I}_n$ whose associated Kottwitz triple is $(\gamma_0, \gamma, \delta)$ has cardinality either* 0 *or* $\left|\ker^1(\mathbb{Q}, I_0)\right|$. □

Lemma 5.2.3 ([Kot92, p. 441]). *Let $(\gamma_0, \gamma, \delta)$ be an $\mathbb{F}_{p^n}$-Kottwitz triple. If $TO_\delta(\phi_n) \neq 0$, then δ satisfies condition (3) in Theorem 4.2.8.* □

5.3 The formula

We now assemble what we have done to obtain the unstabilized Langlands–Kottwitz formula.

Theorem 5.3.1 ([Kot92, (19.5)]). *Fix K^p and $g \in G(\mathbb{A}_f^p)$. Let $f^{p,\infty}$ be the characteristic function of $K^p g^{-1} K^p$. When $j \in \mathbb{N}$ is large enough, we have*

$$T(j, f^{p,\infty}) \left|\ker^1(\mathbb{Q}, G)\right|^{-1} = \sum_{(\gamma_0,\gamma,\delta)} c(\gamma_0, \gamma, \delta) O_\gamma(f^p) TO_\delta(\phi_n) \operatorname{Tr}_\xi(\gamma_0), \tag{5.3.1}$$

where the summation is over $\mathbb{F}_{p^n}$-Kottwitz triples whose Kottwitz invariants are trivial, with $n = rj$.

Proof. This follows from Lemma 2.5.1, (2.5.2), Theorem 4.2.8, Lemma 5.1.1, Lemma 5.2.1, Lemma 5.2.2, Lemma 5.2.3, and the following identity:

$$\left|\ker^1(\mathbb{Q}, G)\right| \left|\ker(\ker^1(\mathbb{Q}, G_{\gamma_0}) \to \ker^1(\mathbb{Q}, G))\right| = \ker^1(\mathbb{Q}, G_{\gamma_0}) \tag{5.3.2}$$

which is valid for any semisimple element $\gamma_0 \in G(\mathbb{Q})$. We explain the truth of (5.3.2). In fact, the pointed sets $\ker^1(\mathbb{Q}, G_{\gamma_0})$ and $\ker^1(\mathbb{Q}, G)$ have canonical structures of abelian groups (see e.g. [Kot84]), so (5.3.2) is equivalent to the surjectivity of $\ker^1(\mathbb{Q}, G_{\gamma_0}) \to \ker^1(\mathbb{Q}, G)$. But $\ker^1(\mathbb{Q}, Z_G)$ already maps onto $\ker^1(\mathbb{Q}, G)$, as shown in [Kot92, § 7]. □

Remark 5.3.2. The assumption that j is large enough is only needed for the validity of (2.5.2). If we had defined $T(j,g) = T(j, f^{p,\infty})$ to be the right hand side of (2.5.2), then Theorem 5.3.1 would be true for all $j \in \mathbb{N}_{\geq 1}$.

Remark 5.3.3. By linearity, Theorem 5.3.1 holds for arbitrary

$$f^{p,\infty} \in \mathcal{H}_{K^p} = \mathcal{H}(G(\mathbb{A}_f^p)//K^p, L(\xi)).$$

Of course then the condition that j is large enough depends on $f^{p,\infty}$.

Remark 5.3.4. As noted in the beginning of § 2.4, the quantity

$$T(j, f^{p,\infty}) \left|\ker^1(\mathbb{Q}, G)\right|^{-1}$$

is the trace on the cohomology of $\mathrm{Sh}_{K^pK_p}(G,X)_{\overline{E}}$ analogous to $T(j,g)$.

6 Stabilization

6.1 The goal

Let $f^{p,\infty} \in \mathcal{H}_{K^p} = \mathcal{H}(G(\mathbb{A}_f^p)//K^p, L(\xi))$. We fix a field embedding $L(\xi) \to \mathbb{C}$ and think of $f^{p,\infty}$ as taking values in $\mathbb{C}$. Denote by $T'(j, f^{p,\infty})$ the right hand side of (5.3.1). Namely

$$T'(j, f^{p,\infty}) := \sum_{(\gamma_0,\gamma,\delta)} c(\gamma_0,\gamma,\delta) O_\gamma(f^p) TO_\delta(\phi_n) \operatorname{Tr}_\xi(\gamma_0) \tag{6.1.1}$$

where the summation is over $\mathbb{F}_{p^n}$-Kottwitz triples with trivial Kottwitz invariant. Our next goal is to describe the stabilization of $T'(j, f^{p,\infty})$. We may and shall assume that ξ is irreducible without loss of generality. Denote by $\chi \in X^*(Z_G)$ the central character of ξ.

One may think of $T'(j, f^{p,\infty})$ as the Shimura-variety analogue of T_e, the elliptic part of Arthur–Selberg trace formula for G. The stabilization of T_e is the main topic of [Har11], and is of the following form (see [Har11, Theorem 6.2]):

$$T_e(\phi) = \sum_{(H,s,\eta)\in\mathfrak{E}} i(G,H) ST_e^{H,*}(\phi^H). \tag{6.1.2}$$

Strictly speaking, in the setting of [Har11], G is assumed to be anisotropic, while our G is only anisotropic modulo center. However for this motivational discussion we ignore this point. Here, as well as in what follows, we replace the notation ξ used in [Har11] by η. The stabilization of $T'(j, f^{p,\infty})$ is structurally very similar to (6.1.2).

We recall briefly the ingredients in (6.1.2), which will also appear in the stabilization of $T'(j, f^{p,\infty})$. The notion of an elliptic endoscopic triple for G over $\mathbb{Q}$ or $\mathbb{Q}_v$ is discussed in [Har11, § 5]. We let $\mathfrak{E}$ (resp. $\mathfrak{E}_v$) denote a fixed set of representatives of the isomorphism classes of elliptic endoscopic triples for G over $\mathbb{Q}$ (resp. $\mathbb{Q}_v$). Each element of $\mathfrak{E}$ (resp. $\mathfrak{E}_v$) is a triple (H,s,η) consisting of

- a semi-simple element $s \in \widehat{G}$
- a quasi-split reductive group H over $\mathbb{Q}$ (resp. $\mathbb{Q}_v$)
- an embedding $\eta : \widehat{H} \to \widehat{G}$ such that $\eta(\widehat{H})$ centralizes s

satisfying extra properties. Since G^{der} is simply connected, we can further extend η to an L-embedding ${}^L H \to {}^L G$, by [Lan79, Proposition 1]. We fix such a choice and still denote it by η.[8] For example, an explicit presentation of $\mathfrak{E}, \mathfrak{E}_v$ and the embeddings ${}^L H \to {}^L G$ in special cases can be found in [Mor10b, § 2.3] and [CHL11, §1.3].

Remark 6.1.1. For our specific $G = \mathrm{GU}(V, [\cdot, \cdot])$, we know that H^{der} is simply connected for all (H, s, η) in $\mathfrak{E}$ or $\mathfrak{E}_v$. This will greatly simplify the theory.

For each $(H, s, \eta) \in \mathfrak{E}$, the constant $i(G, H)$ is defined as

$$\tau(G)\tau(H)^{-1}\lambda(H, s, \eta),$$

where $\tau(G)$ and $\tau(H)$ are the Tamagawa numbers and $\lambda(H, s, \eta)$ is the order of the outer automorphism group of the triple (H, s, η). See [Har11, § 5] for details.

Let $(H, s, \eta) \in \mathfrak{E}$. The notions of (G, H)-regular semisimple elements in $H(\mathbb{Q})$, and stable orbital integrals of them, are discussed in [Har11, § 5]. Recall from [Har11, p. 30] or [Kot90, p. 189] that the *(G, H)-regular elliptic part of the stable trace formula for H* is the distribution

$$C_c^\infty(H(\mathbb{A})) \ni h \mapsto ST_e^{H,*}(h) := \tau(H) \sum_{\gamma_H} SO_{\gamma_H}(h), \tag{6.1.3}$$

where

- γ_H runs through a set of representatives of the stable conjugacy classes in $H(\mathbb{Q})$ that are elliptic and (G, H)-regular.
- each $SO_{\gamma_H}(h)$ is the stable orbital integral of h over γ_H, defined for instance on p. 21 of [Har11].

At this point we have described all the ingredients on the right hand side of (6.1.2) except ϕ^H. The function $\phi^H : H(\mathbb{A}) \to \mathbb{C}$ is defined to be any *Langlands–Shelstad transfer* of ϕ.

The desired stabilization of $T'(j, f^{p,\infty})$ is of the same form as (6.1.2), except that the test function on each H is defined differently.

Theorem 6.1.2 ([Kot90, Theorem 7.2]).

$$T'(j, f^{p,\infty}) = \sum_{(H,s,\eta)\in\mathfrak{E}} i(G, H) ST_e^{H,*}(f_H), \tag{6.1.4}$$

---

[8] In general, when G^{der} is not simply connected, the notion of an *endoscopic datum* in the sense of [LS87] refers to a suitable quadruple $(H, \mathcal{H}, s, \eta : \mathcal{H} \to {}^L G)$. Our endoscopic triple (H, s, η) equipped with the chosen extension of η amounts to an endoscopic datum $(H, \mathcal{H}, s, \eta : \mathcal{H} \to {}^L G)$ equipped with an isomorphism $\mathcal{H} \cong {}^L H$. Such an object is sometimes called an *extended endoscopic datum* in the literature, e.g. [Kal11].

where for each $(H, s, \eta) \in \mathfrak{E}$ *the function* $f_H : H(\mathbb{A}) \to \mathbb{C}$ *is a product* $f_H = f_{H,\infty} f_{H,p} f_H^{p,\infty}$. *The factor* $f_H^{p,\infty} \in C_c^\infty(H(\mathbb{A}_f^p))$ *is any Langlands–Shelstad transfer of* $f^{p,\infty}$ *(with respect to certain normalizations to be specified below). The factors* $f_{H,p} \in C_c^\infty(H(\mathbb{Q}_p))$ *and* $f_{H,\infty} \in C^\infty(H(\mathbb{Q}_p))$ *are to be explicitly specified below. The meanings of* $i(G, H)$ *and* $ST_e^{H,*}(\cdot)$ *are the same as in* (6.1.2).

Remark 6.1.3. The main point here is that the functions $f_{H,p}, f_{H,\infty}$ are specified in an explicit way that is **independent** of $f^{p,\infty}$.

Remark 6.1.4. The function $f_{H,\infty}$ is not compactly supported, but it has compact support modulo center and transforms under the split component of the center by a character. In particular the same definition (6.1.3) of $ST_e^{H,*}(f_H)$ still makes sense.

Remark 6.1.5. Even if we drop the (G, H)-regular condition and allow all elliptic elements in the summation (6.1.3), the equality (6.1.4) still holds. In fact, the function $f_{H,\infty}$ to be specified below will have the property that the resulting f_H satisfies $SO_{\gamma_H}(f_H) = 0$ for all elliptic non-(G, H)-regular γ_H. See the remark below [CHL11, (4.3.5)] and Remark 6.3.5.

Remark 6.1.6. As we mentioned in the introduction, Kottwitz's proof of Theorem 6.1.2 was conditional on certain hypotheses in local harmonic analysis, including the existence of Langlands–Shelstad transfer for the statement to even make sense. These hypotheses have been proved now, as we shall see later.

The rest of this chapter is devoted to explaining the definition of f_H and the proof of Theorem 6.1.2. For brevity we write T' for $T'(j, f^{p,\infty})$. In the following we do not need to assume that we are in the special case of unitary Shimura varieties. All we need is that $(G, X) = (G, h^{-1})$ is a Shimura datum, satisfying the two simplifying conditions in Remark 2.1.1 and the simplifying condition in Remark 6.1.1. (It is also not necessary to assume that G is anisotropic over $\mathbb{Q}$.) Of course in this generality we do not necessarily know the truth of the unstabilized Langlands–Kottwitz formula,[9] but we may and will take (6.1.1) as the definition of T' and still prove Theorem 6.1.2.

6.2 Preliminary steps

Definition 6.2.1. For H a reductive group over $\mathbb{Q}$ or $\mathbb{Q}_v$, we denote by A_H the maximal split torus in the center of H, over $\mathbb{Q}$ or $\mathbb{Q}_v$ respectively.

In particular, by Remark 2.1.1 we have $(A_G)_{\mathbb{R}} = A_{G_{\mathbb{R}}}$.

Lemma 6.2.2. *Let* $(\gamma_0, \gamma, \delta)$ *be an* $\mathbb{F}_{p^n}$*-Kottwitz triple such that* $\alpha(\gamma_0, \gamma, \delta)$ *is trivial. The quantity* $c(\gamma_0, \gamma, \delta)$ *defined in* (5.1.2) *is equal to*

$$\tau(G)\, |\mathfrak{K}(\gamma_0/\mathbb{Q})|\, \mathrm{vol}(A_G(\mathbb{R})^0 \backslash I_0^{comp}(\mathbb{R}))^{-1}.$$

Here $I_0 := G_{\gamma_0}$ *and* I_0^{comp} *is the compact-mod-center inner form of* $I_{0,\mathbb{R}}$. *Note that* I_0^{comp} *exists because* γ_0 *is* $\mathbb{R}$*-elliptic, and is unique as an isomorphism class of inner forms of* $I_{0,\mathbb{R}}$.

[9] See §7 for a discussion of the known cases.

The Haar measure on $I_0^{comp}(\mathbb{R})$ being used in the definition of $\mathrm{vol}(A_G(\mathbb{R})^0\backslash I_0^{comp}(\mathbb{R}))$ *is characterized by its transfer to $I_0(\mathbb{R})$; we normalize the latter such that its product with the chosen Haar measure on $I_0(\mathbb{A}_f)$ is the Tamagawa measure on $I_0(\mathbb{A})$.*

Proof. Let I be the inner form of I_0 as in §5.1. Then I_0^{comp} can be realized as $I_{\mathbb{R}}$. We have

$$\mathrm{vol}(I(\mathbb{Q})\backslash I(\mathbb{A}_f)) = \tau(I)\,\mathrm{vol}(A_G(\mathbb{R})^0\backslash I_0^{comp}(\mathbb{R}))^{-1}.$$

Moreover $\tau(I) = \tau(I_0)$ by [Kot88]. Hence we have

$$c(\gamma_0,\gamma,\delta) = \left|\ker(\ker^1(\mathbb{Q}, I_0) \to \ker^1(\mathbb{Q}, G))\right| \tau(I_0)\,\mathrm{vol}(A_G(\mathbb{R})^0\backslash I_0^{comp}(\mathbb{R}))^{-1}.$$

By [Kot84] and [Kot88], we have

$$\tau(G) = \left|\ker^1(\mathbb{Q}, G)\right|^{-1} \left|\pi_0(Z(\widehat{G})^\Gamma)\right|,$$

and similarly for G replaced by I_0. Thus it suffices to prove

$$\frac{\left|\ker(\ker^1(\mathbb{Q}, I_0) \to \ker^1(\mathbb{Q}, G))\right| \left|\ker^1(\mathbb{Q}, G)\right| \left|\pi_0(Z(\widehat{I_0})^\Gamma)\right|}{\left|\ker^1(\mathbb{Q}, I_0)\right| \left|\pi_0(Z(\widehat{G})^\Gamma)\right| |\mathfrak{K}(I_0/\mathbb{Q})|} = 1.$$

But this follows from the exact sequence

$$\ker^1(\mathbb{Q}, I_0) \to \ker^1(\mathbb{Q}, G) \to \mathfrak{K}(I_0/\mathbb{Q})^D \to X^*(\pi_0(Z(\widehat{I_0})^\Gamma)) \to X^*(\pi_0(Z(\widehat{G})^\Gamma)) \to 0,$$

cf. [Kot86, p. 395] for the dual version. The exactness at $X^*(\pi_0(Z(\widehat{G})^\Gamma))$ follows from the assumption that γ_0 is elliptic. □

Consequently we have

$$T' = \sum_{\substack{(\gamma_0,\gamma,\delta)\\ \alpha(\gamma_0,\gamma,\delta)=1}} \tau(G)\,|\mathfrak{K}(\gamma_0/\mathbb{Q})|\,\mathrm{vol}(A_G(\mathbb{R})^0\backslash I_0^{comp}(\mathbb{R}))^{-1} O_\gamma(f^{p,\infty}) TO_\delta(\phi_n)\,\mathrm{Tr}_\xi(\gamma_0).$$

We now insert the global Kottwitz sign

$$1 = e(I)$$

into each summand, where I is the inner form of I_0 associated to (γ_0,γ,δ) as in § 5.1. We have a product formula

$$e(I) = \prod_v e(I_{\mathbb{Q}_v})$$

with the local Kottwitz signs (see [Kot83]), where

- The sign $e(I_{\mathbb{R}}) = e(I_0^{comp})$ depends only on γ_0.
- For a finite place $v \neq p$, the sign $e(I_{\mathbb{Q}_v}) = e(G_{\gamma_v})$ depends only on the v-component γ_v of γ. We write $e(\gamma_v)$ for it. Almost all of them are equal to 1. Write $e(\gamma)$ for $\prod_{v\neq p,\infty} e(\gamma_v)$.
- The sign $e(I_{\mathbb{Q}_p}) = e(G_\delta^\sigma)$ depends only on δ. We write $e(\delta)$ for it.

Denote

$$\bar{\nu}(\gamma_0) = \bar{\nu}(I_0) := \mathrm{vol}(A_G(\mathbb{R})^0 \backslash I_0^{comp}(\mathbb{R}))e(I_0^{comp}). \tag{6.2.1}$$

Thus we have

$$T' = \sum_{\substack{(\gamma_0,\gamma,\delta) \\ \alpha(\gamma_0,\gamma,\delta)=1}} \tau(G)\, |\mathfrak{K}(\gamma_0/\mathbb{Q})|\, \bar{\nu}(\gamma_0)^{-1} e(\gamma)e(\delta)O_\gamma(f^{p,\infty})TO_\delta(\phi_n)\,\mathrm{Tr}_\xi(\gamma_0).$$

Note that in the above formula, each summand makes sense even if $\alpha(\gamma_0,\gamma,\delta) \neq 1$. Therefore by Fourier inversion on the finite abelian group $\mathfrak{K}(\gamma_0)$, we have

$$T' = \tau(G) \sum_{\gamma_0} \sum_{\kappa \in \mathfrak{K}(\gamma_0/\mathbb{Q})} \bar{\nu}(\gamma_0)^{-1}\,\mathrm{Tr}_\xi(\gamma_0) \sum_{(\gamma,\delta)} e(\gamma)e(\delta)O_\gamma(f^{p,\infty})TO_\delta(\phi_n)\langle\alpha(\gamma_0,\gamma,\delta),\kappa\rangle, \tag{6.2.2}$$

where

- γ_0 runs through a set of representatives for the $\mathbb{R}$-elliptic stable conjugacy classes in $G(\mathbb{Q})$.
- γ runs through a set of representatives for the conjugacy classes in $G(\mathbb{A}_f^p)$ that are stably conjugate to γ_0.
- δ runs through a set of representatives for the σ-conjugacy classes in $G(L_n)$ of which γ_0 is an n-th norm, and which satisfy the condition (3) in Definition 4.1.1.
- In the summation there are only finitely many nonzero terms.

The identity (6.2.2) is [Kot90, (4.2)]. Note the similarity between (6.2.2) and [Har11, (4.8)], but also note that the places ∞ and p play very different roles in the two situations.

6.3 The archimedean place

Let $(H, s, \eta) \in \mathfrak{E}_\infty$. We assume that *an elliptic maximal torus of* $G_\mathbb{R}$ *comes from* H. We explain the meaning of this assumption.

Throughout the whole theory of endoscopy for the group $G_\mathbb{R}$, we need to fix a quasi-split reductive group G^* over $\mathbb{R}$ together with an inner twisting $\psi : G^* \to G_\mathbb{R}$. For any maximal torus T' of $H_\mathbb{C}$, there is a canonical $G^*(\mathbb{C})$-conjugacy class $\mathscr{E}_{T'}$ of embeddings $T' \to G^*_\mathbb{C}$. Now suppose $T_H \subset H$ is a maximal torus defined over $\mathbb{R}$. By an *admissible embedding* of T_H into $G_\mathbb{R}$, we mean an embedding $j : T_H \hookrightarrow G_\mathbb{R}$ defined over $\mathbb{R}$, which is the composition of an element in $\mathscr{E}_{T_{H,\mathbb{C}}}$ with ψ. In general T_H may not have any admissible embedding into $G_\mathbb{R}$. For more details see [LS87, § 1] or [Ren11, § 2].

We can now explain the meaning of our assumption: We require that there exists a maximal torus T_H of H defined over $\mathbb{R}$, and an admissible embedding $j : T_H \hookrightarrow G_\mathbb{R}$, such that $T_G := j(T_H)$ is an **elliptic** maximal torus of $G_\mathbb{R}$. In particular, T_H is an elliptic maximal torus of H and $A_H \cong A_{G_\mathbb{R}}$. Note that the existence of elliptic maximal tori in H is already a non-trivial assumption on H in general. We fix such a triple (T_H, T_G, j).

The definition of $f_{H,\infty}$ depends on the choice of (T_H, T_G, j) as above, as well as the choice of a Borel $B_{G,H}$ of $G_{\mathbb{C}}$ containing $T_{G,\mathbb{C}}$. Let $(T_H, T_G, j, B_{G,H})$ be fixed. In the following we denote this quadruple simply by $(j, B_{G,H})$. Then $(j, B_{G,H})$ determines an embedding of root systems

$$j_* : R(H_{\mathbb{C}}, T_{H,\mathbb{C}}) \to R(G_{\mathbb{C}}, T_{G,\mathbb{C}}),$$

and determines a Borel B_H of $H_{\mathbb{C}}$ containing T_H by pulling back the $B_{G,H}$-positive roots along j_*. Via j_* we view the Weyl group $\Omega_H := \Omega(H_{\mathbb{C}}, T_{H,\mathbb{C}})$ of $H_{\mathbb{C}}$ as a subgroup of the Weyl group $\Omega := \Omega(G_{\mathbb{C}}, T_{G,\mathbb{C}})$ of $G_{\mathbb{C}}$. We also have a subset Ω_* of Ω, consisting of elements $\omega \in \Omega$ such that $(\omega \circ j, B_{G,H})$ still determines the Borel B_H. The multiplication induces a bijection

$$\Omega_H \times \Omega_* \to \Omega$$

known as the *Kostant decomposition*.

Let ξ^* be the contragredient representation of ξ. Let φ_{ξ^*} be the elliptic Langlands parameter of $G_{\mathbb{R}}$ corresponding to ξ^*, so that the L-packet of φ_{ξ^*} is the packet of (essentially) discrete series representations of $G(\mathbb{R})$ whose central and infinitesimal characters are the same as those of ξ^*. Let $\Phi_H(\varphi_{\xi^*})$ be the set of equivalence classes of elliptic Langlands parameters φ_H of H such that $\eta \circ \varphi_H$ is equivalent to φ_{ξ^*}, where η is the fixed L-embedding ${}^LH \to {}^LG$. As on p. 185 of [Kot90], we have a bijection

$$\Phi_H(\varphi_{\xi^*}) \xrightarrow{\sim} \Omega_*$$
$$\varphi_H \mapsto \omega_*(\varphi_H),$$

characterized by the condition that φ_H is *aligned* with

$$(\omega_*(\varphi_H)^{-1} \circ j, B_{G,H}, B_H).$$

We refer to [Kot90, pp. 184–185] for the meaning of the last condition. For general expositions of the Langlands–Shelstad theory of L-packets of discrete series representations over $\mathbb{R}$, we refer the reader to [Ada11] or [Taï14, § 4.2.1] and the references therein.

For any $\varphi_H \in \Phi_H(\varphi_{\xi^*})$, define the *averaged pseudo-coefficient*

$$f_{\varphi_H} := |\Pi(\varphi_H)|^{-1} \sum_{\pi \in \Pi(\varphi_H)} f_\pi,$$

where $\Pi(\varphi_H)$ is the L-packet of φ_H, and $f_\pi \in C^\infty(H(\mathbb{R}))$ is a normalized pseudo-coefficient for π, constructed in [CD85]. Note that the normalization of f_π depends on the choice of a Haar measure on $H(\mathbb{R})$. We also know that each f_π, and hence f_{φ_H}, transform under $A_H(\mathbb{R})^0 \cong A_{G_{\mathbb{R}}}(\mathbb{R})^0$ by the central character χ^{-1} of ξ^*.

Definition 6.3.1. Fix the choice of $(j, B_{G,H})$ as above and fix a Haar measure on $H(\mathbb{R})$. Define

$$f_{H,\infty} := (-1)^{q(G_{\mathbb{R}})} \langle \mu_{T_G}, s \rangle_j \sum_{\varphi_H \in \Phi_H(\varphi_{\xi^*})} \det(\omega_*(\varphi_H)) f_{\varphi_H}.$$

Here μ_{T_G} is the Hodge cocharacter of any $G(\mathbb{R})$-conjugate of h that factors through T_G. The pairing $\langle \mu_{T_G}, s\rangle_j \in \mathbb{C}^\times$ is defined as follows: We use j to view μ_{T_G} as a cocharacter of T_H, and view s as an element of $Z(\widehat{H})$. Then we use the canonical pairing

$$X_*(T_H) \times Z(\widehat{H}) \to X^*(Z(\widehat{H})) \times Z(\widehat{H}) \to \mathbb{C}^\times$$

to define $\langle \mu_{T_G}, s\rangle_j$. The integer $q(G_\mathbb{R})$ is half the real dimension of the symmetric space of $G(\mathbb{R})$.

On the other hand, it is a result of Shelstad (reviewed by Kottwitz on p. 184 of [Kot90]) that for the chosen $(j, B_{G,H})$, there is a normalization of the transfer factors between H and $G_\mathbb{R}$ (see [Har11]) denoted by $\Delta_{j,B_{G,H}}$, such that for all (G, H)-regular elements $\gamma_H \in T_H(\mathbb{R})$ the values $\Delta_{j,B_{G,H}}(\gamma_H, j(\gamma_H))$ are given by an explicit formula. We do not record this explicit formula, but only remark that it involves the Langlands correspondence for tori over $\mathbb{R}$.

Remark 6.3.2. In practice, it is sometimes necessary to compare the normalization $\Delta_{j,B_{G,H}}$ with other normalizations. For example, we obtain another normalization by realizing $G_\mathbb{R}$ as a pure or rigid inner form of $G^*_\mathbb{R}$ and by choosing a Whittaker datum on $G^*_\mathbb{R}$, see the introduction of [Kal16]. In principle, such a comparison can be done by comparing the shapes of the associated endoscopic character identities. See [Zhu18] for a special case of such a computation.

Definition 6.3.1 is motivated by the following computation of stable orbital integrals:

Proposition 6.3.3 ([Kot90, (7.4)]). *Let $\gamma_H \in H(\mathbb{R})$ be a semi-simple (G, H)-regular element. If γ_H is not elliptic (i.e. not in any elliptic maximal torus), then $SO_{\gamma_H}(f_{H,\infty}) = 0$. Suppose γ_H is elliptic. Then*

$$SO_{\gamma_H}(f_{H,\infty}) = \langle \mu_{T_G}, s\rangle_j \, \Delta_{j,B_{G,H}}(\gamma_H, \gamma_0)\bar{v}(\gamma_0)^{-1}\,\mathrm{Tr}_\xi(\gamma_0).$$

Here $\gamma_0 \in G(\mathbb{R})$ is any element such that γ_H is an image of γ_0 (see [Har11, § 5] for the notion of "image"). The term $\bar{v}(\gamma_0)$ is defined as in (6.2.1), *with respect to the Haar measure transferred from the one on $H_{\gamma_H}(\mathbb{R})$ (where H_{γ_H} is an inner form of I_0^{comp}) that is used to define SO_{γ_H}.* □

Remark 6.3.4. Note that γ_0 in the above proposition exists. In fact, since all elliptic maximal tori of H are conjugate under $H(\mathbb{R})$, we may first conjugate γ_H into T_H and then map it to G via j, and the image is an example of γ_0.

Remark 6.3.5. As a special case of an observation of Morel [Mor11, Remarque 3.2.6], we have $SO_{\gamma_H}(f_{H,\infty}) = 0$ for any semi-simple $\gamma_H \in H(\mathbb{R})$ that is not (G, H)-regular.

6.4 Away from p, ∞

Away from p, ∞, the ingredients we need from local harmonic analysis are the same as those in the stabilization of T_e discussed in [Har11], namely the Langlands–Shelstad Transfer Conjecture and the Fundamental Lemma. They are now proved theorems thanks to the work of Ngô [Ngô10], Waldspurger [Wal97] [Wal06], Cluckers-Loeser [CL10], and Hales

[Hal95]. See [Har11, § 6] for more historical remarks. In the following we state these two proved conjectures and their adelic consequence. The statements get simplified by the fact that G and all its local and global elliptic endoscopic groups H have simply connected derived groups in our specific case.

Theorem 6.4.1 (Langlands–Shelstad Transfer Conjecture). *Let v be a finite place of $\mathbb{Q}$ and let $(H, s, \eta) \in \mathfrak{E}_v$. Fix a normalization of the transfer factors between H and $G_{\mathbb{Q}_v}$. Fix Haar measures on $G(\mathbb{Q}_v)$ and $H(\mathbb{Q}_v)$. For any $f \in C_c^\infty(G(\mathbb{Q}_v))$, there exists $f^H \in C_c^\infty(H(\mathbb{Q}_v))$, called a* Langlands–Shelstad transfer *of f, satisfying the following properties. For any semi-simple (G, H)-regular element $\gamma_H \in H(\mathbb{Q}_v)$, we have*

$$SO_{\gamma_H}(f^H) = \begin{cases} 0, & \gamma_H \text{ is not an image from } G(\mathbb{Q}_v) \\ \Delta(\gamma_H, \gamma) O_\gamma^\kappa(f), & \gamma_H \text{ is an image of a semisimple } \gamma \in G(\mathbb{Q}_v) \end{cases} \tag{6.4.1}$$

where in the second situation,

- *the element*

$$\kappa \in \mathfrak{K}(G_\gamma/\mathbb{Q}_v) = \pi_0(Z(\widehat{G_\gamma})^{\Gamma_v})/\pi_0(Z(\widehat{G})^{\Gamma_v})$$

  is the natural image of s (see [Kot86]). The κ-orbital integral O_γ^κ is defined as in [Kot86, § 5] or [Har11, (4.9)].

- *$SO_{\gamma_H}(f^H)$ and $O_\gamma^\kappa(f)$ are defined with respect to the fixed Haar measures on $G(\mathbb{Q}_v)$ and $H(\mathbb{Q}_v)$, and compatible Haar measures on $G_\gamma(\mathbb{Q}_v)$ and $H_{\gamma_H}(\mathbb{Q}_v)$.*

- *$\Delta(\gamma_H, \gamma) \in \mathbb{C}$ is the transfer factor.* □

Theorem 6.4.2 (Fundamental Lemma). *In the situation of Theorem 6.4.1, assume G is unramified over $\mathbb{Q}_v$. Also assume (H, s, η) is unramified, meaning that H is unramified over $\mathbb{Q}_v$ and that $\eta : {}^L H \to {}^L G$ is unramified at v, in the sense that the restriction of η to the inertia group at v (as a subgroup of ${}^L H$) is trivial. Fix Haar measures on $G(\mathbb{Q}_v)$ and $H(\mathbb{Q}_v)$ giving volume 1 to hyperspecial subgroups. Let K (resp. K_H) be an arbitrary hyperspecial subgroup of $G(\mathbb{Q}_v)$ (resp. $H(\mathbb{Q}_v)$). Then 1_{K_H} is a Langlands–Shelstad transfer of 1_K as in Theorem 6.4.1, for the canonical unramified normalization of transfer factors defined in [Hal93].* □

We now state an adelic consequence of the above theorems. Let $(H, s, \eta) \in \mathfrak{E}$. Then it is unramified at almost all places. Suppose for almost all places v at which G and (H, s, η) are unramified, that the transfer factors at v are normalized under the canonical unramified normalization.

Corollary 6.4.3. *For any $f \in C_c^\infty(G(\mathbb{A}_f^p))$, there exists $f^H \in C_c^\infty(H(\mathbb{A}_f^p))$ such that the $\mathbb{A}_f^p$-analogue of* (6.4.1) *holds. Here the meaning of an adelic (G, H)-regular element is as on pp. 178–179 of [Kot90], and all the orbital integrals are defined with respect to adelic Haar measures.* □

Remark 6.4.4. The original Langlands–Shelstad Transfer Conjecture is only stated for G-regular elements. The more general conjecture with (G, H)-regular elements is reduced to that version by Langlands–Shelstad [LS90, § 2.4]. The same remark applies to the Fundamental Lemma.

Definition 6.4.5. Let $(H, s, \eta) \in \mathfrak{E}$. Assume that an elliptic maximal torus of $G_{\mathbb{R}}$ comes from $H_{\mathbb{R}}$. Also assume that the reductive group H is unramified over $\mathbb{Q}_p$ (but we do not assume that η is unramified over $\mathbb{Q}_p$). Fix the Haar measure on $G(\mathbb{A}_f^p)$ giving volume 1 to K^p. Fix the Haar measure on $H(\mathbb{A}_f^p)$ such that its product with the chosen Haar measure on $H(\mathbb{R})$ in Definition 6.3.1 and the Haar measure on $H(\mathbb{Q}_p)$ giving volume 1 to hyperspecial subgroups is equal to the Tamagawa measure on $H(\mathbb{A})$. Normalize the transfer factors between H and G at all places away from p and ∞ in a way that satisfies the hypothesis in Corollary 6.4.3. We define $f_H^{p,\infty} \in C_c^\infty(H(\mathbb{A}_f^p))$ to be a Langlands–Shelstad transfer of $f^{p,\infty}$ as in Corollary 6.4.3.

6.5 The place p

At the place p, the ingredient needed from local harmonic analysis is a variant of the Fundamental Lemma. It is a special case of the Twisted Fundamental Lemma, in the situation known as the "unstable cyclic base change". We shall state it in the form needed here, after recalling some facts. Again, the statement is simplified by the fact that G and all its endoscopic groups H over $\mathbb{Q}_p$ have simply connected derived groups.

Let k be a finite extension of $\mathbb{Q}_p$. Let H_1 be an unramified reductive group over k. Then by the Satake isomorphism we know that the Hecke algebras $\mathcal{H}(H_1(k)//K)$, for different hyperspecial subgroups K of $H_1(k)$, are canonically isomorphic to each other. We denote them by the common notation $\mathcal{H}^{\mathrm{ur}}(H_1(k))$, called the *unramified Hecke algebra*. Moreover, let H_2 be another such reductive group and let ${}^L H_1 \to {}^L H_2$ be an unramified L-morphism. Then we have an associated algebra map $\mathcal{H}^{\mathrm{ur}}(H_2(k)) \to \mathcal{H}^{\mathrm{ur}}(H_1(k))$, defined using the Satake isomorphism. We refer the reader to [Car79], [Bor79], [HR10], [ST16] for expositions of the Satake isomorphism.

Fix $(H, s, \eta) \in \mathfrak{E}_p$. Assume temporarily the following two conditions:

(1) (H, s, η) is unramified.

(2) The element s, when viewed as an element of $Z(\widehat{H})$, is fixed by Γ_p.

Let $R := \mathrm{Res}_{L_n/\mathbb{Q}_p} G$. Then R is unramified over $\mathbb{Q}_p$. On pp. 179-180 of [Kot90], Kottwitz uses $\eta : {}^L H \to {}^L G$ to construct an unramified L-morphism ${}^L H \to {}^L R$, which will serve as a component of a *twisted endoscopic datum*. We then obtain a map

$$b : \mathcal{H}^{\mathrm{ur}}(R(\mathbb{Q}_p)) = \mathcal{H}^{\mathrm{ur}}(G(L_n)) \to \mathcal{H}^{\mathrm{ur}}(H(\mathbb{Q}_p)).$$

Theorem 6.5.1 (Fundamental Lemma for Unstable Cyclic Base Change). *Let $(H, s, \eta) \in \mathfrak{E}_p$ be an element that satisfies the above two temporary assumptions. For any $f \in \mathcal{H}^{\mathrm{ur}}(G(L_n))$ and for any semi-simple (G, H)-regular element $\gamma_H \in H(\mathbb{Q}_p)$, we have*

$$SO_{\gamma_H}(b(f)) = \sum_{\delta} \langle \beta(\gamma_0, \delta), s \rangle \Delta(\gamma_H, \gamma_0) e(\delta) TO_\delta(f), \qquad (6.5.1)$$

where

- $\gamma_0 \in G(\mathbb{Q}_p)$ *is any semi-simple element such that* γ_H *is an image of it. It always exists because* $G_{\mathbb{Q}_p}$ *is quasi-split.*
- δ *runs through a set of representatives for the* σ*-conjugacy classes in* $G(L_n)$ *such that* γ_0 *is an n-th norm of* δ.
- $\beta(\gamma_0, \delta)$ *is the character* $Z(\widehat{G_{\gamma_0}})^{\Gamma_p} Z(\widehat{G}) \to \mathbb{C}^\times$ *defined in the same way as* β_p *at the end of § 4.1. The pairing* $\langle \beta(\gamma_0, \delta), s \rangle$ *is defined in the obvious way.*
- $\Delta(\gamma_H, \gamma_0)$ *is the transfer factor under the canonical unramified normalization.*
- SO_{γ_H} *and* TO_δ *are defined with respect to the Haar measures on* $H(\mathbb{Q}_p)$ *and* $G(L_n)$ *giving volume* 1 *to hyperspecial subgroups, and compatible Haar measures on* $H_{\gamma_H}(\mathbb{Q}_p)$ *and* $G_\delta^\sigma(\mathbb{Q}_p)$.

Proof. This is [Kot90, (7.3)] (with a more general function f here). In [Kot90] it is shown that [Kot90, (7.3)] is equivalent to [Kot90, (7.2)]. In [Mor10b, Appendix A], Kottwitz reduces [Kot90, (7.2)] to a **special case** of the usual form of the Twisted Fundamental Lemma, and he also reduces the (G, H)-regular case to the G-regular case. In [Mor10b, § 9], Morel shows how the desired statement can be further reduced to the Twisted Fundamental Lemma for the unit element in the unramified Hecke algebra.

The general Twisted Fundamental Lemma is known for the unit element in the unramified Hecke algebra for large p, thanks to the work of Ngô [Ngô10], Waldspurger [Wal06] [Wal08], Cluckers-Loeser [CL10] and others. By the recent work of Lemaire, Moeglin, and Waldspurger [LMW15] [LW15], it is also known for general elements of the unramified Hecke algebra, including the case of small p. □

Corollary 6.5.2. *Let* $(H, s, \eta) \in \mathfrak{E}_p$. *We assume that the reductive group* H *is unramified over* $\mathbb{Q}_p$, *but we drop the temporary assumptions (1) (2) above. Fix an arbitrary normalization of transfer factors between* H *and* $G_{\mathbb{Q}_p}$. *For any* $f \in \mathcal{H}^{\mathrm{ur}}(G(L_n))$, *there exists a function* $b'(f) \in \mathcal{H}^{\mathrm{ur}}(H(\mathbb{Q}_p))$, *such that the conclusion of Theorem 6.5.1 holds for* $b'(f)$.

Proof. This follows from Theorem 6.5.1, see [Kot90, p. 181]. □

Definition 6.5.3. Let $(H, s, \eta) \in \mathfrak{E}$. Assume that H is unramified over $\mathbb{Q}_p$ and that an elliptic maximal torus of $G_{\mathbb{R}}$ comes from $H_{\mathbb{R}}$. Take the normalization of transfer factors between $H_{\mathbb{Q}_p}$ and $G_{\mathbb{Q}_p}$ such that its product with the normalization $\Delta_{j,B_{G,H}}$ at ∞ and the normalization away from p, ∞ in Definition 6.4.5 satisfies the global product formula (see [LS87, § 6]). We define $f_{H,p}$ to be $b'(\phi_n)$ as in Corollary 6.5.2, where ϕ_n is defined in § 5.1.

6.6 The final step

Let $(H, s, \eta) \in \mathfrak{E}$. If it does not satisfy the assumptions in Definition 6.5.3, we define $f_H := 0$. Otherwise we define $f_H := f_{H,\infty} f_H^{p,\infty} f_{H,p}$, with $f_{H,\infty}$, $f_H^{p,\infty}$, $f_{H,p}$ defined in Definitions 6.3.1, 6.4.5, 6.5.3 respectively.

Consider a pair (γ_0, κ) appearing on the right hand side of (6.2.2). Assume that the summand indexed by it is non-zero. Recall from [Kot86, § 9] or [Har11, § 5] that (γ_0, κ)

determines a set of *endoscopic quadruples* $\pi^{-1}(\gamma_0, \kappa)$. Each element of $\pi^{-1}(\gamma_0, \kappa)$ is an *endoscopic quadruple* of the form (H, s, η, γ_H), with $(H, s, \eta) \in \mathfrak{E}$ and $\gamma_H \in H(\mathbb{Q})$ a semi-simple (G, H)-regular element that is an image of γ_0. In particular, γ_H is $\mathbb{R}$-elliptic, and an elliptic maximal torus of $G_{\mathbb{R}}$ comes from $H_{\mathbb{R}}$. Moreover it can be checked that $H_{\mathbb{Q}_p}$ is necessarily unramified, as on p. 189 of [Kot90]. Assembling Proposition 6.3.3, Corollary 6.4.3, Corollary 6.5.2, and using the global product formula of transfer factors, it is easy to see that $SO_{\gamma_H}(f_H)$ is equal to the summand in (6.2.2) indexed by (γ_0, κ). Then by the same counting argument as in [Har11, § 5] and the following easy observation, we know that the right hand side of (6.2.2) is equal to the right hand side of (6.1.4), and Theorem 6.1.2 is proved. The observation is that if $\delta \in G(L_n)$ is an element that violates condition (3) in Definition 4.1.1, then $TO_\delta(\phi_n) = 0$.

Q.E.D. for Theorem 6.1.2.

7 Final remarks

7.1 What is after stabilization?

The next step after Theorem 6.1.2 may be called "destabilization". This amounts to expanding the right hand side of (6.1.4) spectrally, in terms of automorphic representations, and then rewriting everything in terms of G alone, so that the endoscopic groups H's no longer appear. In this way, one would obtain a description of the cohomology of the Shimura variety in terms of the conjectural relation between Galois representations and automorphic representations. In fact, a suitable implementation of the destabilization step could be used to construct Galois representations attached to automorphic representations.

Kottwitz outlines in [Kot90, Part II] the destabilization, assuming Arthur's multiplicity conjectures and some related hypotheses on the stable trace formula. For the special case of unitary Shimura varieties, a lot has been known with the aid of the base change of automorphic representations from the unitary group to GL_n. We refer the reader to [Mor10b], [Shi11], and [CHL11].

For cases beyond the unitary groups, the hypotheses made in [Kot90, Part II] have been largely verified for quasi-split classical groups, by Arthur [Art13] (for symplectic and special orthogonal groups) and Mok [Mok15] (for unitary groups). Extensions of these results to non-quasi-split inner forms are available in certain cases, see [KMSW14] and [Taï15], but the general case is still open. We remark that for the theory of Shimura varieties it is usually insufficient to understand only quasi-split groups, as the existence of a Shimura datum often forces the group to be non-quasi-split already at infinity. In [Zhu18], the results from [Art13] and [Taï15] are applied to the problem of destabilization for some special cases of orthogonal Shimura varieties, where the special orthogonal groups in question have signature $(n, 2)$ at infinity and are not quasi-split.

7.2 What is known beyond PEL type?

An important circle of ideas that is closely related to Kottwitz's work [Kot92] is the Langlands–Rapoport conjecture [LR87] (later corrected by Reimann [Rei97]). This conjecture gives a group theoretic description of the reduction of the Shimura variety at hyperspecial primes. It is essentially shown in [LR87] that this conjecture implies the unstabilized Langlands–Kottwitz formula (5.3.1) (cf. Remark 5.3.4). Recently, a modified version of the Langlands–Rapoport conjecture has been proved by Kisin [Kis17] for all abelian-type Shimura varieties. See [Kis17] for more historical remarks and references. In an ongoing project [KSZ], it is expected that the results in [Kis17] can be refined in such a way as to imply the generalization of the unstabilized Langlands–Kottwitz formula for all abelian-type Shimura varieties. Also the stabilization procedure is carried out in [KSZ] in this generality, with the usual group theoretic assumptions removed. Thus the expected main result of [KSZ] will generalize the stabilized formula (6.1.4) from PEL type to abelian type.

7.3 What is beyond the compact and good reduction case?

Kottwitz's work has also been generalized to some non-compact Shimura varieties, where the intersection cohomology of the Baily–Borel compactification turns out to be the correct object to compare with Arthur–Selberg trace formulas. See [LR92], [Lau97], [Mor10a], [Mor10b], [Mor11], and [Zhu18]. In an orthogonal direction of generalization, the bad reduction of Shimura varieties is an extremely active and fruitful field of study, with profound arithmetic consequences which are not obtainable by studying good reductions only. See e.g. [HT01], [Rap05], [Hai05], [Shi11], [Man], [Shi], [Sch], and the references therein.

References

[Ada11] Jeffrey Adams. Discrete series and characters of the component group. In *On the stabilization of the trace formula*, volume 1 of *Stab. Trace Formula Shimura Var. Arith. Appl.*, pages 369–387. Int. Press, Somerville, MA, 2011.

[Art13] James Arthur. *The endoscopic classification of representations*, volume 61 of *American Mathematical Society Colloquium Publications*. American Mathematical Society, Providence, RI, 2013. Orthogonal and symplectic groups.

[Bor79] A. Borel. Automorphic L-functions. In *Automorphic forms, representations and L-functions (Proc. Sympos. Pure Math., Oregon State Univ., Corvallis, Ore., 1977), Part 2*, Proc. Sympos. Pure Math., XXXIII, pages 27–61. Amer. Math. Soc., Providence, R.I., 1979.

[Car79] P. Cartier. Representations of p-adic groups: a survey. In *Automorphic forms, representations and L-functions (Proc. Sympos. Pure Math., Oregon State Univ., Corvallis, Ore., 1977), Part 1*, Proc. Sympos. Pure Math., XXXIII, pages 111–155. Amer. Math. Soc., Providence, R.I., 1979.

[Car86] Henri Carayol. Sur les représentations l-adiques associées aux formes modulaires de Hilbert. *Ann. Sci. École Norm. Sup. (4)*, 19(3):409–468, 1986.

[CD85] Laurent Clozel and Patrick Delorme. Pseudo-coefficients et cohomologie des groupes de Lie réductifs réels. *C. R. Acad. Sci. Paris Sér. I Math.*, 300(12):385–387, 1985.

[CHL11] Laurent Clozel, Michael Harris, and Jean-Pierre Labesse. Construction of automorphic Galois representations, I. In *On the stabilization of the trace formula*, volume 1 of *Stab. Trace Formula Shimura Var. Arith. Appl.*, pages 497–527. Int. Press, Somerville, MA, 2011.

[CL10] Raf Cluckers and François Loeser. Constructible exponential functions, motivic Fourier transform and transfer principle. *Ann. of Math. (2)*, 171(2):1011–1065, 2010.

[Del79] Pierre Deligne. Variétés de Shimura: interprétation modulaire, et techniques de construction de modèles canoniques. In *Automorphic forms, representations and L-functions (Proc. Sympos. Pure Math., Oregon State Univ., Corvallis, Ore., 1977), Part 2*, Proc. Sympos. Pure Math., XXXIII, pages 247–289. Amer. Math. Soc., Providence, R.I., 1979.

[Fuj97] Kazuhiro Fujiwara. Rigid geometry, Lefschetz-Verdier trace formula and Deligne's conjecture. *Invent. Math.*, 127(3):489–533, 1997.

[GN] Alain Genestier and Bao Châu Ngô. Chapter in this volume.

[Hai02] Thomas J. Haines. On connected components of Shimura varieties. *Canad. J. Math.*, 54(2):352–395, 2002.

[Hai05] Thomas J. Haines. Introduction to Shimura varieties with bad reduction of parahoric type. In *Harmonic analysis, the trace formula, and Shimura varieties*, volume 4 of *Clay Math. Proc.*, pages 583–642. Amer. Math. Soc., Providence, RI, 2005.

[Hal93] Thomas C. Hales. A simple definition of transfer factors for unramified groups. In *Representation theory of groups and algebras*, volume 145 of *Contemp. Math.*, pages 109–134. Amer. Math. Soc., Providence, RI, 1993.

[Hal95] Thomas C. Hales. On the fundamental lemma for standard endoscopy: reduction to unit elements. *Canad. J. Math.*, 47(5):974–994, 1995.

[Har11] Michael Harris. An introduction to the stable trace formula. In *On the stabilization of the trace formula*, volume 1 of *Stab. Trace Formula Shimura Var. Arith. Appl.*, pages 3–47. Int. Press, Somerville, MA, 2011.

[HR10] Thomas J. Haines and Sean Rostami. The Satake isomorphism for special maximal parahoric Hecke algebras. *Represent. Theory*, 14:264–284, 2010.

[HT01] Michael Harris and Richard Taylor. *The geometry and cohomology of some simple Shimura varieties*, volume 151 of *Annals of Mathematics Studies*. Princeton University Press, Princeton, NJ, 2001. With an appendix by Vladimir G. Berkovich.

[Kal11] Tasho Kaletha. Endoscopic character identities for depth-zero supercuspidal L-packets. *Duke Math. J.*, 158(2):161–224, 2011.

[Kal16] Tasho Kaletha. Rigid inner forms of real and p-adic groups. *Ann. of Math. (2)*, 184(2):559–632, 2016.

[Kis17] Mark Kisin. Mod p points on Shimura varieties of abelian type. *J. Amer. Math. Soc.*, 30(3):819–914, 2017.

[KMSW14] T. Kaletha, A. Minguez, S. W. Shin, and P.-J. White. Endoscopic Classification of Representations: Inner Forms of Unitary Groups. *ArXiv e-prints*, September 2014.

[Kot82] Robert E. Kottwitz. Rational conjugacy classes in reductive groups. *Duke Math. J.*, 49(4):785–806, 1982.

[Kot83] Robert E. Kottwitz. Sign changes in harmonic analysis on reductive groups. *Trans. Amer. Math. Soc.*, 278(1):289–297, 1983.

[Kot84] Robert E. Kottwitz. Stable trace formula: cuspidal tempered terms. *Duke Math. J.*, 51(3):611–650, 1984.

[Kot85] Robert E. Kottwitz. Isocrystals with additional structure. *Compositio Math.*, 56(2):201–220, 1985.

[Kot86] Robert E. Kottwitz. Stable trace formula: elliptic singular terms. *Math. Ann.*, 275(3):365–399, 1986.

[Kot88] Robert E. Kottwitz. Tamagawa numbers. *Ann. of Math. (2)*, 127(3):629–646, 1988.

[Kot90] Robert E. Kottwitz. Shimura varieties and λ-adic representations. In *Automorphic forms, Shimura varieties, and L-functions, Vol. I (Ann Arbor, MI, 1988)*, volume 10 of *Perspect. Math.*, pages 161–209. Academic Press, Boston, MA, 1990.

[Kot92] Robert E. Kottwitz. Points on some Shimura varieties over finite fields. *J. Amer. Math. Soc.*, 5(2):373–444, 1992.

[KSZ] Mark Kisin, Sug Woo Shin, and Yihang Zhu. The stable trace formula for certain Shimura varieties of abelian type. in preparation.

[Lan79] R. P. Langlands. Stable conjugacy: definitions and lemmas. *Canad. J. Math.*, 31(4):700–725, 1979.

[Lau97] Gérard Laumon. Sur la cohomologie à supports compacts des variétés de Shimura pour $GSp(4)_{\mathbb{Q}}$. *Compositio Math.*, 105(3):267–359, 1997.

[LMW15] B. Lemaire, C. Moeglin, and J.-L. Waldspurger. Le lemme fondamental pour l'endoscopie tordue: réduction aux éléments unités. *ArXiv e-prints*, June 2015.

[LR87] R. P. Langlands and M. Rapoport. Shimuravarietäten und Gerben. *J. Reine Angew. Math.*, 378:113–220, 1987.

[LR92] Robert P. Langlands and Dinakar Ramakrishnan, editors. *The zeta functions of Picard modular surfaces*. Université de Montréal, Centre de Recherches Mathématiques, Montreal, QC, 1992.

[LS87] R. P. Langlands and D. Shelstad. On the definition of transfer factors. *Math. Ann.*, 278(1-4):219–271, 1987.

[LS90] R. Langlands and D. Shelstad. Descent for transfer factors. In *The Grothendieck Festschrift, Vol. II*, volume 87 of *Progr. Math.*, pages 485–563. Birkhäuser Boston, Boston, MA, 1990.

[LW15] B. Lemaire and J.-L. Waldspurger. Le lemme fondamental pour l'endoscopie tordue: le cas où le groupe endoscopique non ramifié est un tore. *ArXiv e-prints*, November 2015.

[Man] Elena Mantovan. Chapter in this volume.

[Mil05] J. S. Milne. Introduction to Shimura varieties. In *Harmonic analysis, the trace formula, and Shimura varieties*, volume 4 of *Clay Math. Proc.*, pages 265–378. Amer. Math. Soc., Providence, RI, 2005.

[Mok15] Chung Pang Mok. Endoscopic classification of representations of quasi-split unitary groups. *Mem. Amer. Math. Soc.*, 235(1108):vi+248, 2015.

[Mor10a] Sophie Morel. The intersection complex as a weight truncation and an application to Shimura varieties. In *Proceedings of the International Congress of Mathematicians. Volume II*, pages 312–334, New Delhi, 2010. Hindustan Book Agency.

[Mor10b] Sophie Morel. *On the cohomology of certain noncompact Shimura varieties*, volume 173 of *Annals of Mathematics Studies*. Princeton University Press, Princeton, NJ, 2010. With an appendix by Robert Kottwitz.

[Mor11] Sophie Morel. Cohomologie d'intersection des variétés modulaires de Siegel, suite. *Compos. Math.*, 147(6):1671–1740, 2011.

[Ngô10] Bao Châu Ngô. Le lemme fondamental pour les algèbres de Lie. *Publ. Math. Inst. Hautes Études Sci.*, (111):1–169, 2010.

[Nic] Marc-Hubert Nicole. Chapter in this volume.

[Pin92a] Richard Pink. On l-adic sheaves on Shimura varieties and their higher direct images in the Baily-Borel compactification. *Math. Ann.*, 292(2):197–240, 1992.

[Pin92b] Richard Pink. On the calculation of local terms in the Lefschetz-Verdier trace formula and its application to a conjecture of Deligne. *Ann. of Math. (2)*, 135(3):483–525, 1992.

[Rap05] Michael Rapoport. A guide to the reduction modulo p of Shimura varieties. *Astérisque*, (298):271–318, 2005. Automorphic forms. I.

[Rei97] Harry Reimann. *The semi-simple zeta function of quaternionic Shimura varieties*, volume 1657 of *Lecture Notes in Mathematics*. Springer-Verlag, Berlin, 1997.

[Ren11] David Renard. Endoscopy for real reductive groups. In *On the stabilization of the trace formula*, volume 1 of *Stab. Trace Formula Shimura Var. Arith. Appl.*, pages 95–141. Int. Press, Somerville, MA, 2011.

[Roz] Sandra Rozensztajn. Chapter in this volume.

[RZ88] Harry Reimann and Thomas Zink. Der Dieudonnémodul einer polarisierten abelschen Mannigfaltigkeit vom CM-Typ. *Ann. of Math. (2)*, 128(3):461–482, 1988.

[Sch] Peter Scholze. Chapter in this volume.

[Shi] Sug Woo Shin. Chapter in this volume.

[Shi11] Sug Woo Shin. Galois representations arising from some compact Shimura varieties. *Ann. of Math. (2)*, 173(3):1645–1741, 2011.

[ST16] Sug Woo Shin and Nicolas Templier. Sato-Tate theorem for families and low-lying zeros of automorphic L-functions. *Invent. Math.*, 203(1):1–177, 2016. Appendix A by Robert Kottwitz, and Appendix B by Raf Cluckers, Julia Gordon and Immanuel Halupczok.

[Taï14] O. Taïbi. Dimensions of spaces of level one automorphic forms for split classical groups using the trace formula. *ArXiv e-prints*, June 2014.

[Taï15] O. Taïbi. Arthur's multiplicity formula for certain inner forms of special orthogonal and symplectic groups. *ArXiv e-prints*, October 2015.

[Var05] Yakov Varshavsky. A proof of a generalization of Deligne's conjecture. *Electron. Res. Announc. Amer. Math. Soc.*, 11:78–88, 2005.

[Wal97] J.-L. Waldspurger. Le lemme fondamental implique le transfert. *Compositio Math.*, 105(2):153–236, 1997.

[Wal06] J.-L. Waldspurger. Endoscopie et changement de caractéristique. *J. Inst. Math. Jussieu*, 5(3):423–525, 2006.

[Wal08] J.-L. Waldspurger. L'endoscopie tordue n'est pas si tordue. *Mem. Amer. Math. Soc.*, 194(908):x+261, 2008.

[Zhu18] Yihang Zhu. The stabilization of the Frobenius–Hecke traces on the intersection cohomology of orthogonal Shimura varieties. *ArXiv e-prints*, January 2018.

INTEGRAL CANONICAL MODELS OF SHIMURA VARIETIES: AN UPDATE

MARK KISIN

Department of Mathematics, Harvard University
(email: kisin@math.harvard.edu)

1 Shimura varieties

The aim of these notes is to provide an introduction to the subject of integral canonical models of Shimura varieties, as well as to explain some more recent developments, since [Ki 3] and [Ki 4] were written. Most of these notes are an updated version of [Ki 3], and their aim is to sketch a proof of the existence of such models for Shimura varieties of Hodge and, more generally, abelian type. In the final section we explain some recent developments on models for Shimura varieties with parahoric level structuredue to Pappas and the author.

It is a pleasure to thank B. Conrad, O. Gabber, B. Moonen and A. Vasiu for helpful comments regarding various versions of these notes.

1.1 Shimura data

We recall the definition of a Shimura datum and the associated Shimura variety [De 1, §2.1]. Let G be a connected reductive group over $\mathbb{Q}$ and X a $G(\mathbb{R})$-conjugacy class of maps of algebraic groups over $\mathbb{R}$

$$h : \mathbb{S} = \mathrm{Res}_{\mathbb{C}/\mathbb{R}}\,\mathbb{G}_m \to G_{\mathbb{R}}.$$

On $\mathbb{R}$-points such a map induces a map of real groups $\mathbb{C}^\times \to G(\mathbb{R})$.

The pair (G, X) is required to satisfy the following conditions:

(1) Let $\mathfrak{g}$ denote the Lie algebra of $G_{\mathbb{R}}$. We require that the composite

$$\mathbb{S} \to G_{\mathbb{R}} \to G_{\mathbb{R}}^{\mathrm{ad}} \to \mathrm{GL}(\mathfrak{g})$$

defines a Hodge structure of type $(-1,1), (0,0), (1,-1)$. This means that under the action of $\mathbb{C}^\times$ on $\mathfrak{g}_{\mathbb{C}} = \mathfrak{g} \otimes_{\mathbb{R}} \mathbb{C}$ we have a decomposition

$$\mathfrak{g}_{\mathbb{C}} = V^{-1,1} \oplus V^{0,0} \oplus V^{1,-1}$$

where $z \in \mathbb{C}^\times$ acts on $V^{p,q}$ via $z^{-p}\bar{z}^{-q}$.

(2) Conjugation by $h(i)$ induces a Cartan involution of $G_{\mathbb{R}}^{\mathrm{ad}}$ (note that $\mathrm{ad}h(-1) = 1$ on $\mathfrak{g}$ so $h(i)$ induces an involution of $G_{\mathbb{R}}^{\mathrm{ad}}$). This means that we require the real form of G defined by the involution $g \mapsto h(i)\bar{g}h(i)^{-1}$ to be compact.

(3) G^{ad} has no factor defined over $\mathbb{Q}$ whose real points form a compact group.

The author was partially supported by NSF grant DMS-1601054

Parts of this chapter first appeared in M. Kisin, Integral canonical models of Shimura varieties (Journees Arithmetiques 2007), *J. Th. Nombres Bordeaux* 21(2) (2009).

The second condition implies that for any $h_0 \in X$ the stabilizer $K_\infty \subset G^{\text{ad}}(\mathbb{R})$ (acting by conjugation) of h_0 is compact and $G^{\text{ad}}(\mathbb{R})/K_\infty \xrightarrow{\sim} X$ has a complex structure.

A pair (G, X) satisfying the above conditions is called a *Shimura datum*. A morphism $(G_1, X_1) \to (G_2, X_2)$ of Shimura data is a map of groups $G_1 \to G_2$, which induces a map $X_1 \to X_2$.

Now let $K \subset G(\mathbb{A}_f)$ be a compact open subgroup, where $\mathbb{A}_f$ denotes the finite adeles over $\mathbb{Q}$. Then a theorem of Baily-Borel asserts that

$$\mathrm{Sh}_K(G, X) = G(\mathbb{Q})\backslash X \times G(\mathbb{A}_f)/K$$

has a natural structure of an algebraic variety over $\mathbb{C}$. Results of Shimura, Deligne, Milne and others imply that $\mathrm{Sh}_K(G, X)$ has a model over a number field $E = E(G, X)$ - the reflex field - which does not depend on K [Mi 2, §4,5]. We will again denote by $\mathrm{Sh}_K(G, X)$ this algebraic variety over $E(G, X)$.

Fix a prime p. We will sometimes consider the pro-variety

$$\mathrm{Sh}_{K_p}(G, X) = \varprojlim \mathrm{Sh}_K(G, X),$$

where K runs through compact open subgroups of the form $K_p K^p$, with $K_p \subset G(\mathbb{Q}_p)$ fixed, and $K^p \subset G(\mathbb{A}_f^p)$ compact open. Here $\mathbb{A}_f^p$ denotes the adeles with trivial component at p.

A morphism of Shimura data $(G_1, X_1) \to (G_2, X_2)$ induces a morphism of the corresponding Shimura varieties $\mathrm{Sh}_{K_1}(G_1, X_1) \to \mathrm{Sh}_{K_2}(G_2, X_2)$, provided the compact open subgroups are chosen so that K_1 maps into K_2.

1.2 Examples

(1) Let $G = \mathrm{GL}_2$, and let X be the $\mathrm{GL}_2(\mathbb{R})$ orbit (or equivalently the $\mathrm{PGL}_2(\mathbb{R})$-orbit) of

$$h_0 : \mathbb{C}^\times \to G(\mathbb{R}); \quad a + ib \mapsto \begin{pmatrix} a & -b \\ b & a \end{pmatrix}.$$

Then the map $\mathrm{ad}(g) \cdot h_0 \mapsto g \cdot i$ identifies X with the upper and lower half planes $\mathcal{H}^\pm$, compatibly with the action of $\mathrm{GL}_2(\mathbb{R})$. Here $\mathrm{GL}_2(\mathbb{R})$ acts on $\mathcal{H}^\pm$ in the usual way, by Möbius transformations.

(2) Fix a $\mathbb{Q}$-vector space V with a perfect alternating pairing ψ. Take $G = \mathrm{GSp}(V, \psi)$ the corresponding group of symplectic similitudes, and let $X = S^\pm$ be the Siegel double space, defined as the set of maps $h : \mathbb{S} \to G_{\mathbb{R}}$ such that

(1) The $\mathbb{C}^\times$ action on $V_{\mathbb{R}}$ gives rise to a Hodge structure of type $(-1, 0), (0, -1)$:

$$V_{\mathbb{C}} \xrightarrow{\sim} V^{-1,0} \oplus V^{0,-1}.$$

(2) $(x, y) \mapsto (x, h(i)y)$ is (positive or negative) definite on $V_{\mathbb{R}}$.

If $V_{\mathbb{Z}} \subset V$ is a $\mathbb{Z}$-lattice, and $h \in S^\pm$, then $V^{-1,0}/V_{\mathbb{Z}}$ is an abelian variety, and $\mathrm{Sh}_K(G, X) = \mathrm{Sh}_K(\mathrm{GSp}, S^\pm)$ has an interpretation as a moduli space for abelian varieties.

(3) One may use families of abelian varieties to study a more general class of Shimura varieties. A *Siegel embedding* $(G, X) \hookrightarrow (\mathrm{GSp}(V, \psi), S^{\pm})$ is an injective map $G \hookrightarrow \mathrm{GSp}(V, \psi)$ which induces a (necessarily injective) map $X \hookrightarrow S^{\pm}$. A Shimura datum (G, X) which admits a Siegel embedding is called of *Hodge type*. The resulting Shimura varieties $\mathrm{Sh}_K(G, X)$ carry families of abelian varieties equipped with certain Hodge cycles.

(4) There is a larger class of Shimura varieties which are related to moduli of abelian varieties in a more indirect manner. For any (G, X) the *adjoint* Shimura datum is the pair $(G^{\mathrm{ad}}, X^{\mathrm{ad}})$ where X^{ad} is the $G^{\mathrm{ad}}(\mathbb{R})$ orbit of a map $h^{\mathrm{ad}} : \mathbb{S} \to G^{\mathrm{ad}}(R)$ induced by some $h \in X$. A Shimura datum (G, X) is called of *abelian type* if there exists a central isogeny $G_1^{\mathrm{der}} \to G^{\mathrm{der}}$ inducing an isomorphism $(G_1^{\mathrm{ad}}, X_2^{\mathrm{ad}}) \xrightarrow{\sim} (G^{\mathrm{ad}}, X^{\mathrm{ad}})$.

Problems for Shimura varieties of abelian type can often be reduced to the case of Hodge type, and ultimately to the moduli of principally polarized abelian varieties. For example, this is how Deligne [De 1], following Shimura, constructed canonical models for Shimura varieties of abelian type.

This class includes Shimura varieties for orthogonal groups. These include moduli of polarized $K3$ surfaces as a special case.

2 Integral Canonical models

2.1 The Langlands-Milne conjecture

In this section we explain a conjecture on the existence of integral models with good reduction for Shimura varieties with hyperspecial level structure.

2.1.1

A compact open subgroup $K_p \subset G(\mathbb{Q}_p)$ is called *hyperspecial* if there exists a reductive group $\mathcal{G}$ over $\mathbb{Z}_p$ extending $G_{\mathbb{Q}_p}$ and such that $K_p = \mathcal{G}(\mathbb{Z}_p)$. A hyperspecial subgroup is a maximal compact open subgroup. Such subgroups exist if G is quasi-split at p and split over an unramified extension.

Fix a finite prime p, and a compact open subgroup $K = K_p K^p \subset G(\mathbb{A}_f)$, with K_p, K^p as in 1.1 above. Write $E = E(G, X)$, and let $\mathcal{O} \subset E$ be its ring of integers. For a prime $\lambda | p$ of $E(G, X)$ denote by $\mathcal{O}_{(\lambda)}$ the localization of $\mathcal{O}$ at λ.

The following was conjectured in a rough form by Langlands [La] and made precise by Milne [Mi 1].

Conjecture 2.1.1 (Langlands-Milne). *If K_p is hyperspecial, then for $\lambda | p$, $\mathrm{Sh}_K(G, X)$ has an integral canonical model over $\mathcal{O}_{(\lambda)}$.*

2.1.2

The conjecture is a statement about the tower of Shimura varieties

$$\mathrm{Sh}_{K_p}(G, X) = \varprojlim \mathrm{Sh}_{K'}(G, X)$$

where $K' = K'_p K'^p$ runs over compact open subgroups with $K'_p = K_p$. The group $G(\mathbb{A}_f^p)$ of points in the ring of finite adeles with trivial p-component acts on the pro-scheme $\mathrm{Sh}_{K_p}(G, X)$. The term *integral canonical model* in the conjecture means, in particular, an extension of the tower of E-schemes $\mathrm{Sh}_{K'}(G, X)$ with its $G(\mathbb{A}_f^p)$-action to a tower of *smooth* $\mathcal{O}_{(\lambda)}$ schemes with $G(\mathbb{A}_f^p)$-action.

On its own the condition in the previous paragraph is vacuous, since it is satisfied by the tower $\mathrm{Sh}_{K_p}(G, X)$ itself! We need another condition which expresses the integrality of the extension. If the Shimura varieties $\mathrm{Sh}_{K'}(G, X)$ happen to be proper, then we can simply insist that the extension consist of a tower of proper $\mathcal{O}_{(\lambda)}$-schemes as in [La].

In the non-proper case Milne observed that one can still formulate an extension property by using the whole tower: Namely we can require that the tower $\mathrm{Sh}_{K_p}(G, X)$ satisfy the valuative criterion with respect to any discrete valuation ring R of mixed characteristic $(0, p)$. That is, any $R[1/p]$-valued point of the tower extends to an R-valued point. (We remark however, that even when $\mathrm{Sh}_{K'}(G, X)$ is proper, Milne's condition does not obviously imply that the integral canonical models are proper. For the models whose construction will be explained below this was proved by Madapusi-Pera [MP]).

2.1.3

We can see why this might be a reasonable definition if we look at the example of principally polarized abelian varieties (PPAV's), and take for R the strict henselisation of $\mathbb{Z}_{(p)}$ in a fixed algebraic closure $\bar{\mathbb{Q}}$ of $\mathbb{Q}$ (to check the valuative criterion one may always replace R by a strict henselisation). The moduli theoretic description of the moduli space of PPAV's gives rise to a natural integral model of it which is smooth at primes where K_p is maximal. An $R[1/p]$-valued point of the tower gives rise to a PPAV $\mathcal{A}$ over $R[1/p]$ together with a basis of the ℓ-adic Tate module $T_\ell\mathcal{A}$ where $\ell \neq p$. The action of $\mathrm{Gal}(\bar{\mathbb{Q}}/R[1/p])$ on $T_\ell\mathcal{A}$ is trivial, and so in particular the action of an inertia subgroup at p is trivial. This implies that $\mathcal{A}$ has good reduction, so the point extends to an R-valued point of the tower.

2.1.4

We have already explained most of the features of the precise conjecture. The only difference is that Milne's version of the extension property is formulated for a more general class of schemes than discrete valuation rings (see also [Mo, §3.5]). This has the effect that one is able to show that if the integral canonical model exists then it is unique.

For PEL Shimura varieties, this conjecture was established in by Zink and Langlands-Rapoport [Zi], [LR]. The conjecture for Shimura varieties of abelian type is claimed in Vasiu's papers [Va 1], [Va 2], as well as the more recent [Va 4], [Va 5]. Here we will explain the approach of [Ki 3] to the case of abelian type.

2.2 Examples

(1) Take $(G, X) = (\mathrm{GSp}, S^\pm)$, the Siegel Shimura datum defined by the symplectic space (V, ψ) as above. If we fix the $\mathbb{Z}$-lattice $V_\mathbb{Z} \subset V$, then we may consider the tower $\mathrm{Sh}_{K_p}(G, X)$ where K_p is the maximal compact open subgroup which leaves $V_{\mathbb{Z}_p} = V_\mathbb{Z} \otimes_\mathbb{Z} \mathbb{Z}_p \subset V \otimes \mathbb{Q}_p$ stable.

The group K_p is hyperspecial if and only if ψ induces a perfect, $\mathbb{Z}_p$-valued pairing on $V_{\mathbb{Z}_p}$, in which case it may be identified with the $\mathbb{Z}_p$-points of the group of symplectic similitudes defined by $(V_{\mathbb{Z}_p}, \psi)$, which is a reductive group over $\mathbb{Z}_p$.

Any choice of $V_{\mathbb{Z}}$ gives rise to an interpretation of $\mathrm{Sh}_K(G,X)$ as a moduli space for polarized abelian varieties, and hence to a model $\mathscr{S}_K(G,X)$ for $\mathrm{Sh}_K(G,X)$ over the ring of integers of E. The $G(\mathbb{A}_f^p)$-action extends to this model, and it will satisfy the extension property for discrete valuation rings, as explained above. On the other hand Milne's stronger form of the extension property requires more work, even in the smooth case. See [Mo, 3.6] and the references therein.

However the varieties $\mathrm{Sh}_K(G,X)$ will be *smooth* over $\mathcal{O}_{(\lambda)}$ if and only if the degree of the polarization in the moduli problem is prime to p. This corresponds to the condition that ψ induces a perfect pairing on $V_{\mathbb{Z}_p}$.

(2) Another example is given by Hilbert modular varieties, for which $G = \mathrm{Res}_{F/\mathbb{Q}}\,\mathrm{GL}_2$ for F a totally real field. If K_p is maximal compact, then it is conjugate to $\prod_{\lambda|p} \mathrm{GL}_2(\mathcal{O}_{F,\lambda})$. This is hyperspecial if and only if F is unramified at p. The corresponding integral canonical models were constructed by Deligne-Pappas [DP], and are indeed smooth if and only if F is unramified at p.

2.3 Results

The theorem below was proved in [Ki 3] with some restrictions when $p = 2$. These were removed by Kim-Madapusi Pera.

Theorem 2.3.1. *If (G,X) is of Hodge type, and $K_p \subset G(\mathbb{Q}_p)$ is hyperspecial, then the tower* $\mathrm{Sh}_{K_p}(G,X)$ *admits a canonical integral model.*

If $(G,X) \hookrightarrow (\mathrm{GSp}S^{\pm})$ is any Siegel embedding, then the corresponding extension of $\mathscr{S}_{K_p}(G,X)$ is given by taking the normalization of the closure of $\mathrm{Sh}_{K_p}(G,X)$ *in* $\mathscr{S}_{K'_p}(\mathrm{GSp}, S^{\pm})$ *for a suitable choice of lattice $V_{\mathbb{Z}} \subset V$ and $K'_p \subset \mathrm{GSp}(\mathbb{Q}_p)$ the stabilizer of $V_{\mathbb{Z}_p}$.*

Using the theorem one can deduce the existence of integral canonical models for Shimura varieties of abelian type - see 1.2(4) above.

Corollary 2.3.2. *Suppose that K_p is hyperspecial, and that (G,X) is of abelian type. Then* $\mathrm{Sh}_{K_p}(G,X)$ *has an integral canonical model.*

2.4 Remarks

(1) Deligne has given an explicit description of the $\mathbb{Q}$-simple Shimura data which satisfy the condition of the corollary. They include the cases when G is of type A, B, C and certain cases of type D [De 1, 2.3.10].

(2) As remarked above, Madapusi Pera has proved that when $\mathrm{Sh}_K(G,X)$ is proper - which is equivalent to the condition that G is anisotropic over $\mathbb{Q}$, then $\mathscr{S}_K(G,X)$ is proper.

3 Proof of the results

3.1 The set up

We now sketch some of the key arguments from [Ki 3] which go into the proof of Theorem 2.3.1.

3.1.1

We put ourselves in the situation of the theorem. Let $\mathcal{G}$ be a reductive group over $\mathbb{Z}_p$, which extends G and such that $K_p = \mathcal{G}(\mathbb{Z}_p)$. If $V_{\mathbb{Z}_p} \subset V_{\mathbb{Q}_p}$ is a $\mathbb{Z}_p$-lattice, denote by $\mathrm{GL}(V_{\mathbb{Z}_p})$ the $\mathbb{Z}_p$-group scheme of automorphisms of $V_{\mathbb{Z}_p}$. If $V_{\mathbb{Z}_p}$ is stable by K_p, then we obviously have

$$K_p \subset \mathrm{GL}(V_{\mathbb{Z}_p})(\mathbb{Z}_p) = \mathrm{Aut}_{\mathbb{Z}_p} V_{\mathbb{Z}_p}.$$

Unfortunately this does not quite imply that $G \hookrightarrow \mathrm{GSp}(V, \psi) \subset \mathrm{GL}(V)$ extends to an embedding $\mathcal{G} \subset \mathrm{GL}(V_{\mathbb{Z}_p})$. However one can choose $V_{\mathbb{Z}_p}$ so that we do in fact have such embedding (cf. [PY, Cor. 1.3], [Va 2, 3.1.2.1]).

3.1.2

Choose a $\mathbb{Z}$-lattice $V_{\mathbb{Z}} \subset V_{\mathbb{Q}}$ such that $V_{\mathbb{Z}} \otimes_{\mathbb{Z}} \mathbb{Z}_p = V_{\mathbb{Z}_p}$. We will consider $K' = K'_p K'^p \subset \mathrm{GSp}(\mathbb{A}_f)$ compact open, as above, so that $K \subset K'$, K' leaves $V_{\mathbb{Z}} \otimes \prod_\ell \mathbb{Z}_\ell$ stable, and K'_p is the stabilizer of $V_{\mathbb{Z}_p}$ in $\mathrm{GSp}(\mathbb{Q}_p)$. The moduli theoretic interpretation of $\mathrm{Sh}_{K'}(\mathrm{GSp}, S^{\pm})$ then gives rise to a natural extension of this scheme to a $\mathbb{Z}_{(p)}$-scheme $\mathscr{S}_{K'}(\mathrm{GSp}, S^{\pm})$.

The compact open subgroups K and K' can be so arranged that $i : \mathrm{Sh}_K(G, X) \to \mathrm{Sh}_{K'}(\mathrm{GSp}, S^{\pm})$ is a closed embedding (cf. [De 1, 1.15]). We assume this and write $\mathscr{S}_K(G, X)$ for the normalization of the closure of $\mathrm{Sh}_K(G, X)$ in $\mathscr{S}_{K'}(\mathrm{GSp}, S^{\pm})$.

3.1.3

Since the tower $\varprojlim_{L'} \mathscr{S}_{L'}(\mathrm{GSp}, S^{\pm})$ with $L'_p = K'_p$ fixed, is an integral canonical model for $\mathrm{Sh}_{K'_p}(\mathrm{GSp}, S^{\pm})$, one sees that the tower $\varprojlim_L \mathrm{Sh}_L(G, X)$ with $L_p = K_p$ has an action of $G(\mathbb{A}_f^p)$ and satisfies the extension property. (At least as far as it was explained in § 2). We have to show that $\mathscr{S}_K(G, X)$ is smooth over primes $\lambda | p$. We will describe it in a formal neighborhood of a closed point $\bar{x}$ of its mod λ fibre.

3.2 The key lemma

3.2.1

Let $x \in \mathrm{Sh}_K(G, X)$ be a closed point with residue field $\kappa(x)$, specializing to $\bar{x}$ in the mod λ fibre of $\mathscr{S}_K(G, X)$, and $\mathcal{A}_x$ the corresponding abelian variety over $\kappa(x)$. We will again denote by λ an extension of our chosen place of E to $\kappa(x)$. The p-adic Tate module $T_p\mathcal{A}_x$ may be canonically identified with $V_{\mathbb{Z}_p}$ up to the action of K_p on $V_{\mathbb{Z}_p}$. Fix such an identification. Deligne's theorem that a Hodge cycle on $\mathcal{A}_x$ is absolutely Hodge [De 2] implies that the representation

$$\rho_x : G_x := \mathrm{Gal}(\bar{\kappa}(x)/\kappa(x)) \to \mathrm{Aut}_{\mathbb{Z}_p} T_p\mathcal{A}_x$$

factors through $G(\mathbb{Q}_p)$ (this also follows from the earlier results of Piatetski-Shapiro and Borovoi), and hence through $G(\mathbb{Q}_p) \cap \mathrm{Aut}_{\mathbb{Z}_p} V_{\mathbb{Z}_p} = K_p$.

3.2.2

It will be convenient to make the following convention. If R is a ring and M is a free R-module we write $M^{\otimes}$ for the direct sum of all the R-modules formed from M by taking tensor products, symmetric powers, exterior powers and duals in all possible combinations. A collection of tensors $(s_\alpha) \subset M^{\otimes}$ defines a closed subgroup of $\mathrm{GL}(M)$, namely the closed subgroup which fixes each s_α. Note that $M^{\otimes} \xrightarrow{\sim} M^{*\otimes}$ so that a tensor in the left hand side may be regarded in the right hand side.

Note that the closure of G in $\mathrm{GL}(V_{\mathbb{Z}_{(p)}})$ (here $V_{\mathbb{Z}_{(p)}} = V_{\mathbb{Z}} \otimes_{\mathbb{Z}} \mathbb{Z}_{(p)}$) is a group whose base change to $\mathbb{Z}_p$ is $\mathcal{G}$. In particular, this group is reductive, and one can show that it is defined by a family of tensors $(s_\alpha) \subset V_{\mathbb{Z}_{(p)}}^{\otimes}$. Since the action of G_x on $T_p\mathcal{A}_x \xrightarrow{\sim} V_{\mathbb{Z}_p}$ factors through $K_p = \mathcal{G}(\mathbb{Z}_p)$, the s_α are all Galois invariant.

Let $\mathbb{D}_x$ be the Dieudonné module of $\mathcal{A}_x$. If $\kappa(\bar{x})$ denotes the residue field of $\bar{x}$, then $\mathbb{D}_x$ is a finite free $W(\kappa(\bar{x}))$-module. Let $G_{x,\lambda} \subset G_x$ denote a decomposition group at λ.

The p-adic comparison isomorphism is a canonical isomorphism

$$V_{\mathbb{Z}_p}^* \otimes_{\mathbb{Z}_p} B_{\mathrm{cris}} \xrightarrow{\sim} \mathbb{D}_x \otimes B_{\mathrm{cris}}$$

which is compatible with the actions of $G_{x,\lambda}$, the Frobenius φ on the two sides, as well as with filtrations after tensoring by $\kappa(x)_\lambda$. Since the s_α are $G_{x,\lambda}$ invariant, they are in $\mathbb{D}_x^{\otimes} \otimes_{\mathbb{Z}_p} \mathbb{Q}_p$. when regarded as elements of the right hand side of this isomorphism. They are of course invariant by φ and lie in $\mathrm{Fil}^0(\mathbb{D}_x^{\otimes} \otimes_{W(\kappa(\bar{x}))} \kappa(x)_\lambda)$.

Key Lemma 3.2.3. $(s_\alpha) \subset \mathbb{D}_x^{\otimes}$ *(not just after inverting p) and this collection of tensors defines a reductive subgroup of* $\mathrm{GL}(\mathbb{D}_x)$.

3.3 From the key lemma to integral canonical models

3.3.1

It was known to experts that a statement like the key lemma should allow the construction of integral canonical models for Shimura varieties of Hodge type. The point is that using the collection of tensors in the lemma, one can use a construction of Faltings [Fa] to define a deformation problem involving p-divisible groups. This can then be shown to describe the complete local ring of $\mathrm{Sh}_K(G, X)$ at $\bar{x}$. We sketch the argument here (cf. [Va 2, §5] and [Mo, 5.8]).

3.3.2

Let H be the p-divisible group of the mod p reduction $\mathcal{A}_{\bar{x}}$ of $\mathcal{A}_x$. This is the p-divisible group over $\kappa(\bar{x})$ attached to $\mathbb{D}_x$. Let $\hat{U} = \mathrm{Spf}\, R$ be the versal deformation space of H, and $\hat{H}$ a versal p-divisible group over $\hat{U}$. Then $\hat{U}$ is formally smooth, and we fix a lift of Frobenius on R.

The Lie algebra of the universal vector extension of $\hat{H}$ gives rise to a vector bundle $\mathbb{D}(\hat{H})$ over $\hat{U}$, which is equipped with a a connection, an action of Frobenius and a filtration.

A construction of Faltings [Fa, p136] provides a closed, formally smooth, formal subscheme $\hat{U}_{\mathcal{G}} \subset U$ over which the tensors (s_α) extend to parallel, Frobenius invariant sections of $\mathrm{Fil}^0(\mathbb{D}(\hat{H})^\otimes)$. Here the formal smoothness of $\hat{U}_{\mathcal{G}}$ relies on the fact that $\mathcal{G}$ is reductive.

3.3.3

The composite of the inclusions $G \subset \mathrm{GSp}(V_{\mathbb{Q}}, \psi) \subset \mathrm{GL}(V_{\mathbb{Q}})$ of reductive groups over $\mathbb{Q}$ is defined by the tensors $s_\alpha \in V_{\mathbb{Z}_{(p)}}^\otimes \subset V_{\mathbb{Q}}^\otimes$. Let $\mathcal{V}$ denote the de Rham cohomology of the universal polarized abelian variety over $\mathrm{Sh}_K(G, X) \subset \mathrm{Sh}_{K'}(\mathrm{GSp}, S^\pm)$. Then the s_α define tensors in the de Rham cohomology of the universal abelian variety over $S^\pm$, and since they are invariant by $G(\mathbb{Q})$, their restrictions to X descend to $s_{\alpha,\mathrm{dR}} \in \mathrm{Fil}^0(\mathcal{V}^\otimes)$.

Let $\hat{M}$ be the formal neighborhood of $\bar{x} \in \mathscr{S}_K(G, X)$, and let $\hat{N} \subset \hat{M}$, be the irreducible component which contains the point x. One can show that the restrictions of $s_{\alpha,\mathrm{dR}}$ to $\hat{N}$ define parallel, Frobenius invariant sections in $\mathrm{Fil}^0(\mathbb{D}(\hat{H})^\otimes)$, which extend $s_\alpha \in \mathbb{D}_x$. The fact that these sections are Frobenius invariant follows from a result of Blasius and Wintenberger [Mo, 5.6.3].

3.3.4

One can show that a map $\hat{N} \to \hat{U}$ which induces the p-divisible group of the abelian variety over $\hat{N}$ factors through $\hat{U}_{\mathcal{G}}$. This plausible since over both spaces the tensors (s_α) extend to parallel, Frobenius invariant sections in Fil^0. However the proof of this result actually requires the Key Lemma.

The map $\varepsilon : \hat{N} \to \hat{U}_{\mathcal{G}}$ is a closed embedding, because a deformation of H determines a deformation of $\mathcal{A}_{\bar{x}}$, and if the given polarization of $\mathcal{A}_{\bar{x}}$ lifts to this deformation, then it does so uniquely. On the other hand one can check that $\hat{\mathcal{N}}$ and $\hat{U}_{\mathcal{G}}$ have the same dimension. Since $\hat{U}_{\mathcal{G}}$ is formally smooth, we find that ε is an isomorphism.

4 Proof of the key lemma

4.1 Classification of crystalline representations

4.1.1

We recall some of the results of [Ki 1] regarding the classification of crystalline representations and p-divisible groups.

Let k be a perfect field of characteristic p, $W = W(k)$ its ring of Witt vectors and $K_0 = W(k)[1/p]$. Let K be a finite totally ramified extension of K_0, and $\mathcal{O}_K$ its ring of integers. Fix an algebraic closure $\bar{K}$ of K, and set $G_K = \mathrm{Gal}(\bar{K}/K)$.

We denote by $\mathrm{Rep}_{G_K}^{\mathrm{cris}}$ the category of crystalline G_K-representations, and by $\mathrm{Rep}_{G_K}^{\mathrm{cris}\,\circ}$ the category of G_K-stable lattices which span a representation in $\mathrm{Rep}_{G_K}^{\mathrm{cris}}$. For V a crystalline representation recall Fontaine's functors

$$D_{\mathrm{cris}}(V) = (B_{\mathrm{cris}} \otimes_{\mathbb{Q}_p} V)^{G_K} \quad \textit{and} \quad D_{\mathrm{dR}}(V) = (B_{\mathrm{dR}} \otimes_{\mathbb{Q}_p} V)^{G_K}.$$

Fix a uniformiser $\pi \in K$, and let $E(u) \in W(k)[u]$ be the Eisenstein polynomial for π. We set $\mathfrak{S} = W[\![u]\!]$ equipped with a Frobenius φ which acts as the usual Frobenius on W and sends u to u^p.

Let $\mathrm{Mod}^{\varphi}_{/\mathfrak{S}}$ denote the category of finite free $\mathfrak{S}$-modules $\mathfrak{M}$ equipped with a Frobenius semi-linear isomorphism

$$\varphi^*(\mathfrak{M})[1/E(u)] \xrightarrow{\sim} \mathfrak{M}[1/E(u)].$$

Note that this definition differs slightly from that of [Ki 1], where we insisted that the above map be induced by a map $\varphi^*(\mathfrak{M}) \to \mathfrak{M}$. This is related to the fact that in *loc. cit* we considered crystalline representations with non-negative Hodge-Tate weights, whereas here we will allow arbitrary Hodge-Tate weights.

Let $\mathcal{O}_{\mathcal{E}}$ denote the p-adic completion of $\mathfrak{S}_{(p)}$. If $\mathfrak{M}$ is in $\mathrm{Mod}^{\varphi}_{/\mathfrak{S}}$ then $M = \mathcal{O}_{\mathcal{E}} \otimes_{\mathfrak{S}} \mathfrak{M}$ is a finite free $\mathcal{O}_{\mathcal{E}}$-module equipped with an isomorphism $\varphi^*(M) \xrightarrow{\sim} M$.

Theorem 4.1.2. *There exists a fully faithful tensor functor*

$$\mathfrak{M} : \mathrm{Rep}^{\mathrm{cris}\,\circ}_{G_K} \to \mathrm{Mod}^{\varphi}_{/\mathfrak{S}}.$$

If L is in $\mathrm{Rep}^{\mathrm{cris}\,\circ}_{G_K}$, $V = L \otimes_{\mathbb{Z}_p} \mathbb{Q}_p$, and $\mathfrak{M} = \mathfrak{M}(L)$, then

(1) *There are canonical isomorphisms*

$$D_{\mathrm{cris}}(V) \xrightarrow{\sim} \varphi^*(\mathfrak{M}/u\mathfrak{M})[1/p] \quad \textit{and} \quad D_{\mathrm{dR}}(V) \xrightarrow{\sim} \mathfrak{M} \otimes_{\mathfrak{S}} K.$$

Here the first isomorphism is compatible with Frobenius and in the second the map $\mathfrak{S} \to K$ is given by sending $u \mapsto \pi$.

(2) *There is a canonical isomorphism*

$$\mathcal{O}_{\widehat{\mathcal{E}^{\mathrm{ur}}}} \otimes_{\mathbb{Z}_p} L \xrightarrow{\sim} \mathcal{O}_{\widehat{\mathcal{E}^{\mathrm{ur}}}} \otimes_{\mathfrak{S}} \mathfrak{M},$$

where $\mathcal{O}_{\widehat{\mathcal{E}^{\mathrm{ur}}}}$ is the completion of the strict henselisation of $\mathcal{O}_{\mathcal{E}}$.

(3) *If G is a p-divisible group over $\mathcal{O}_K$ and $L = (T_pG)^*$ is the dual of its p-adic Tate module, then $\mathfrak{M}$ is stable by φ, and there is a canonical φ-equivariant isomorphism*

$$\mathbb{D}(G) \xrightarrow{\sim} \varphi^*(\mathfrak{M}/u\mathfrak{M}).$$

The theorem follows result follows immediately from [Ki 1], with the restriction that $p > 2$ if G is not connected in (3). This restriction was independently removed by Kim [Kim], Liu [Li], and Lau [Lau].

4.2 Reductive groups

4.2.1

We now prove the key lemma. We will apply the above theory with $K = \kappa(x)_\lambda$ and $L = V^*_{\mathbb{Z}_p}$. Recall that $\mathcal{G} \subset \mathrm{GL}(V^*_{\mathbb{Z}_p})$ denotes the reductive group defined by (s_α). We may view the tensors s_α as morphisms $s_\alpha : \mathbf{1} \to L^\otimes$ in $\mathrm{Rep}^{\mathrm{cris}\,\circ}_{G_K}$. Applying the functor $\mathfrak{M}$ of the theorem, we obtain morphisms $\tilde{s}_\alpha : \mathbf{1} \to \mathfrak{M}^\otimes$ in $\mathrm{Mod}^{\varphi}_{/\mathfrak{S}}$.

Note that the theorem immediately implies the first part of the key lemma, since specializing the $(\tilde{s}_\alpha)$ at $u = 0$ produces the tensors $s_\alpha \in \mathbb{D}_x[1/p]^\otimes$, which lie in $\mathbb{D}_x^\otimes \xrightarrow{\sim} \varphi^*(\mathfrak{M}/u\mathfrak{M})^\otimes$ by construction.

We will show that the $(\tilde{s}_\alpha)$, define a reductive subgroup of $\mathrm{GL}(\mathfrak{M})$. This will complete the proof of Theorem 2.3.1. Denote this closed subgroup by $\mathcal{G}_\mathfrak{S} \subset \mathrm{GL}(\mathfrak{M})$. It suffices to prove the statement after making the faithfully flat, formally étale base extension $\mathfrak{S} \to W(k^{\mathrm{sep}})[\![u]\!]$. Hence we will assume from now on that the residue field k is separably closed.

In fact we will prove the following stronger statement.

Proposition 4.2.2. *Let L be in $\mathrm{Rep}_{G_K}^{\mathrm{cris}\circ}$ and $\mathfrak{M} = \mathfrak{M}(L)$. Let $(s_\alpha) \subset L^\otimes$ be a finite collection of G_K-invariant tensors defining a reductive $\mathbb{Z}_p$-subgroup $\mathcal{G}$ of* $\mathrm{GL}(L)$*, and let $(\tilde{s}_\alpha)$ be the corresponding tensors in $\mathfrak{M}^\otimes$.*

Let $\mathfrak{M}' = L \otimes_W \mathfrak{S}$. If k is separably closed, there is an isomorphism $\mathfrak{M} \xrightarrow{\sim} \mathfrak{M}'$ which takes each tensor $\tilde{s}_\alpha$ to s_α. In particular, the subgroup $\mathcal{G}_\mathfrak{S} \subset \mathrm{GL}(\mathfrak{M})$ defined by $(\tilde{s}_\alpha)$ is isomorphic to $\mathcal{G} \times_{\mathrm{Spec}\,\mathbb{Z}_p} \mathrm{Spec}\,\mathfrak{S}$.

Proof. Let $P \subset \underline{\mathrm{Hom}}_\mathfrak{S}(\mathfrak{M}, \mathfrak{M}')$ be the subscheme of isomorphisms between $\mathfrak{M}$ and $\mathfrak{M}'$ which take $\tilde{s}_\alpha$ to s_α. The fibres of P are either empty or a torsors under $\mathcal{G}$. We claim that P is a $\mathcal{G}$-torsor. That is, P is flat over $\mathfrak{S}$ with non-empty fibres. The claim implies the proposition since a torsor under a reductive group is étale locally trivial, while the ring $\mathfrak{S}$ is strictly henselian, so any $\mathcal{G}$ torsor over $\mathfrak{S}$ is trivial.

To prove the claim we proceed in several steps. For R a $\mathfrak{S}$-algebra, we set $P_R = P \times_{\mathrm{Spec}\,\mathfrak{S}} \mathrm{Spec}\,R$.

Step 1: $P_{\mathfrak{S}_{(p)}}$ is a $\mathcal{G}$-torsor. Since $\mathcal{O}_{\widehat{\mathcal{E}^{\mathrm{ur}}}}$ is faithfully flat over $\mathcal{O}_\mathcal{E}$ and $\mathcal{O}_\mathcal{E}$ is faithfully flat over $\mathfrak{S}_{(p)}$, it suffices to show that $P_{\mathcal{O}_{\widehat{\mathcal{E}^{\mathrm{ur}}}}}$ is a $\mathcal{G}$-torsor. However the isomorphism in (2) of the Theorem 4.1.2 shows that $P_{\mathcal{O}_{\widehat{\mathcal{E}^{\mathrm{ur}}}}}$ is a trivial $\mathcal{G}$-torsor.

Step 2: P_{K_0} is a $\mathcal{G}$-torsor, where we regard K_0 as a $\mathfrak{S}$-algebra via $u \mapsto 0$. This follows from (1) of the theorem in 4.1, which implies the existence of a canonical isomorphism

$$B_{\mathrm{dR}} \otimes L \xrightarrow{\sim} B_{\mathrm{dR}} \otimes \mathfrak{M}/u\mathfrak{M}.$$

Step 3: $P_{\mathfrak{S}[1/pu]}$ is a $\mathcal{G}$-torsor. Let $U \subset \mathrm{Spec}\,\mathfrak{S}[1/up]$ denote the maximal open subset over which P is flat with non-empty fibres. By Step 1, we know this subset is non-empty, since it contains the generic point. In particular, the complement of U in $\mathrm{Spec}\,\mathfrak{S}[1/up]$ contains finitely many closed points.

Let $x \in \mathrm{Spec}\,\mathfrak{S}[1/up]$ be a closed point. If $x \notin U$, we consider two cases. If $|u(x)| < |\pi|$, then since the s_α are Frobenius invariant, we have $P_{\mathfrak{S}[1/p]} \xrightarrow{\sim} \varphi^*(P_{\mathfrak{S}[1/p]})$ in a formal neighbourhood of x. Hence $P_\mathfrak{S}[1/p]$ cannot be a $\mathcal{G}$-torsor at $\varphi(x)$, since φ is a faithfully flat map on $\mathfrak{S}$. Repeating the argument we find $\varphi(x), \varphi^2(x), \dots . \notin U$, which gives a contradiction.

Similarly, if $|u(x)| \geq |\pi|$ consider a sequence of points $x_0, x_1, \dots$ with $x_0 = x$, and $\varphi(x_{i+1}) = x_i$. For $i \geq 1$, we have $P_{\mathfrak{S}[1/p]} \xrightarrow{\sim} \varphi^*(P_{\mathfrak{S}[1/p]})$ in a formal neighbourhood of x_i, so we find that $x_i \notin U$ for $i \geq 1$.

Step 4: $P_{\mathfrak{S}[1/p]}$ is a $\mathcal{G}$-torsor. By Step 3, it suffices to show that the restriction of P to $K_0[\![u]\!]$ is a $\mathcal{G}$-torsor. For any $\mathfrak{N}$ in $\mathrm{Mod}^{\varphi}_{/\mathfrak{S}}$ there is a unique, φ-equivariant isomorphism

$$\mathfrak{N} \otimes_{\mathfrak{S}} K_0[\![u]\!] \xrightarrow{\sim} K_0[\![u]\!] \otimes_{K_0} \mathfrak{N}/u\mathfrak{N}[1/p]$$

lifting the identity map on $\mathfrak{N}/u\mathfrak{N} \otimes_{\mathcal{O}_{K_0}} K_0$, which is functorial in $\mathfrak{N}$ (see, for example, [Ki 1, 1.2.6]). Applying this to $\mathfrak{M}$ and the morphisms $\tilde{s}_\alpha$ shows that the restriction of P to $K_0[\![u]\!]$ is isomorphic to $P_{K_0} \otimes_{K_0} K_0[\![u]\!]$, which is a $\mathcal{G}$-torsor by Step 2.

Step 5: P is a $\mathcal{G}$-torsor. Let U be the complement of the closed point in $\operatorname{Spec} \mathfrak{S}$. By Steps 1 and 4 we know that $P|_U$ is a $\mathcal{G}$-torsor. By a result of Colliot-Thélène and Sansuc [CS, Thm. 6.13] P extends to a $\mathcal{G}$-torsor over $\mathfrak{S}$ and, as we remarked above, any such torsor is trivial. Hence $P|_U$ is trivial, and there is an isomorphism $\mathfrak{M}|_U \xrightarrow{\sim} \mathfrak{M}'|_U$ taking $\tilde{s}_\alpha$ to s_α. Since any vector bundle over U has a canonical extension to $\mathfrak{S}$, obtained by taking its global sections, this isomorphism extends to $\mathfrak{S}$. This implies that P is the trivial $\mathcal{G}$-torsor, and completes the proof of the proposition and of the key lemma. □

Note that the proof of the proposition implies the following result.

Corollary 4.2.3. *Suppose that k is separably closed or finite. Then for any $i \geq 0$ there is an isomorphism $L \otimes_{\mathbb{Z}_p} W \xrightarrow{\sim} \varphi^{i*}(\mathfrak{M}/u\mathfrak{M})$ which takes $s_\alpha \in L$ to the corresponding tensor in $\varphi^{i*}(\mathfrak{M}/u\mathfrak{M}) \subset D_{\mathrm{cris}}(L \otimes \mathbb{Q}_p)$.*

Proof. Define P as in the proof of the proposition (without assuming k separably closed). Then we saw that P is a $\mathcal{G}$-torsor. If k is separably closed or finite a torsor under a reductive W-group is necessarily trivial [Sp, 4.4]. Hence P is trivial, as is $\varphi^{i*}(P)$ for $i \geq 0$. □

When L is dual to the Tate module of a p-divisible group and $\varphi^*(\mathfrak{M}/u\mathfrak{M})$ is replaced by the Dieudonné module of the p-divisible group, then the corollary is a conjecture of Milne [Mi 3]. This conjecture follows from the corollary and theorem of 4.1. Vasiu [Va 3] has also claimed a proof of the conjecture.

5 Parahoric Level

We now discuss extensions of the above results where we drop the assumption that K_p is hyperspecial, and assume only that K_p is parahoric. This means that $K_p = \mathcal{G}^\circ(\mathbb{Z}_p)$ where $\mathcal{G}^\circ$ is a parahoric group scheme. That is, $\mathcal{G}^\circ$ is the connected stabilizer of a point x in the building $B(G, \mathbb{Q}_p)$ in the sense of Bruhat-Tits. It is a smooth group scheme with generic fiber G. For the remainder of this article, we assume that K_p is parahoric and that $p > 2$.

5.1 Conjectures

5.1.1

We now fix a sufficiently small compact open $K^p \subset G(\mathbb{A}_f^p)$ and set $K = K_p K^p$. We also fix a prime $\lambda | p$ of E. When K_p is parahoric, one does not expect that $\mathrm{Sh}_K(G, X)$ should have a smooth integral model. Indeed, one can sometimes prove that this is impossible, by showing that the $\mathrm{Gal}(\bar{E}/E)$-action on the cohomology of $\mathrm{Sh}_K(G, X)$ is ramified at λ.

Motivated by the work of a number of authors, and in particular Rapoport-Zink [RZ] and Pappas [Pa] in the PEL case, it was conjectured by Rapoport [Ra] that there should exist integral models of $\mathrm{Sh}_K(G, X)$ over $\mathcal{O}_\lambda$, which are modeled by certain, group theoretically defined, *local models* $\mathrm{M}^{\mathrm{loc}}_{G,X}$.

To explain some of the properties of $\mathrm{M}^{\mathrm{loc}}_{G,X}$, we denote by X_μ the E-scheme of parabolics in G associated to a cocharacter in the conjugacy class of μ_h for $h \in X$. Then X_μ is naturally a G-homogeneous space over E. One requires that $\mathrm{M}^{\mathrm{loc}}_{G,X}$ should be a projective $\mathcal{O}_{E_\lambda}$ scheme, equipped with an action of $\mathcal{G}^\circ$, together with a G-equivariant identification $\mathrm{M}^{\mathrm{loc}}_{G,X} \otimes E \xrightarrow{\sim} X_\mu$. There is also a precise group theoretic description of $\mathrm{M}^{\mathrm{loc}}_{G,X}(k)$ for k an algebraic closure of $\kappa(\lambda)$ - see [Pa, Conj. 2.1] - as well as the structure of the $\mathcal{G}^\circ$ orbits of the special fibre.

For G which split over a tamely ramified extension of $\mathbb{Q}_p$, local models with these properties were constructed by Pappas-Zhu [PZ]; this was extended to a larger class of groups by Levin [Le]. The definition of $\mathrm{M}^{\mathrm{loc}}_{G,X}$ is as an orbit closure in an affine Grassmannian. It is shown in [KP, 2.3.16] that for (G, X) of Hodge type,. $\mathrm{M}^{\mathrm{loc}}_{G,X}$ may also be identified with the closure of X_μ in a usual Grassmannian.

5.1.2

Rapoport conjectures that there is an integral model $\mathscr{S}_K(G, X)$ of $\mathrm{Sh}_K(G, X)$, and a smooth morphism of stacks

$$\mathscr{S}_K(G, X) \to [\mathcal{G}^{c,\circ} \backslash \mathrm{M}^{\mathrm{loc}}_{G,X}]$$

where $\mathcal{G}^{c,\circ}$ is the quotient of $\mathcal{G}^\circ$ by the closure of the maximal $\mathbb{R}$-split, $\mathbb{Q}$-anisotropic torus in the center of G. (For Hodge type Shimura data this torus is trivial and $\mathcal{G}^{c,\circ} = \mathcal{G}^\circ$.) More explicitly, this means that there is a *local model diagram*

$$\begin{array}{ccc} & \widetilde{\mathscr{S}_K} & \\ \swarrow^{\pi} & & \searrow^{q} \\ \mathscr{S}_K(G, X) & & \mathrm{M}^{\mathrm{loc}}_{G,X} \end{array}$$

where π is a $\mathcal{G}^{c,\circ}$-torsor and q is smooth and $\mathcal{G}^{c,\circ}$-equivariant. In order to make the conjecture above non-vacuous, one should insist that the special fibre of $\mathscr{S} = \mathscr{S}_K(G, X)$ is non-empty. Although there is no known analogue of the extension property which would characterize $\mathscr{S}$ as in the hyperspecial case, one can at least ask that for R a mixed characteristic discrete valuation ring, the map $\mathscr{S}(R) \to \mathscr{S}(R[1/p])$ is a bijection. In the context of parahoric level structure, we will refer to this condition as the *extension property*.

5.2 Results

We give a sample of the results of [KP]. We denote by $\mathcal{G}^{\mathrm{ad},\circ}$ the connected stabilizer of the image of x in the building of the adjoint group $B(G^{\mathrm{ad}}, \mathbb{Q}_p)$.

Theorem 5.2.1. *Let (G, X) be of abelian type, and suppose that G splits over a tamely ramified extension of $\mathbb{Q}_p$. The E_λ-scheme $\mathrm{Sh}_{K_p}(G, X)$ admits a $G(\mathbb{A}_f^p)$-equivariant extension to a flat $\mathcal{O}_{E_\lambda}$-scheme $\mathscr{S}_{K_p}(G, X)$, satisfying the extension property. Any sufficiently small compact open $K^p \subset G(\mathbb{A}_f^p)$ acts freely on $\mathscr{S}_{K_p}(G, X)$, and the quotient*

$$\mathscr{S}_K(G, X) := \mathscr{S}_{K_p}(G, X)/K^p$$

is a finite $\mathcal{O}_{E_\lambda}$-scheme extending $\mathrm{Sh}_{K_p}(G, X)_E$.

Suppose that $p \nmid |\pi_1(G^{\mathrm{der}})|$, and that either $(G^{\mathrm{ad}}, X^{\mathrm{ad}})$ has no factor of type $D^{\mathbb{H}}$, or that G is unramified over $\mathbb{Q}_p$, then there is a smooth morphism

$$\mathscr{S}_K(G, X) \to [\mathcal{G}^{\mathrm{ad},\circ} \backslash \mathrm{M}^{\mathrm{loc}}_{G,X}].$$

Using the properties of local models established by Pappas-Zhu [PZ], one deduces

Corollary 5.2.2. *Let (G, X) be of abelian type, and suppose that G splits over a tamely ramified extension of $\mathbb{Q}_p$. If $\kappa/\kappa(\lambda)$ is a finite extension, and $z \in \mathscr{S}_K(G, X)(\kappa)$, then there exists $w \in \mathrm{M}^{\mathrm{loc}}_{G,X}(\kappa')$, with $\kappa'/\kappa(\lambda)$ a finite extension, and an isomorphism of strict henselizations*

$$\mathcal{O}^{\mathrm{sh}}_{\mathscr{S}_K(G,X),z} \xrightarrow{\sim} \mathcal{O}^{\mathrm{sh}}_{\mathrm{M}^{\mathrm{loc}}_{G,X},w}.$$

The special fibre $\mathscr{S}_K(G, X) \otimes \kappa(\lambda)$ is reduced, and the strict henselizations of the local rings on $\mathscr{S}_K(G, X) \otimes \kappa(\lambda)$ have irreducible components which are normal and Cohen-Macaulay.

5.2.3

We remark that for (G, X) of abelian type the results of [KP] (for example the two results above) relate $\mathscr{S}_{K_p}(G, X)$ and the local model of an auxiliary *Hodge type* Shimura datum. It has been shown by Haines-Richarz that this local model agrees with the one for (G, X) [HR], if we assume the normality of the Schubert varieties appearing in the generic and special fibers of the local model for $(G^{\mathrm{ad}}, X^{\mathrm{ad}})$; the latter is known to hold for example when the residual characteristic does not divide the order of $\pi_1(G^{\mathrm{ad}})$.

The proofs of these results are much more technically involved than those of [Ki 3]. Among the points which require new ideas are the construction of a suitable symplectic embedding of G, the proof of the generalization of the Key Lemma, and the construction of the required p-divisible group over $\mathrm{M}^{\mathrm{loc}}_{G,X}$, which uses displays in place of Faltings' construction. For the analogue of the Key Lemma one proves

Theorem 5.2.4. *Suppose that G splits over a tamely ramified extension of $\mathbb{Q}_p$, and contains no factors of type E_8. If $D^\times \subset \operatorname{Spec} \mathfrak{S}$ denotes the complement of the closed point, then any $\mathcal{G}^\circ$-torsor on $D^\times$ is trivial, and so extends to* $\operatorname{Spec} \mathfrak{S}$.

5.2.5

This Theorem allows one to prove the analogue of the Key Lemma, along the lines explained in the hyperspecial case, above, but with the result of Colliot-Thélène and Sansuc in Step 5, replaced by Theorem 5.2.4; the case of type E_8 is of course not needed for applications to Shimura varieties. The proof uses, among other things, results of Gille and Bayer-Fluckiger - Parimala on Serre's conjecture II, which, together with results of Kato and Gabber, implies that any G-torsor on Spec Frac$\mathfrak{S}$ is trivial.

Theorem 5.2.4 has been extended by Anschütz to a number of other cases, including all cases when G which is semi-simple and simply connected or a torus [An]. It seems reasonable to conjecture it holds for any reductive G.

References

[An] J. Anschütz, *Extending torsors on the punctured* Spec (A_{inf}), preprint (2018), 36 pages.

[BT] F. Bruhat, J. Tits, *Groupes réductifs sur un corps local. II. Schémas en groupes. Existence d'une donnée radicielle valuée*, Pub. Math IHES **60** (1984), 197–376.

[CS] J-L. Colliot-Thélène, J-J Sansuc, *Fibrés quadratiques et composantes connexes réelles*, Math. Ann **244** (1979), 105–134.

[De 1] P. Deligne, *Variètés de Shimura: interprétation modulaire, et techniques de construction de modèles canoniques*, Automorphic forms, representations and L-functions (Corvallis 1977), Proc. Sympos. Pure Math XXXIII, Amer. Math. Soc, pp. 247–289, 1979.

[De 2] P. Deligne, *Hodge cycles on abelian varieties*, Hodge cycles motives and Shimura varieties, Lecture notes in math. **900**, pp. 9–100, 1982.

[DP] P. Deligne, G. Pappas, *Singularités des espaces de modules de Hilbert, en les caractéristiques divisant le discriminant*, Compositio Math. **90** (1994), 59–79.

[Fa] G. Faltings, *Integral crystalline cohomology over very ramified valuation rings*, J. AMS. **12(1)** (1999), 117–144.

[HR] T. Haines, T. Richarz, *Normality and Cohen-Macaulayness of parahoric local models*, preprint (2019), 23 pages.

[Ki 1] M. Kisin, *Crystalline representations and F-crystals*, Algebraic geometry and number theory, Progr. Math 253, Birkhäuser Boston, pp. 459–496, 2006.

[Ki 2] M. Kisin, *Modularity of 2-adic Barsotti-Tate representations*, Invent. Math. **178(3)** (2009), 587–634.

[Ki 3] M. Kisin, *Integral models for Shimura varieties of abelian type*, J. AMS. **23(4)** (2010), 967–1012.

[Ki 4] M. Kisin, *Integral canonical models of Shimura varieties (Journees Arithmetiques 2007)*, J. Th. Nombres Bordeaux **21(2)** (2009), 301–312.

[Kim] W. Kim, *The classification of p-divisible groups over 2-adic discrete valuation rings*, Math. Res. Lett. **19(1)** (2012).

[KM] W. Kim, K. Madapusi Pera, *2-adic integral canonical models*, Forum Math. Sigma **4** (2016), e28, 34 pp.

[KP] M. Kisin, G. Pappas, *Integral models for Shimura varieties with parahoric level structure*, Pub. Math IHES **128** (2018), 121–218.

[La] R. Langlands, *Some contemporary problems with origins in the Jugendtraum*, Mathematical developments arising from Hilbert problems (De Kalb 1974), Proc. Sympos. Pure Math. XXVIII, Amer. Math. Soc, pp. 401–418, 1976.

[Lau] E. Lau, *Relations between Dieudonné displays and crystalline Dieudonné theory*, Algebra Number Theory **8(9)** (2014), 2201–2262.

[Le] B. Levin, *Local models for Weil-restricted groups.*, Compos. Math. **152(12)** (2016), 2563–2601.

[Li] T. Liu, *The correspondence between Barsotti-Tate groups and Kisin modules when $p = 2$*, J. Thèor. Nombres Bordeaux **25(3)** (2013), 661–676.

[LR] R. Langlands, M. Rapoport, *Shimuravarietäten und Gerben*, J. Reine Angew. Math **378** (1987), 113–220.

[Mi 1] J. Milne, *The points on a Shimura variety modulo a prime of good reduction*, The zeta functions of Picard modular surfaces, (1992), pp. 151–253.

[Mi 2] J. Milne, *Canonical models of (mixed) Shimura varieties and automorphic vector bundles*, Automorphic forms, Shimura varieties. and L-functions I (Ann Arbor 1988), Perspectives in Math. 10, pp. 284–414, 1990.

[Mi 3] J. Milne, *On the conjecture of Langlands and Rapoport*, available at **arxiv.org** (1995), 31 pages.

[Mo] B. Moonen, *Models of Shimura varieties in mixed characteristics*, Galois representations in arithmetic algebraic geometry (Durham, 1996), London Math. Soc. Lecture Note Ser. 254, Cambridge Univ. Press, pp. 267–350, 1998.

[MP] K. Madapusi Pera, *Toroidal compactifications of integral models of Shimura varieties of Hodge type*, Annales d'École Normale (to appear), 105 pages.

[Sp] T. Springer Automorphic forms, representations and L-functions (Corvallis 1977), Proc. Sympos. Pure Math XXXIII, Amer. Math. Soc, 1979.

[Pa] G. Pappas, *On the arithmetic moduli schemes of PEL Shimura varieties*, J. Algebraic Geom. **9** (2000(3)), 577–605.

[Pa 2] G. Pappas, *Arithmetic models for Shimura varieties*, Proceedings of the ICM, Rio, Brazil, 2018.

[PY] G. Prasad; J-K. Yu, *On quasi-reductive group schemes. With an appendix by Brian Conrad*, J. Algebraic Geom. **15** (2006), 507–549.

[PZ] G. Pappas, X. Zhu, *Local models of Shimura varieties and a conjecture of Kottwitz*, Invent. Math. **194(1)** (2013), 147–254.

[Ra] M. Rapoport, *A guide to the reduction modulo p of Shimura varieties*, Automorphic forms. I., Astérisque, 298, pp. 271–318, 2005.

[RZ] M. Rapoport, T. Zink, *Period spaces for p-divisible groups*, Annals of Mathematics Studies, 141, Princeton University Press, pp. 324 pp,1996.

[Va 1] A. Vasiu, *Integral Canonical Models of Shimura Varieties of Preabelian Type*, Asian J. Math (1999), 401–518.

[Va 2] A. Vasiu, *Integral Canonical Models of Shimura Varieties of Preabelian Type (Fully corrected version.)*, available at **arxiv.org** (2003), 135 pages.

[Va 3] A. Vasiu, *A motivic conjecture of Milne*, available at **arxiv.org** (2003), 46 pages.

[Va 4] A. Vasiu, *Good reduction of Shimura varieties in arbitrary unramified mixed characteristic I*, available at **arxiv.org** (2007), 48 pages.

[Va 5] A. Vasiu, *Good reduction of Shimura varieties in arbitrary unramified mixed characteristic I*, available at **arxiv.org** (2007), 33 pages.

[Zi] T. Zink, *Über die schlechte Reduktion einiger Shimuramannigfaltigkeiten*, Compositio Math. **45(1)** (1982), 15–107.

THE NEWTON STRATIFICATION

ELENA MANTOVAN

Contents

1 Introduction

The Newton stratification of moduli spaces for abelian varieties in positive characteristic naturally arise in the study of Shimura varieties of PEL type. This stratification is defined by the loci where the equivalence class of the F-isocrystals attached to the abelian varieties is constant. In the following, we recall the notion of F-isocrystals with G-structure and the definition of the Newton stratification of the reduction modulo p of Shimura varieties. We discuss the geometry of the Newton strata and of the associated Oort foliations and their relation to the Rapoport-Zink spaces constructed by Rapoport and Zink, and to some moduli spaces for abelian varieties in positive characteristic, called Igusa varieties. In particular, we discuss a formula, in the appropriate Grothendieck group, computing the l-adic cohomology of the Newton strata in terms of those of the Igusa varieties and the Rapoport-Zink spaces. Finally, in the last section, we focus on a special class of Shimura varieties, which is closely related to and includes that studied by Harris and Taylor in [9]. For these Shimura varieties, the associated Rapoport-Zink spaces are easily related to Lubin-Tate spaces and Drinfeld modular varieties. As the geometry and cohomology of these latter spaces are well understood by the work of Boyer ([3]), we deduce many interesting results on the geometry and cohomology of these Shimura varieties.

2 F-isocrystals with G-structure

The notion of F-isocrystal with G-structure is due to Kottwitz. In this section, we recall some of his definitions and results from [14] and [15], as well as results from the work of Rapoport and Richartz in [34].

2.1 Notations

Let k be a finite field of characteristic p (e.g. $k = k(v)$), $\overline{k}$ an algebraic closure of k, and L the fraction field of the Witt ring $W(\overline{k})$, $L = W(\overline{k})[\frac{1}{p}]$. The Frobenius automorphism on $\overline{k}$, $x \mapsto x^p$, induces the Frobenius automorphism of L over $\mathbb{Q}_p$ which we denote by σ. For any connected reductive group G over $\mathbb{Q}_p$, we also denote by σ the corresponding automorphism of $G(L)$. We say that two elements b, b' of $G(L)$ are *σ-conjugate* if there exists $g \in G(L)$ such that $b' = gb\sigma(g)^{-1}$. We define $B(G)$ to be the set of σ-conjugacy classes of $G(L)$ and write $[b] \in B(G)$ for the class of an element $b \in G(L)$.

2.2 Definitions

An *F-isocrystal* is a finite dimensional vector space V over L together with a σ-linear automorphism $\Phi : V \to V$ (called the Frobenius morphism). An *F-isocrystal with G-structure*, for G a connected reductive group over $\mathbb{Q}_p$, is an exact faithful tensor functor from the category of finite dimensional p-adic representations of G to that of F-isocrystals. In the cases of interest to us (e.g. $G = GL(V)$ or $G = GSp(V)$), any such functor N is uniquely determined by its value on the natural representation of G. Therefore, in these cases, an F-isocrystal with G-structure can be identified with a classical F-isocrystal endowed with a G-structure, i.e. an F-isocrystal of the form (V_L, Φ), where $V_L = V \otimes_{\mathbb{Q}_p} L$, for V a finite dimensional $\mathbb{Q}_p$-vector space together with a G-structure, $G \subset GL(V)$, and where the Frobenius morphism Φ commutes with the G-structure on V_L.

To each element $b \in G(L)$ we associate an F-isocrystal with G-structure N_b, via $N_b(W, \rho) = (W_L, \rho(b)(\mathrm{id}_W \otimes \sigma))$. It follows from the definition that any such functor N is defined by a unique $b \in G(L)$, and that if b, b' are σ-conjugate in $G(L)$ then the corresponding functors $N_b, N_{b'}$ are isomorphic. Thus, the set $B(G)$ can be identified with the set of isomorphism classes of F-isocrystals with G-structure.

E.g., for $G = GL(V)$, $B(G)$ can be regarded as the set of isomorphism classes of h-dimensional F-isocrystals, where $h = \dim_{\mathbb{Q}_p} V$.

2.3 Newton points

The Dieudonné-Manin description of the category of F-isocrystals makes it possible to give a simple classification of the σ-conjugacy classes in $G(L)$. In particular, we recall that the category of F-isocrystals is semisimple, and that the simple objects are parameterized by rational numbers (called the *slopes*). Let $\mathbb{D}$ be the diagonalizable pro-algebraic group over $\mathbb{Q}_p$ with character group $\mathbb{Q}$. For any finite dimensional p-adic representation (W, ρ) of G, and $b \in G(L)$, the slope decomposition of the associated F-isocrystal $N_b(W, \rho)$ gives a homomorphism $\nu_{\rho,b} : \mathbb{D} \to GL(W)$ defined over L. Thus, for each $\rho : G \to GL(W)$, we get a map $\nu_\rho : G(L) \to \mathrm{Hom}_L(\mathbb{D}, GL(W))$, via $b \mapsto \nu_{\rho,b}$. We fix $b \in G(L)$, and consider the elements $\nu_{\rho,b}$ as ρ varies. The functoriality of the construction implies that there exists a unique element $\nu_b \in \mathrm{Hom}_L(\mathbb{D}, G)$ (called the *slope homomorphism* of b) such that $\rho \circ \nu_b = \nu_{\rho,b}$ for all ρ. This defines a map

$$\nu : G(L) \to \mathrm{Hom}_L(\mathbb{D}, G)$$

by $b \mapsto \nu_b$ ([14], Section 4).

For each $b \in G(L)$, the corresponding slope homomorphism ν_b can be characterized as follows ([14], Section 4). It is the unique element in $\mathrm{Hom}_L(\mathbb{D}, G)$ for which there exists an integer $n \geq 1$, an element $c \in G(L)$ and a unit $u \in \mathbb{Z}_p^*$ such that $n\nu_b \in \mathrm{Hom}_L(\mathbb{G}_m, G)$, $Int(c) \circ (n\nu_b)$ is defined over the fixed field of σ^n on L, and $c(b\sigma)^n c^{-1} = c((n\nu_b)(up))c^{-1}\sigma^n$ as elements in $G(L) \rtimes \langle\sigma\rangle$ (here the inclusion $\mathrm{Hom}_L(\mathbb{G}_m, G) \subset \mathrm{Hom}_L(\mathbb{D}, G)$ is the one induced by the $\mathbb{Q}_p$-homomorphism $\mathbb{D} \to \mathbb{G}_m$ corresponding to the usual inclusion $\mathbb{Z} \subset \mathbb{Q}$, and $Int(c)$ denotes the inner automorphism $x \to cxc^{-1}$).

These properties imply that the mapping $b \mapsto \nu_b$ satisfies the conditions $\sigma(b) \mapsto \sigma(\nu_b)$ and $gb\sigma(g)^{-1} \mapsto Int(g) \circ \nu_b$, for all $g \in G(L)$. In particular, the $G(L)$-conjugacy class of the homomorphism ν_b (which we denote by $[\nu_b]$) depends only on the σ-conjugacy class of b, and is fixed by σ (the latter property follows from the equality $\nu_b = Int(b) \circ \sigma(\nu_b)$). Thus, the map $b \mapsto \nu_b$ induces a map

$$\bar{\nu} : B(G) \to \mathcal{N}(G)$$

by $[b] \mapsto \bar{\nu}_{[b]} = [\nu_b]$, where $\mathcal{N}(G) = (Int(G(L))\backslash \mathrm{Hom}_L(\mathbb{D}, G))^{\langle\sigma\rangle}$ is the set of the σ-invariant $G(L)$-conjugacy classes of homomorphisms $\mathbb{D}_L \to G_L$. We call $\mathcal{N}(G)$ the *set of Newton points of G*. The map $\bar{\nu}$ (the definition of which is due to Rapoport and Richartz in [34], Section 1) is called the *Newton map of the group G*, and defines a natural transformation of set-valued functors on the category of connected reductive groups. We remark that, in the cases of interest to us, the Dieudonné-Manin classification of F-isocrystals implies that the Newton map is injective.

E.g., for $G = GL(V)$, the slope decomposition of the F-isocrystal $N_b(V)$ associated to an element $b \in G(L)$ is a decomposition $V_L = \oplus_1^r V_i$ into Φ-stable subspaces of dimension $h_i = \dim_L V_i$, for which there exist $\mathcal{O}_L$-lattices $M_i \subset V_i$ with $\Phi^{h_i} M_i = \pi^{d_i} M_i$, where $d_i = \lambda_i h_i \in \mathbb{Z}$, for some uniquely determined rational numbers $\lambda_1 > \lambda_2 > \cdots > \lambda_r$ (each V_i is called the *isotypical component* of slope λ_i in V). Then, the associated slope homomorphism $\nu_b : \mathbb{D}_L \to G_L$ is equal to $\nu_b = \oplus_i \lambda_i \cdot \mathrm{id}_{V_i}$. Classically, the data of the slopes $\lambda_1, \ldots, \lambda_r$ and their multiplicities $h_1, \ldots, h_r$ are used to form a convex polygon with integral vertices, starting at $(0, 0)$ and ending at (h, d), where $h = \dim_L V$ and $d = \sum_i d_i$ (called respectively the *height* and *dimension* of the F-isocrystal $N_b(V)$). It follows by the construction that this polygon, which is called *Newton polygon* of the F-isocrystal $N_b(V)$, uniquely determines the isomorphism class of the corresponding F-isocrystal.

2.4 The group J_b

To each $b \in G(L)$ (or equivalently, to each pair (G, b), $b \in G(L)$), we associate a connected reductive group J_b over $\mathbb{Q}_p$,

$$J_b(R) = \{g \in G(L \otimes_{\mathbb{Q}_p} R) | g = b\sigma(g)b^{-1}\}$$

for any $\mathbb{Q}_p$-algebra R. Then, $J_b(\mathbb{Q}_p)$ is the group of automorphisms of the F-isocrystal with G-structure N_b associated with b. If $b' = hb\sigma(h)^{-1}$ is a σ-conjugate of b in $G(L)$, then the inner automorphism $Int(h)$ on $G(L)$ induces a $\mathbb{Q}_p$-isomorphism $J_b \to J_{b'}$. Thus the isomorphism class of J_b depends only on the σ-conjugacy class of b, $[b] \in B(G)$.

We say that an element $b \in G(L)$ is *basic* if its slope homomorphism $\nu_b \in \mathrm{Hom}_L(\mathbb{D}_m, G)$ factors through the center $Z = Z(G)$ of G. We remark that if b is basic, then ν_b is necessarily defined over $\mathbb{Q}_p$. We say that a σ-conjugacy class $[b] \in B(L)$ is *basic* if it consists of basic elements, and write $B(G)_b \subset B(G)$ for the set of basic classes in $B(G)$. We recall that an element $b \in G(L)$ is basic if and only if the corresponding connected reductive group J_b is an inner twist of G ([14], Section 5).

Let us assume for simplicity that G is a quasi-split connected reductive group (a similar description of $B(G)$ for G any connected reductive group is also possible). Then, any element $b \in G(L)$ is σ-conjugate to a basic element of some Levi subgroup of G. Equivalently, each $[b] \in B(G)$ is in the image of $B(M)_b$, for some Levi subgroup M of G, under the map $B(M) \to B(G)$ induced by the inclusion $M(L) \subset G(L)$. Furthermore, for any Levi subgroup M of G and $b \in B(M)_b$, the centralizer $\mathrm{Cent}_G(\nu_b)$ of ν_b in G is a Levi subgroup containing M. We say that b is *G-regular* if $\mathrm{Cent}_G(\nu_b)$ is equal to M, and write $B(M)_{br}$ for the set of G-regular basic elements of $B(M)$. Then, each $[b] \in B(G)$ is in the image of $B(M)_{br}$, for some Levi subgroup M of G. Moreover, given any two Levi subgroups M_1, M_2 of G and two basic G-regular elements $b_i \in M_i(L)$, $i = 1, 2$, b_1, b_2 have the same image in $B(G)$ if and only if there exists $g \in G(\mathbb{Q}_p)$ such that $Int(g)(M_1) = M_2$ and $Int(g)(b_1) = b_2$. (We remark that since $g \in G(\mathbb{Q}_p)$ the isomorphism $Int(g) : M_1 \to M_2$ induces a map $B(M_1) \to B(M_2)$ which extends to be the identity on $B(G)$.)

Let M be a Levi subgroup of G and b is a basic G-regular element of $M(L)$. It is a consequence of the definitions that the connected reductive group associated with the pair (M, b) is the same as the group J_b associated with the pair (G, b), $b \in G(L)$. Thus, in particular, J_b is an inner twist of M. Then, it follows from the previous results that, for each $b \in G(L)$, the corresponding connected reductive group J_b is an inner twist of a Levi subgroup of G ([14], Section 6).

More explicitly, for each $[b] \in B(G)$, $b \in G(L)$, since the $G(L)$-conjugacy class of the slope morphism ν_b is fixed by σ, there exists $g \in G(L)$ such that $Int(g) \circ \nu_b$ is defined over $\mathbb{Q}_p$ ([14], Section 6). Thus, after replacing b by $gb\sigma(g)^{-1}$, we may assume that ν_b is defined over $\mathbb{Q}_p$. We write $M_b = \mathrm{Cent}_G(\nu_b)$. It is immediate that $b \in M_b(L)$, and b is basic and G-regular in $M_b(L)$. Thus, in particular, the group J_b is an inner form of M_b.

E.g., for $G = GL(V)$, the Levi subgroup M_b associated to an element $b \in G(L)$ is the stabilizer in G of the decomposition $V_L = \oplus_i V_i$ underlying the slope decomposition of the F-isocrystal $N_b(V)$. I.e. it is the Levi subgroup corresponding to the partition $(h_1, \ldots, h_r)$ of $h = \dim_{\mathbb{Q}_p} V$, determined by the L-dimensions of the isotypical components of $N_b(V)$, $h_i = \dim_L V_i$. In particular, $b \in G(L)$ is basic if the corresponding F-isocrystal $N_b(V)$ has a unique isotypical component (in which case the F-isocrystal is called *isoclinic*).

2.5 Bruhat–Tits order

Let G be a reductive connected group, and consider $\mathcal{N}(G)$ the set of Newton points of G. If $A \subset G$ is a maximal torus with Weyl group Ω, then $\mathcal{N}(G) = (X_*(A)_{\mathbb{Q}}/\Omega)^{\Gamma}$, for Γ the absolute Galois group of $\mathbb{Q}_p$ ([34], Section 1). We define a partial order $\preceq$ on $\mathcal{N}(G)$, via $\bar{\nu} \preceq \bar{\nu}'$ if the orbit $\Omega \cdot \bar{\nu}$ under the Weyl group in $X_*(A)_{\mathbb{R}}$ lies in the convex hull of the orbit $\Omega \cdot \bar{\nu}'$.

Under the Newton map $\bar{\nu} : B(G) \to \mathcal{N}(G)$, the partial order on $\mathcal{N}(G)$ defines a partial order on the set $B(G)$, via $[b] \preceq [b']$ if and only if $\bar{\nu}_{[b]} \preceq \bar{\nu}_{[b']}$. We write $[b] \prec [b']$ if $[b] \preceq [b']$ and $[b] \neq [b']$. The order $\preceq$ on $B(G)$ is called the *Bruhat–Tits order*. Under the Bruhat–Tits order, the set of basic elements $B(G)_b$ is equal to the set of minimal elements. Moreover, for all $[b] \in B(G)$ the set $X_{[b]} = \{[b'] \in B(G) | [b'] \preceq [b]\}$ is finite ([34], Section 2). We remark that, for any representation $\rho : G \to GL(W)$, the corresponding map $B(G) \to B(GL(W))$ (resp. $\mathcal{N}(G) \to \mathcal{N}(GL(W))$) preserves the order $\preceq$, i.e. if $[b] \preceq [b']$ in $B(G)$ then $\rho([b]) \preceq \rho([b'])$ in $B(GL(W)$ (resp. if $\bar{\nu} \preceq \bar{\nu}'$ in $\mathcal{N}(G)$ then $\rho(\bar{\nu}) \preceq \rho(\bar{\nu}')$ in $\mathcal{N}(GL(W))$) ([34], Section 2).

E.g., for $G = GL(V)$, the Bruhat–Tits order on $B(G)$ can be described as follows. For any $[b], [b'] \in B(G)$, let $\bar{\nu}_{[b]}, \bar{\nu}_{[b]}$ be the corresponding Newton points in $\mathcal{N}(G)$. As described above, we regard $\bar{\nu}_{[b]}$ and $\bar{\nu}_{[b']}$ as two convex polygons with integral vertices starting at $(0, 0)$. Then, $[b] \preceq [b']$ (resp. $\bar{\nu}_{[b]} \preceq \bar{\nu}_{[b']}$) if and only if the Newton polygon $\bar{\nu}_{[b]}$ lies above the Newton polygon $\bar{\nu}_{[b']}$ and has the same end-point. (We warn that the opposite partial order on the set of Newton polygons is also sometimes used in the literature.)

2.6 Hodge points

Let (G, μ) be a pair consisting of a reductive connected group over $\mathbb{Q}_p$ and a conjugacy class of cocharacters of G. (Such pairs arise naturally in the context of Shimura varieties, with μ minuscule). We assume for simplicity that G is quasi-split and μ unramified (i.e. μ defined over L).

To μ we associate an element $\bar{\mu} \in \mathcal{N}(G)$ as follows. The class μ determines an element in $Int(G(L))\backslash\mathrm{Hom}_L(\mathbb{D}, G)$, via pullback under the natural morphism $\mathbb{D} \to \mathbb{G}_m$. For each μ, there exists an integer $r \geq 1$ such that this element (which we still denote by μ) is fixed under the action of σ^r (e.g. r equal to the degree over $\mathbb{Q}_p$ of the field of definition of μ). We define

$$\bar{\mu} = \frac{1}{r} \sum_{1}^{r} \sigma^i(\mu).$$

Its definition is independent of the choice of r, and it implies that $\bar{\mu}$ is σ-invariant. Thus, $\bar{\mu} \in \mathcal{N}(G)$ and is called the *Hodge point of* μ.

We say that an element $[b] \in B(G)$ is *μ-admissible* if the corresponding Newton point $\bar{\nu}_{[b]} \in \mathcal{N}(G)$ satisfies the condition $\bar{\nu}_{[b]} \preceq \bar{\mu}$. We write $B(G, \mu) \subset B(G)$ for the set of μ-admissible elements ([15], Section 4). Then, it follows from the above discussion that the set $B(G, \mu)$ is finite, and that it contains a unique basic element. If we regard $B(G, \mu)$ as a partially ordered set under the restriction of the Bruhat–Tits order of $B(G)$, then $B(G, \mu)$ has a unique minimum (which is its unique basic element), and in the cases of interest to us, a unique maximum (called the *μ-ordinary* element), which is the unique element mapping to the Hodge point of μ under the Newton map.

E.g., for $G = GL(V)$, a conjugacy class of cocharacters μ of G is uniquely determine by its weights and their multiplicities, i.e. by integral numbers $w_1 < w_2 \cdots < w_s$ and multiplicities $t_1, \dots t_s$, for some $s \geq 1$. Similarly to the construction of Newton polygons, these data can be used to form a convex polygon with integral vertices, starting at $(0, 0)$. This polygon is called the *Hodge polygon* of μ and the set $B(G,\mu)$ can be identified with the set of all Newton polygons which lie above the Hodge polygon of μ and have the same end-point. If the class of cocharacters μ is minuscule (i.e. if μ has weights in $\{0,1\}$), then it is uniquely determined by the multiplicities $(h-d,d)$ of the weights $(0,1)$ (for $h=\dim_{\mathbb{Q}_p} V \geq d \geq 0$). It follows that the Hodge polygon of μ is the convex polygon of slopes in $\{0,1\}$, starting at $(0,0)$ and ending at (h,d), with breakpoint $(h-d,0)$. Thus, an element $[b] \in B(G)$ is μ-admissible if and only if the corresponding F-isocrystals N_b of height h has dimension d, and slopes in the closed interval $[0,1]$. Then, $B(G,\mu)$ is equal to the set of F-isocrystals of Barsotti–Tate groups of height h and dimension d. Equivalently, it can be identified with the set of isogeny classes of Barsotti–Tate groups over $\bar{k}$, of height h and dimension d. The unique basic element of $B(G,\mu)$ corresponds to the isoclinic F-isocrystal of slope d/h. The μ-ordinary element of $B(G,\mu)$ corresponds to the ordinary F-isocrystal, i.e. the F-isocrystal of slopes 0 and 1.

3 Newton strata and Oort's foliations

In this section, we explain how the geometry in characteristic p of PEL type Shimura varieties can be discussed in terms of the F-isocrystals associated to the p-divisible part of the corresponding abelian varieties.

A first study of the behavior in families of the Newton polygons of F-crystals associated with Barsotti–Tate groups is due to Grothendieck. His results were later improved by Katz and are known as Grothendieck's Specialization Theorem ([12], Theorem 2.3.1). A converse to this theorem and its application to the study of the Newton polygon stratification of moduli spaces of abelian varieties in positive characteristic (which is defined by the loci where the isogeny class of the p-divisible part of the abelian varieties is constant) are due to Oort in [27]. Subsequently, in the work of Rapoport and Richartz in [34], these ideas have been reformulated in and applied to the study of Shimura varieties. More recently, in [28], Oort has obtained new results on the geometry of the Newton polygon strata of moduli spaces of abelian varieties in positive characteristic p by introducing two foliations on the strata. The first one (called the *central foliation*) is defined by the loci where the isomorphism class of the p-divisible part of the abelian varieties is constant. The second one (called the *isogeny foliation*) is defined by the loci where the p-isogeny class of the abelian varieties is constant. (The latter, as we will explain in the next section, is closely related to the construction of moduli spaces of p-divisible group due to Rapoport and Zink in [36]). In [18] and [19], these constructions of Oort have been generalized in and applied to the study of Shimura varieties. In the following, we recall these notions and results following respectively the work in [34] and [19].

3.1 Shimura varieties

We briefly recall notations from the previous chapters in this book by Nicole [25] and Rozensztajn [37]. Let (G,X) be a PEL-type Shimura datum as in [25] (Section 2.2) and [37] (Section 1.1). Assume p is an unramified prime of good reduction for (G,X) (as in [37], Section 1.2), and $K_{p,0} \subset G(\mathbb{Q}_p)$ is hyperspecial (in [37], Section 1.3, $K_{p,0}$ is denoted by C_0).

For $K^p \subset G(\mathbb{A}_f^p)$ a sufficiently small open compact subgroup, let S_{K^p} over $\mathcal{O}_{E_v}$ be the canonical integral model of the Shimura variety $Sh_K \otimes_E E_v$, of level $K = K^p K_{p,0} \subset G(\mathbb{A}_f)$ ([37], Section 3), and denote by $\bar{S}_{K^p}$ over $k(v)$ its reduction modulo p. We write A for the universal abelian variety over $\bar{S}_{K^p}$, and $H = A[p^\infty]$ for its p-divisible group. Then, H is a Barsotti–Tate group with $G_{\mathbb{Q}_p}$-structure, namely in this case a Barsotti–Tate group endowed with a quasi-polarization and an action of $\mathcal{O}_B \otimes \mathbb{Z}_p$, the p-adic completion of $\mathcal{O}_B$ in $B_{\mathbb{Q}_p}$.

For any $h \in X$, let $\mu_h : \mathbb{G}_{m,\mathbb{C}} \to G_{\mathbb{C}}$ denote the cocharacter $\mu_h(z) = h_{\mathbb{C}}(z, 1)$. After choosing a local embedding at p of the algebraic closure $\bar{\mathbb{Q}}$ of $\mathbb{Q}$ in $\mathbb{C}$, $\nu : \bar{\mathbb{Q}} \to \bar{\mathbb{Q}}_p$, X determines a conjugacy class of cocharacters of $G_{\mathbb{Q}_p}$, namely the class of $\mu_{\bar{\mathbb{Q}}_p} = \nu \circ \mu_h : \mathbb{G}_m \to G_{\bar{\mathbb{Q}}_p}$. The conditions satisfied by the Shimura datum imply that the cocharacter $\mu_{\bar{\mathbb{Q}}_p}$ is minuscule ([25], Section 2.2). A Barsotti–Tate group H with $G_{\mathbb{Q}_p}$-structure is said to be *compatible* with $\mu_{\bar{\mathbb{Q}}_p}$ if the induced $G_{\mathbb{Q}_p}$-structure on its Lie algebra, $\mathrm{Lie}(H)$, satisfies Kottwitz's *determinant condition* ([37], Section 2.2). It follows that the universal Barsotti–Tate group H over $\bar{S}_{K^p}$ is a $\mu_{\bar{\mathbb{Q}}_p}$-compatible Barsotti–Tate group with $G_{\mathbb{Q}_p}$-structure.

3.2 Newton strata

To each geometric closed point x of $\bar{S}_{K^p}$, we associate H_x the fiber of H at x, $\mathbb{D}H_x$ the (covariant) Dieudonné module of H_x, and $N_x = \mathbb{D}H_x \otimes_{\mathbb{Z}_p} \mathbb{Q}_p$ the F-isocrystal of H_x. We write $k_x \supset \overline{k(v)}$ for the field of definition of x. For any point x in $\bar{S}_{K^p}$, the corresponding F-isocrystal N_x determines a unique element $b_x \in B(G_{\mathbb{Q}_p})$ (here, we simply write b in place of $[b]$ for an element in $B(G)$). It is important to remark that, since the $G_{\mathbb{Q}_p}$-structure on H_x is compatible with $\mu_{\bar{\mathbb{Q}}_p}$, the corresponding element $b_x \in B(G_{\mathbb{Q}_p})$ is $\mu_{\bar{\mathbb{Q}}_p}$-admissible, i.e. $b_x \in B(G_{\mathbb{Q}_p}, \mu_{\bar{\mathbb{Q}}_p})$.

The generalization of Katz-Grothendieck's Specialization Theorem to this context (which is due to Rapoport-Richartz in [34], Theorem 3.6) implies that for each $b \in B(G_{\mathbb{Q}_p})$ the subset

$$\bar{S}_{K^p}(\leq b) = \{x \in \bar{S}_{K^p} | b_x \leq b\}$$

is a Zariski-closed subspace of $\bar{S}_{K^p}$. We define $\bar{S}_{K^p}(\leq b)$ as a closed subscheme of $\bar{S}_{K^p}$, endowed with the induced reduced structure. It follows from the definition that if $b \leq b'$ then $\bar{S}_{K^p}(\leq b)$ is contained in $\bar{S}_{K^p}(\leq b')$. Thus, they form a stratification by closed subschemes of $\bar{S}_{K^p}$, indexed by the elements in $B(G_{\mathbb{Q}_p})$. We call this the *Newton stratification of* $\bar{S}_{K^p}$ and, for each $b \in B(G)$, we call $\bar{S}_{K^p}(\leq b)$ the *closed Newton stratum attached to* b. For each $b \in B(G)$, we define the *Newton stratum attached to* b as

$$\bar{S}_{K^p}(b) = \bar{S}_{K^p}(\leq b) - \bigcup_{b' < b} \bar{S}_{K^p}(\leq b').$$

Then, $\bar{S}_{K^p}(b)$ is a locally closed reduced subscheme of $\bar{S}_{K^p}$, which is open inside the corresponding closed Newton stratum $\bar{S}_{K^p}(\leq b)$.

We remark that since H is a $\mu_{\bar{\mathbb{Q}}_p}$-compatible Barsotti–Tate group with $G_{\mathbb{Q}_p}$-structure then the strata $\bar{S}_{K^p}(b)$ are empty for all $b \in B(G_{\mathbb{Q}_p}) - B(G_{\mathbb{Q}_p}, \mu_{\bar{\mathbb{Q}}_p})$. Thus, the Newton stratification of $\bar{S}_{K^p}$ is actually indexed by the elements $b \in B(G_{\mathbb{Q}_p}, \mu_{\bar{\mathbb{Q}}_p})$. The converse of this statement, i.e. that the strata $\bar{S}_{K^p}(b)$ are non-empty for all $b \in B(G_{\mathbb{Q}_p}, \mu_{\bar{\mathbb{Q}}_p})$, is also true. In the classical case of $G = GL_n$ or $G = GSp_{2n}$, this was conjectured by Manin in [17] and proved first by Honda in [10] and later by Oort in [26]. For PEL-type Shimura varieties, this was conjectured by Fargues ([5], Conjecture 3.1.1) and by Rapoport ([33], Conjecture 7.1), and proved by Viehmann and Wedhorn ([40], Theorem 6). Furthermore, the converse of the generalization of Katz-Grothendieck's Specialization Theorem to this context implies that, for each $b \in B(G_{\mathbb{Q}_p}, \mu_{\bar{\mathbb{Q}}_p})$, the closed Newton stratum attached to b is equal to the Zariski-closure of the corresponding Newton stratum, or equivalently, that for each $b \in B(G_{\mathbb{Q}_p}, \mu_{\bar{\mathbb{Q}}_p})$ the stratum $\bar{S}_{K^p}(b)$ is open and dense in $\bar{S}_{K^p}(\leq b)$. In the case of $G = GL_n$ or $G = GSp_{2n}$, this was conjectured by Grothendieck in [6] and proved by Oort in [26]. For PEL-type Shimura varieties, this was proposed by Rapoport ([33], Question 7.3), and proved by Hamacher ([7], Theorem 1.1). We refer to the chapter in this book by Viehmann [39] for an overview of recent results on the geometry of the Newton stratification.

It follows from the property of the Bruhat–Tits order on $B(G_{\mathbb{Q}_p}, \mu_{\bar{\mathbb{Q}}_p})$ that there is a unique Newton stratum which is closed in $\bar{S}_{K^p}$, the stratum corresponding to the basic element in $B(G_{\mathbb{Q}_p}, \mu_{\bar{\mathbb{Q}}_p})$, and that there is a unique Newton stratum which is open in $\bar{S}_{K^p}$, the stratum corresponding to the $\mu_{\bar{\mathbb{Q}}_p}$-ordinary element in $B(G_{\mathbb{Q}_p}, \mu_{\bar{\mathbb{Q}}_p})$. We call them respectively the *basic Newton stratum* and the *$\mu_{\bar{\mathbb{Q}}_p}$-ordinary Newton stratum*. We recall that in [41] Wedhorn proved that the $\mu_{\bar{\mathbb{Q}}_p}$-ordinary Newton stratum is dense in $\bar{S}_{K^p}$ (and thus, in particular, non-empty).

If we consider the *classical* Newton stratification of $\bar{S}_{K^p}$ (a.k.a. the Newton polygon stratification), which is defined by looking at the isogeny class of the Barsotti–Tate groups underlying the H_x's (i.e., with our notations, $\bar{S}_{K^p}(\leq \rho(b)) = \{x \in \bar{S}_{K^p} | \rho(b_x) \leq \rho(b)\}$, for $\rho : G \to GL(V)$ the standard representation), then it follows from the functorial property of the Bruhat–Tits order on $B(G)$ that the Newton polygon stratification is coarser than or equal to the Newton stratification . While in general the corresponding inclusions are strict, in the cases of interest to us, the two stratifications are actually the same (see [34], Theorem 3.8).

It is an immediate consequence of the definition that, as the level K^p varies, the action of $G(\mathbb{A}_f^p)$ on the Shimura varieties respects the Newton stratifications. Furthermore, for each $b \in B(G_{\mathbb{Q}_p}, \mu_{\bar{\mathbb{Q}}_p})$ and any $K_1^p \subset K^p$, the Newton stratum of $\bar{S}_{K_1^p}$ attached to b is equal to the pullback of the corresponding Newton stratum of $\bar{S}_{K^p}$, under the natural projection between Shimura varieties $\bar{S}_{K_1^p} \to \bar{S}_{K^p}$.

3.3 The cohomology of Shimura varieties

Let l be a prime number, $l \neq p$, and $\mathcal{L}_\rho$ the étale l-adic local system on the Shimura varieties, associated with a representation $\rho \in \mathrm{Rep}_{\mathbb{C}}(G)$ ([25], Sections 4.2, and 5). For all (sufficiently small) open compact subgroups $K^p \subset G(\mathbb{A}_f^p)$ and $K_p \subset K_{p,0}$, we write $f_{K_p} = f_{K^pK_p} : Sh_{K^pK_p} \to Sh_{K^pK_{p,0}}$ for the natural projection between the Shimura varieties, and $R^q\psi(f_{K_p*}(\mathcal{L}_\rho))$ for the nearby cycle sheaves over $\bar{S}_{K^p}$ of $f_{K_p*}(\mathcal{L}_\rho)$, for all $q \geq 0$. With abuse of notation, we also write $R^q\psi(f_{K_p*}(\mathcal{L}_\rho))$ for their restriction to the Newton strata $\bar{S}_{K^p}(b)$, for all $b \in B(G_{\mathbb{Q}_p}, \mu_{\bar{\mathbb{Q}}_p})$. For all $t, q \geq 0$, we define

$$H_c^t(\bar{S}(b), R^q\psi(\mathcal{L}_\rho)) = \varinjlim_{K^p, K_p} H_c^t(\bar{S}_{K^p}(b), R^q\psi(f_{K_p*}(\mathcal{L}_\rho)).$$

These cohomology spaces naturally inherit an action of $G(\mathbb{A}_f) \times W_{E_v}$, and as l-adic representations of $G(\mathbb{A}_f) \times W_{E_v}$, they are admissible. We define

$$H_c(\bar{S}(b), R\psi(\mathcal{L}_\rho)) = \sum_{t,q} (-1)^{t+q} H_c^t(\bar{S}(b), R^q\psi(\mathcal{L}_\rho)) \in \mathrm{Groth}(G(\mathbb{A}_f) \times W_{E_v}).$$

We remark that, if the Shimura varieties considered are proper, then it follows from the theory of nearby cycles and the above properties of the Newton stratification that the following equality holds in $\mathrm{Groth}(G(\mathbb{A}_f) \times W_{E_v})$:

$$H_{\text{ét}}(Sh, \mathcal{L}_\rho) = \sum_{b \in B(G_{\mathbb{Q}_p}, \mu_{\bar{\mathbb{Q}}_p})} H_c(\bar{S}(b), R\psi(\mathcal{L}_\rho)).$$

3.4 Central foliation

Let $\mathbb{X}$ over $\overline{k(v)}$ be a $\mu_{\bar{\mathbb{Q}}_p}$-compatible Barsotti–Tate group with $G_{\mathbb{Q}_p}$-structure, and write $b = b_{\mathbb{X}}$ for the corresponding element in $B(G_{\mathbb{Q}_p}, \mu_{\bar{\mathbb{Q}}_p})$. We say that a ($\mu_{\bar{\mathbb{Q}}_p}$-compatible) Barsotti–Tate group with $G_{\mathbb{Q}_p}$-structure Y, defined over a field $k \supset \overline{k(v)}$, is *geometrically isomorphic* to $\mathbb{X}$ (and write $Y \cong_g \mathbb{X}$) if there exists a field extension k' of k over which the two become isomorphic. We define the *central leaf* associated to $\mathbb{X}$ to be

$$C_{\mathbb{X}} = \{x \in \bar{S}_{K^p} | H_x \cong_g \mathbb{X}\}.$$

It follows from the definition that $C_{\mathbb{X}} \subset \bar{S}_{K^p}(b)$. Furthermore, it is a closed subspace of $\bar{S}_{K^p}(b)$, and if endowed with the induced reduced subscheme structure then it is smooth ([27], Sections 2 and Section 3; [19], Proposition 1). We remark that the latter property is tautological. (Indeed, given such scheme structure, Serre-Tate theory implies that the complete local ring of $C_{\mathbb{X}}$ at any closed geometric point depends only on the geometric isomorphism class of $\mathbb{X}$, and in particular it is independent of the point. Since the smooth locus of a non-empty reduced scheme is non-empty, we conclude that $C_{\mathbb{X}}$ is smooth.)

It follows from the definition that any two central leaves $C_{\mathbb{X}}, C_{\mathbb{X}'}$ of $\bar{S}_{K^p}(b)$ either are disjoint or coincide, and that any closed geometric point x of $\bar{S}_{K^p}(b)$ is contained in exactly one central leaf (which we denote by C_x). Moreover, if $\mathbb{X}, \mathbb{X}'$ are two $\mu_{\bar{\mathbb{Q}}_p}$-compatible Barsotti–Tate groups with $G_{\mathbb{Q}_p}$-structure, with $b = b_{\mathbb{X}} = b_{\mathbb{X}'}$ (thus $\mathbb{X}$ and $\mathbb{X}'$ are isogenous), then there exists a scheme T and two finite surjective morphisms $C_{\mathbb{X}} \twoheadleftarrow T \twoheadrightarrow C_{\mathbb{X}'}$. In particular, $\dim C_{\mathbb{X}} = \dim C_{\mathbb{X}'}$ ([27], Section 2 and Section 3; [19], Section 5). Because of these properties, the central leaves are said to form a foliation (called the *central foliation*) of $\bar{S}_{K^p}(b)$.

We remark that, for any $b \in B(G_{\mathbb{Q}_p}, \mu_{\bar{\mathbb{Q}}_p})$, if the Newton stratum $\bar{S}_{K^p}(b)$ is not empty then there exists a $\mu_{\bar{\mathbb{Q}}_p}$-compatible Barsotti–Tate group with $G_{\mathbb{Q}_p}$-structure $\mathbb{X}_0$ over $\overline{k(v)}$, with $b_{\mathbb{X}_0} = b$, such that the corresponding central leaf $C_{\mathbb{X}_0}$ is not empty, and in such case, all central leaves $C_{\mathbb{X}}$ are not empty for all $\mu_{\bar{\mathbb{Q}}_p}$-compatible Barsotti–Tate groups with $G_{\mathbb{Q}_p}$-structure $\mathbb{X}$ over $\overline{k(v)}$, with $b_{\mathbb{X}} = b$.

We also remark that if we consider the *original* central foliations of the Newton strata, which is defined by fixing the isomorphism class of the Barsotti–Tate groups underlying the

H_x's, then it follows from the definition that it is coarser than or equal to the corresponding above central foliations. In particular, the central leaves (as defined above) are unions of connected components of the corresponding original ones ([19], Proposition 1).

Finally, it is an immediate consequence of the definition that, as the level K^p varies, the action of $G(\mathbb{A}_f^p)$ on the Shimura varieties respects the central foliations. Furthermore, for each a $\mu_{\bar{\mathbb{Q}}_p}$-compatible Barsotti–Tate group with $G_{\mathbb{Q}_p}$-structure $\mathbb{X}$ and any $K_1^p \subset K^p$, the associated central leaf of $\bar{S}_{K_1^p}(b)$ $(b = b_{\mathbb{X}})$ is equal to the pullback of the corresponding central leaf of $\bar{S}_{K^p}$, under the natural projection $\bar{S}_{K_1^p}(b) \to \bar{S}_{K^p}(b)$.

3.5 Isogeny foliation

Let x be a closed geometric point of $\bar{S}_{K^p}(b)$ defined over $\overline{k(v)}$. We write A_x for the associated abelian variety with G-structure, the fiber of A at x. We say that an abelian variety with G-structure B, defined over a field $k \supset \overline{k(v)}$, is *geometrically p-isogenous* to A_x (and write $B \sim_g A_x$) if there exist a field extension k' of k and a p-power isogeny $A_x \otimes_{\overline{k(v)}} k' \to B \otimes_k k'$ of abelian varieties with the G-structures. We define the *isogeny leaf* of x to be

$$I_x = \{y \in \bar{S}_{K^p} | A_y \sim_g A_x\}.$$

It follows from the definition that $I_x \subset \bar{S}_{K^p}(b)$. Furthermore, it is a closed subspace of $\bar{S}_{K^p}(b)$. We regard I_x as a subscheme of $\bar{S}_{K^p}(b)$ via the induced reduced scheme structure ([27], Section 4).

It follows from the definition that any two isogeny leaves $I_x, I_{x'}$ of $\bar{S}_{K^p}(b)$ either are disjoint or coincide, and that any closed geometric point x of $\bar{S}_{K^p}(b)$ is contained in exactly one isogeny leaf. Moreover, if x, x' are two closed geometric point of $\bar{S}_{K^p}(b)$, then there exists a scheme J and two finite surjective morphisms $I_x \twoheadleftarrow J \twoheadrightarrow I_{x'}$. In particular, $\dim I_x = \dim I_{x'}$ ([27], Section 4; [19], Section 5). Thus, the isogeny leaves form a foliation (called the *isogeny foliation*) of $\bar{S}_{K^p}(b)$.

Finally, it is an immediate consequence of the definition that, as the level K^p varies, the isogeny foliation is respected by the action of $G(\mathbb{A}_f^p)$ on the Shimura varieties, i.e. for all $g \in G(\mathbb{A}_f^p) : g(I_x) = I_{g(x)}$, for any geometric closed point x of $\bar{S}_{K^p}$.

4 Igusa varieties and Rapoport-Zink spaces

In this section, we introduce the notions of Igusa variety and Rapoport-Zink space, and explain their relation to the central and isogeny foliations of the Newton strata of Shimura varieties of PEL type.

A crucial ingredient in the study of the bad reduction of modular curves is due to Igusa. In [11], Igusa introduced some new moduli spaces for elliptic curves (now called *Igusa curves*), which exist exclusively in positive characteristic, and proved that up to a finite inseparable morphism they can be identified with the smooth components of the reduction of the corresponding modular curves. Many years later, in [9], Harris and Taylor generalized Igusa's ideas in the context of some simple Shimura varieties. They defined some new varieties, which they called *Igusa varieties*, as smooth covering spaces of the Newton strata of Shimura varieties with good reduction, and applied their new construction to the study of the cases of bad reduction. In particular, similarly to the case of modular curves, the smooth components of the bad reduction of a simple Shimura variety can be identified, up to a finite inseparable morphism, to a *compactification* of the Igusa variety defined over the $\mu_{\bar{\mathbb{Q}}_p}$-ordinary Newton stratum of the corresponding Shimura variety with good reduction ([20], Proposition 16). In [18] and [19], Harris-Taylor's new notion of Igusa variety is extended to the context of Shimura varieties of PEL type and applied to the study of their bad reductions. In this generality, the Igusa varieties arise as smooth covering spaces of the central leaves of the Newton strata of the Shimura varieties with good reduction. In particular, even in the case corresponding to the $\mu_{\bar{\mathbb{Q}}_p}$-ordinary Newton stratum, their dimension is in general strictly smaller than the dimension of the Shimura varieties. (We will see in the last section that, in the case of the simple Shimura varieties considered in [9], each central foliation consists of a unique leaf, equal to the whole Newton stratum.) This difference in dimension is explained by the existence of the isogeny foliation (which in the case of simple Shimura varieties is zero dimensional). As it was already mentioned, the isogeny foliation is closely related to certain moduli spaces of p-divisible groups with additional structure constructed by Rapoport and Zink in [36] as local analogues of Shimura varieties. More precisely, Rapoport-Zink's p-adic uniformization theorem completely describes the geometry of the isogeny leaves in terms of these new spaces.

4.1 Slope filtration

Let $\mathbb{X}$ be a Barsotti–Tate group and $N = N_{\mathbb{X}}$ the associated F-isocrystal. We write $N = \oplus_i N^i$ for the slope decomposition of N, each N^i isoclinic of slope λ_i. We also order the slope of N in decreasing order: $1 \geq \lambda_1 > \lambda_2 > \cdots > \lambda_r \geq 0$. We call a *slope filtration* of $\mathbb{X}$ a filtration $0 = \mathbb{X}_0 \subset \mathbb{X}_1 \subset \mathbb{X}_2 \cdots \subset \mathbb{X}_r = \mathbb{X}$ of $\mathbb{X}$ by Barsotti–Tate subgroups satisfying the condition that for each $i \in \{0, \dots, r\}$ the corresponding subquotient $\mathbb{X}^i = \mathbb{X}_i/\mathbb{X}_{i-1}$ is an isoclinic Barsotti–Tate group of slope λ_i (thus, for each i, N^i can be identified with the F-isocrystal associated to $\mathbb{X}^i$). It follows from the definition that if a slope filtration exists then it is unique. The notion of slope filtration for Barsotti–Tate groups is due to Grothendieck who in [6] proved that any Barsotti–Tate group defined over a field of positive characteristic admits a slope filtration and that this filtration is split in the case when the field is perfect. Grothendieck's result was later extended by Katz ([12]) to families of Barsotti–Tate groups over a smooth curve in characteristic p, and more recently by Zink ([42]) over regular schemes and by Oort and Zink ([31]) over normal schemes. In [31], Oort and Zink also discuss counterexamples to more general statements.

Let $\mathbb{X}$ over $\overline{k(v)}$ be a $\mu_{\bar{\mathbb{Q}}_p}$-compatible Barsotti–Tate group with $G_{\mathbb{Q}_p}$-structure, $N = N_{\mathbb{X}}$ the associated F-isocrystal with $G_{\mathbb{Q}_p}$-structure, $b = b_{\mathbb{X}}$ the corresponding element in $B(G_{\mathbb{Q}_p}, \mu_{\bar{\mathbb{Q}}_p})$, and $1 \geq \lambda_1 > \lambda_2 > \cdots > \lambda_r \geq 0$ the slopes of $\mathbb{X}$. We write $\mathbb{X} = \oplus_i \mathbb{X}^i$ for the slope decomposition of $\mathbb{X}$, each $\mathbb{X}^i$ an isoclinic Barsotti–Tate group of slope λ_i defined over $\overline{k(v)}$. Then, it follows from the definition that $\mathbb{X}$ has a natural M_b-structure (for M_b the centralizer of the slope morphism ν_b of $b = b_{\mathbb{X}}$). Indeed, the action of the maximal order $\mathcal{O}_B \otimes \mathbb{Z}_p$ of $B_{\mathbb{Q}_p}$ on $\mathbb{X}$ respects the slope decomposition of $\mathbb{X}$, i.e. each isoclinic component $\mathbb{X}^i$ inherits a structure of $\mathcal{O}_B \otimes \mathbb{Z}_p$-module. Moreover, the quasi-polarization of $\mathbb{X}$, $\ell : \mathbb{X} \to \mathbb{X}^\vee$, naturally decomposes as $\ell = \oplus_i \ell^i$, the direct sum of $\mathcal{O}_B \otimes \mathbb{Z}_p$-equivariant isomorphisms $\ell^i : \mathbb{X}^i \to (\mathbb{X}^{r+1-i})^\vee$ satisfying the conditions $\ell^i = c \cdot (\ell^{r+1-i})^\vee$ for all $i \in \{0, \ldots, r\}$ and a constant $c \in \mathbb{Z}_{(p)}^\times$ independent of i.

In the following we will assume that $\mathbb{X}$ is *completely slope divisible.* The notion of complete slope divisibility is due to Zink in [42]; we recall this definition. Let $s > 0$ and $t_1, \ldots t_r$ be integers such that $s \geq t_1 > t_2 > \cdots > t_r \geq 0$. A Barsotti–Tate group $\mathbb{X}$ is said to be completely slope divisible with respect to these integers if $\mathbb{X}$ has a filtration by Barsotti–Tate subgroups $0 = \mathbb{X}_0 \subset \mathbb{X}_1 \subset \mathbb{X}_2 \cdots \subset \mathbb{X}_r = \mathbb{X}$ such that for each $i \in \{1, \ldots r\}$ the quasi isogenies $p^{-t_i} F^s : \mathbb{X}_i \to \mathbb{X}_i^{(p^s)}$ are isogenies, and the induced morphisms $p^{-t_i} F^s : \mathbb{X}_i/\mathbb{X}_{i-1} \to (\mathbb{X}_i/\mathbb{X}_{i-1})^{(p^s)}$ are isomorphisms (here, $F : \mathbb{X} \to \mathbb{X}^{(p)}$ denotes the relative Frobenius of $\mathbb{X}$). Note that the last condition implies that for each i the Barsotti–Tate subquotient $\mathbb{X}^i = \mathbb{X}_i/\mathbb{X}_{i-1}$ is isoclinic of slope $\lambda_i = t_i/s$, and that the filtration is a slope filtration. In [31], Oort and Zink proved that if $\mathbb{X}$ is a Barsotti–Tate group defined over an algebraically closed field, then $\mathbb{X}$ is completely slope divisible if and only if it is isomorphic to the direct sum of isoclinic Barsotti–Tate groups defined over finite fields. On the other hand, we recall that an element $b \in B(G_{\mathbb{Q}_p})$ is said to be *decent* if it contains an element of $G(L)$ defined over an unramified finite field extension of $\mathbb{Q}_p$. In [14] Kottwitz proved that if $G_{\mathbb{Q}_p}$ is connected then every element in $B(G_{\mathbb{Q}_p})$ is decent. The above result of Oort and Zink implies that if $b \in B(G_{\mathbb{Q}_p}, \mu_{\bar{\mathbb{Q}}_p})$ then b is decent if and only if there exists a completely slope divisible $\mu_{\bar{\mathbb{Q}}_p}$-compatible Barsotti–Tate group with $G_{\mathbb{Q}_p}$-structure defined over $\overline{k(v)}$ associated with b.

4.2 Igusa varieties

Let $b \in B(G_{\mathbb{Q}_p}, \mu_{\bar{\mathbb{Q}}_p})$, and $\mathbb{X}$ a completely slope divisible $\mu_{\bar{\mathbb{Q}}_p}$-compatible Barsotti–Tate group with $G_{\mathbb{Q}_p}$-structure associated with b (the above observations imply that under our assumptions such a Barsotti–Tate group always exists). For $K^p \subset G(\mathbb{A}_f^p)$ a sufficiently small open compact subgroup, let $C = C_{\mathbb{X}} \subset \bar{S}_{K^p}(b)$ denote the central leaf and Newton stratum associated respectively with $\mathbb{X}$ and b inside the reduction of the Shimura variety of level K^p. It is an easy application of Zink's work to our context that the restriction of the universal family of Barsotti–Tate groups H over C admits a slope filtration ([18], Section 3.2.3). Moreover, if we write $H^1, \ldots, H^r$ for the Barsotti–Tate subquotients associated with the slope filtration of H over C, then it follows from the definition that the Barsotti–Tate group $H^{\mathrm{sp}} = \oplus_i H^i$ has a natural M_b-structure and that for each point x of C the Barsotti–Tate group with M_b-structure H_x^{sp} is geometrically isomorphic to $\mathbb{X}$. In particular, for each $i \in \{0, \ldots, r\}$, the Barsotti–Tate group H_x^i is geometrically isomorphic to $\mathbb{X}^i$.

For any integer $m \geq 1$, we define $\mathrm{Ig}_m = \mathrm{Ig}_{\mathbb{X},m}$ the *Igusa variety of level m attached to* $\mathbb{X}$ as the universal space over $C = C_{\mathbb{X}}$ for the existence of an isomorphism of truncated Barsotti–Tate groups with M_b-structure $j_m : \mathbb{X}[p^m]_C \cong H^{\mathrm{sp}}[p^m]$. (By a homomorphism of truncated Barsotti–Tate groups with M_b-structure we mean a homomorphism of the corresponding finite flat group schemes which commutes with the induced M_b-structures and which extends étale locally to any depth $m' \geq m$, for m the depth of the truncation.) It follows from the definition that for each $m \geq 1$ the corresponding Igusa variety $\mathrm{Ig}_m \to C$ is naturally endowed with an action of the group of isomorphisms of the m-th truncation of $\mathbb{X}$. We denote this group by $\Gamma_m = \Gamma_{\mathbb{X},m}$ and regard it as the quotient of $\Gamma = \Gamma_{\mathbb{X}}$, the group of automorphisms of $\mathbb{X}$, by the subgroup of those automorphisms which induce the identity on $\mathbb{X}[p^m]$. For each $m \geq 1$, the Igusa variety $\mathrm{Ig}_m \to C$ is finite étale and Galois over C, with Galois group Γ_m. In particular, it is a smooth scheme over $\overline{k(v)}$ ([19], Proposition 4). In the following, we sometimes write $\mathrm{Ig}_0 = C$ and $\mathrm{Ig}_{m,K^p} = \mathrm{Ig}_{\mathbb{X},m,K^p}$ for the Igusa variety of level m, when we wish to include the level of the corresponding Shimura variety in the notation.

It follows from the moduli interpretation of the Igusa varieties that, as the level m varies, they form a projective system. Furthermore, if we regard the Igusa varieties as endowed with an action of Γ (via the natural projections $\Gamma \twoheadrightarrow \Gamma_m$, for all $m \geq 1$), then the projective system naturally inherits an action of Γ. We regard Γ as a subgroup of the group of the quasi-self-isogenies of $\mathbb{X}$, which we identify with $J_b(\mathbb{Q}_p)$. Let $\rho \in J_b(\mathbb{Q}_p)$. If ρ^{-1} is an isogeny of $\mathbb{X}$, we define $f_i = f_i(\rho)$ and $e_i = e_i(\rho)$ to be respectively the maximal and minimal positive integers such that $\oplus_i \mathbb{X}^i[p^{f_i}] \subset \ker(\rho^{-1}) \subset \oplus_i \mathbb{X}^i[p^{e_i}]$ (thus $f_i \leq e_i$, for all i). We define $S = S_{\mathbb{X}}$ as the set of elements in $J_b(\mathbb{Q}_p)$ whose inverse is an isogeny of $\mathbb{X}$ satisfying the conditions $f_{i-1} \geq e_i$, for all $i \geq 2$. It follows that S is a submonoid of $J_b(\mathbb{Q}_p)$ which contains Γ. Moreover, the action of Γ on the tower of Igusa varieties can be extended to the monoid S, each $\rho \in S$ defining a compatible system of morphisms $\rho : \mathrm{Ig}_m \to \mathrm{Ig}_{m-e}$, for $e = e_1(\rho)$ and any $m \geq e$ ([19], Lemma 5).

As we also allow the level K^p to vary, then the corresponding Igusa varieties Ig_{m,K^p} form a projective system indexed by both the levels m and K^p. Moreover, for each $g \in G(\mathbb{A}_f^p)$ and $K_1^p \subset g^{-1}K^p g$, the morphism $g : C_{K_1^p} \to C_{K^p}$ canonically lifts to morphisms $g : \mathrm{Ig}_{m,K_1^p} \to \mathrm{Ig}_{m,K^p}$ for all $m \geq 1$, defining an action of $G(\mathbb{A}_f^p)$ on the projective system of Igusa varieties. It is easy to see that the action of $G(\mathbb{A}_f^p)$ commutes with the previously defined action of $S \subset J_b(\mathbb{Q}_p)$.

4.3 The cohomology of Igusa varieties

Let l be prime number, $l \neq p$, and $\mathcal{L}_\rho$ the étale l-adic local system on the tower of Shimura varieties associated with a representation $\rho \in \mathrm{Rep}_{\mathbb{C}}(G)$. With abuse of notations, we also denote by $\mathcal{L} = \mathcal{L}_\rho$ the pullback of $\mathcal{L}_\rho$ to the tower of Igusa varieties, under the natural morphisms $\mathrm{Ig}_{\mathbb{X},m,K^p} \to C_{\mathbb{X},K^p} \subset \bar{S}_{K^p}(b) \subset \bar{S}_{K^p}$. For all $t \geq 0$, we define

$$H_c^t(\mathrm{Ig}_{\mathbb{X}}, \mathcal{L}) = \varinjlim_{m,K^p} H_c^t(\mathrm{Ig}_{\mathbb{X},m,K^p}, \mathcal{L}).$$

These cohomology groups naturally inherit an action of $S_{\mathbb{X}} \times G(\mathbb{A}_f^p)$ and it is not hard to see that this action uniquely extends to an action of $J_b(\mathbb{Q}_p) \times G(\mathbb{A}_f^p)$. Moreover, as l-adic representations of $J_b(\mathbb{Q}_p) \times G(\mathbb{A}_f^p)$, they are admissible ([19], Proposition 7). We define

$$H_c(\mathrm{Ig}_{\mathbb{X}}, \mathcal{L}) = \sum_t (-1)^t H_c^t(\mathrm{Ig}_{\mathbb{X}}, \mathcal{L}) \in \mathrm{Groth}(J_b(\mathbb{Q}_p) \times G(\mathbb{A}_f^p)).$$

4.4 Rapoport-Zink spaces

In [36], Rapoport and Zink associate to each class $b \in B(G_{\mathbb{Q}_p}, \mu_{\bar{\mathbb{Q}}_p})$ a tower of moduli spaces of $\mu_{\bar{\mathbb{Q}}_p}$-compatible Barsotti–Tate groups with $G_{\mathbb{Q}_p}$-structure in the category of rigid analytic spaces. For each $b \in B(G_{\mathbb{Q}_p}, \mu_{\bar{\mathbb{Q}}_p})$, the associated tower of Rapoport–Zink spaces is a projective system of rigid analytic spaces defined over L, indexed by the set of open compact subgroups K_p of $G(\mathbb{Q}_p)$. This projective system is naturally endowed with an action of $G(\mathbb{Q}_p) \times J_b(\mathbb{Q}_p)$ and with a non-effective decent datum to E_v. Moreover, similarly to the case of the corresponding Shimura varieties, in the case of $K_p = K_{p,0}$ maximal compact open of $G(\mathbb{Q}_p)$, the corresponding Rapoport–Zink space admits a smooth integral model in the category of formal schemes over $\mathcal{O}_L$. We recall Rapoport and Zink's construction in more detail.

Let $\mathbb{X}$ over $\overline{k(v)}$ be a $\mu_{\bar{\mathbb{Q}}_p}$-compatible Barsotti–Tate group with $G_{\mathbb{Q}_p}$-structure associated with b. For $K_p = K_{p,0}$, we write $\mathcal{M}_b = \mathcal{M}_{b,\mathbb{X}}$ for the corresponding Rapoport–Zink formal scheme over $\mathcal{O}_L$. The set-valued functor on the category of formal $\mathcal{O}_L$-schemes associated to $\mathcal{M}_b$ is defined as follows. For any $\mathcal{O}_L$-scheme S where p is locally nilpotent, $\mathcal{M}_b(S)$ is the set of isomorphism classes of pairs (H, β), where H is a $\mu_{\bar{\mathbb{Q}}_p}$-compatible Barsotti–Tate group with $G_{\mathbb{Q}_p}$-structure defined over S, and β is a quasi-isogeny $\beta : \mathbb{X} \times_{\overline{k(v)}} \bar{S} \to H \times_S \bar{S}$, defined over $\bar{S} = Z(p) \subset S$. After identifying $J_b(\mathbb{Q}_p)$ with the group of quasi-self-isogenies of $\mathbb{X}$, we define an action of $J_b(\mathbb{Q}_p)$ on $\mathcal{M}_b$ by right translations, i.e. for all $\rho \in J_b(\mathbb{Q}_p) : (H, \beta) \mapsto (H, \beta \circ \rho)$. Similarly, we define a non-effective descent datum on $\mathcal{M}_b$, via the σ-semi-linear isomorphism of $\mathcal{M}_b$ arising from the Frobenius of $\mathbb{X}$, $(H, \beta) \mapsto (H, F^{-1} \circ \beta)$. We remark that the descent datum and the action of $J_b(\mathbb{Q}_p)$ commute with each other.

Let H denote the universal family of Barsotti–Tate groups over $\mathcal{M}_b$, we define $\mathcal{M}_{b,K_{p,0}}$ to be the rigid analytic fiber over L of $\mathcal{M}_b$ and T_pH the Tate module of H over $\mathcal{M}_{b,K_{p,0}}$, regarded as a locally constant étale $\mathbb{Z}_p$-sheaf over $\mathcal{M}_{b,K_{p,0}}$. For each $K_p \subset K_{p,0}$, the *Rapoport–Zink space of level K_p attached to b*, $\mathcal{M}_{b,K_p}$, is defined as the universal space over $\mathcal{M}_{b,K_{p,0}}$ for the existence of a K_p-level structure on T_pH (i.e. the covering space parameterizing the classes modulo K_p of trivialization of T_pH). For each K_p, the space $\mathcal{M}_{b,K_p}$ is finite étale and Galois over $\mathcal{M}_{b,K_{p,0}}$. In particular, it is a smooth rigid analytic space of dimension $D_b = \dim \mathcal{M}_{b,K_{p,0}}$ ([36], Section 5.34). It also follows from the definition that the action of $J_b(\mathbb{Q}_p)$ and the descent datum on $\mathcal{M}_{b,K_{p,0}}$ canonically lift to $\mathcal{M}_{b,K_p}$. Moreover, as the level K_p varies, these covers naturally form a projective system endowed with the obvious action of $G(\mathbb{Q}_p)$.

We remark that although the definition depends on the choice of a Barsotti–Tate group $\mathbb{X}$ associated with b, the isomorphism class of the corresponding Rapoport–Zink spaces depends only on the class $b \in B(G_{\mathbb{Q}_p}, \mu_{\bar{\mathbb{Q}}_p})$. We also point out that the Rapoport–Zink spaces are only locally of finite type. In fact, they naturally arise as a direct limit of subspaces of finite type. More precisely, we may describe $\mathcal{M}_b$ as $\mathcal{M}_b = \varinjlim_{n,d} \mathcal{M}^{n,d}_{b,\mathbb{X}}$, where for each pair of positive integers n, d, we denote by $\mathcal{M}^{n,d}_{b,\mathbb{X}}$ the formal subscheme of $\mathcal{M}_b$ where the universal quasi-isogeny $\beta : \mathbb{X} \times_{\overline{k(v)}} \mathcal{M}^{\mathrm{red}}_b \to H \times_{\mathcal{M}_b} \mathcal{M}^{\mathrm{red}}_b$, defined over the reduced fiber of $\mathcal{M}_b$, satisfies the condition that $p^n\beta$ is a well-defined isogeny of degree less than or equal to p^d (the latter is a closed condition in $\mathcal{M}^{\mathrm{red}}_b$). We remark that the truncated Rapoport–Zink spaces $\mathcal{M}^{n,d}_{b,\mathbb{X}}$ are not stable under either the action of $J_b(\mathbb{Q}_p)$ or the descent datum. Moreover, their definition depends on the choice of a Barsotti–Tate group $\mathbb{X}$ associated with b.

4.5 The cohomology of Rapoport–Zink spaces

Let l be a prime number, $l \neq p$. For all $t \geq 0$, we define

$$H^t_c(\mathcal{M}_b, \bar{\mathbb{Q}}_l) = \varinjlim_{K_p} H^t_c(\mathcal{M}_{b,K_p} \times_L \hat{\bar{E}}_v, \bar{\mathbb{Q}}_l).$$

It follows from the definition that the above cohomology groups are naturally endowed with commuting actions of $G(\mathbb{Q}_p)$, $J_b(\mathbb{Q}_p)$ and of the Weil group W_{E_v}. As representations of $J_b(\mathbb{Q}_p) \times G(\mathbb{Q}_p) \times W_{E_v}$, they are smooth but it is not known whether they are also admissible. On the other hand, for any admissible representation ρ of $J_b(\mathbb{Q}_p)$, the groups $\mathrm{Ext}^i_{J_b(\mathbb{Q})}(H^t_c(\mathcal{M}_b, \bar{\mathbb{Q}}_l), \rho)$, regarded as smooth representations of $G(\mathbb{Q}_p)\times W_{E_v}$, are admissible and vanish for almost all $i, t \geq 0$ ([5], Corollary 4.4.14; [18], Section 8.2). (Here, $\mathrm{Ext}^i_{J_b(\mathbb{Q}_p)}(\cdot,\cdot)$ are the derived functors on the category of smooth $J(\mathbb{Q}_p)$-representations.) It follows that to each $b \in B(G_{\mathbb{Q}_p}, \mu_{\bar{\mathbb{Q}}_p})$ we can associate a functor $\mathcal{E}_b : \mathrm{Groth}(J_b(\mathbb{Q}_p)) \to \mathrm{Groth}(G(\mathbb{Q}_p) \times W_{E_v})$ defined as

$$\mathcal{E}_b(\rho) = \sum_{i,t\geq 0} (-1)^{i+t} \mathrm{Ext}^i_{J_b(\mathbb{Q})}(H^t_c(\mathcal{M}_b, \bar{\mathbb{Q}}_l), \rho)(-D_b).$$

In the following we will also denote by $\mathcal{E}_b$ the functor from $\mathrm{Groth}(J_b(\mathbb{Q}_p) \times G(\mathbb{A}^p_f))$ to $\mathrm{Groth}(G(\mathbb{A}_f) \times W_{E_v})$ obtained by extending the above functor by the identity on $\mathrm{Groth}(G(\mathbb{A}^p_f))$.

Alternatively, using Berkovich's theory of nearby cycle sheaves for formal schemes ([1],[2]), the l-adic cohomology of the Rapoport–Zink spaces can be computed in terms of the cohomology of the reduced fiber $\mathcal{M}^{\mathrm{red}}_b$ with coefficient in the appropriate nearby cycle sheaves. For each $K_p \subset K_{p,0}$, we write $g_{K_p} : \mathcal{M}_{b,K_p} \to \mathcal{M}_{b,K_{p,0}}$ for the natural projection among Rapoport–Zink spaces, and $R^q\psi(g_{K_p *}(\bar{\mathbb{Q}}_l))$ for the nearby cycle sheaves over $\mathcal{M}^{\mathrm{red}}_b$ of $g_{K_p *}(\bar{\mathbb{Q}}_l)$, for all $q \geq 0$. Then, the following equality holds in $\mathrm{Groth}(G(\mathbb{Q}_p) \times W_{E_v})$, for each $\rho \in \mathrm{Groth}(J_b(\mathbb{Q}_p))$,

$$\mathcal{E}_b(\rho) = \sum_{i,p,q\geq 0} (-1)^{i+p+q} \varinjlim_{K_p} \mathrm{Tor}^i_{J_b(\mathbb{Q}_p)}(H^p_c(\mathcal{M}^{\mathrm{red}}_b, R^q\psi(g_{K_p *}(\bar{\mathbb{Q}}_l))), \rho)$$

([18], Theorem 8.7). (Here, the interchange between Tor-groups and Ext-groups reflects the fact that we are considering cohomology with compact support.)

4.6 p-adic uniformization

Finally, we recall Rapoport's and Zink's p-adic uniformization Theorem ([36], Chapter 6). Let K^p be a level away from p, $\bar{S}_{K^p}$ the reduction in positive characteristic of the corresponding Shimura variety, and $\bar{S}_{K^p}(b)$ the Newton stratum associated to an element $b \in B(G_{\mathbb{Q}_p}, \mu_{\bar{\mathbb{Q}}_p})$. For any closed geometric point x of $\bar{S}_{K^p}(b)$, we write I_x and $(A_x, \lambda_x, \iota_x; \bar{\eta}_x)$ respectively for the corresponding isogeny leaf in $\bar{S}_{K^p}(b)$ and polarized abelian variety with G-structure and K^p-level structure over $\overline{k(v)}$. For simplicity, we choose $\mathcal{M}_b = \mathcal{M}_{b,\mathbb{X}}$ for $\mathbb{X} = A_x[p^\infty]$. Let $\mathcal{M}^{\mathrm{red}}_b$ denote the reduced fiber of $\mathcal{M}_b$, and for any closed geometric points y of $\mathcal{M}_b^{\mathrm{red}}$ write (H_y, β_y) for the fiber at y of the universal family (H, β) over $\mathcal{M}_b$. We define a morphism of $\overline{k(v)}$-schemes

$$\Phi_x : \mathcal{M}_b^{\mathrm{red}} \to I_x \subset \bar{S}_{K^p}$$

as $y \mapsto (A_x/(\ker(p^n\beta_y)), \lambda', \iota'; \bar{\eta}')$, where the additional structures on the abelian variety $A_x/(\ker(p^n\beta_y))$ are induced by the those of A_x, via pushforward under the isogeny $A_x \twoheadrightarrow A_x/(\ker(p^n\beta_y))$ (here, n is a sufficiently large positive integer depending on y). The morphism Φ_x is well-defined and surjective onto I_x. Moreover, using Serre-Tate theory, it is easy to see that Φ_x can be canonically extended to a formally étale morphism of formal $\mathcal{O}_L$-schemes $\Phi^{/x}$, from $\mathcal{M}_b$ to the formal completion of S_{K^p} along the isogeny leaf I_x, $(S_{K^p})^{/I_x}$. Finally, it follows from the definition of the Rapoport–Zink covers that the rigid analytic fiber of $\Phi^{/x}$ canonically lifts to a $G(\mathbb{Q}_p)$-equivariant projective system of morphisms $\Phi^{/x}_{K_p}$, from $\mathcal{M}_{b,K_p}$ to the rigid covers of $(S_{K^p})^{/I_x}$ corresponding to the Shimura varieties $Sh_{K^pK_p} \to Sh_{K^pK_{p,0}}$, for all $K_p \subset K_{p,0}$.

5 The quasi-product structure of the Newton strata

In this section, we explain how the geometry and cohomology of the Newton strata in the reduction of the Shimura varieties can be described in terms of those of the corresponding Rapoport-Zink spaces and Igusa varieties.

In [28] (Theorem 5.3) Oort proved that the central and isogeny foliations define an almost product structure on the Newton strata of Siegel varieties. More precisely, he proved that, for each Newton stratum W in the reduction of a Siegel variety, there exist two schemes T and J of finite type over $\bar{\mathbb{F}}_p$ and a finite surjective $\bar{\mathbb{F}}_p$-morphism $\Phi : T \times J \to W$ such that for each closed geometric point x of J, $\Phi(T \times \{x\})$ is a central leaf in W and every central leaf in W can be obtained in this way, and for each closed geometric point y of T, $\Phi(\{y\} \times J)$ is an isogeny leaf in W and every isogeny leaf in W can be obtained in this way. In [18], [19], these ideas are adapted to the context of Shimura varieties. In particular, in the cases of interest to us, for each Newton stratum $\bar{S}_{K^p}(b)$ in the reduction of the Shimura variety $\bar{S}_{K^p}$, for K^p a sufficiently small open compact subgroup of $G(\mathbb{A}_f^p)$ and $b \in B(G_{\mathbb{Q}_p}, \mu_{\bar{\mathbb{Q}}_p})$, the schemes T and J in Oort's theorem can be chosen equal respectively to an Igusa variety $\mathrm{Ig}_{m,\mathbb{X}}$ and the reduced fiber of a truncated Rapoport–Zink space $\mathcal{M}^{n,d}_{b,\mathbb{X}}$, for some sufficiently large integers m, n, d and any choice of a completely slope divisible Barsotti–Tate group $\mathbb{X}$ associated with b. Further more, in these cases, for each closed geometric point x' in $\mathrm{Ig}_{m,\mathbb{X}}$ the restriction of the morphism Φ to the space $\{x'\} \times \mathcal{M}^{n,d}_{b,\mathbb{X}}$ agrees (up to a purely inseparable morphism) with the morphism Φ_x in Rapoport's and Zink's p-adic uniformization theorem, for x the image of x' in $\bar{S}_{K^p}(b)$. These observations lead to a group-equivariant description of the geometry and cohomology of the Newton strata in terms of those of the corresponding products of Igusa varieties and Rapoport-Zink spaces. We recall the main results of [18], [19].

5.1 Quasi-product structure

Let K^p a sufficiently small open compact subgroup of $G(\mathbb{A}_f^p)$, and $\bar{S}_{K^p}$ the corresponding Shimura variety modulo p. For each $b \in B(G_{\mathbb{Q}_p}, \mu_{\bar{\mathbb{Q}}_p})$, let $\mathbb{X}$ be a complete slope divisible

$\mu_{\bar{\mathbb{Q}}_p}$-compatible Barsotti–Tate group with $G_{\mathbb{Q}_p}$-structure associated with b, and for all m, n, d denote by $\mathrm{Ig}_{m,\mathbb{X}}$ and $\bar{\mathcal{M}}^{n,d}_{b,\mathbb{X}}$ the corresponding Igusa varieties and the reduced fibers of the truncated Rapoport–Zink spaces. There exists a system of finite (and almost all surjective) $\overline{k(v)}$-morphisms

$$\pi_N : \mathrm{Ig}_{m,\mathbb{X}} \times \bar{\mathcal{M}}^{n,d}_{b,\mathbb{X}} \to \bar{S}_{K^p}(b)$$

indexed by quadruples of positive integers m, n, d, N satisfying the conditions $m \geq d$ and $N \geq \frac{d}{\delta \log_p q}$, for $\delta = \delta_b \in \mathbb{Q}$ a constant depending only on $b \in B(G)$. As the indexes m, n, d, N vary, these morphisms are compatible under the projections among the Igusa varieties and the inclusion among the reduced truncated Rapoport–Zink spaces, up to a purely inseparable morphism of $\bar{S}_{K^p}(b)$ (namely a power of the q-Frobenius morphism $F_{\bar{S}} = F_{\bar{S}_{K^p}(b)}$ of $\bar{S}_{K^p}(b)$, the actual power depending on the variation of the index N). Furthermore, the resulting compatible system of morphism is also invariant under the action of $S_{\mathbb{X}} \subset J_b(\mathbb{Q}_p)$ on the product of the Igusa varieties by the Rapoport–Zink spaces, and equivariant for the descent data to E_v. Finally, as we also allow the level $K^p \subset G(\mathbb{A}^p_f)$ to vary, the resulting morphisms form a $G(\mathbb{A}^p_f)$-equivariant projective system. We refer to [18] (Section 4) and [19] (Section 5) for the details of the above construction. Here, we simply recall its main properties and its application to the computation of the cohomology of the Newton strata.

5.2 Fiber at a point

Let x be a closed geometric point of $\bar{S}_{K^p}$. For each triple of integers m, n, d, with $m \geq d$, we consider the morphisms $\pi_N : \mathrm{Ig}_{m,\mathbb{X}} \times \bar{\mathcal{M}}^{n,d}_{b,\mathbb{X}} \to \bar{S}_{K^p}(b)$ as N varies. It follows from the compatibility among the various π_N's that, for each $N \geq N_0 = [\frac{d}{\delta \log_p q}] + 1$, there is an equality of finite sets $\pi_N^{-1}(F_{\bar{S}}^N(x)) = \pi_{N_0}^{-1}(F_{\bar{S}}^{N_0}(x))$. Moreover, as m varies, the sets $\pi_N^{-1}(F_{\bar{S}}^N(x))$ form an inverse system under the projection maps among the Igusa varieties, and the corresponding limits form a direct system under the inclusions among the reduced truncated Rapoport–Zink spaces, as n, d vary. We call *the fiber at* x the resulting set

$$\Pi^{-1}(x) = \varinjlim_{n,d} \left(\varprojlim_{m} \pi_N^{-1}(Fr^{fN} x) \right),$$

endowed with the topology of direct limit of inverse limits of discrete sets. The topological space $\Pi^{-1}(x)$ inherits a continuous action of the monoid $S_{\mathbb{X}}$, which it is easy to see extends uniquely to a continuous action of the group $J_b(\mathbb{Q}_p)$.

Alternatively, we can describe the fiber at x, $\Pi^{-1}(x)$, as follows. We define the topological space $\mathrm{Ig}_{\mathbb{X}}(\overline{k(v)}) = \varprojlim_m \mathrm{Ig}_{m,\mathbb{X}}(\overline{k(v)})$, or equivalently

$$\mathrm{Ig}_{\mathbb{X}}(\overline{k(v)}) = \{(B, \lambda, i, \bar{\mu}^p; j) | (B, \lambda, i, \bar{\mu}^p) \in \bar{S}_{K^p}(b)(\overline{k(v)}) \text{ and } j : \mathbb{X} \cong B[p^\infty]\},$$

endowed with the topology of inverse limit of discrete sets. In the following, we simply write (B, j) (resp. B) in place of $(B, \lambda, i, \bar{\mu}^p; j)$ (resp. $(B, \lambda, i, \bar{\mu}^p)$) for a point of $\mathrm{Ig}_{\mathbb{X}}(\overline{k(v)})$ (resp. of $\bar{S}_{K^p}(b)(\overline{k(v)})$). We regard $\mathcal{M}_b(\overline{k(v)})$ as endowed with the discrete topology and the product $\mathrm{Ig}_{\mathbb{X}}(\overline{k(v)}) \times \mathcal{M}_b(\overline{k(v)})$ with the product topology. The action of the monoid $S_{\mathbb{X}}$ on the product of Igusa varieties and Rapoport–Zink spaces induces a continuous action of $S_{\mathbb{X}}$ on the topological space $\mathrm{Ig}_{\mathbb{X}}(\overline{k(v)}) \times \bar{\mathcal{M}}_b(\overline{k(v)})$, namely for all $\rho \in S_{\mathbb{X}} : ((B, j), (H, \beta)) \mapsto \big((B/j(\ker \rho^{-1}), j \circ \rho), (H, \beta \circ \rho)\big)$. It is easy to see that this action extends to a continuous action of $J_b(\mathbb{Q}_p)$. We define a map

$$\Pi : \mathrm{Ig}_{\mathbb{X}}(\overline{k(v)}) \times \bar{\mathcal{M}}_b(\overline{k(v)}) \to \bar{S}_{K^p}(b)(\overline{k(v)})$$

as $((B, j), (H, \beta)) \mapsto B/j(\ker(p^n \beta)$, where the abelian variety $B/j(\ker(p^n \beta))$ is endowed with the additional structures induced by those of B (for n a sufficiently large integer depending on β). It is easy to see that the map Π is continuous for the discrete topology on $\bar{S}_{K^p}(b)(\overline{k(v)})$, and $J_b(\mathbb{Q}_p)$-invariant. In particular, for each $x \in \bar{S}_{K^p}(b)(\overline{k(v)})$, the preimage under Π of x is naturally endowed with a continuous action of $J_b(\mathbb{Q}_p)$. Moreover, it follows from the construction that it is a principle homogeneous $J_b(\mathbb{Q}_p)$-space and that it can be identified (as a topological $J_b(\mathbb{Q}_p)$-space) with the fiber at x, $\Pi^{-1}(x)$.

5.3 The cohomology of the Newton strata

Let l be a prime number, $l \neq p$. The existence of the morphisms π_N enable us to compare the l-adic nearby cycles sheaves of the Shimura varieties, when restricted to the Newton strata, with those of the corresponding Rapoport–Zink spaces, when restricted to the truncated subspaces, after pulling back to the common covers $\mathrm{Ig}_{m,\mathbb{X}} \times \bar{\mathcal{M}}^{n,d}_{b,\mathbb{X}}$. Indeed, it follows from the above construction, that, for each quadruple of sufficiently large integers m, n, d, N, the pullbacks over $\mathrm{Ig}_{m,\mathbb{X}} \times \bar{\mathcal{M}}^{n,d}_{b,\mathbb{X}}$ of the l-adic nearby cycles sheaves defined over $\bar{S}_{K^p}(b)$ and $\bar{\mathcal{M}}_b$, pulled back respectively via the morphism π_N and the second projection $\mathrm{Ig}_{m,\mathbb{X}} \times \bar{\mathcal{M}}^{n,d}_{b,\mathbb{X}} \to \bar{\mathcal{M}}^{n,d}_{b,\mathbb{X}} \subset \bar{\mathcal{M}}_b$, are isomorphic ([19], Proposition 21). By combining this observation with a $J_b(\mathbb{Q}_p)$-equivariant version of the Künneth formula ([18], Section 5), we compute the l-adic cohomology of the Newton strata in terms of the l-adic cohomology of the corresponding Igusa varieties and Rapoport–Zink spaces. More precisely, let $\mathcal{L}_\rho$ be the étale l-adic local system on the Shimura varieties, associated with a representation $\rho \in \mathrm{Rep}_{\mathbb{C}}(G)$. Then, for each $b \in B(G_{\mathbb{Q}_p}, \mu_{\bar{\mathbb{Q}}_p})$, the following equality holds in $\mathrm{Groth}(G(\mathbb{A}_f) \times W_{E_v})$:

$$H_c(\bar{S}(b), R\psi(\mathcal{L}_\rho)) = \mathcal{E}_b(H_c(\mathrm{Ig}_{\mathbb{X}}, \mathcal{L}_\rho)).$$

We recall that, in the case of b basic, an analogous formula computing the cohomology of the basic Newton stratum in terms of that of the corresponding basic Rapoport–Zink space was obtained by Fargues in [5] (Corollary 4.6.3), as an application of Rapoport's and Zink's p-adic uniformization theorem.

6 The signature $(1, n-1)$ case and Lubin-Tate spaces

We discuss in further details the geometry of the reduction modulo p of a special class of PEL type Shimura varieties, which contains the class of simple Shimura varieties considered in [9]. The Shimura varieties in this class parameterize abelian varieties whose local geometry is completely controlled by a one-dimensional Barsotti–Tate $\mathcal{O}_K$-module, for K a local p-adic field. In this case, we allow p to be ramified in the reflex field E, and K

to be ramified over $\mathbb{Q}_p$. The Rapoport–Zink spaces associated with this class of Shimura varieties are closely related to the formal moduli spaces for one-dimensional formal Lie groups constructed by Lubin and Tate in [16], and to the modular varieties constructed by Drinfeld in [4].

6.1 One-dimensional Barsotti–Tate $\mathcal{O}_K$-modules

Let $\mathcal{O}_K$ be the ring of integers of a p-adic field K. A *one-dimensional Barsotti–Tate $\mathcal{O}_K$-module* is a one-dimensional Barsotti–Tate group together with a faithful action of $\mathcal{O}_K$. If H is a one-dimensional Barsotti–Tate $\mathcal{O}_K$-module defined over a $\mathcal{O}_K$-scheme S where p is locally nilpotent, we say that H is *compatible* if the two actions of $\mathcal{O}_K$ on the Lie algebra of H (one from the action of $\mathcal{O}_K$ on H and one from the structure morphism $\mathcal{O}_K \to \mathcal{O}_S$) coincide. In the following we assume all one-dimensional Barsotti–Tate $\mathcal{O}_K$-modules to be compatible. We say that a one-dimensional Barsotti–Tate $\mathcal{O}_K$-module is *ind-étale* (resp. *formal*) if the underlying Barsotti–Tate group is ind-étale (resp. formal).

Let k be an algebraically closed field of characteristic p containing the residue field of K. It follows from the definition that if H is a one-dimensional Barsotti–Tate $\mathcal{O}_K$-module, then the height h of the underlying Barsotti–Tate group is divisible by the degree $[K : \mathbb{Q}_p]$. We define the height of H as $h(H) = \frac{h}{[K:\mathbb{Q}_p]}$. For each integer $g \geq 1$, there is a unique up to isomorphism formal one-dimensional Barsotti–Tate $\mathcal{O}_K$-module $\Sigma_g = \Sigma_{K,g}$ over k of height g. Furthermore, every one-dimensional Barsotti–Tate $\mathcal{O}_K$-module H over k is of the form $\Sigma_g \times (K/\mathcal{O}_K)^r$, for some non-negative integers g and r, with $g + r$ equal to the height of H. We call r the *p-rank* of H, and denote it by $r = \mathrm{rk}(H)$. It follows from the definitions that if H is a one-dimensional Barsotti–Tate $\mathcal{O}_K$-module over k, of height n and p-rank r, then the Newton polygon of the Barsotti–Tate group underlying H is the convex polygon which starts at $(0, 0)$, ends at $(n[K : \mathbb{Q}_p], 1)$, and has a unique break-point at $(r[K : \mathbb{Q}_p], 0)$. In particular, any two one-dimensional Barsotti–Tate $\mathcal{O}_K$-modules over k are isogenous if and only if they are isomorphic, i.e. if and only if they have the same height and p-rank.

6.2 The Shimura datum

We recall the specific assumptions on the moduli data. Let $(B, *, V, \langle,\rangle, h)$ be a Shimura datum of PEL type (as in [37], Section 1.1). We assume B to be central over a number field F. The positivity condition on $*$ implies that F is either a CM field or a totally real field. We assume F is a CM field and fix $\Phi \subset \mathrm{Hom}(F, \mathbb{C})$ a CM type of F. Let G be the unitary similitude group defined over $\mathbb{Q}$ associated with the Shimura datum. The similitude factor c defines a character $c : G \to \mathbb{G}_m$, and we write $G_1 = \ker(c)$. For each $\tau \in \Phi$, we write (p_τ, q_τ) for the signature of G_1 at τ, $p_\tau + q_\tau = n$, i.e.

$$G_{1/\mathbb{R}} \cong \prod_{\tau \in \Phi} U(p_\tau, q_\tau).$$

In the following we assume that there exists a place $\tau_0 \in \Phi$ such that $(p_{\tau_0}, q_{\tau_0}) = (1, n-1)$, and $(p_\tau, q_\tau) = (0, n)$ for all $\tau \in \Phi - \{\tau_0\}$.

Let F^+ be the maximal totally real subfield of F. For simplicity, we assume that F is of the form $F = F^+\mathcal{K}$, for $\mathcal{K}$ a quadratic imaginary extension of $\mathbb{Q}$, and that the CM type Φ

is induced by a CM type of $\mathcal{K}$. Then, under all these assumptions, the reflex field of the Shimura datum is simply $E = \tau_0(F)$.

Let p be a prime number which is split in $\mathcal{K}$ (but not necessarily unramified in F). After choosing a local embedding $\nu : \bar{\mathbb{Q}} \to \bar{\mathbb{Q}}_p$, to each $\tau \in \Phi$, $\tau : F \to \bar{\mathbb{Q}}$, we associate the embedding $\nu \circ \tau : F \to \bar{\mathbb{Q}}_p$. We write $v_1, \ldots, v_r$ for the places of F corresponding to these embeddings as τ varies in Φ, and $v = v_1$ for the place corresponding to τ_0. Then, the set of places of F dividing p is exactly $\{v_1, \ldots, v_r, v_1^c, \ldots, v_r^c\}$. We assume B is split at p. Then $B_{\mathbb{Q}_p}$ is of the form $B_{\mathbb{Q}_p} \cong \prod_{i=1}^r (M_n(F_{v_i}) \times M_n(F_{v_i^c})^{\mathrm{opp}})$, and $G(\mathbb{Q}_p) \cong \prod_{i=1}^r GL_n(F_{v_i}) \times \mathbb{Q}_p^\times$. We denote by ϵ the idempotent in $\mathcal{O}_B \otimes \mathbb{Z}_p$ corresponding to the matrix of $M_n(F_v)$ with entry 1 in $(1,1)$ and 0 elsewhere.

6.3 Shimura varieties of signature (1,n-1)

Let $K^p \subset G(\mathbb{A}_f^p)$ be a sufficiently small open compact subgroup, and S_{K^p} the integral model of the Shimura variety Sh_K, for $K = K^p K_{p,0}$. We write A for the universal abelian variety over S_{K^p}, and $A[p^\infty]$ for its p-divisible group. Then it follows from our assumptions that the Barsotti–Tate group $A[p^\infty]$ decomposes as

$$A[p^\infty] = \prod_{i=1}^r (A[v_i^\infty] \times A[(v_i^c)^\infty]),$$

where the polarization of A induces an isomorphism of Barsotti–Tate $\mathcal{O}_{F_{v_i}}$-modules between $A[v_i^\infty]$ and $A[(v_i^c)^\infty]$, for all $i \geq 1$, the Barsotti–Tate $\mathcal{O}_{F_{v_i}}$-modules $A[v_i^\infty]$ are ind-étale, for all $i > 1$, and for $i = 1$ the Barsotti–Tate group $H = \epsilon A[v^\infty]$ is a compatible one-dimensional $\mathcal{O}_{F_v}$-module.

6.4 p-rank strata

For each closed geometric point x of $\bar{S}_{K^p}$, we write H_x (resp. A_x) for the fiber at x of H (resp. A). We regard the height and p-rank of H_x, as x varies, as $\mathbb{Z}$-valued functions on $\bar{S}_{K^p}$. Then, the height is constant and equal to n, while the p-rank is lower semicontinuous. I.e., for each integer $i \geq 0$, the subspace

$$\bar{S}_{K^p}^{\leq i} = \{x \in \bar{S}_{K^p} \,|\, \mathrm{rk}(H_x) \leq i\}$$

is a closed subspace of $\bar{S}_{K^p}$. (This result is due to Messing in [24], Proposition II.4.9.) We call the associated stratification of $\bar{S}_{K^p}$ by closed reduced subschemes the *p-rank stratification* of $\bar{S}_{K^p}$, and for each integer i, we define *the i-th p-rank stratum* of $\bar{S}_{K^p}$ as the locally closed reduced subscheme

$$\bar{S}_{K^p}^{(i)} = \bar{S}_{K^p}^{\leq i} - \bar{S}_{K^p}^{\leq i-1} = \{x \in \bar{S}_{K^p} \,|\, \mathrm{rk}(H_x) = i\}.$$

We remark that the stratum $\bar{S}_{K^p}^{(i)}$ is empty unless $i \in \{0, \ldots, n-1\}$, in which case it is smooth of pure dimension i ([9], Corollary III.4.4). It also follows from the work of Messing ([24], Proposition II.4.9) that, for each i, $0 \leq i \leq n-1$, the restriction of H over the stratum $\bar{S}_{K^p}^{(i)}$ admits a unique filtration $0 \subset H_0 \subset H$, where H_0 is a formal one-dimensional Barsotti–Tate

$\mathcal{O}_{F_v}$-module of height $n-i$ and $H_1 = H/H_0$ is an ind-étale Barsotti–Tate $\mathcal{O}_{F_v}$-module of height i. This filtration is called the *formal-ind-étale* filtration of H.

We remark that, for each closed geometric point x of $\bar{S}_{K^p}$, the isogeny and isomorphism classes of A_x only depend on H_x, namely on its p-rank. Thus, in particular, the Newton stratification of $\bar{S}_{K^p}$ coincides with the p-rank stratification, the basic (resp. μ-ordinary) stratum equal to the 0-th (resp. $(n-1)$-th) p-rank stratum. Moreover, each Newton stratum has a unique central leaf, which is equal to the whole stratum, and the slope filtration of H defined over it agrees with the formal-ind-étale filtration.

6.5 Lubin-Tate spaces

It follows from the above discussion, that the theory of Rapoport-Zink spaces for this class of Shimura varieties can also be described in terms of the appropriate one-dimensional Barsotti–Tate groups. More precisely, for each integer i, $0 \leq i \leq n-1$, we write $\mathcal{M}_{i,n}$ for the formal Rapoport–Zink space parameterizing one-dimensional Barsotti–Tate $\mathcal{O}_{F_v}$-modules, of height n and p-rank i. We write $\breve{E}$ and $\mathcal{O}_{\breve{E}}$ for the maximal unramified extension of the local reflex field E_v and its ring of integers (in the unramified case $\breve{E} = L$). Then, $\mathcal{M}_{i,n}$ is by definition a formal $\mathcal{O}_{\breve{E}}$-scheme, and it follows from the above considerations that its reduced fiber is a zero-dimensional k-scheme. These are the Rapoport-Zink spaces associated with the group $G' = \mathrm{Res}_{F_v/\mathbb{Q}_p}(GL_n)$ (under our current assumptions, $G(\mathbb{Q}_p) \cong \prod_{i=1}^r GL_n(F_{v_i}) \times \mathbb{Q}_p^\times$.)

The geometry and cohomology of these Rapoport–Zink spaces and their rigid analytic covers is easily described in terms of those of the Lubin-Tate spaces ([16]) and of Drinfeld varieties ([4]). To make the connection, we recall the following two results of Messing. Let S be an $\mathcal{O}_{F_v}$-scheme where p is locally nilpotent. There is an equivalence of categories between the category of formal Barsotti–Tate $\mathcal{O}_{F_v}$-modules over S and that of formal Lie groups over S, together with an action of $\mathcal{O}_{F_v}$, which satisfy the condition that multiplication by p on the Lie group is an epimorphism with finite and locally free kernel ([24], Corollary II.4.5.). Let H be a Barsotti–Tate $\mathcal{O}_{F_v}$-module over S. If H has constant height n and constant p-rank r, then it admits a unique filtration $0 \subset H_0 \subset H$, where H_0 is a formal Barsotti–Tate $\mathcal{O}_{F_v}$-module over S of constant height $n-r$, and $H_1 = H/H_0$ is an ind-étale Barsotti–Tate $\mathcal{O}_{F_v}$-module over S of constant height r ([24], Proposition II.4.9).

For $i = 0$, the Rapoport–Zink space $\mathcal{M}_{0,n}$ parameterizes formal one-dimensional Barsotti–Tate $\mathcal{O}_{F_v}$-modules, of height n. It follows from the fact that any quasi-isogeny of height 0 between formal one-dimensional Barsotti–Tate $\mathcal{O}_{F_v}$-modules over an algebraically closed field is an isomorphism, that $\mathcal{M}(\overline{k(v)}) = \mathbb{Z}$ (the bijection given by the height of the quasi-isogeny). Furthermore, the infinitesimal deformation functor for formal one-dimensional Barsotti–Tate $\mathcal{O}_{F_v}$-modules of height n, $\mathcal{D}_{0,n}$ (which was studied by Lubin and Tate in [16], and by Drinfeld in [4]), is represented by the formal $\mathcal{O}_{\breve{E}}$-scheme $Spf(\mathcal{O}_{\breve{E}})[[T_1,\dots,T_{n-1}]]$. We conclude that the Rapoport–Zink space $\mathcal{M}_{0,n}$ is (non-canonically) isomorphic to a disjoint union of copies of $Spf(\mathcal{O}_{\breve{E}})[[T_1,\dots,T_{n-1}]]$, indexed by $\mathbb{Z}$ ([36], Proposition 3.79).

For $i > 0$, it follows from the fact that over an algebraically closed field any quasi-isogeny between one-dimensional $\mathcal{O}_{F_v}$-modules splits as the product of a quasi-isogeny between their ind-étale parts by a quasi-isogeny between their formal parts, that

$\mathcal{M}_{i,n}(\overline{k(v)}) = GL_i(F_v)/GL_i(\mathcal{O}_{F_v})\times\mathbb{Z}$. Furthermore, the infinitesimal deformation functor for one-dimensional $\mathcal{O}_{F_v}$-modules of height n and p-rank i, $\mathcal{D}_{i,n}$ (which was studied by Drinfeld in [4]), is also represented by $Spf(\mathcal{O}_{\breve{E}})[[T_1,\dots,T_{n-1}]]$. We conclude that the Rapoport–Zink space $\mathcal{M}_{i,n}$ is (non-canonically) isomorphic to a disjoint union of copies of $Spf(\mathcal{O}_{\breve{E}})[[T_1,\dots,T_{n-1}]]$, indexed by $GL_i(F_v)/GL_i(\mathcal{O}_{F_v}) \times \mathbb{Z}$. Moreover, it follows from the existence and uniqueness of the formal-ind-étale filtration of Barsotti–Tate $\mathcal{O}_{F_v}$-modules, that the Rapoport–Zink space $\mathcal{M}_{i,n}$ admits a surjective homomorphism θ onto a disjoint union of copies of $\mathcal{M}_{0,n-i}$, indexed by $GL_i(F_v)/GL_i(\mathcal{O}_{F_v})$. In the above coordinates, the corresponding homomorphism of $\mathcal{O}_{\breve{E}}$-algebras $\theta^* : \mathcal{O}_{\breve{E}}[[T_1,\dots,T_{n-i-1}]] \to \mathcal{O}_{\breve{E}}[[T_1,\dots,T_{n-1}]]$ is defined by $T_s \mapsto T_s$, for all $s \in \{1,\dots,n-i-1\}$. In particular, the map θ is formally smooth of relative dimension i.

6.6 Adding level structure

Let $K_p \subset K_{p,0} = GL_n(\mathcal{O}_{F_v})$ be an open compact subgroup of $GL_n(F_v)$.

For each i, $0 \le i \le n-1$, we write $\mathcal{M}_{i,n,K_{p,0}}$ for the rigid analytic fiber of $\mathcal{M}_{i,n}$, and $\mathcal{M}_{i,n,K_p} \to \mathcal{M}_{i,n,K_{p,0}}$ for the associated finite étale cover, of level K_p. We also denote by $\mathcal{D}_{i,n,K_p} \to \mathcal{D}_{i,n,K_{p,0}}$ the rigid analytic Drinfeld cover of level K_p (a.k.a. the rigid analytic Lubin-Tate cover in the case of $i = 0$). It follows from the universal properties of the covers that, for each $i \ge 0$, the above description in the case of level $K_{p,0}$ extends to the cases of higher levels $K_p \subset K_{p,0}$. I.e., there exists a $GL_n(F_v)$-equivariant system of isomorphisms between $\mathcal{M}_{i,n,K_p}$ and disjoints union of $GL_i(F_v)/GL_i(\mathcal{O}_{F_v})\times\mathbb{Z}$-copies of $\mathcal{D}_{i,n,K_p}$, indexed by the open compact subgroups K_p.

For simplicity, we now assume K_p is of the form

$$K_{p,m} = \{A \in GL_n(\mathcal{O}_{F_v}) | A \equiv \mathbb{I} \mod v^m\},$$

for some $m \ge 0$. For each $m \ge 0$, let $\mathcal{D}_{i,n,m}$ be the formal Drinfeld space of level m (a.k.a. the formal Lubin-Tate space of level m in the case of $n = 0$). The space $\mathcal{D}_{i,n,m}$ is defined as the formal $\mathcal{O}_{\breve{E}}$-scheme representing the infinitesimal deformation functor for one-dimensional $\mathcal{O}_{F_v}$-modules of height n and p-rank i, endowed with a Drinfeld structure of level m (its construction is due to Drinfeld in [4]). It follows from the definition that the rigid analytic fiber of $\mathcal{D}_{i,n,m}$ can be canonically identified with $\mathcal{D}_{i,n,K_{p,m}}$. Moreover, for all $m' \ge m \ge 0$, the projections $\mathcal{D}_{i,n,K_{p,m'}} \to \mathcal{D}_{i,n,K_{p,m}}$ canonically extend to homomorphisms of formal $\mathcal{O}_{\breve{E}}$-schemes $\mathcal{D}_{i,n,m'} \to \mathcal{D}_{i,n,m}$. In [4] (Proposition 4.3) Drinfeld proved that the formal schemes $\mathcal{D}_{i,n,m}$ are regular and flat and that the homomorphisms $\mathcal{D}_{i,n,m'} \to \mathcal{D}_{i,n,m}$ are finite and flat, of degree $\#GL_n(\mathcal{O}_{F_v}/v^{m'})/\#GL_n(\mathcal{O}_{F_v}/v^m)$.

For $i > 0$, we write $\mathcal{D}_{0,n-i,m}$ for the corresponding formal Lubin-Tate space of level m, for each $m \ge 0$. Let H be the universal one-dimensional Barsotti–Tate $\mathcal{O}_{F_v}$-module over $\mathcal{D}_{i,n,0}$, $0 \subset H_0 \subset H$ denote the formal-ind-étale filtration of H, and $H_1 = H/H_0$. Étale locally on $\mathcal{D}_{i,n,m}$, the datum of a Drinfeld structure of level m on H determines a direct summand N of $\mathcal{O}_{F_v}^n$ modulo $K_{p,m}$ (each N of rank $n-i$), and two classes of isomorphisms $N \cong T_pH_0$ and $\mathcal{O}_{F_v}^n/N \cong T_pH_1$. After choosing an isomorphism between the 2-flag $0 \subset N \subset \mathcal{O}_{F_v}^n$ and the standard 2-flag of ranks $(n-i, n)$, the latter data can be identified with structures of level m on H_0 and H_1, respectively. Thus, for each $m \ge 1$, the restriction of θ to $\mathcal{D}_{i,n,0}$

(which we denote by $\theta_0 : \mathcal{D}_{i,n,0} \to \mathcal{D}_{0,n-i,0}$) canonically lifts to a $GL_n(\mathcal{O}_{F_v})$-equivariant homomorphism of formal $\mathcal{O}_{\breve{E}}$-schemes

$$\theta_m : \mathcal{D}_{i,n,m} \to \coprod_{P_{n-i,n}(F_v)\backslash GL_n(F_v)/K_{p,m}} \left(\coprod_{GL_i(\mathcal{O}_{F_v}/v^m)} \mathcal{D}_{0,n-i,m} \right),$$

where $P_{n-i,n}$ denotes the parabolic subgroup of GL_n which stabilizes the standard 2-flag of dimensions $(n-i, n)$. For all $m \geq 1$, the homomorphisms θ_m are surjective, finite and flat. Moreover, as the level m varies, they form a $GL_n(F_v)$-equivariant projective system, under the natural projections among the Drinfeld and Lubin-Tate spaces and epimorphisms among the indexing sets.

6.7 The supercuspidal part of cohomology

It is an immediate consequence of the latter construction that, for each $i > 0$, the l-adic cohomology groups with compact supports of the tower of Drinfeld spaces $\mathcal{D}_{i,n,K_p}$, and thus also that of the associated tower of Rapoport–Zink spaces $\mathcal{M}_{i,n,K_p}$, regarded as representations of $GL_n(F_v)$, are parabolically induced by $P_{n-i,n}(F_v)$, and thus in particular contain no supercuspidal representations of $GL_n(F_v)$. (This simple but important observation is due to Boyer in [3], and reappeared in the work of Harris and Taylor in [9].)

This observation, together with the formula in section 5.3, implies that, for any étale l-adic local system $\mathcal{L}_\rho$ associated with a representation $\rho \in \mathrm{Rep}_{\mathbb{C}}(G)$, and for each $i > 0$, the cohomology groups of the i-th p-rank strata $\bar{S}^{(i)}_{K_p}$, with compact supports and coefficients in the vanishing cycles sheaves of the pushforwards of $\mathcal{L}_\rho$ (as defined in section 3.3), contain no supercuspidal representations of $G(\mathbb{Q}_p)$. Thus, in particular, in the case when the Shimura varieties are proper (i.e. in the case when $[F^+ : \mathbb{Q}] \geq 2$), for any étale l-adic local system $\mathcal{L}_\rho$, only the cohomology groups of the basic strata (i.e. the 0-th p-rank strata) contribute, in the sense of the formula in section 3.3, to the supercuspidal part of the cohomology of the Shimura varieties.

6.8 Harris–Viehmann conjecture

We conclude by observing that the above phenomena are expected to occur in much larger generality. Indeed, the Harris–Viehmann conjecture for Rapoport–Zink spaces, together with the formula of section 5.3, implies that, for any Shimura datum (G, h) of PEL type which is unramified at p, and any étale l-adic local system $\mathcal{L}_\rho$ associated with an algebraic representation ρ of G, the corresponding cohomology groups of the Newton strata of the mod p reduction of the Shimura varieties defined by (G, h) (section 3.3) contain no supercuspidal representations of $G(\mathbb{Q}_p)$ except in the case of the basic Newton strata.

This conjecture was first proposed by Harris ([8], Conjecture 5.2), and later modified by Viehmann ([35], Conjecture 8.4), who together with Rapoport also generalized the statement to include local Shimura varieties. In the case of an unramified local datum (G, μ) with G quasi-split, the Harris–Viehmann conjecture predicts that for any non-basic element $b \in B(G, \mu)$, the l-adic cohomology with compact supports of the Rapoport–Zink spaces attached to the triple (b, G, μ) is parabolically induced from that of the associated basic Rapoport–Zink spaces. More precisely, for each $b \in B(G, \mu)$, and any Levi subgroup M of G containing an inner form of J_b (e.g., for $M = M_b$ as introduced in section 2.4), the Harris–Viehmann conjecture predicts the following equalities in $\mathrm{Groth}(G(\mathbb{Q}_p) \times W_{E_v})$, for all $\rho \in \mathrm{Groth}(J_b(\mathbb{Q}_p))$:

$$\mathcal{E}_{(b,G,\mu)}(\rho) = \mathrm{Ind}_{P(\mathbb{Q}_p)}^{G(\mathbb{Q}_p)}\left(\sum_{\mu' \in I_{b,\mu,M}} \mathcal{E}_{(b,M,\mu')}(\rho)\right),$$

where $I_{b,\mu,M}$ is the set of M-conjugacy classes μ' of cocharacters of M which are G-conjugate to μ, with $b \in B(M, \mu')$, and P is the unique parabolic with Levi subgroup M which, after base changed to L, contains to the parabolic P_b corresponding to the slope filtration of the F-isocrystal $N_b(V)$. In particular, for $M = M_b$, we have $P = P_b$ and $b \in B(M_b, \mu')$ basic, for all $\mu' \in I_{b,\mu,M_b}$. In [35] (Lemma 8.1), Rapoport and Viehmann show that the set $I_{b,\mu,M}$ is finite, non-empty, and consists of a single element if G is spilt. In [35] (Example 8.3), they compute an example with $I_{b,\mu,M}$ of size 2.

For $G = GL_n$, and $\mu = (1, 0, \ldots, 0)$, this conjecture was proved by Harris and Taylor in [9], their approach following that of Boyer in [3] where the analogous result for Drinfeld modular varieties is proved. The same approach was later extended in [21] to establish new cases of the Harris–Viehmann conjecture, under the further assumption that $b \in B(G, \mu)$ is Hodge–Newton reducible, and some restrictions on the Levi M. To be precise, the condition for Hodge–Newton reducibility used in [21] was the one originally given in [23], and was later relaxed by Shen in [38].

References

[1] V. Berkovich, *Vanishing cycles for formal schemes.* Invent. Math. 115 (1994), no. **3**, 539–571.

[2] V. Berkovich, *Vanishing cycles for formal schemes. II.* Invent. Math. 125 (1996), no. **2**, 367–390.

[3] P. Boyer, *Mauvaise réduction des variétés de Drinfeld et correspondance de Langlands locale. (French) [Bad reduction of Drinfeld varieties and local Langlands correspondence]* Invent. Math. 138 (1999), no. 3, 573–629.

[4] V. Drinfeld, *Elliptic modules.* Mat. Sb. (N.S.) 94 (136)(197), 594–627, 656.

[5] L. Fargues, *Cohomologie d'espaces de modules de groupes p-divisibles et correspondences de Langlands locales.* Variétés de Shimura, espaces de Rapoport-Zink et correspondances de Langlands locales. Astérisque **291** (2004), 1–199.

[6] A. Grothendieck, *Groupes de Barsotti–Tate et cristaux de Dieudonné.* Sém. Math. Sup. Univ. Montréal. Presses Univ. Montréal, 1974.

[7] P. Hamacher, *The geometry of Newton strata in the reduction modulo p of Shimura varieties of PEL type.* Duke Math. J. 164 (2015), no. 15, 2809–2895.

[8] M. Harris, *Local Langlands correspondences and vanishing cycles on Shimura varieties.* European Congress of Mathematics, Vol. I (Barcelona, 2000), 407–427, Progr. Math., 201, Birkhäuser, Basel, 2001.

[9] M. Harris, R. Taylor, *On the geometry and cohomology of some simple Shimura varieties.* Volume **151**, Annals of Math. Studies, Princeton University Press, 2001.

[10] T. Honda, *Isogeny classes of abelian varieties over finite fields.* J. Math. Soc. Japan 20 1968 83–95.

[11] J. Igusa, *Kroneckerian model of fields of elliptic modular functions.* Amer. J. Math. **81**, 1959, 561–577.
[12] N. Katz, *Slope Filtration of F-crystals.* Journées de Géométrie Algébrique de Rennes (Rennes, 1978), Vol. I, pp. 113–163, Astérisque, **63**, Soc. Math. France, Paris, 1979.
[13] R. Kottwitz, *Points on some Shimura varieties over finite fields.* J. Amer. Math. Soc. **5** (1992), no. 2, 373–444.
[14] R. Kottwitz, *Isocrystals with additional structure.* Compositio Math. **56** (1985), no. 2, 201–220.
[15] R. Kottwitz, *Isocrystals with additional structure. II.* Compositio Math. **109** (1997), no. 3, 255–339.
[16] J. Lubin, J. Tate, *Formal moduli for one-parameter formal Lie groups.* Bull. Soc. Math. France **94** (1966), 49–59.
[17] Manin, Yu. I. *Theory of commutative formal groups over fields of finite characteristic.* (Russian) Uspehi Mat. Nauk **18** (1963) no. 6 (114), 3–90; Russian Math. Surveys **18** (1963), 1–80.
[18] E. Mantovan, *On certain unitary group Shimura varieties.* Variétés de Shimura, espaces de Rapoport-Zink et correspondances de Langlands locales. Astérisque **291** (2004), 200–331.
[19] E. Mantovan, *On the cohomology of certain PEL-type Shimura varieties.* Duke Math. J. **129** (2005), no. 3, 573–610.
[20] E. Mantovan, *A compactification of Igusa varieties.* Math. Ann. **340** (2008), no. 2, 265–292.
[21] E. Mantovan, *On non-basic Rapoport–Zink spaces.* Ann. Sci. École Norm. Sup. (4) 41 (2008), no. 5, 671-716.
[22] E. Mantovan, *l-Adic Étale Cohomology of PEL Type Shimura Varieties with Non-Trivial Coefficients.* Win- Women in Numbers: Research Directions in Number Theory. Fields Institute Communications **60** (2011), 61–83.
[23] E. Mantovan, E. Viehmann, *On the Hodge–Newton filtration for p-divisible $\mathcal{O}$-modules.* Math. Z. **266** (2010), no. 1, 193–205.
[24] W. Messing, *The crystals associated to Barsotti-Tate groups: with applications to abelian schemes.* Lecture Notes in Mathematics, Vol. 264. Springer-Verlag, Berlin-New York, 1972.
[25] M.-H. Nicole, *Unitary Shimura varieties.* This volume.
[26] F. Oort, *Newton polygons and formal groups: conjectures by Manin and Grothendieck.* Ann. of Math. (2) **152** (2000), no. 1, 183–206.
[27] F. Oort, *Moduli of abelian varieties and Newton polygons.* C. R. Acad. Sci. Paris Sér. I Math. **312** (1991), no. 5, 385–389.
[28] F. Oort. *Foliations in moduli spaces of abelian varieties.* J. Amer. Math. Soc. **17** (2004), no. 2, 267–296.
[29] F. Oort. *A stratification of a moduli space of abelian varieties.* Moduli of abelian varieties (Texel Island, 1999), 345-416, Progr. Math., 195, Birkhäuser, Basel, 2001.
[30] F. Oort. *Minimal p-divisible groups.* Ann. of Math. (2) 161 (2005), no. 2, 1021-1036.
[31] F. Oort, Th. Zink. *Families of p-divisible groups with constant Newton polygon.* Documenta Mathematica **7** (2002), 183–201.
[32] M. Rapoport, *Non-Archimedean period domains.* Proceeding of the International Congress of Mathematicians, Vol. 1, 2 (Zürich, 1994), 423–434.
[33] M. Rapoport, *A guide to the reduction modulo p of Shimura varieties.* Astérisque **298** (2005), 271–318.
[34] M. Rapoport, Richartz, M. *On the classification and specialization of F-isocrystals with additional structure.* Compositio Math. **103** (1996), no. 2, 153–181.
[35] M. Rapoport, E. Viehmann, *Towards a theory of local Shimura varieties* Münster Journal of Mathematics **7** (2014), on the occasion of P. Schneider's 60th birthday, p. 273-326.

[36] M. Rapoport, Th. Zink, *Period spaces for p-divisible groups.* Annals of Mathematics Studies, **141**, Princeton University Press, Princeton, NJ, 1996.

[37] S. Rozensztajn, *Integral models of Shimura varieties of PEL type.* This volume.

[38] X. Shen, *On the Hodge–Newton filtration for p-divisible groups with additional structures.* Int. Math. Res. Notices (2014), no. 13, 3582–3631.

[39] E. Viehmann, *On the geometry of the Newton stratification.* This volume

[40] E. Viehmann, T. Wedhorn, *Ekedahl-Oort and Newton strata for Shimura varieties of PEL type.* Math. Ann. **356** (2013), 1493–1550.

[41] T. Wedhorn, *Ordinariness in good reductions of Shimura varieties of PEL-type.* Ann. Sci. École Norm. Sup. (4) **32** (1999), no. 5, 575–618.

[42] Th. Zink, *On the slope filtration.* Duke Math. J. Vol. **109** (2001), 79–95.

ON THE GEOMETRY OF THE NEWTON STRATIFICATION

EVA VIEHMANN

Technische Universität München, Fakultät für Mathematik - M11, Boltzmannstr. 3,
85748 Garching bei München, Germany
Email address: viehmann@ma.tum.de

Abstract

We give an overview over recent results on the global structure and geometry of the Newton stratification of the reduction modulo p of Shimura varieties of Hodge type with hyperspecial level structure. More precisely, we discuss non-emptiness, dimensions, and closure relations of Newton strata. We also explain the group-theoretic description and methods leading to their proofs.

1 Introduction

One of the key invariants of an abelian variety in characteristic p is the Newton polygon describing the isogeny class of its p-divisible group. It is a central tool in the study of the reduction modulo p of Shimura varieties of PEL type in almost all of the main breakthroughs in this area (as for example the computation of the Hasse-Weil ζ-function by Kottwitz in [Kot2], or the proof of the local Langlands correspondence for GL_n by Harris and Taylor [HT]).

However, even very basic questions on the geometry of the induced stratification of the fiber at p of a given Shimura variety of PEL type remained open. For example, except for special cases, one did not even know the set of strata. For several years there is now a trend towards applying more group-theoretic methods (building on the classical works of Kottwitz [Kot1]) to address these questions. In this overview we report on recent developments in this direction. We describe the set of strata, discuss their dimensions and the closure relations. The group-theoretic methods underlying the proofs do not make a difference between groups associated with Shimura varieties of PEL type or not. Therefore, the natural context to apply them is the most general one where we still have a good theory of integral models of the Shimura varieties and a translation between points in the special fiber and suitable elements of the corresponding group. Thus, instead of Shimura varieties of PEL type we also discuss the recent generalizations to Hodge type Shimura varieties. Besides, we briefly report on the parallel theory for the function field case, i.e. for moduli spaces of global and local G-shtukas, or for loop groups.

The author was partially supported by ERC starting grant 277889 "Moduli spaces of local G-shtukas".

2 Points in the reduction of Shimura varieties of Hodge type

2.1 The Siegel case

Let $\mathcal{A}_g$ be the moduli space of principally polarized abelian varieties of dimension g over $\mathrm{Spec}\,(\mathbb{F}_p)$ for some fixed characteristic $p > 0$. It is the fiber at p of a Shimura variety for the group GSp_{2g}. Let k be an algebraically closed field of characteristic p. Then the points of $\mathcal{A}_g(k)$ correspond to pairs (A, λ) where A is an abelian variety over k and λ a principal polarization of A. An important invariant of an abelian variety in characteristic p is its p-divisible group $A[p^\infty]$. It is equipped with a quasi-polarization λ induced by λ on A. By Dieudonné theory this datum corresponds bijectively to the Dieudonné module $(M, F, \langle\cdot,\cdot\rangle)$ where M is a free $W(k)$-module of rank $2g$, where $\langle\cdot,\cdot\rangle$ is a symplectic pairing induced by λ and where F is a Frobenius-linear map $M \to M$ with $pM \subset F(M)$, and satisfying a compatibility condition with λ. Trivializing M in such a way that $\langle\cdot,\cdot\rangle$ is identified with the standard pairing we can write $F = b\sigma$ for some $b \in \mathrm{GSp}_{2g}(W(k)[1/p])$ and where σ is the Frobenius of $W(k)[1/p]$ over $\mathbb{Q}_p$. The element b is well-defined up to base change, i.e. up to replacing it by $g^{-1}b\sigma(g)$ for some $g \in \mathrm{GSp}_{2g}(W(k))$.

A similar, but more tedious description is available for all Shimura varieties of PEL type with good reduction at p, compare [VW], 1 and 7.

2.2 Shimura varieties of Hodge type

Let now $\mathcal{D} = (G, X)$ be a Shimura datum of Hodge type, and let p be a prime. Let $K = K^pK_p \subset G(\mathbb{A}_f^p)G(\mathbb{Q}_p)$ be a compact open subgroup. We always assume that we are in the case of good reduction, i.e. that K_p is hyperspecial. In other words we assume that G extends to a reductive group over $\mathbb{Z}_p$ and that $K_p = G_{\mathbb{Z}_p}(\mathbb{Z}_p)$. This implies in particular that G is unramified at p, i.e. quasi-split and split over an unramified extension of $\mathbb{Q}_p$.

Then the corresponding Shimura variety $Sh_K(G, X)$ is a moduli space of abelian varieties with certain Hodge cycles associated with the Shimura datum. By [Ki1] it has an integral model at the prime p which is obtained by taking an embedding into a suitable Siegel moduli space, and taking the normalization of the closure in an integral model of that space. We denote its special fiber by $\mathcal{S}_K(G, X)$. Associated with closed points of $\mathcal{S}_K(G, X)$ we thus still have an abelian variety, but there is no explicit moduli theoretic interpretation of $\mathcal{S}_K(G, X)$.

Let $h : \mathbb{S} \to G_{\mathbb{R}}$ be an element of the conjugacy class X of the Shimura datum. Let $\psi = \psi_h$ be the cocharacter defined on R-points (for any $\mathbb{C}$-algebra R) by

$$R^\times \to (R \times c^*(R))^\times = (R \otimes_{\mathbb{R}} \mathbb{C})^\times = \mathbb{S}(R) \to G(R).$$

Here c denotes complex conjugation. Via a fixed isomorphism $\overline{\mathbb{Q}}_p \cong \mathbb{C}$ we view ψ as a p-adic cocharacter. Then let $\mu = \sigma(\psi^{-1})$. Note that our use of the letter μ differs slightly from that of [Ki2], whom we follow for the next construction.

Let k be an algebraically closed field of characteristic p and let $\mathcal{O}_L = W(k)$. Let $x \in \mathcal{S}_K(G, X)(k)$. Let $\mathcal{A}_x$ be the fiber at x of the universal abelian variety $\mathcal{A}$, and let $\mathcal{G}_x$ be its p-divisible group. Let $g = \dim \mathcal{A}$. Following [Ki2], 1.4.1 (or [VW], 7.2 for the PEL case) there is a trivialization of the Dieudonné module $\mathbb{D}(\mathcal{G}_x)(\mathcal{O}_L)$ of $\mathcal{G}_x$, i.e. an isomorphism

$$\mathcal{O}_L^{2g} \to \mathbb{D}(\mathcal{G}_x)(\mathcal{O}_L)$$

and such that it maps the Hodge tensors on $\mathcal{A}_x$ to certain fixed F-invariant tensors defining a subgroup $G_{\mathcal{O}_L} \cong G_{\mathbb{Z}_p} \otimes_{\mathbb{Z}_p} \mathcal{O}_L \subsetneq \mathrm{GL}_{2g,\mathcal{O}_L}$. By fixing one such isomorphism we can write $F = b\sigma$ where σ is the Frobenius on $\mathcal{O}_L$ and where $b \in G_{\mathbb{Z}_p}(\mathcal{O}_L)\mu(p)G_{\mathbb{Z}_p}(\mathcal{O}_L) \subset G(L)$. The element b is well-defined up to σ-conjugation by elements of $G_{\mathbb{Z}_p}(\mathcal{O}_L)$. For the Siegel case, this coincides with the class constructed in the previous paragraph.

For $c \in G(L)$ let

$$[\![c]\!] := \{g^{-1}c\sigma(g) \mid g \in G(\mathcal{O}_L)\}.$$

Let $C(G,\mu) = \{[\![c]\!] \mid c \in G(\mathcal{O}_L)\mu(p)G(\mathcal{O}_L)\}$. Then by the considerations above we have a natural map

$$\Upsilon : \mathcal{S}_K(G,X)(k) \to C(G,\mu).$$

3 The set of Newton points

Before introducing the Newton stratification on the special fiber of a Shimura variety we discuss in this section the index set for such stratifications. This set of Newton points has an abstract group-theoretic definition independent of the context of the previous section. Thus for this section we are in the following, more general setting.

Let F be a local field of residue characteristic p, let $\mathcal{O}_F$ be its ring of integers, and ϵ a uniformizer. Let $L \supset \mathcal{O}_L$ be the completion of the maximal unramified extension and its ring of integers, and let σ denote the Frobenius of L over F.

Let G be a reductive group over $\mathcal{O}_F$, and fix a Borel subgroup B and a maximal torus T, both defined over F.

3.1 σ-conjugacy classes and their Newton points

For $b \in G(L)$ let

$$[b] = \{g^{-1}b\sigma(g) \mid g \in G(L)\}$$

denote its σ-conjugacy class, and let $B(G)$ be the set of σ-conjugacy classes for $b \in G(L)$. Kottwitz [Kot1] has given a classification of $B(G)$ generalizing the Dieudonné-Manin classification of isocrystals by Newton polygons. He assigns to an element of $B(G)$ two invariants. One, the Newton point, is the direct analog of the Newton polygon. It is defined by assigning to every $b \in G(L)$ a homomorphism $\nu_b : \mathbb{D} \to G$. Here $\mathbb{D}$ is the pro-algebraic torus with character group $\mathbb{Q}$. Varying b in $[b]$ leads to conjugate homomorphisms. The $G(L)$-conjugacy class of ν_b is stable under the action of σ. The Newton point associated with $[b]$ is then defined to be this conjugacy class, an element $\nu([b]) \in (G\backslash X_*(G)_{\mathbb{Q}})^{\Gamma}$. Often, one represents it by its unique representative in $\mathcal{N}(G) := (W\backslash X_*(T)_{\mathbb{Q}})^{\Gamma}$ where W is the Weyl group of G. Choosing dominant representatives we can also identify $\mathcal{N}(G)$ with $X_*(T)^{\Gamma}_{\mathbb{Q},\mathrm{dom}}$. We call $\mathcal{N}(G)$ the Newton cone, and denote the map $B(G) \to \mathcal{N}(G)$ by ν.

The second classifying invariant of $[b]$ is the image under the so-called Kottwitz map

$$\kappa_G : B(G) \to \pi_1(G)_{\Gamma}$$

(see also Rapoport and Richartz [RR] for the reformulation). Here, $\pi_1(G)$ is Borovoi's fundamental group, defined as the quotient of $X_*(T)$ by the coroot lattice, and we are taking coinvariants under the absolute Galois group of F. As G is an unramified reductive group over F, the map κ_G has the following explicit description. By the Cartan decomposition every $b \in G(L)$ is in $G(\mathcal{O}_L)\mu(\epsilon)G(\mathcal{O}_L)$ for some $\mu \in X_*(T)_{\mathrm{dom}}$. Then $\kappa_G([b])$ is the image of μ under the projection map to $\pi_1(G)_\Gamma$.

The images of $\nu([b])$ and $\kappa_G([b])$ in $\pi_1(G)_\Gamma \otimes_{\mathbb{Z}} \mathbb{Q}$ coincide. Thus for groups where $\pi_1(G)_\Gamma$ is torsion free, the Newton point alone already determines an element of $B(G)$.

Example 3.1. (1) For the group GL_n we choose the upper triangular matrices as Borel subgroup and let T be the diagonal torus. Then we have $X_*(T)_{\mathbb{Q}} \cong \mathbb{Q}^n$ and $\nu = (\nu_i)$ is dominant if and only if $\nu_i \geq \nu_{i+1}$ for all i. The Newton point $\nu = (\nu_i)$ of $[b]$ coincides with the classical Newton point of the isocrystal $(L^n, b\sigma)$ for any representative $b \in [b]$. Let p_ν be the polygon associated with ν, that is the graph of the continuous, piecewise linear function $[0, n] \to \mathbb{R}$ mapping 0 to 0 and with slope ν_i on $[i-1, i]$. Then $\kappa_G([b]) \in \pi_1(G)_\Gamma \cong \mathbb{Z}$ is equal to $\sum \nu_i$, i.e. to the second coordinate of the endpoint of the Newton polygon.

(2) Let us now consider the group PGL_n. Here we have $X_*(T) \cong \mathbb{Z}^n/\mathbb{Z} \cdot (1, \ldots, 1)$. Let $b_1 \in \mathrm{PGL}_n(L)$ with $b_1(e_i) = e_{i+1}$ if $i < n$, and $b_1(e_n) = \epsilon e_1$. Then $(b_1\sigma)^n = b_1^n\sigma^n = \epsilon\sigma^n$. Hence for every integer l the Newton point of b_1^l is constant and thus equal to $(0, \ldots, 0)$ in $X_*(T)_{\mathbb{Q}}$. However, we have $\pi_1(G)_\Gamma \cong \mathbb{Z}/n\mathbb{Z}$ and $\kappa_{\mathrm{PGL}_n}(b_1^l) \equiv l \pmod{n}$. Thus $[b_1^l] = [1]$ if and only if $n|l$.

Let $\mathcal{N}(G)_{\mathbb{Z}} \subset \mathcal{N}(G)$ be the image of ν. It is called the Newton lattice. This set also has an intrinsic, group-theoretic description, see [C]. For our standard example $G = \mathrm{GL}_n$ we obtain that $\nu \in \mathcal{N}(\mathrm{GL}_n)_{\mathbb{Z}}$ if and only if all break points of p_ν as well as its endpoint have integral coordinates.

3.2 $B(G)$ as ranked poset

In this section we explain Chai's theory of the ordering and lengths of chains in the set of Newton points, [C].

There is a natural partial ordering on the Newton lattice $\mathcal{N}(G)_{\mathbb{Z}}$, induced by the partial ordering on $X_*(T)_{\mathbb{Q},\mathrm{dom}}$ and given by $\nu \preceq \nu'$ if and only if $\nu' - \nu$ is a non-negative rational linear combination of positive coroots. A characterization not using our choice of B (and thus showing that it is independent of that choice) is the following. Let $\nu, \nu' \in W\backslash X_*(T)_{\mathbb{Q}}$ and $W \cdot \nu, W \cdot \nu'$ the corresponding orbits in $X_*(T)_{\mathbb{R}}$. Then $\nu \preceq \nu'$ if and only if the convex hull of $W \cdot \nu'$ contains the convex hull of $W \cdot \nu$. On the set $B(G)$ this induces a partial ordering given by $[b'] \preceq [b]$ if and only if $\nu([b']) \preceq \nu([b])$ and $\kappa_G([b]) = \kappa_G([b'])$.

For the group GL_n we have $(\nu_i) = \nu \preceq \nu' = (\nu_i')$ if and only if $\sum_{i=1}^l \nu_i \leq \sum_{i=1}^l \nu_i'$ for all l with equality for $l = n$. For the associated polygons this means that the polygon for ν' lies on or above the polygon of ν and that they have the same endpoints. Figure 7.1 shows the ordering on the set of $\nu \in \mathcal{N}(\mathrm{GL}_4)_{\mathbb{Z}}$ with $\nu \preceq (1, 1, 0, 0)$. The arrows point from larger to smaller elements. For better readability we include the relevant points with integral coordinates lying on or below the polygons. An additional important combinatorial observation then is the following: Each polygon is the convex hull of the set of integral points lying on or below the polygon, and for neighboring polygons, these sets differ by one element which is a breakpoint of the larger of the two polygons. (Compare also Example 3.5 below.)

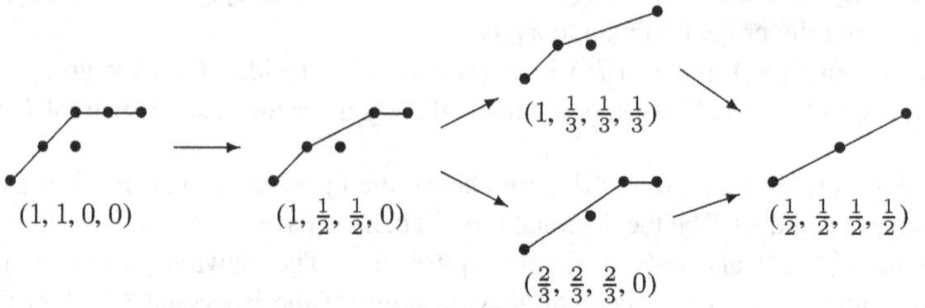

Figure 7.1.

Definition 3.2. For $[b] \in B(G)$ let $\nu = \nu([b])$ be its Newton point. Let $\mathcal{N}(G)_{\preceq\nu}$ be the image in $\mathcal{N}(G)$ of $\{[b'] \in B(G) \mid [b'] \preceq [b]\}$. By [C], Prop. 4.4 this only depends on ν.

For $\nu \in \mathcal{N}(G)$ and $\nu' \in \mathcal{N}(G)_{\preceq\nu}$ a chain between ν' and ν is a sequence $\nu' = \nu_0 \preceq \cdots \preceq \nu_n = \nu$ with $\nu_i \in \mathcal{N}(G)_{\preceq\nu}$ and $\nu_i \neq \nu_{i+1}$ for all i. We call n the length of the chain. A chain between ν' and ν is called maximal if it is not a proper subsequence of another chain between ν' and ν.

Before we can compute the lengths of maximal chains we need some more notation

Definition 3.3. Let l be the number of Galois orbits of absolute fundamental weights of G. For $j = 1, \ldots, l$ let $\underline{\omega}_j$ be the sum of all elements of the corresponding orbit.

For $b \in G(L)$ the defect of b is defined as $\mathrm{rk}\, G - \mathrm{rk}_F J_b$. Here J_b is the reductive group over F with $J_b(A) = \{g \in G(A \otimes_F L) \mid gb = b\sigma(g)\}$ for every F-algebra A.

Theorem 3.4 (Chai, Kottwitz). *Let $\nu \in \mathcal{N}(G)_{\mathbb{Z}}$.*

(1) *The length of chains between two given elements ν', ν'' with $\nu' \preceq \nu'' \in \mathcal{N}(G)_{\preceq\nu}$ is bounded above, in particular there are maximal chains between ν' and ν''.*

(2) *The set $\mathcal{N}(G)_{\preceq\nu}$ is ranked, i.e. for any $\nu' \preceq \nu''$ in $\mathcal{N}(G)_{\preceq\nu}$ every maximal chain between ν' and ν'' has the same length, which is independent of ν and denoted by length$([\nu', \nu''])$. We have*

$$\mathit{length}([\nu', \nu'']) = \sum_{j=1}^{l} \lfloor\langle \nu'', \underline{\omega}_j\rangle\rfloor - \lfloor\langle \nu', \underline{\omega}_j\rangle\rfloor.$$

Here $\lfloor x \rfloor$ denotes the greatest integer less or equal to x.

(3) *If in the above context $\nu'' \in X_*(T)$, then*

$$\mathit{length}([\nu', \nu'']) = \langle \rho, \nu'' - \nu'\rangle + \frac{1}{2}\mathrm{def}_G(b')$$

where $[b'] \in B(G)$ with $\nu_G([b']) = \nu'$.

(4) $\mathcal{N}(G)_{\leq\nu}$ *is a finite set. Every non-empty subset of* $\mathcal{N}(G)_{\leq\nu}$ *has a unique supremum and a unique infimum in* $\mathcal{N}(G)_{\leq\nu}$.

References for the proof. (1) and (4) are due to Chai, [C] Thm. 7.4. The formula in (2) is shown essentially by the same proof as given for [C], Theorem 7.4. It corrects the formula in loc. cit. in two respects: One needs to use sums over Galois orbits of absolute fundamental weights, as was pointed out to us by P. Hamacher (compare [Ha2], Prop. 3.11). Furthermore, the combinatorics of counting chain lengths gives the above formula instead of the one of [C], Theorem 7.4. The proof and formula given by Chai are correct for the case $\nu \in X_*(T)_{\mathrm{dom}}$ and then imply our formula for general ν. (3) is due to Kottwitz [Kot4], and Hamacher [Ha2], 3.3. □

Example 3.5. For the group GL_n there is the following explicit reformulation of the length formula: In this case $l = n$, the $\underline{\omega}_j$ are the fundamental weights $(1, \dots, 1, 0, \dots, 0)$ with multiplicities $j, n - j$. Thus $\langle \nu, \underline{\omega}_j \rangle$ is the value of the Newton polygon at the point j. Assume for the moment that the polygon lies above the first coordinate axis, then $\lfloor \langle \nu, \underline{\omega}_j \rangle \rfloor$ is the number of points of the form (j, l) with $0 \leq l \in \mathbb{Z}$ lying on or below p_ν. Altogether (and without the additional assumption) we obtain that the length of each maximal chain of Newton points between ν' and ν'' (with $\nu' \preceq \nu''$) is equal to the number of points with integral coordinates lying on or below ν'' and strictly above ν'.

3.3 The Newton stratification

We now consider stratifications induced by this invariant, or rather by the pair of invariants (ν, κ_G).

Recall the Cartan decomposition $G(L) = \bigcup_{\mu \in X_*(T)_{\mathrm{dom}}} G(\mathcal{O}_L)\mu(\epsilon)G(\mathcal{O}_L)$.

Definition 3.6. For $\mu \in X_*(T)_{\mathrm{dom}}$ let $B(G, \mu)$ denote the set of $[b] \in B(G)$ with $[b] \leq [\mu(\epsilon)]$.

Theorem 3.7 (Katz, Kottwitz). *Let* $b \in G(L)$ *and let* $\mu \in X_*(T)_{\mathrm{dom}}$ *with* $b \in G(\mathcal{O}_L)\mu(\epsilon)G(\mathcal{O}_L)$. *Then* $[b] \in B(G, \mu)$.

Remark 3.8. Therefore for every μ we obtain a natural map $C(G, \mu) \to B(G, \mu)$ with $[\![b]\!] \mapsto [b]$. Note that $B(G, \mu)$ is not defined as the set of σ-conjugacy classes of elements of $G(\mathcal{O}_L)\mu(\epsilon)G(\mathcal{O}_L)$. Therefore it is a priori not clear if this map is surjective, this only follows from Theorem 3.9 below.

The theorem applies in particular to the case where $b \in G(L)$ is a representative of $\Upsilon(x) \in C(G, \mu)$ for some point x in the reduction of a Shimura variety of Hodge type, and μ is the cocharacter associated with the Shimura variety.

The theorem has originally been proved by Katz [Ka] for F-isocrystals (i.e. the case of $G = \mathrm{GL}_n$) who refers to it as Mazur's inequality. The group theoretic generalization is due to Kottwitz [Kot3]. In both references this is only stated for $F = \mathbb{Q}_p$, but the same proof also shows the analog for function fields.

We have the following converse.

Theorem 3.9 (Kottwitz-Rapoport [KR], Lucarelli [Lu], Gashi [Ga]). *Let $\mu \in X_*(T)_{\mathrm{dom}}$ and let $[b] \in B(G, \mu)$. Then there is an $x \in G(\mathcal{O}_L)\mu(\epsilon)G(\mathcal{O}_L)$ with $[x] = [b]$.*

We now discuss a first important geometric property of Newton strata. We use the notion of an F-isocrystal with G-structure which is defined by Rapoport and Richartz as an exact faithful tensor functor $(\mathrm{Rep}_{\mathbb{Q}_p} G) \to (F-\mathrm{Isoc})$. For the precise definition and explanations compare [RR], 3.1–3.3. In particular, this theorem applies to the reduction modulo p of Shimura varieties of Hodge type.

Theorem 3.10 (Grothendieck, Rapoport-Richartz). *Let S be a scheme of characteristic p. Let X be an F-isocrystal with G-structure over S. Let $[b] \in B(G)$. Then*

$$S_{\leq [b]}(k) = \{s \in S(k) \mid [X_s] \preceq [b]\}$$

defines a closed subscheme $S_{\leq [b]}$ of S, the closed Newton stratum associated with $[b]$. *The* Newton stratum associated with $[b]$ *is the open subscheme of $S_{\leq [b]}$ defined by*

$$S_{[b]}(k) = \{s \in S(k) \mid [X_s] = [b]\}.$$

Remark 3.11. (1) This theorem is known as Grothendieck's specialization theorem. Grothendieck's original proof [Gr] is for isocrystals (or p-divisible groups) without additional structure. The group theoretic generalization is due to Rapoport and Richartz, [RR].

(2) Assume that the map κ_G has the constant value $\kappa_G([b])$ in all closed points of S. This is for example the case if S is connected. Then for $\nu = \nu([b])$ we also write $\mathcal{N}_\nu$ instead of $S_{[b]}$ (and similarly for $S_{\leq [b]}$).

(3) There is the following analog in the function field case. Let LG be the loop group of G, i.e. the ind-scheme over $\mathbb{F}_q = \mathcal{O}_F/\epsilon\mathcal{O}_F$ representing the functor assigning to an $\mathbb{F}_q$-algebra R the set $G(R[\![\epsilon]\!][\frac{1}{\epsilon}])$. Let $X \in LG(S)$ for S as in the theorem. Then the same statement as in Theorem 3.10 holds for the induced decomposition of S. This analog follows from essentially the same proof as Theorem 3.10, compare also [HV1], Theorem 7.3.

Definition 3.12. Let $\mathcal{S}_K(G, X)$ be a Hodge type Shimura variety and let μ be the associated cocharacter. Then the μ-ordinary Newton stratum is the one associated with the unique maximal element $[\mu(p)]$ of $B(G, \mu)$ (compare Theorem 3.4 (4)). It is open in $\mathcal{S}_K(G, X)$. The basic Newton stratum is the one associated with the unique minimal element of $B(G, \mu)$. It is closed in $\mathcal{S}_K(G, X)$.

In the following sections we explain converses of the two preceding theorems.

4 Non-emptiness of strata

By Theorem 3.7 the set of Newton strata in a Shimura variety of Hodge type with hyperspecial level structure is indexed by the set $B(G, \mu)$ where G and μ are the group and coweight determined by the Shimura datum. However, it is a priori not clear if for each $[b] \in B(G, \mu)$ the corresponding stratum $\mathcal{N}_{[b]} \subseteq \mathcal{S}_K(G, X)$ is indeed non-empty. To show such a statement, the task is to construct an abelian variety with additional structure whose rational Dieudonné module (with additional structure) is a given one.

The first result in this direction valid for a large class of Shimura varieties is the following theorem.

Theorem 4.1 (Viehmann-Wedhorn [VW], Thm. 1.6). *Let G, μ be associated with a Shimura datum of PEL type. Let $[b] \in B(G, \mu)$. Then the Newton stratum associated with $[b]$ in the special fiber of the associated Shimura variety is non-empty.*

Remark 4.2. This proves a conjecture of Rapoport, [Ra], Conj. 7.1. Note that non-emptiness of Newton strata does not follow directly from Theorem 3.9, as we do not know if the map $\Upsilon : \mathcal{S}_K(G, X)(k) \to C(G, \mu)$ is surjective. Also, it cannot be directly deduced from Kottwitz's [Kot2] resp. Kisin's [Ki2] computation of the ζ-function of the Shimura variety. It is a priori not clear from that formula that the contribution of a given Newton stratum is non-zero.

There are several special cases where non-emptiness of Newton strata has been shown already some time ago.

In [We], Wedhorn shows that for PEL type Shimura varieties (as usual assumed to have good reduction) the μ-ordinary Newton stratum is dense in the Shimura variety and in particular non-empty. Using Kottwitz's description of the points in the good reduction of Shimura varieties of PEL type one can show that the basic Newton stratum is also non-empty (see Fargues [Far], 3.1.8 or Kottwitz [Kot2], 18).

Using a different language, the non-emptiness question has also been adressed by Vasiu in [Va2].

Strategy of proof of Theorem 4.1. In addition to the Newton stratification we consider the Ekedahl-Oort stratification. The Ekedahl-Oort invariant associates with an abelian variety with PEL structure its p-torsion (with induced additional structure). Group-theoretically, and on geometric points in the special fiber of the Shimura variety, this means the following: Let $x \in \mathcal{S}_K(G, X)(k)$ for some algebraically closed field k. Let $[\![g_x]\!] = \Upsilon(x) \in C(G, \mu)$. Let $G_1 = \{g \in G(\mathcal{O}_L) \mid g \equiv 1 \text{ in } G(k)\}$. Then the Ekedahl-Oort invariant corresponds to considering the G_1-double coset of $[\![g_x]\!]$ (see [Vi3]). Note that this is well-defined as G_1 is a normal subgroup of $G(\mathcal{O}_L)$. The set of possible values of this invariant is finite, and in bijection with a certain subset ${}^{\mu}W$ of the Weyl group of G. The Ekedahl-Oort invariant induces a stratification of the Shimura variety.

The above definition of the Ekedahl-Oort invariant has a strong focus on the group theoretic background. There are alternative definitions, for example using G-zips [MW] that rather use the p-divisible groups directly.

The first main ingredient in the proof of Theorem 4.1 is to show that for every element $[b] \in B(G, \mu)$ there is an element of ${}^{\mu}W$ such that the corresponding Ekedahl-Oort stratum is contained in the Newton stratum of $[b]$. Strictly speaking, in [VW] we showed a slightly weaker statement by modifying the Shimura datum. However, in the meantime, Nie [N] has shown that also the stronger statement above holds, which leads to a simplification of the proof. It is therefore enough to show that all Ekedahl-Oort strata are non-empty.

The second main ingredient is to use flatness of the morphism from $\mathcal{S}_K(G, X)$ to the stack of Ekedahl-Oort invariants (or equivalently: G-zips) to show that all Ekedahl-Oort strata are non-empty if and only if the same holds for the one corresponding to the unique closed point in this stack. This 'minimal' Ekedahl-Oort stratum is known to be contained in the basic Newton stratum of the Shimura variety.

Finally, one uses that non-emptiness of the basic Newton stratum implies that also the minimal Ekedahl-Oort stratum is non-empty to complete the proof. □

Let $\mathcal{A}$ be an abelian variety over an algebraically closed field of characteristic p, let $\mathcal{A}[p^\infty]$ be its p-divisible group, and X a p-divisible group isogenous to $\mathcal{A}[p^\infty]$. Dividing by the kernel of such an isogeny one can directly construct an abelian variety $\mathcal{B}$ isogenous to $\mathcal{A}$ and with $\mathcal{B}[p^\infty] = X$. A version of this argument including the PEL structure (see [VW], 11) leads to the following integral version of Theorem 4.1.

Theorem 4.3 ([VW], Theorem 1.6(2)). *Let $\mathscr{D}$ be a PEL Shimura-datum and let $\mathcal{S}_K(G, X)$ be the corresponding Shimura variety. Then for every p-divisible group $\mathcal{G}$ with $\mathscr{D}$-structure over an algebraically closed field k in characteristic p there is a k-valued point of $\mathcal{S}_K(G, X)$ whose attached p-divisible group with $\mathscr{D}$-structure is isomorphic to $\mathcal{G}$. In group-theoretic terms: The map $\mathcal{S}_K(G, X)(k) \to C(G, \mu)$ is surjective.*

Remark 4.4. Since the proof of the above Theorem 4.1, the question has attracted significant attention. Several people have generalized Theorem 4.1 further to other classes of Shimura varieties of Hodge type, and/or have given other strategies to prove non-emptiness of Newton strata.

(1) Scholze and Shin [SS], Corollary 8.4 give a different proof of Theorem 4.1 for certain compact unitary group Shimura varieties. They use the description of points on the Shimura variety by Kottwitz triples and stabilization of the trace formula to show non-emptiness in this case.

(2) Kret [Kr] uses Kisin's formula for the number of mod p points of Shimura varieties of Hodge type (see [Ki2]), together with stabilization of the twisted trace formula. In this way he shows non-emptiness of Newton strata for Shimura varieties of Hodge type where the associated group is of type (A), and under assumption of the stabilization of the trace formula also for types (B) and (C).

(3) Koskivirta [Kos] uses quasi-affineness of Ekedahl-Oort strata (which he reproves using generalized Hasse invariants) and flatness of the projection of the special fiber of a Shimura variety to a corresponding stack of G-zips to show that if $\mathcal{S}_K(G, X)$ is a Shimura variety of Hodge type that is projective, then all Ekedahl-Oort strata are non-empty. Using a result of Nie [N] he can then in the same way as in the original proof of Theorem 4.1 deduce that also all Newton strata are non-empty.

(4) Lee gives two proofs of the generalization of Theorem 4.1 to Shimura varieties of Hodge type, see [Le]. In his first proof he constructs a point in the given Newton stratum by realizing it as the image of a point of a special sub-Shimura variety. For the second proof he uses that by work of Kisin [Ki2], the points mod p of a Shimura

variety of Hodge type are in bijection with Kottwitz triples. The Newton point can be read off easily from one of the components of the Kottwitz triple. Lee's approach is then to show directly that for each σ-conjugacy class a suitable Kottwitz triple exists.

(5) C.-F. Yu [Yu2] announced a proof of non-emptiness of the basic locus of Shimura varieties having what he calls an enhanced integral model. This kind of integral model is defined by an abstract set of axioms, for example satisfied by the integral models for Shimura varieties of Hodge type. He then also uses the strategy of constructing points via special sub-Shimura varieties, using a Lemma shown by Langlands and Rapoport.

(6) In a talk given in Oberwolfach in August 2015, Kisin has announced a proof (in joint work with Madapusi Pera and Shin) for all Shimura varieties of Hodge type, also using special sub-Shimura varieties, and a lemma of Langlands-Rapoport from [LR].

5 Closure relations and dimensions

5.1 The Grothendieck conjecture on closures of Newton strata

In 1970 ([Gr], letter to Barsotti, p. 150), Grothendieck conjectured the following converse to his specialization theorem (compare Theorem 3.10).

Conjecture 5.1. *Let $\mathbb{X}_0$ be a p-divisible group over a field k of characteristic p, and let ν be its Newton polygon. Let $\nu' \in \mathcal{N}(G)_{\mathbb{Z}}$ with $\nu \preceq \nu' \preceq \mu$. Here $\mu = (1, \dots, 1, 0, \dots, 0)$ with multiplicities the codimension and dimension of $\mathbb{X}_0$ is the Hodge polygon of $\mathbb{X}_0$. Then there is a p-divisible group $\mathbb{X}$ over* Spec $(k[[t]])$ *with special fiber $\mathbb{X}_0$ and such that the generic fiber has Newton polygon ν'.*

This conjecture has been shown by Oort in 2002, see Remark 5.3(1) below.

Using the bijection between deformations of an abelian variety in characteristic p and deformations of its p-divisible group, (as well as liftability properties in the spirit of Theorem 4.3) one can translate this into a statement on the closures of Newton strata in a suitable Shimura variety. The natural generalization to Shimura varieties of Hodge type is the following theorem.

Theorem 5.2 (Hamacher, [Ha3], Theorem 1.2). *Let $\mathcal{S}_K(G, X)$ be a Shimura variety of Hodge type with hyperspecial level structure at p. Let $[b] \in B(G, \mu)$ where μ is the cocharacter associated with the Shimura datum, and let $\nu = \nu([b])$. Then*

$$\overline{\mathcal{N}_\nu} = \bigcup_{\nu' \preceq \nu} \mathcal{N}_{\nu'}.$$

Remark 5.3. Before Hamacher's proof, a number of partial results were already known.

(1) Grothendieck's original conjecture corresponds to the case that $G_{\mathbb{Q}_p} \cong \mathrm{GL}_h$. This has been shown by Oort [O1]. There, Oort also proved Theorem 5.2 for the Siegel modular variety (i.e. for $G = \mathrm{GSp}_{2n}$), or equivalently for deformations of principally polarized p-divisible groups.

(2) Wedhorn used explicit deformations to prove in [We] that the μ-ordinary Newton stratum (i.e. the one corresponding to the unique maximal element of $B(G, \mu)$ with respect to $\preceq$) of a Shimura variety of PEL type with good reduction is dense. This was generalized to Shimura varieties of Hodge type by Wortmann [Wo], using Ekedahl-Oort strata.

(3) The general statement of Theorem 5.2 (together with weaker forms) was asked for by Rapoport in [Ra], Questions 7.3.

(4) In a small number of other special cases, this statement was known due to explicit calculations, for example for the Shimura variety studied by Harris and Taylor in their proof of the local Langlands correspondence for GL_n.

(5) Hamacher himself first proved the case of PEL Shimura varieties in [Ha2], Theorem 1.1.

Remark 5.4. Parts of Oort's proof of the original Grothendieck conjecture have also been generalized, however, in general this has not led to a proof of Theorem 5.2. Let us recall briefly the two main steps in Oort's approach. The first step is to show that one can deform a given p-divisible group to a p-divisible group with the same Newton point, but a-invariant ≤ 1. Here, the a-invariant of a p-divisible group $\mathbb{X}$ over a field k is defined as $\dim_k \mathrm{Hom}(\alpha_p, \mathbb{X})$. The main difficulty is here to consider the case where the isocrystal of the p-divisible group is simple (and then to use induction on the number of simple summands for the general case). Main ingredients in this step of the proof (in joint work of de Jong and Oort, [JO]) are de Jong and Oort's purity theorem for the Newton stratification (a weak form of Theorem 5.10 below), and combinatorial arguments on some stratification of the Rapoport-Zink moduli space of p-divisible groups isogenous to $\mathbb{X}$. A key intermediate result is that the Rapoport-Zink moduli space is equi-dimensional, and to compute its dimension.

In a second step Oort uses a Cayley-Hamilton type argument to explicitly compute the Newton polygon of deformations of p-divisible groups of a-invariant 1 to conclude. This second part of the argument has been generalized by Yu [Yu1].

However, it seems to be hard to extend this to a proof of the whole statement, simply because there is (for this particular purpose at least) no good analogue of the a-invariant on moduli spaces of p-divisible groups with additional structure. For general Newton strata on Shimura varieties of PEL type the locus where the a-invariant is at most 1 is in general no longer dense, it can even be empty.

5.2 An analog for function fields

The proof of Theorem 5.2 is modelled along the lines of an analogous statement for the function field case, i.e. for Newton strata in loop groups [Vi2].

To formulate this statement let G be a split connected reductive group over $\mathbb{F}_q$, and let $B \supseteq T$ be a Borel subgroup and a split maximal torus of G. We denote by LG the loop group of G, i.e. the ind-scheme over $\mathbb{F}_q$ representing the sheaf of groups for the fpqc-topology whose sections for an $\mathbb{F}_q$-algebra R are given by $LG(R) = G(R((z)))$, see [Fal], Definition 1. Let L^+G be the sub-group scheme of LG with $L^+G(R) = G(R[[z]])$.

Let S be a scheme and $Y \in LG(S)$. Recall from Remark 3.11(3) that for $[b] \in B(G)$ the locus $\mathcal{N}_{[b]} \subseteq S$ where Y is in $[b]$ defines a locally closed reduced subscheme of S.

On the set of dominant coweights $X_*(T)_{\mathrm{dom}}$ we consider the partial ordering induced by the Bruhat order, i.e. $\mu \preceq \mu'$ if and only if $\mu' - \mu$ is a non-negative integral linear combination of positive coroots. Note that this is slightly finer than the ordering induced by $\preceq$ on $X_*(T)_{\mathbb{Q},\mathrm{dom}}$, as we do not allow rational linear combinations.

Theorem 5.5 ([Vi2], Thm. 2). *Let $\mu_1 \preceq \mu_2 \in X_*(T)$ be dominant coweights. Let*

$$S_{\mu_1,\mu_2} = \bigcup_{\mu_1 \preceq \mu' \preceq \mu_2} L^+G\mu'(z)L^+G.$$

Let $[b] \in B(G, \mu_2)$. Then the Newton stratum $\mathcal{N}_{[b]} = [b] \cap S_{\mu_1,\mu_2}$ is non-empty and pure of codimension

$$\mathrm{length}([\nu([b]), \mu_2]) = \langle \rho, \mu_2 - \nu([b]) \rangle + \frac{1}{2}\mathrm{def}(b)$$

in S_{μ_1,μ_2}. The closure of $\mathcal{N}_{[b]}$ in S_{μ_1,μ_2} is the union of all $\mathcal{N}_{[b']}$ for $[b'] \preceq [b]$.

Here the defect $\mathrm{def}(b)$ is as in Definition 3.3 defined as $\mathrm{rk}\, G - \mathrm{rk}_{\mathbb{F}_q((z))} J_b$.

Note that one still needs to define the notions of codimension and of the closure of the infinite-dimensional schemes $\mathcal{N}_b$. Both of these definitions use that there is an open subgroup H of L^+G such that the Newton point of an element of S_{μ_1,μ_2} only depends on its image in the finite-dimensional scheme $S_{\mu_1,\mu_2}/H$, compare [Vi2], 4.3.

Remark 5.6. Theorem 5.5 is not precisely the analog of Theorem 5.2:

(1) In this context we can consider the Newton stratification on arbitrary (unions of) double cosets. The double coset is the replacement of the Hodge polygon in the Shimura variety context. There, one is limited to the case of minuscule μ.

(2) For the closure relations we could also consider the Newton stratification on the whole loop group, but this yields a much weaker statement: There it is very easy to see that the closure of a stratum is a union of strata (just because the σ-conjugation action of $G(L)$ is transitive on each Newton stratum), and it consists of the conjectured strata because of [VW], Cor. 1.9. There is no good analogue for the whole loop group of the statement about codimensions.

(3) Theorem 5.5 is only proven for split reductive groups. The reason is that at the time, the dimension formula and theory of local G-shtukas that is used in the proof was not yet developed in greater generality. We expect that essentially the same proof shows the theorem for all unramified groups. (Which is also justified by the fact that the analog for Shimura varieties in Theorem 5.2 is shown in that generality.) On the other hand, the assumption that K_p is a hyperspecial maximal compact subgroup cannot easily be removed.

(4) The direct analog of Theorem 5.2 in the function field context would be a statement about moduli spaces of shtukas with additional structure. The present statement is rather the generalized analog of Grothendieck's conjecture in the sense that it

considers deformations of the local data. In the same way as for abelian varieties and p-divisible groups it should be possible to use the correspondence between deformations of global shtukas and the corresponding local shtukas to translate Theorem 5.5 into a statement about moduli spaces of global shtukas with additional structure.

5.3 Outline of the proof and dimensions of Newton strata

We will spend the rest of this section to outline some of the joint main ingredients of the proofs of Theorems 5.2 and 5.5. For simplicity we formulate statements only in the context of Shimura varieties. For the corresponding statements in the function field context compare [Vi2]. The strategy of proof differs from Oort's proof of the Grothendieck conjecture in the sense that it avoids the use of the a-invariant. Instead we use a stronger version of the purity theorem to derive the closure relations directly from the computation of dimensions of suitable moduli spaces.

The proof of Theorem 5.2 at the same time proves a formula for the dimensions of Newton strata which is interesting in its own right. It is the counterpart of the statement about codimensions in Theorem 5.5.

Theorem 5.7 ([Ha3], Thm. 1.2). *Let $Sh_K(G,X)$ be a Shimura variety of Hodge type with hyperspecial level structure at p, and let $[b] \in B(G,\mu)$ be in the associated set of σ-conjugacy classes. Then the Newton stratum $\mathcal{N}_{[b]}$ in $Sh_K(G,X)$ is equidimensional of dimension*

$$\langle \rho, \mu + \nu([b]) \rangle - \frac{1}{2}\mathrm{def}(b).$$

The translation between statements about (co)dimensions and statements about closure relations is done using purity properties. The following definition and Lemma 5.12 are an abstract version of the key idea behind the proofs of 5.5, 5.2 and 5.7.

Definition 5.8. Let G be a reductive group over $\mathbb{F}_q$, B a Borel subgroup, $T \subset B$ a maximal torus, and let $\mu \in X_*(T)_{\mathrm{dom}}$ be given. Let S be a scheme, and for every $[b] \in B(G,\mu)$ let $S_{[b]}$ be a locally closed subscheme such that S is the disjoint union of the $S_{[b]}$ and for every $[b]$, the subscheme $\bigcup_{[b']\preceq[b]} S_{[b']}$ is closed. Then we say that this decomposition of S satisfies strong purity if the following condition holds. Let $[b] \in B(G,\mu)$ such that $S_{[b]} \neq \emptyset$. Let $[b'] \preceq [b]$ such that $\mathrm{length}([\nu[b'], \nu([b])]) = 1$. Let I be the closure in S of an irreducible component of $S_{[b]}$. Let $I^{[b']}$ be the complement in I of $I \cap \bigcup_{[b'']\preceq[b']} S_{[b'']}$. Then for every such I and $[b']$ we have that $I^{[b']}$ is an affine I-scheme.

The decomposition satisfies weak purity if for every I as above, $I \cap S_{[b]}$ is an affine I-scheme.

Example 5.9. For the group GL_n we have the following explicit description: Let ν, ν' be the Newton points of $[b]$ and $[b']$ with $[b], [b']$ as in the definition above. Then the integral points lying on or below p_ν are the same as those for $p_{\nu'}$ except for one breakpoint y of ν which does no longer lie on ν'. The subscheme $I^{[b']}$ is now the union of all Newton strata of I where the Newton polygon (which automatically lies below ν) contains the given point y.

Theorem 5.10. *The Newton stratification on the reduction modulo p of Hodge type Shimura varieties satisfies strong purity.*

Remark 5.11. The first version of a theorem along these lines was shown by de Jong and Oort [JO], who showed that the Newton stratification for p-divisible groups without additional structure satisfies a weaker version of weak purity where instead of affineness we require that the complement of $I \cap \mathcal{N}_{[b]}$ in I is either empty or pure of codimension 1. Vasiu [Va1] showed that this same stratification also satisfies weak purity as defined above. Finally, Hartl and the author proved a group-theoretic version of weak purity in the function field case (i.e. for moduli of local G-shtukas), see [HV1].

Only several years later, Yang [Ya] introduced the concept of strong purity (under a different name) and observed that de Jong and Oort's statement can be generalized to prove (the same weakened version of) strong purity for the Newton stratification on families of p-divisible groups. After this idea, the same generalizations as for weak purity were made, in the function field case in [Vi2], for Shimura varieties of Hodge type in [Ha3].

Lemma 5.12. *Let S be an irreducible scheme and let $S = \bigcup_{[b]\in B(G,\mu)} S_{[b]}$ be a decomposition satisfying strong purity. Let $[b_\eta] \in B(G, \mu)$ be the generic σ-conjugacy class. Let $[b_0] \in B(G, \mu)$ be such that $S_{[b_0]}$ is non-empty and that the codimension of every irreducible component of $S_{[b_0]}$ in S is at least length$([\nu([b_0]), \nu([b_\eta])])$. Then for every $[b'] \in B(G, \mu)$ with $\nu([b_0]) \preceq \nu([b']) \preceq \nu([b_\eta])$ we have*

(1) $S_{[b_0]} \subset \overline{S_{[b']}}$.

(2) In particular, $S_{[b']} \neq \emptyset$.

(3) Every irreducible component I of $S_{[b_0]}$ is of codimension length$([\nu([b_0]), \nu([b_\eta])])$ in S.

Proof. The proof is based on the fact that if $I^{[b']}$ is an affine I-scheme, then its complement in I is either empty or pure of codimension 1. Using that, the lemma is shown in the same combinatorial way as the more explicit statements for Newton strata in loop groups [Vi2] or Shimura varieties [Ha2], Prop. 5.13. □

In view of Lemma 5.12, Theorem 5.10 implies that in order to prove Theorems 5.2 and 5.7 it is enough to show that for every element in $B(G, \mu)$, the codimension of each irreducible component of the corresponding Newton stratum is at least length$([\nu([b]), \mu])$.

From the almost product structure on Newton strata (see [O2] for the case of $G = \mathrm{GL}_n$ or GSp_{2n}, [Ma1], 4 for the general case for PEL type, and [Ha3] for Hodge type) we obtain for every Newton stratum $\mathcal{N}_\nu$ in $\mathcal{S}_K(G, X)$ with $\nu = \nu([b])$ associated with a p-divisible group with additional structure $\underline{\mathbb{X}}$ that

$$\dim \mathcal{N}_\nu = \dim X_\mu(b) + \dim C_{\underline{\mathbb{X}}}.$$

Here, $C_{\underline{\mathbb{X}}}$ is the central leaf associated with $\underline{\mathbb{X}}$. The first summand is the dimension of the Rapoport-Zink space associated with $\underline{\mathbb{X}}$ or equivalently of the affine Deligne-Lusztig variety $X_\mu(b)$. Note that central leaves and the above decomposition are essentially a special case of the foliation structure on Newton strata using Igusa varieties explained in [Ma2], 5.

For the two summands we have

Theorem 5.13 ([GHKR],[Vi1],[Ha1], Thm. 1.1). *Let* $[b] \in B(G, \mu)$. *Then*

$$\dim X_\mu(b) = \langle \rho, \mu - \nu \rangle - \frac{1}{2}\mathrm{def}(b).$$

Another proof of this result for unramified groups G in the arithmetic context has recently been given by Zhu [Z],3.

Theorem 5.14 ([O3], [Ha3], Prop. 3.9). *Let* $[b] \in B(G, \mu)$ *be the class corresponding to* $\underline{\mathbb{X}}$. *Then*

$$\dim C_{\underline{\mathbb{X}}} = \langle 2\rho, \nu([b]) \rangle.$$

Altogether we obtain

$$\begin{aligned} \dim \mathcal{N}_\nu &= \dim X_\mu(b) + \dim C_{\underline{\mathbb{X}}} \\ &= \langle \rho, \mu + \nu \rangle - \frac{1}{2}\mathrm{def}(b) \\ &= \dim \mathcal{S}_K(G, X) - \langle \rho, \mu - \nu \rangle - \frac{1}{2}\mathrm{def}(b) \\ &= \dim \mathcal{S}_K(G, X) - \mathrm{length}([\nu, \mu]). \end{aligned}$$

Thus the codimension of every irreducible component of $\mathcal{N}_\nu$ is at least length$([\nu, \mu])$, and Lemma 5.12 finishes the proof of Theorems 5.2 and 5.7.

Acknowledgments

I thank P. Hamacher for helpful discussions. The author was partially supported by ERC starting grant 277889 "Moduli spaces of local G-shtukas".

References

[C] C.-L. Chai, *Newton polygons as lattice points*, Amer. J. Math. **122** (2000), 967–990.

[Fal] G. Faltings, *Algebraic loop groups and moduli spaces of bundles*, J. Eur. Math. Soc. **5** (2003), 41–68.

[Far] L. Fargues, *Cohomologie des espaces de modules de groupes p-divisibles et correspondances de Langlands locales*, in: Variétés de Shimura, espaces de Rapoport-Zink et correspondances de Langlands locales, Astérisque **291** (2004), 1–199.

[Ga] Q. Gashi, *On a conjecture of Kottwitz and Rapoport*, Ann. Sci. École Norm. Sup. **43** (2010), 1017–1038.

[GHKR] U. Görtz, Th. Haines, R. Kottwitz, D. Reuman, *Dimensions of some affine Deligne-Lusztig varieties*, Ann. Sci. École Norm. Sup. **39** (2006), 467–511.

[GHKR2] U. Görtz, Th. Haines, R. Kottwitz, D. Reuman, *Affine Deligne-Lusztig varieties in affine flag varieties*, Compositio Math. **146** (2010), 1339–1382.

[Gr] A. Grothendieck, *Groupes de Barsotti-Tate et cristaux de Dieudonné*, Séminaire de Mathématiques Supérieures, No. 45 (Eté, 1970). Les Presses de l'Université de Montréal, 1974.

[Ha1] P. Hamacher, *The dimension of affine Deligne-Lusztig varieties in the affine Grassmannian of unramified groups*, IMRN 2015, 12804–12839.

[Ha2] P. Hamacher, *The geometry of Newton strata in the reduction modulo p of Shimura varieties of PEL type*, Duke Math. J. **164** (2015), 2809–2895.

[Ha3] P. Hamacher, *The almost product structure of Newton strata in the deformation space of a Barsotti-Tate group with crystalline Tate tensors*, Math. Z. **287** (2017), 1255–1277.

[HT] M. Harris, R. Taylor, The geometry and cohomology of some simple Shimura varieties. With an appendix by Vladimir G. Berkovich. Annals of Mathematics Studies, **151**. Princeton University Press, Princeton, NJ, 2001. viii+276 pp.

[HV1] U. Hartl, E. Viehmann, *The Newton stratification on deformations of local G-shtukas*, J. reine angew. Mathematik (Crelle's Journal) **656** (2011), 87–129.

[HV2] U. Hartl, E. Viehmann, *Foliations in deformation spaces of local G-shtukas*, Adv. Math. **229** (2012), 54–78.

[JO] A. J. de Jong, F. Oort, *Purity of the stratification by Newton polygons*, J. Amer. Math. Soc. **13** (2000), 209–241.

[Ka] N. M. Katz, *Slope filtration of F-crystals*, Astérisque **63** (1979), 113–164.

[Ki1] M. Kisin, *Integral models for Shimura varieties of abelian type*, J.A.M.S. **23** (2010), 967–1012.

[Ki2] M. Kisin, *Mod p points on Shimura varieties of abelian type*, J.A.M.S. **30** (2017), 819–914.

[Kos] J.-S. Koskivirta, *Sections of the Hodge bundle over Ekedahl-Oort strata of Shimura varieties of Hodge type*, Journal of Algebra **449**, 446–459.

[Kot1] R. E. Kottwitz, *Isocrystals with additional structure*, Compositio Math. **56** (1985), 201–220.

[Kot2] R. E. Kottwitz, *Points on some Shimura varieties over finite fields*, J. Amer. Math. Soc. **5** (1992), 373–444.

[Kot3] R. E. Kottwitz, *On the Hodge-Newton decomposition for split groups*, IMRN **26** (2003), 1433–1447.

[Kot4] R. E. Kottwitz, *Dimensions of Newton strata in the adjoint quotient of reductive groups*, Pure Appl. Math. Q. **2** (2006), 817–836.

[KR] R. E. Kottwitz, M. Rapoport, *On the existence of F-crystals*, Comment. Math. Helv. **78** (2003), 153–184.

[Kr] A. Kret, *The trace formula and the existence of PEL type Abelian varieties modulo p*, preprint, 2012, arXiv:1209.0264

[LR] R. Langlands, M. Rapoport, *Shimuravarietäten und Gerben*, J. reine angew. Math. (1987) **378**, 113–220.

[Le] D. U. Lee, *Non-emptiness of Newton strata of Shimura varieties of Hodge type*, Algebra and Number Theory **12** (2018), 259–283.

[Lu] C. Lucarelli, *A converse to Mazur's inequality for split classical groups*, J. Inst. Math. Jussieu **3** (2004), 165–183.

[Ma1] E. Mantovan, *On certain unitary group Shimura varieties*. In: Variétés de Shimura, espaces de Rapoport-Zink et correspondances de Langlands locales, Astérisque **291** (2004), 201–331.

[Ma2] E. Mantovan, *The Newton stratification*, this volume.

[MW] B. Moonen, T. Wedhorn, *Discrete invariants of varieties in positive characteristic*, Int. Math. Res. Notices 2004, 3855–3903.

[N] S. Nie, *Fundamental elements of an affine Weyl group*, Math. Ann. **362** (2015), 485–499.

[O1] F. Oort, *Newton polygon strata in the moduli space of abelian varieties*, in: Moduli of abelian varieties (Texel Island, 1999), 417–440, Progr. Math., **195**, Birkhäuser, Basel, 2001.

[O2] F. Oort, *Foliations in moduli spaces of abelian varieties*, J. Amer. Math. Soc. **17** (2004), 267–296.

[O3] F. Oort, *Foliations in moduli spaces of abelian varieties and dimension of leaves*, in: Algebra, arithmetic, and geometry: in honor of Yu. I. Manin. Vol. II, 465–501, Progr. Math., 270, Birkhäuser Boston, Inc., Boston, MA, 2009.

[Ra] M. Rapoport, *A guide to the reduction modulo p of Shimura varieties*, Astérisque **298** (2005), 271–318.

[RR] M. Rapoport, M. Richartz, *On the classification and specialization of F-isocrystals with additional structure*, Compositio Math. **103** (1996), 153–181.

[RZ] M. Rapoport, Th. Zink, *Period spaces for p-divisible groups*, Princeton Univ. Press, 1996.

[SS] P. Scholze, S. W. Shin, *On the cohomology of compact unitary group Shimura varieties at ramified split places*, J. Amer. Math. Soc. **26** (2013), 261–294.

[Va1] A. Vasiu, *Crystalline boundedness principle*, Ann. Sci. École Norm. Sup. **39** (2006), 245–300.

[Va2] A. Vasiu, *Manin problems for Shimura varieties of Hodge type*, J. Ramanujan Math. Soc. **26** (2011), 31–84.

[Vi1] E. Viehmann, *The dimension of some affine Deligne-Lusztig varieties*, Ann. Sci. École Norm. Sup. **39** (2006), 513–526.

[Vi2] E. Viehmann, *Newton strata in the loop group of a reductive group*, American Journal of Mathematics **135** (2013), 499–518.

[Vi3] E. Viehmann, *Truncations of level 1 of elements in the loop group of a reductive group*, Annals of Math. **179** (2014), 1009–1040.

[VW] E. Viehmann, T. Wedhorn, *Ekedahl-Oort and Newton strata for Shimura varieties of PEL type*, Math. Ann. **356** (2013), 1493–1550.

[We] T. Wedhorn, *Ordinariness in good reductions of Shimura varieties of PEL-type*, Ann. scient. Éc. Norm. Sup., **32**, (1999), 575–618.

[Wo] D. Wortmann, *The μ-ordinary locus for Shimura varieties of Hodge type*, preprint, 2013, arxiv:1310.6444.

[Ya] Y. Yang, *An improvement of de Jong–Oort's purity theorem*, Münster J. Math. **4** (2011), 129–140.

[Yu1] C.-F. Yu, *On the slope stratification of certain Shimura varieties*, Math. Z. **251** (2005), 859 –873.

[Yu2] C.-F. Yu, *Non-emptiness of the basic locus of Shimura varieties*, Oberwolfach reports (2015).

[Z] X. Zhu, *Affine Grassmannians and the geometric Satake in mixed characteristic*, Annals of Math. **185** (2017), 403–492.

CONSTRUCTION OF AUTOMORPHIC GALOIS REPRESENTATIONS: THE SELF-DUAL CASE

SUG WOO SHIN

This survey paper grew out of the author's lecture notes for the Clay Summer School in 2009. Proofs are often omitted or only sketched, but a number of references are given to help the reader find more details if he or she wishes. The article aims to be a guide to the outline of argument for constructing Galois representations from the so-called regular algebraic conjugate self-dual automorphic representations, implementing recent improvements in the trace formula and endoscopy.

Warning. Although it is important to keep track of various constants (e.g. sign matters!), twists by characters, etc in proving representation-theoretic identities, I often ignore them in favor of simpler notation and formulas. Thus many identities and statements in this article should be taken with a grain of salt and not cited in academic work. Correct and precise statements can be found in original papers. I should also point out that significant progress has been made, especially in the case of *non-self-dual* automorphic representations ([HLTT16], [Sch15]) since the original version of these notes was written. I have kept updates minimal (mostly taking place in introduction, adding references to some recent developments) and do not discuss these exciting new results here. The interested readers are referred to survey papers such as [Wei16, Mor16, Sch14, Sch16, Car].

Acknowledgments

I heartily thank the Clay Mathematics Institute for its generous support toward the 2009 summer school and choosing an exciting location as Hawaii. I am grateful to the participants for their interest and wonderful questions.

1 Introduction

1.1 Reading list

This incomplete list of survey articles is only intended to suggest some starting points for further learning. The reader should not feel compelled to go through too many of them before starting to read the present article.

- [Tay04] is a nice survey of various topics on Galois representations.
- local and global Weil groups, Weil-Deligne groups [Tat79]; L-groups, morphisms of L-groups (L-morphisms) and Satake isomorphisms [Bor79].

- representation theory of GL_n over p-adic fields or archimedean fields with a view toward the local Langlands correspondence [Kud94], [Kna94]
- [Art05] may be a good place where one can start to learn the Arthur-Selberg trace formula. Arthur also wrote several short survey papers on the trace formula, which one may find very helpful.
- (conjectural formulation of) endoscopy for unitary groups: [Rog92, §2], [Mok15]
- [BR94] would be helpful in that it reviews numerous concepts that constantly show up in the study of cohomology of Shimura varieties.
- Surveys on various topics rotating around the trace formula and endoscopy can be found in the Paris book project volume I [CHLN11].
- We do not do enough justice to the history of the subject though we have a few remarks in §1.5 below. The best source is papers by Langlands on the subject, found at:
  http://publications.ias.edu/rpl/section/26

The following are research papers and books where one can seriously learn some of the advanced topics that are important to this article. I do not claim by any means that this list is even nearly complete but let me add that there has been recent progress on extending some key constructions like Newton stratification, Igusa varieties, and Rapoport-Zink spaces to the setup for Shimura varieties of Hodge type (and sometimes abelian type) by Hamacher, Howard-Pappas, and W. Kim among others.

- Base change for unitary groups: [Lab11], [Mok15]; for GL_n: [AC89]
- PEL-type Shimura varieties: [Kot92b] (esp. §5), [HT01, Ch III], [Mil05] (esp. §8), [Lan13]
- Newton stratification (in the case of interest): [HT01, III.4], [Man]
- Igusa varieties: [HT01, Ch IV], [Man05]
- Rapoport-Zink spaces: [RZ96], [Far04]
- Stabilization of (elliptic part of) the trace formula: [Kot86], [Kot90]

Finally, §1.5 contains some research papers that are directly related to the main theorem to be discussed.

1.2 Notation

- $\mathbb{A}$ is the adèle ring over $\mathbb{Q}$. If S is a finite set of places of $\mathbb{Q}$ then $\mathbb{A}^S$ is the restricted product of $\mathbb{Q}_v$ over $v \notin S$. In particular, $\mathbb{A}^\infty$ is the ring of finite adèles. If F is a finite extension of $\mathbb{Q}$, $\mathbb{A}_F := \mathbb{A} \otimes_{\mathbb{Q}} F$.
- $\mathrm{Irr}(G(K))$ is the set of isomorphism classes of irreducible smooth representations of $G(K)$, when G is a connected reductive group over a p-adic field K.

- $\mathrm{Irr}(G(\mathbb{A}_F))$ is the set of isomorphism classes of irreducible admissible representations of $G(\mathbb{A}_F)$, when G is a connected reductive group over a number field F. Similarly $\mathrm{Irr}(G(\mathbb{A}_F^S))$ is defined.
- $\mathrm{Groth}(G)$ is the Grothendieck group of admissible and/or continuous representations of G, where G can be a p-adic Lie group, a finite adélic group, a Galois group, and so on. The precise definition is found in [HT01, pp.23–25].
- $\boxplus$ signifies the irreducible parabolic induction either for smooth representations of a p-adic group or for automorphic representations.

1.3 Main theorem

Let F be a number field. There is a famous conjecture of Langlands, complemented by an observation of Clozel, Fontaine and Mazur, which goes as follows. For the notion of a compatible family of Galois representations (for varying primes l and field isomorphisms $\overline{\mathbb{Q}}_l \simeq \mathbb{C}$), see [Tay04, §1] or [BLGGT14, §5].

Conjecture 1.1. *There is a bijection between the following two sets consisting of isomorphism classes.*

$$\left\{ \begin{array}{c} \text{cuspidal automorphic} \\ \text{reps } \Pi \text{ of } GL_m(\mathbb{A}_F) \\ \text{algebraic at } \infty \end{array} \right\} \longleftrightarrow \left\{ \begin{array}{c} \text{compatible systems of} \\ \text{irred. continuous reps} \\ \rho_{l,\iota_l} : \mathrm{Gal}(\overline{F}/F) \to GL_m(\overline{\mathbb{Q}}_l) \\ \text{unram. at all but fin. many places,} \\ \text{de Rham at } l \end{array} \right\}$$

such that

$$\mathrm{WD}(\rho_{l,\iota_l}(\Pi)|_{\mathrm{Gal}(\overline{F}_v/F_v)})^{\mathrm{F-ss}} \simeq \iota_l^{-1}\mathscr{L}_{F_v}(\Pi_v) \tag{1.1}$$

at every finite place v of F.

The algebraicity at ∞ is reviewed in Definition 1.7 below. The "de Rham" property is a technical condition from l-adic Hodge theory. For our purpose it suffices to remark that it is the counterpart of the algebraicity condition for Π at ∞. The functor $\mathrm{WD}(\cdot)$ assigns a Weil-Deligne representation to an l-adic local Galois representation, and the superscript "F-ss" means the Frobenius semisimplification of a Weil-Deligne representation. (See [Tay04, §1] for definitions.) The notation $\mathscr{L}_{F_v}$ denotes a "geometric" normalization (e.g. [Shi11, §2.3]) of the local Langlands correspondence for $GL_m(F_v)$, which was established by Harris-Taylor ([HT01]) and Henniart ([Hen00]) about 10 years ago.

Remark 1.2. In order to uniquely determine the bijection, it suffices to require (1.1) at all but finitely many places v by the strong multiplicity one theorem (on the automorphic side) and the Cebotarev density theorem (on the Galois side).

It is customary to call each direction of the arrow in Conjecture 1.1 as

$\longrightarrow$ construction of Galois representations

$\longleftarrow$ modularity (or automorphy) of Galois representations.

When $m = 1$, Conjecture 1.1 is a consequence of class field theory. The case $m = 2$ with totally real F is discussed in Tilouine's lectures and will not be discussed in my lectures. (This case is separated because Hilbert modular varieties and Shimura curves are used when $m = 2$ while unitary Shimura varieties are used when $m > 2$.)

When $m \geq 3$, the best known case of the above conjecture is the following theorem due to various people. (See §1.5 below for major contributions.) The analogue over totally real fields can be deduced from this theorem (cf. [BLGHT11, Thm 1.1], [BLGGT14, Thm 2.1.1]).

Theorem 1.3. *Assume $m \geq 3$. If*

- *F is a CM field,*
- *Π is a cuspidal automorphic representation of $GL_m(\mathbb{A}_F)$,*
- *$\Pi^\vee \simeq \Pi \circ c$, and*
- *Π_∞ is regular and algebraic,*

then for each prime l and ι_l there exists a semisimple continuous representation

$$\rho_{l,\iota_l}(\Pi) : \mathrm{Gal}(\overline{F}/F) \to GL_m(\overline{\mathbb{Q}}_l),$$

which is unramified at all but finitely many places and de Rham at l, such that (1.1) *holds at every finite place v of F.*

Remark 1.4. It is clear that $\rho_{l,\iota_l}(\Pi)$ is unique up to isomorphism by Cebotarev and Brauer-Nesbitt theorems. We do not know whether $\rho_{l,\iota_l}(\Pi)$ is irreducible in general even though it is expected, unless Π is square-integrable at a finite prime ([TY07, Cor 1.3]). See [BLGGT14, Thm 5.5.2] and [PT15, Thm D] for some partial results. (The dictionary is that the cuspidality of Π should correspond to the irreducibility of $\rho_{l,\iota_l}(\Pi)$.)

Remark 1.5. Roughly speaking, the conditions on Π in the theorem mean that Π comes from a cohomological automorphic representation of a unitary group via quadratic base change.

Remark 1.6. The information about Π at infinite places is encoded by $\rho_{l,\iota_l}(\Pi)$ in the image of complex conjugation (if F has a real place) and the Hodge-Tate numbers at l-adic places. There is also a sign for ρ_{l,ι_l}. See [BLGGT14, Thm 2.1.1] for complete statements (also [CLH16] and [BC11] for complex conjugation and sign; [CLH16] builds on earlier results by Taylor and Taïbi).

My goal is to explain the ideas for the proof of Theorem 1.3.

1.4 Conditions on Π_∞

Let us recall some basic terminology regarding Π_∞. The skimming reader should feel free to skip this subsection.

Let F be any number field. Let $\Pi_\infty = \prod_{v|\infty} \Pi_v$ be a representation of $GL_m(F \otimes_{\mathbb{Q}} \mathbb{R}) = \prod_{v|\infty} GL_m(F_v)$. Let $\phi_v : W_{F_v} \to GL_m(\mathbb{C})$ be the L-parameter for Π_v, where $W_{\mathbb{C}} = \mathbb{C}^\times$ and $W_{\mathbb{R}}$ contains $W_{\mathbb{C}}$ as an index two subgroup. Thus we can write

$$\phi_v|_{W_{\mathbb{C}}} \simeq \chi_{v,1} \oplus \cdots \oplus \chi_{v,m}$$

for characters $\chi_{v,i} : \mathbb{C}^\times \to \mathbb{C}^\times$.

Definition 1.7 ([Clo90, Def 1.8]). We say Π_∞ is **algebraic** if there exist $a_{v,i}, b_{v,i} \in \mathbb{Z}$ for all $v|\infty$ and $1 \le i \le m$ such that for all $z \in \mathbb{C}$,

$$\chi_{v,i}(z) = z^{a_{v,i}+\frac{m-1}{2}} \bar{z}^{b_{v,i}+\frac{m-1}{2}}.$$

Remark 1.8. Buzzard and Gee [BG14] defines C-algebraic and L-algebraic representations. The former coincides with algebraic representations above. To define L-algebraic ones, one simply removes $\frac{m-1}{2}$ from the exponent in the above definition.

For simplicity, we restrict ourselves to a CM field F. (See Clozel's article for the general case.) Assume that Π_∞ is algebraic, and reorder indices so that $a_{v,1} \ge \cdots \ge a_{v,m}$ for each v.

Definition 1.9 ([Clo90, Def 3.12], [Shi11, §7.1]). An algebraic representation Π_∞ is said to be **regular algebraic** if $a_{v,1} > \cdots > a_{v,m}$ (or equivalently if Π_∞ has the same infinitesimal character as an algebraic representation Ξ of $(R_{F/\mathbb{Q}}GL_m) \times_{\mathbb{Q}} \mathbb{R}$). A regular algebraic Π_∞ is **slightly regular** if there exist $v|\infty$ and an odd i such that

$$a_{v,i} > a_{v,i+1} + 1.$$

(This may be an unfortunate name. The name indicates that the highest weight of Ξ is slightly regular.)

1.5 Methods to construct Galois representations

A natural abundant source of Galois representations is the l-adic cohomology of varieties over number fields. To make a connection with automorphic representations, it is the best to work with Shimura varieties, which come with canonical models over number fields as well as Hecke correspondences which are also defined over the same number fields. In any successful method (except the $m = 1$ case covered by class field theory), the basic starting point is to look for the desired Galois representation attached to a given automorphic representation in the l-adic cohomology of a well-chosen Shimura variety, possibly under some technical conditions. Then one tries to relax those conditions by various methods.

The fundamental idea goes back to 1970s and is due to Langlands, who laid out the program to study the zeta function and cohomology of Shimura varieties by describing the mod p points of Shimura varieties and then comparing the Grothendieck-Lefschetz fixed point formula with the Arthur-Selberg trace formula. Langlands understood the role of endoscopy in this context early on; when the group G is sandwiched between SL_2 and GL_2, he noticed in particular that the zeta function of a Shimura variety was factorized into the L-functions pertaining to not only G but a smaller "endoscopic" group, cf. [Lan79]. For the early history, the reader is referred to Langlands's papers in 1970s; see the last paragraph of [Lan77] for the author's guidance to some of them. (The interested reader may also read his commentary to [Lan77] in 1995 on his IAS website.) Kottwitz substantially contributed to flesh out Langlands's ideas and carried out the program for many PEL-type Shimura varieties in [Kot90, Kot92b, Kot92a], to name a few. Essentially any construction of automorphic Galois representations ultimately relies on this method initiated by Langlands and furthered by Kottwitz, which often goes by the "Langlands–Kottwitz method".

To discuss variants of this method and fine technical points, we list some key considerations in the construction of Galois representations.

(1) good reduction of Shimura varieties

(1′) bad reduction of Shimura varieties (involving nearby cycles, relation with a moduli space of p-divisible groups, etc)

(2) counting points on the special fibers of Shimura varieties at good primes

(2′) counting points on Igusa varieties at primes of bad reduction

(3) Arthur-Selberg trace formula (including twisted trace formula) when endoscopy is trivial

(3′) stable trace formula and nontrivial endoscopy

(4) compactification, boundary contributions in the counting point formula and Arthur-Selberg trace formula

(5) congruences; p-adic approximation

Let us remark on various approaches to Theorem 1.3 when $m \geq 3$.[1] As for (1) and (1)′, every work makes use of certain unitary PEL-type Shimura varieties. Naturally it is essential to consider the case of unitary (similitude) groups in the trace formula method alluded to above.

(i) Collaboration of many people ([LR92]): $m = 3$. (1), (2), (3), (4).

(ii) Clozel ([Clo91]), Kottwitz ([Kot92a], [Kot92b]): (1), (2), (3).

(iii) Harris-Taylor ([HT01]), Taylor-Yoshida ([TY07]): (1)′, (2)′, (3).

(iv) Morel ([Mor10]): (1), (2), (3)′, (4).

(v) Clozel-Harris-Labesse ([CHL11]): (1), (2), (3)′.

(vi) Shin ([Shi11]): (1)′, (2)′, (3)′.

(vii) Chenevier-Harris ([CH13]): (5).

Note that only (iii) and (vi) prove (1.1) at ramified places. There (1)′ and (2)′ are crucial.

[1] We apologize for suppressing the rich history when $m = 2$ for Galois representations associated to (Hilbert) modular forms. Also when $m = 4$, there has been much earlier work coming from Shimura varieties for GSp_4.

In (ii) and (iii), Theorem 1.3 was proved under the additional assumption that Π_w is square-integrable at some place w. Later (iv), (v) and (vi) established the theorem without the square-integrability assumption but under a mild condition on Π_∞ when m is even (see Hypothesis 4.9). The last condition was removed by (vii) at the cost of proving (1.1) only up to semisimplification (thus losing the precise information on the monodromy operator). The full version of (1.1) was established by Caraiani [Car12, Car14].

It is worth noting that only special cases of the fundamental lemma were available when (i), (ii) and (iii) were carried out. The improvements (iv)-(vii) have become possible largely thanks to the proof of the fundamental lemma by Laumon, Ngô, Waldspurger and others.

Finally we mention that there is another approach to the bad reduction of Shimura varieties and the counting point formula, shedding some new light on the bad reduction of Shimura varieties and its interaction with representation theory. It may not be unreasonable to say that the flavor of this approach is somewhere between (1)+(2) and (1)′+(2)′, if compared with the cited work above. See nice surveys by Rapoport [Rap05] and Haines [Hai05] as well as some research papers [HR12, Sch13a, Sch13b, SS13]. In particular the last two papers simplified the proof (by Harris-Taylor and Henniart) of the local Langlands correspondence for general linear groups and the proof of Theorem 1.3. See [Sch] for a survey on some of these ideas.

From the next section I will mainly follow the approach of (vi), which improves on the method of (iii) especially in the aspect of the counting point formula, its comparison with the Arthur-Selberg trace formula via the stable trace formula. The reader is strongly encouraged to refer to other references as well.

1.6 The Ramanujan-Petersson conjecture

Here is a vast generalization of the conjecture (proved by Deligne) that Ramanujan's τ-function satisfies the bound $|\tau(p)| \leq 2p^{11/2}$ for all primes p. (This conjecture is a special case of the conjecture below when $m = 2$, $F = \mathbb{Q}$ and Π corresponds to the cuspform Δ of weight 12 and level 1.)

Conjecture 1.10 (Ramanujan-Petersson). *Let $m \geq 1$, F be a number field and Π be a cuspidal automorphic representation of $GL_m(\mathbb{A}_F)$. Then Π_v is tempered at every finite place v of F.*

Remark 1.11. In fact, it would be reasonable to expect something stronger when Π is algebraic. For instance, one can conjecture that if Π is algebraic then for every $\sigma \in \mathrm{Aut}(\mathbb{C})$ and every finite place v, the σ-twist Π_v^σ is tempered. In other words, Π_v should be "absolutely tempered".

By combining Theorem 1.3 and some facts in the representation theory of GL_m over p-adic fields, we obtain

Theorem 1.12. *Conjecture 1.10 is true for F and Π as in Theorem 1.3.*

To be precise, the theorem is a corollary of Theorem 1.3 (with a semisimplified version of (1.1)) when m is even and Π_∞ is slightly regular (or when Π is square-integrable at a finite prime). Several mathematicians have made contributions to this. Namely, whenever the authors in (i)–(vi) of §1.5 proved the local-global compatibility at v for constructed Galois representations, they obtained the corresponding case of Theorem 1.12. Finally in the missing case where m is even and Π_∞ is not slightly regular, Caraiani [Car12] proved Theorem 1.12 in the course of strengthening (1.1), improving on Clozel's result [Clo13] in the unramified case by a different method.

2 Base change for unitary groups

Throughout the article we assume $n \geq 3$. In this case there are no Shimura varieties attached to GL_n (or its inner forms). The next best thing is to study Shimura varieties associated to unitary (similitude) groups since a unitary group becomes isomorphic to GL_n after quadratic extension of the base field. To analyze the cohomology of those Shimura varieties, it is essential to understand automorphic representations of unitary groups, especially in connection with those of GL_n because the representations of GL_n are better understood and also because we would like to prove a theorem about automorphic representations of GL_n rather than a unitary group. This can be achieved by (automorphic) base change for unitary groups, which will be discussed in this section. For the case of unitary similitude groups, see §3.4.

2.1 Unramified local Langlands correspondence

This subsection is a general background needed for §2 and §3. Details may be found in [Bor79] and [Min11] for instance. Let G be an unramified connected reductive group over a p-adic field F. Recall that G is said to be unramified over F if G has a smooth fiberwise reductive integral model over $\mathcal{O}_F$, or equivalently if G is quasi-split over F and splits over an unramified extension of F. We will fix such a model, thus also a compact subgroup $G(\mathcal{O}_F) \subset G(F)$, often called "hyperspecial". Denote by $\mathscr{H}^{\mathrm{ur}}(G(F)) := C_c^\infty(G(\mathcal{O}_F)\backslash G(F)/G(\mathcal{O}_F))$ the unramified Hecke algebra equipped with the convolution product. Then the following sets are in canonical bijection with each other (e.g. [Min11, 2.6]). This is a consequence of the Satake isomorphism.

(1) $\mathbb{C}$-algebra morphisms $\chi : \mathscr{H}^{\mathrm{ur}}(G(F)) \to \mathbb{C}$.

(2) isomorphism classes of unramified L-parameters $\varphi : W_F \to {}^LG$.

(3) isomorphism classes of unramified (irreducible admissible) representations of $G(F)$.

By definition φ is unramified if $\varphi(I_F) = (1)$, and $\pi \in \mathrm{Irr}(G(F))$ is unramified if π has a nonzero fixed vector under $G(\mathcal{O}_F)$.

One is invited to define canonical maps between them and prove that they are bijections in case G is a torus. The proof in the general case can be reduced, with some work, to the case of tori by considering a maximal torus of G.

It is an extremely important fact that the bijections among (1), (2) and (3) are *functorial* in G. This means the following: If H, G are connected reductive groups over F and $\widetilde{\eta} : {}^LH \to {}^LG$ is an L-morphism, then there is a canonical map (e.g. [Min11, 2.7])

$$\widetilde{\eta}^* : \mathscr{H}^{\mathrm{ur}}(G(F)) \to \mathscr{H}^{\mathrm{ur}}(H(F))$$

characterized by the property that whenever $\chi_H : \mathscr{H}^{\mathrm{ur}}(H(F)) \to \mathbb{C}$ and $\varphi_H : W_F \to {}^L H$ correspond, we have $\chi_H \circ \widetilde{\eta}^*$ and $\widetilde{\eta} \circ \varphi_H$ correspond to each other. The map

$$\widetilde{\eta}_* : \mathrm{Irr}^{\mathrm{ur}}(H(F)) \to \mathrm{Irr}^{\mathrm{ur}}(G(F)) \tag{2.1}$$

corresponding to $\varphi_H \mapsto \widetilde{\eta} \circ \varphi_H$ is basically the "transfer of unramified representations" with respect to $\widetilde{\eta}$. A quick characterization of the map $\widetilde{\eta}_*$ is possible using the following identity: for every $f \in \mathscr{H}^{\mathrm{ur}}(G(F))$,

$$\mathrm{tr} \widetilde{\eta}_* \pi_H(f) = \mathrm{tr} \pi_H(\widetilde{\eta}^* f). \tag{2.2}$$

2.2 Setup

Now we restrict our attention to unitary groups and general linear groups. The following notation will be used.

- F is a CM field with complex conjugation c. Set $F^+ := F^{c=1}$.
- $\vec{n} = (n_1, \dots, n_r)$, $n_i, r \in \mathbb{Z}_{>0}$, $\sum_{i=1}^r n_i = n$.
- $i_{\vec{n}} : GL_{\vec{n}} \hookrightarrow GL_N$ ($N = \sum_i n_i$) is the embedding

$$(A_1, \dots, A_r) \mapsto \begin{pmatrix} A_1 & 0 & \cdots & 0 \\ 0 & A_2 & \cdots & 0 \\ \vdots & \cdots & \cdots & \vdots \\ 0 & \cdots & 0 & A_r \end{pmatrix}.$$

- $GL_{\vec{n}} = \prod_{n=1}^r GL_{n_i}$ (over any base).
- U_n is a quasi-split unitary group in n variables over F^+ (for an n-dimensional F-vector space with a Hermitian pairing). Define

$$U_{\vec{n}} := \prod_{1 \le i \le r} U_{n_i}.$$

We will work under the following hypothesis. (This is not a vacuous condition. For instance, $F = \mathbb{Q}(\zeta_5)$ does not satisfy it.)

Hypothesis 2.1. *$F = EF^+$ for an imaginary quadratic field $E \subset F$.*

Define

$$G^1_{\vec{n}} := R_{F^+/\mathbb{Q}} U_{\vec{n}}, \qquad \mathbb{G}^1_{\vec{n}} := R_{E/\mathbb{Q}}(G^1_{\vec{n}} \times_{\mathbb{Q}} E)$$

where $R_{(\cdot)}$ denotes the Weil restriction of scalars. When $\vec{n} = (n)$, we prefer to write G^1_n and $\mathbb{G}^1_n$ for $G^1_{\vec{n}}$ and $\mathbb{G}^1_{\vec{n}}$, respectively. (These ugly notations anticipate their similitude cousins, which will show up in §3.4. The superscript means that the similitude factor equals 1.) The connected reductive groups $G^1_{\vec{n}}$ and $\mathbb{G}^1_{\vec{n}}$ are defined over $\mathbb{Q}$ and $G^1_{\vec{n}}(\mathbb{Q}) = U_{\vec{n}}(F^+)$, $\mathbb{G}^1_{\vec{n}}(\mathbb{Q}) = G^1_{\vec{n}}(E) = U_{\vec{n}}(F) \simeq GL_{\vec{n}}(F)$. Also note that

$$G^1_{\vec{n}}(\mathbb{A}) = U_{\vec{n}}(\mathbb{A}_{F^+}), \qquad \mathbb{G}^1_{\vec{n}}(\mathbb{A}) = G^1_{\vec{n}}(\mathbb{A}_E) = U_{\vec{n}}(\mathbb{A}_F) \simeq GL_{\vec{n}}(\mathbb{A}_F). \tag{2.3}$$

Let θ denote the action on $\mathbb{G}^1_{\vec{n}}$ induced by $1 \times c$ on $G^1_{\vec{n}} \times_{\mathbb{Q}} E$. Under the above isomorphism, θ is transported to the action $(g_1, \ldots, g_r) \mapsto ({}^t g_1^{-c}, \ldots, {}^t g_r^{-c})$ on $GL_{\vec{n}}(\mathbb{A}_F)$ up to conjugation by $GL_{\vec{n}}(\mathbb{A}_F)$.

Consider the L-groups ${}^L G^1_{\vec{n}} := \hat{G}_{\vec{n}} \rtimes W_{\mathbb{Q}}$ and ${}^L \mathbb{G}^1_{\vec{n}} := \hat{\mathbb{G}}^1_{\vec{n}} \rtimes W_{\mathbb{Q}}$. The dual groups may be identified as

$$\hat{G}^1_{\vec{n}} = GL_{\vec{n}}(\mathbb{C})^{\mathrm{Hom}(F^+, \mathbb{C})}, \qquad \hat{\mathbb{G}}^1_{\vec{n}} = GL_{\vec{n}}(\mathbb{C})^{\mathrm{Hom}(F, \mathbb{C})}, \tag{2.4}$$

equipped with $W_{\mathbb{Q}}$-actions. There is a natural L-morphism

$$\mathrm{BC}_{\vec{n}} : {}^L G^1_{\vec{n}} \to {}^L \mathbb{G}^1_{\vec{n}}$$

which extends the diagonal embedding on the dual group. On the level of dual groups, $\mathrm{BC}_{\vec{n}}$ maps $(g_\sigma) \mapsto (h_\sigma)$ so that $h_\sigma = h_{\sigma^c} = g_\sigma$ for every $\sigma \in \mathrm{Hom}(F^+, \mathbb{C})$.

2.3 Local base change

Fix a finite place v of $\mathbb{Q}$. Consider

$$\pi_v \in \mathrm{Irr}(G^1_{\vec{n}}(\mathbb{Q}_v)) = \mathrm{Irr}(\prod_{w|v} U_{\vec{n}}(F_w)).$$

(i) For any v, when π_v is unramified, let $\phi(\pi_v) : W_{\mathbb{Q}_v} \to {}^L G^1_{\vec{n}}$ be the corresponding unramified parameter. Define $\mathrm{BC}_{\vec{n},*}(\pi_v) \in \mathrm{Irr}^{\mathrm{ur}}(\mathbb{G}^1_{\vec{n}}(\mathbb{Q}_v))$ corresponding to the unramified parameter $\mathrm{BC}_{\vec{n}} \circ \phi(\pi_v)$. Thus obtain a map (whose image sits inside the set of θ-stable representations).

$$\mathrm{BC}_{\vec{n},*} : \mathrm{Irr}^{\mathrm{ur}}(G^1_{\vec{n}}(\mathbb{Q}_v)) \to \mathrm{Irr}^{\mathrm{ur}}(\mathbb{G}^1_{\vec{n}}(\mathbb{Q}_v)). \tag{2.5}$$

(ii) If v *splits in* E, say u and u^c are primes of E above v. There is an isomorphism $\mathbb{G}^1_{\vec{n}}(\mathbb{Q}_v) = G^1_{\vec{n}}(E_v) \simeq G^1_{\vec{n}}(E_u) \times G^1_{\vec{n}}(E_{u^c})$ where θ acts as $(g_1, g_2) \mapsto (g_2^c, g_1^c)$. (Note that $c : E \xrightarrow{\sim} E$ induces $E_u \xrightarrow{\sim} E_{u^c}$.) Define

$$\mathrm{BC}_{\vec{n},*} : \mathrm{Irr}(G^1_{\vec{n}}(\mathbb{Q}_v)) \to \mathrm{Irr}(\mathbb{G}^1_{\vec{n}}(\mathbb{Q}_v)) \tag{2.6}$$

by $\pi \mapsto \pi \otimes \pi$. The image is clearly θ-stable. It is an exercise to check that the map (2.6) restricts to the map (2.5) defined above if π_v is unramified.

(iii) When $v = \infty$, the base change of any L-packet of $G^1_{\vec{n}}(\mathbb{R})$ can be defined as a representation of $\mathbb{G}^1_{\vec{n}}(R) = G^1_{\vec{n}}(\mathbb{C})$, but the details will be omitted. The fact that this is possible should not be surprising as the local Langlands correspondence is known for any real or complex group due to Langlands ([Lan89]). We will be interested in only those L-packets consisting of discrete series representations of $G^1_{\vec{n}}(\mathbb{R})$ and their base change.

Although the "explicit" base change in the above list does not exhaust all cases, it suffices for constructing Galois representations. (To our knowledge this observation was first made by Harris.) This is desirable since the base change for unitary groups is not established in full generality yet.

We will often write BC or $\mathrm{BC}_{\vec{n}}$ for the map $\mathrm{BC}_{\vec{n},*}$ in (2.5) or (2.6) if there is no danger of confusion.

In general the base change along a finite cyclic extension is characterized by a trace identity with respect to the transfer of test functions. The existence of transfer for any endoscopy (twisted or untwisted) is a consequence of the recent proof of the fundamental lemma, thanks to Laumon, Ngô, Waldspurger and others. In particular the transfer of test functions is known for any cyclic base change, which is an instance of twisted endoscopy. (However see Remark 2.4 below.)

Proposition 2.2. (1) *In case (i) and (ii) above, let* $f_v \in C_c^\infty(\mathbb{G}^1_{\vec{n}}(\mathbb{Q}_v))$. *Then there exists* $\phi_v \in C_c^\infty(G^1_{\vec{n}}(\mathbb{Q}_v))$ *with matching orbital integrals. In case (i), if* $f_v \in \mathscr{H}^{\mathrm{ur}}(\mathbb{G}^1_{\vec{n}}(\mathbb{Q}_v))$ *then one can take* $\phi_v = \widetilde{BC_{\vec{n}}}^*(f_v)$ *in the notation of §2.1.*

(2) *The function* ϕ_v *in part (1) satisfies the following in case (ii):*

$$\widetilde{\mathrm{tr}} BC(\pi_v)(f_v) = \mathrm{tr}\pi_v(\phi_v), \qquad \forall \pi_v \in \mathrm{Irr}(G^1_{\vec{n}}(\mathbb{Q}_v)),$$

where $\widetilde{\mathrm{tr}}$ *denotes the (suitably normalized) twisted trace. In case (i), the same identity holds for all unramified* $\pi_v \in \mathrm{Irr}(G^1_{\vec{n}}(\mathbb{Q}_v))$.

Sketch of proof. Part (1) is a consequence of the fundamental lemma as explained above. As for Part (2), (2.2) is the desired identity in case (i). In case (ii) it is shown by a direct computation (which is not hard). □

Remark 2.3. We apologize for the imprecision in the statement of (1). We chose not to write out the precise identity for matching of orbital integrals. Moreover the experienced reader must have noticed an abuse of language. Orbital integrals there really mean stable (twisted) orbital integrals.

Remark 2.4. The second assertion of (1) is usually refereed to as the base change fundamental lemma and due to Kottwitz, Clozel and Labesse and was proved almost 20 years ago.

Remark 2.5. You may have noticed that no properties of unitary groups are used in §2.3. Indeed, local base change can be defined in the same manner as above when

- the quadratic extension $E/\mathbb{Q}$ is replaced with any finite cyclic extension E'/E'',
- $G^1_{\vec{n}}$ is replaced with any connected reductive group H over E'',
- $\mathbb{G}^1_{\vec{n}}$ is replaced with $\mathbb{H} := R_{E'/E''}(H \times_{E''} E')$ and
- v runs over places of E''.

What is nice about the quadratic base change from $G^1_{\vec{n}}$ to $\mathbb{G}^1_{\vec{n}}$ is that the representation theory of $\mathbb{G}^1_{\vec{n}}$ (which is a general linear group) is much more complete than that of other groups. As such, this base change enables us to study representations of $G^1_{\vec{n}}$ through those of $\mathbb{G}^1_{\vec{n}}$. (For instance, one may try to define a local or global L-packet for $G^1_{\vec{n}}$ as the set of those representations of $G^1_{\vec{n}}$ whose base change images are isomorphic.)

2.4 Weak base change (global)

Definition 2.6. Let $\pi \in \mathrm{Irr}(G_n^1(\mathbb{A}))$ and Π be an automorphic representation of $\mathbb{G}_n^1(\mathbb{A})$. We say that Π is a *weak base change* of π and write $\Pi = \mathrm{WBC}(\pi)$ if $\mathrm{BC}(\pi_v) \simeq \Pi_v$ for all but finitely many v. (This makes sense as π_v is unramified and so $\mathrm{BC}(\pi_v)$ is defined for almost all v.)

By the strong multiplicity one theorem for GL_n, the weak base change $\mathrm{WBC}(\pi)$ is unique (up to isomorphism) if it exists (cf. (2.3)). There is a fairly general existence result by Clozel and Labesse.

Proposition 2.7 ([Lab11, Cor 5.3]). *If π_∞ is a discrete series whose infinitesimal character is sufficiently regular, then* $\mathrm{WBC}(\pi)$ *exists. In other words, the conclusion is that there exists an automorphic representation Π of $\mathbb{G}_n^1(\mathbb{A})$ such that* $\mathrm{BC}(\pi_v) \simeq \Pi_v$ *for almost all v.*

Remark 2.8. Actually, in the construction of Galois representations, it is not necessary to use this proposition until the last stage where the p-adic approximation argument ([CH13]) is used. The reverse of (weak) base change, or "descent", often turns out to be more essential. Labesse ([Lab11]) also proves many instances of descent.

We will be mostly interested in the case when

$$\mathrm{WBC}(\pi) = \Pi \simeq \mathrm{Ind}(\Pi_1 \otimes \cdots \otimes \Pi_r) \tag{2.7}$$

such that each Π_i is *cuspidal* and $\Pi_i^\theta \simeq \Pi_i$. (In general, it can happen that some Π_i is discrete but not cuspidal and also that for $i \neq j$, $\Pi_i^\theta \simeq \Pi_j$ and $\Pi_j^\theta \simeq \Pi_i$.) It is built into assumption that the parabolic induction of (2.7) is irreducible. The Π as above can be thought of as coming from the discrete tempered spectrum for $G^1(\mathbb{A}) = U(\mathbb{A}_{F^+})$. For our application, the case $r \leq 2$ is the most important.

Definition 2.9. Suppose that $\mathrm{WBC}(\pi)$ exists and has the form as in (2.7). We say π is *stable* if $r = 1$ and *endoscopic* if $r > 1$.

3 Endoscopy for unitary groups

3.1 Endoscopic groups

Define $\mathcal{E}(G_n^1)$ (resp. $\mathcal{E}^{\mathrm{ell}}(G_n^1)$) to be the set of $G_{\vec{n}}^1$ such that $\sum_{i=1}^r n_i = n$ (resp. $\sum_{i=1}^r n_i = n$ and $r \leq 2$). In other words,

$$\mathcal{E}^{\mathrm{ell}}(G_n^1) = \{G_n^1\} \cup \{G_{n_1,n_2}^1 \mid n_1 + n_2 = n,\ n_1, n_2 > 0\}.$$

The elements of $\mathcal{E}^{\mathrm{ell}}(G_n^1)$ are called the elliptic endoscopic groups for G_n^1 and play a fundamental role in the trace formula for G_n^1. For each $G_{\vec{n}}^1 \in \mathcal{E}^{\mathrm{ell}}(G_n^1)$, fix an L-embedding

$$\tilde{\eta}_{\vec{n}}^1 : {}^L G_{\vec{n}}^1 \to {}^L G_n^1 \tag{3.1}$$

extending $\hat{G}_{\vec{n}}^1 \to \hat{G}_n^1$ given by the block diagonal embedding $i_{\vec{n}}$. The choice of $\tilde{\eta}_{\vec{n}}^1$ is not unique and there is no canonical choice in general. An explicit form of $\tilde{\eta}_{\vec{n}}^1$ will be omitted but the interested reader may find it as an exercise or look up [Rog92, §1.2], cf. [Shi11, §3.2].

3.2 Endoscopic transfer

Let $G^1_{\vec{n}} \in \mathscr{E}(G^1_n)$. (In fact, $G^1_{\vec{n}} \in \mathscr{E}^{\mathrm{ell}}(G^1_n)$ suffices for our later discussion.) There should be a transfer of representations from $G^1_{\vec{n}}$ to G^1_n corresponding to $\widetilde{\eta}^1_{\vec{n}}$. Such a transfer is called an *endoscopic transfer* and an instance of the Langlands functoriality. Although the Langlands functoriality is still largely open, which should involve the description of L-packets among other things, there are some favorable cases (cf. §2.3), as we will see below, where the local transfer of representations can be explicitly worked out.

(i) For any finite v,

$$\widetilde{\eta}^1_{\vec{n},*} : \mathrm{Irr}^{\mathrm{ur}}(G^1_{\vec{n}}(\mathbb{Q}_v)) \to \mathrm{Irr}^{\mathrm{ur}}(G^1_n(\mathbb{Q}_v))$$

is defined as in (2.1).

(ii) If v splits in E,

$$\widetilde{\eta}^1_{\vec{n},*} : \mathrm{Irr}(G^1_{\vec{n}}(\mathbb{Q}_v)) \to \mathrm{Irr}(G^1_n(\mathbb{Q}_v))$$

is defined as a character twist of the parabolic induction, noting that $G^1_{\vec{n}}(\mathbb{Q}_v)$ and $G^1_n(\mathbb{Q}_v)$ are isomorphic to $\prod_w GL_{\vec{n}}(F_w)$ and $\prod_w GL_n(F_w)$, respectively, where w runs over the half of the places dividing v. The character twist depends on $\widetilde{\eta}^1_{\vec{n},*}$ at v, which is not the identity map on $W_{\mathbb{Q}_v}$ (inside ${}^L G^1_{\vec{n}}$) in general.

(iii) If $v = \infty$, the transfer of representations is defined by the Langlands correspondence for $G^1_{\vec{n}}(\mathbb{R})$ and $G^1_n(\mathbb{R})$. (We are especially interested in the discrete series representations.)

We have the analogue of Proposition 2.2 characterizing the endoscopic transfer described above with respect to the transfer of test functions. As explained in the paragraph above Proposition 2.2, the existence of transfer is known.

Proposition 3.1. (1) *In case (i) and (ii) above, let* $f_v \in C^\infty_c(\mathbb{G}^1_{\vec{n}}(\mathbb{Q}_v))$. *Then there exists* $\phi_v \in C^\infty_c(G^1_{\vec{n}}(\mathbb{Q}_v))$ *with matching orbital integrals. Moreover in case (ii), if* $f_v \in \mathscr{H}^{\mathrm{ur}}(\mathbb{G}^1_{\vec{n}}(\mathbb{Q}_v))$ *then one can take* $\phi_v = (\widetilde{\eta}^1_{\vec{n}})^*(f_v)$ *in the notation of §2.1.*

(2) *The function* ϕ_v *in part (1) satisfies the following in case (ii).*

$$\widetilde{\mathrm{tr}}\left(\widetilde{\eta}^1_{\vec{n},*}(\pi_{H,v})\right)(f_v) = \mathrm{tr}\,\pi_{H,v}(\phi_v), \qquad \forall \pi_{H,v} \in \mathrm{Irr}(G^1_{\vec{n}}(\mathbb{Q}_v)).$$

In case (i), the same identity holds for all unramified $\pi_{H,v} \in \mathrm{Irr}(G^1_{\vec{n}}(\mathbb{Q}_v))$.

Sketch of proof. The same remarks in the proof of Proposition 2.2 apply. Remark 2.3 is valid here as well. □

Remark 3.2. The article [CHL11] would be an excellent place to learn details about the material of this subsection.

3.3 Base change and endoscopic transfer

Let $G^1_{\vec{n}} \in \mathscr{E}(G^1_n)$ as before. We would like to understand the interplay between the local base change and the local endoscopic transfer. For this let us look at the following diagram of L-morphisms. The three maps $\widetilde{\eta}_{\vec{n}}$, $\mathrm{BC}_{\vec{n}}$ and BC_n are already defined. It is easy to choose $\widetilde{i}_{\vec{n}}$ explicitly so that the diagram commutes.

$$\begin{array}{ccc} {}^L G^1_{\vec{n}} & \xrightarrow{\widetilde{\eta}^1_{\vec{n}}} & {}^L G^1_n \\ \downarrow{\scriptstyle \mathrm{BC}_{\vec{n}}} & & \downarrow{\scriptstyle \mathrm{BC}_n} \\ {}^L \mathbb{G}^1_{\vec{n}} & \xrightarrow[\widetilde{i}_{\vec{n}}]{} & {}^L \mathbb{G}^1_n \end{array} \tag{3.2}$$

By §2.1, we immediately obtain a commutative diagram at each finite v.

$$\begin{array}{ccc} \mathrm{Irr}^{\mathrm{ur}}(G^1_{\vec{n}}(\mathbb{Q}_v)) & \xrightarrow{\widetilde{\eta}^1_{\vec{n},*}} & \mathrm{Irr}^{\mathrm{ur}}(G^1_n(\mathbb{Q}_v)) \\ \downarrow{\scriptstyle \mathrm{BC}_{\vec{n},*}} & & \downarrow{\scriptstyle \mathrm{BC}_{n,*}} \\ \mathrm{Irr}^{\mathrm{ur}}(\mathbb{G}^1_{\vec{n}}(\mathbb{Q}_v)) & \xrightarrow[\widetilde{i}_{\vec{n},*}]{} & \mathrm{Irr}^{\mathrm{ur}}(\mathbb{G}^1_n(\mathbb{Q}_v)) \end{array} \tag{3.3}$$

If v splits in E or $v = \infty$, there is a similar diagram with $\mathrm{Irr}(\cdot)$ in place of $\mathrm{Irr}^{\mathrm{ur}}(\cdot)$.

The functoriality with respect to $\widetilde{i}_{\vec{n}}$ is simply a parabolic induction up to a twist by a character, noting that $\mathbb{G}^1_{\vec{n}}$ and $\mathbb{G}^1_n$ are essentially $GL_{\vec{n}}$ and GL_n (2.3). (This statement can be made precise and proved.) This leads to the following interesting observation. Suppose that $\pi \in \mathscr{A}(G^1_n(\mathbb{A}))$ is the endoscopic transfer with respect to $\widetilde{\eta}^1_{\vec{n}}$ of $\pi_0 \in \mathscr{A}(G^1_{\vec{n}}(\mathbb{A}))$ at almost all places, where $\vec{n} \neq (n)$. Assume the sufficient regularity of Proposition 2.7 for π and π_0. Then $\mathrm{WBC}(\pi) \in \mathscr{A}(\mathbb{G}^1_n(\mathbb{A}))$ and $\mathrm{WBC}(\pi_0) \in \mathscr{A}(\mathbb{G}^1_{\vec{n}}(\mathbb{A}))$ exist. The commutativity of (3.3) implies that

$$\mathrm{WBC}(\pi) \simeq \mathrm{Ind}^{G^1_n}_{G^1_{\vec{n}}}(\mathrm{WBC}(\pi_0) \otimes \chi) \tag{3.4}$$

for a certain character χ (determined by $\widetilde{\eta}^1_{\vec{n}}$). We have shown that if π is the image of endoscopic transfer from an endoscopic group $G^1_{\vec{n}}$ different from G^1_n, then $\mathrm{WBC}(\pi)$ is endoscopic, namely it is induced from a representation of a proper parabolic subgroup (if it is also the case that $\mathrm{WBC}(\pi)$ is induced from a cuspidal representation). This justifies the terminology of Definition 2.9. The converse is expected to be true and proved in some cases. (The converse says that any endoscopic π arises from some $G^1_{\vec{n}} \neq G^1_n$ via endoscopic transfer.)

3.4 Unitary similitude groups

The results of §2 and §3 carry over with very minor changes to unitary similitude groups $G_{\vec{n}}$ (including the case $\vec{n} = (n)$) sitting inside the following exact sequence where $G_{\vec{n}} \to \mathbb{G}_m$ is the multiplier map.

$$1 \to G^1_{\vec{n}} \to G_{\vec{n}} \to \mathbb{G}_m \to 1$$

Define $\mathbb{G}_{\vec{n}} := R_{E/\mathbb{Q}}(G_{\vec{n}} \times_{\mathbb{Q}} E)$. Observe that

$$\mathbb{G}_{\vec{n}}(\mathbb{A}) = G(\mathbb{A}_E) \simeq GL_1(\mathbb{A}_E) \times GL_{\vec{n}}(\mathbb{A}_F).$$

The extra factor $GL_1(\mathbb{A}_E)$, which did not exist in $\mathbb{G}^1_{\vec{n}}(\mathbb{A})$, is a nuisance but does not increase technical difficulty.

It is the groups $G_{\vec{n}}$ and $\mathbb{G}_{\vec{n}}$ that constantly show up later on. Why do we care about $G_{\vec{n}}$ and $\mathbb{G}_{\vec{n}}$ when it looks simpler to work with $G^1_{\vec{n}}$ and $\mathbb{G}^1_{\vec{n}}$? The main reason is that $G_{\vec{n}}$ naturally occurs in the context of PEL-type Shimura varieties, whose cohomology is an essential input in the construction of Galois representations. Although it is possible to understand the representations of $G_{\vec{n}}$ through those of $G^1_{\vec{n}}$, it seems more satisfactory to deal with $G_{\vec{n}}$ directly. Nevertheless, in the first reading of the subject, it may be harmless to ignore the similitude part and pretend that you are working with unitary groups.

4 Shimura varieties

We keep the previous notation. In particular, F is a CM field, F^+ is the maximal totally real subfield of F, and $F = EF^+$ for some imaginary quadratic field $E \subset F$.

4.1 Choice of unitary group

From here on, we will fix an *odd* integer $n \in \mathbb{Z}_{\geq 3}$. Let G be an inner form of the quasi-split group G_n over $\mathbb{Q}$. Such a G is also a unitary similitude group and fits into an exact sequence

$$1 \to G^1 \to G \to \mathbb{G}_m \to 1$$

so that G^1 is an inner form of G^1_n. When n is odd, there is no obstruction in finding G such that

- G^1 is quasi-split at all finite places and
- $G^1(\mathbb{R}) \simeq U(1, n-1) \times U(0, n)^{[F^+:\mathbb{Q}]-1}$.

(In general, there is a cohomological obstruction for finding a unitary group with prescribed local conditions at all places. The obstruction always vanishes if n is odd, which is not the case if n is even. See [Clo91, §2] for a detailed computation of cohomological obstructions.) The main reason for choosing G^1 to be quasi-split at all finite places is that we want to see in the cohomology of Shimura varieties as many Galois and automorphic representations as possible.[2] In case you wonder why we choose $G^1(\mathbb{R})$ as above, see Remark 4.4 and 5.2.

Remark 4.1. In work of Clozel, Kottwitz and Harris-Taylor, they had an assumption which implies that G^1 is *not* quasi-split at some finite place (but they had the same $G^1(\mathbb{R})$). That assumption was imposed mainly because some techniques in the trace formula were

[2] A general principle is that a quasi-split group has more representations than its non quasi-split inner forms, either locally or globally. For an example, think of the Jacquet-Langlands correspondence between GL_n and its inner form coming from a division algebra.

available only in limited cases at that time (but there was another good reason for Harris-Taylor when they proved a counting point formula). This is why Π was assumed to be square-integrable at a finite place in their proof of Theorem 1.3. The whole point of recent work by several people is to remove the last assumption on Π, so it would be reasonable to appreciate the quasi-split condition on G^1 at finite places.

4.2 *l*-adic cohomology

Let $\mathrm{Sh} = \{\mathrm{Sh}_U\}$ be (the projective system of) Shimura varieties associated to G. (To be precise we also have to choose an $\mathbb{R}$-morphism $R_{\mathbb{C}/\mathbb{R}}\mathbb{G}_m \to G$ but in our case there is a natural choice for this morphism, which we suppress.) Each Sh_U is a smooth quasi-projective variety over F of dimension $n-1$ (if U is sufficiently small). From now on, we make another

Hypothesis 4.2. $F^+ \neq \mathbb{Q}$

so that $G^1(\mathbb{R})$ has at least one compact factor $U(0,n)$. Then it can be shown that Sh_U is projective. Postponing the moduli problem for Sh_U to §5.1, we would like to explain the rough structure of the l-adic étale cohomology of Sh.

Let ξ be an irreducible algebraic representation of G over $\overline{\mathbb{Q}}_l$. The ξ gives rise to a compatible system of smooth $\overline{\mathbb{Q}}_l$-sheaves $\mathcal{L}_\xi$ on Sh_U. (We have seen this in Tilouine's lectures. For a precise construction, refer to [HT01, III.2].) The $\overline{\mathbb{Q}}_l$-vector space

$$H^k(\mathrm{Sh}, \mathcal{L}_\xi) := \varinjlim_{U} H^k(\mathrm{Sh}_U \times_F \overline{F}, \mathcal{L}_\xi)$$

is equipped with a smooth action of $G(\mathbb{A}^\infty) \times \mathrm{Gal}(\overline{F}/F)$. (Smoothness means that each vector has an open stabilizer in $G(\mathbb{A}^\infty)$.) There is a direct sum decomposition

$$H^k(\mathrm{Sh}, \mathcal{L}_\xi) = \bigoplus_{\pi^\infty} \pi^\infty \otimes R_l^k(\pi^\infty)$$

where π^∞ runs over irreducible admissible representations of $G(\mathbb{A}^\infty)$ and $R_l^k(\pi^\infty)$ is a finite dimensional representation of $\mathrm{Gal}(\overline{F}/F)$. (Comparison with Matsushima's formula tells us that the above formula is indeed a direct sum.)

Often it is convenient to consider virtual representations (integral combinations of representations with possibly negative coefficients)

$$H(\mathrm{Sh}, \mathcal{L}_\xi) = \sum_{k\geq 0}(-1)^k H^k(\mathrm{Sh}, \mathcal{L}_\xi), \qquad R_l(\pi^\infty) = \sum_{k\geq 0}(-1)^k R_l^k(\pi^\infty).$$

Write $\mathbb{G} := R_{E/\mathbb{Q}}(G \times_{\mathbb{Q}} E)$. Let $\Pi = \chi \otimes \Pi^1$ be an automorphic representation (not necessarily cuspidal) of $\mathbb{G}(\mathbb{A}) = G(\mathbb{A}_E) \simeq GL_1(\mathbb{A}_E) \times GL_n(\mathbb{A}_F)$. Define a finite sum

$$R_l(\Pi) := \sum_{\mathrm{WBC}(\pi^\infty)=\Pi^\infty} R_l(\pi^\infty).$$

In order to understand $H^k(\mathrm{Sh}, \mathcal{L}_\xi)$ (for instance if one is interested in the L-functions for Sh_U), it would be ideal to describe $R_l(\pi^\infty)$ for each π^∞, or even $R_l^k(\pi^\infty)$ for each $k \in \mathbb{Z}_{\geq 0}$. For the construction of Galois representations, it suffices to describe $R_l(\Pi)$.

Let $\Pi^1 = \boxplus_{i=1}^r \Pi_i^1$ such that each Π_i^1 is θ-stable (cf. (2.7)). We assume that $R_l(\Pi) \neq 0$. (Roughly speaking, this is the case if Π_∞ is the base change of the discrete series representation "determined by" ξ.) Then we expect the following.

(1) If Π^1 is cuspidal ($r = 1$) then $R_l(\Pi)$ should correspond to Π^1 (in the sense of Conjecture 1.1) up to a character twist.

(2) In general, there should exist some i such that $R_l(\Pi)$ corresponds to Π_i^1.

(3) Suppose Π_i^1 is cuspidal for all i. Among $H^k(\mathrm{Sh}, \mathcal{L}_\xi)$, only H^{n-1} contributes to $R_l(\Pi)$. (This is not expected to be true if some θ-stable representation Π_i^1 is discrete but not cuspidal.) So $R_l(\Pi)$ should come with the sign $(-1)^{n-1}$.

Remark 4.3. The same assertion should hold for $R_l(\pi^\infty)$ in place of $R_l(\Pi)$ if $\mathrm{WBC}(\pi^\infty) = \Pi^\infty$. In particular, we should allow a nonzero integral multiplicity for $R_l(\Pi)$ (which is harmless) in general.

Remark 4.4. The above expectation is valid only under the condition on $G^1(\mathbb{R})$ as in §4.1. Let us briefly discuss what happens to (1) and (2) if that condition is given up. If $U(1, n-1)$ is replaced with $U(a, b)$ with $a + b = n$ then $R_l(\Pi)$ should be roughly the a-th exterior power of the Galois representation associated to Π^1 if Π^1 is cuspidal. (Depending on the convention, you get the b-th exterior power, which is dual to the a-th exterior power up to twist.) In the $U(a, b)$ case, if Π^1 is not cuspidal, the description of $R_l(\Pi)$ is more complicated. In fact there is a concrete recipe for predicting Galois representations in the cohomology of PEL-type Shimura varieties which are of unitary or symplectic type in terms of Arthur's parameters ([Kot90, §10]). The above expectation for $R_l(\Pi)$ should be viewed as a special case of the general principle.

4.3 Finding a candidate Galois representation

Let us go back to the setting of Theorem 1.3. Changing notation slightly from the theorem, let Π^1 denote a cuspidal automorphic representation of $GL_m(\mathbb{A}_F)$ which is θ-stable. A big question is where to look for $\rho_l(\Pi^1)$ corresponding to Π^1. One can learn the following recipe from some experience.

<u>m: odd</u>

Take $n := m$ and choose $\psi : \mathbb{A}_E^\times/E^\times \to \mathbb{C}^\times$ so that $\Pi := \psi \otimes \Pi^1$ is a θ-stable representation of $\mathbb{G}_n(\mathbb{A})$. Then $\rho_l(\Pi^1) := R_l(\Pi)$ should work for the main theorem (up to a nonzero multiplicity and a character twist).

<u>m: even</u>

Take $n := m + 1$ and set $\Pi_1 := \Pi^1$. Choose $\psi : \mathbb{A}_E^\times/E^\times \to \mathbb{C}^\times$ and $\Pi_2 : \mathbb{A}_F^\times/F^\times \to \mathbb{C}^\times$ so that Π defined as below is a θ-stable representation of $\mathbb{G}_n(\mathbb{A})$. (In particular, $\Pi_2^\vee = \Pi_2^{-1}$ should hold.)

$$\Pi_M := \psi \otimes \Pi_1 \otimes \Pi_2, \qquad \Pi := \mathrm{Ind}_{\mathbb{G}_{m,1}}^{\mathbb{G}_n}(\Pi_M) = \psi \otimes (\Pi_1 \boxplus \Pi_2). \tag{4.1}$$

Then $\rho_l(\Pi_1) := R_l(\Pi)$ should work (again up to a nonzero multiplicity and a character twist), as long as $R_l(\Pi)$ corresponds to not Π_2 but Π_1. (See the paragraph preceding Remark 4.3.) If Π_1 is slightly regular, then Π_2 (and ψ) can be chosen such that $R_l(\Pi)$ corresponds to Π_1. (The proof involves an explicit sign computation in endoscopy theory, especially for real L-parameters. As this is not very intuitive at first sight, you may regard this as a black box. In case you are curious, the relevant result is Lemma 7.3 of [Shi11].)

The problem is more tractable if we control the set of primes where ramification occurs:

Hypothesis 4.5. *If v is a finite place of $\mathbb{Q}$ such that either v is ramified in F or Π_v is ramified, then v splits in E.*

If v splits in E then $G(\mathbb{Q}_v)$ is isomorphic to a product of general linear groups (cf. (5.1)) in which case almost everything is better understood than other groups, so the ramification causes less difficulty. (If v is not split in E then $G(\mathbb{Q}_v)$ is a v-adic unitary similitude group.)

4.4 What should we do?

To prove Theorem 1.3, we must show that $\rho_l(\Pi^1)$ defined as in §4.3 does correspond to Π^1 at each finite place (excluding those dividing l) via the local Langlands correspondence. This amounts to analyzing $R_l(\Pi)$ in the two cases depending on the parity of m. Indeed, the main theorem will essentially follow from

Theorem 4.6. *Keep the notation of §4.3. At each finite place w not dividing l,*

$$R_l(\Pi)|W_{F_w} \sim \begin{cases} \mathcal{L}_{F_w}(\Pi^1_w), & \text{if } m \text{ is odd,} \\ \mathcal{L}_{F_w}(\Pi_{1,w}) \text{ or } \mathcal{L}_{F_w}(\Pi_{2,w}), & \text{if } m \text{ is even,} \end{cases}$$

where $\sim$ means an isomorphism up to a nonzero multiplicity, a character twist and semisimplification.

In fact we prove the theorem under one more hypothesis:

Hypothesis 4.7. *$p := w|_{\mathbb{Q}}$ splits in E.*

That is to say, we compute $R_l(\Pi)$ only at those places where G is a product of general linear groups. This may seem like a defect, but it turns out that once the theorem is shown under Hypothesis 4.7, the latter can be removed without much difficulty (cf. §4.5).

Remark 4.8. The ambiguity in $\sim$ raises nontrivial issues, but they are resolved after all. The possibility that the multiplicity could be greater than one is handled by Taylor's trick ([HT01, Prop VII.1.8]). The character twist is not a problem as one can always twist back. As for the last issue, Taylor and Yoshida ([TY07]) removed semisimplification by proving the purity of $WD(R_l(\Pi)|W_{F_w})$ by studying Shimura varieties with Iwahori level structure. See their article for further detail.

If m is odd, it is not hard to see that Theorem 4.6 implies Theorem 1.3 (cf. Remark 4.8 above). In case m is even, if

$$R_l(\Pi)|W_{F_w} \sim \mathscr{L}_{F_w}(\Pi_{1,w}) \tag{4.2}$$

(resp. $R_l(\Pi)|W_{F_w} \sim \mathscr{L}_{F_w}(\Pi_{2,w})$) at one w then it is so at every other $w \nmid l$. We would be happy if (4.2) is the case. Unfortunately we are unable to tell whether $\mathscr{L}_{F_w}(\Pi_{1,w})$ or $\mathscr{L}_{F_w}(\Pi_{2,w})$ occurs. (This problem is linked with the computation of the sign e_2 in (6.8).) Nevertheless, if $\mathscr{L}_{F_w}(\Pi_{2,w})$ occurs in the formula for a given $\Pi = \chi \otimes (\Pi_1 \boxplus \Pi_2)$, we can show the following key fact ([Shi11, Lem 7.3]) by some technical sign computation in endoscopy: if we suppose

Hypothesis 4.9. Π_1 *is assumed to be slightly regular when m is even*

then there exists Π_2' such that if we set $\Pi' := \chi \otimes (\Pi_1 \boxplus \Pi_2')$, then $R_l(\Pi')|W_{F_w} \sim \mathscr{L}_{F_w}(\Pi_{1,w})$ for all $w \nmid l$. This suffices for the purpose of deducing Theorem 1.3 under the running hypotheses, as $R_l(\Pi')$ is essentially the desired Galois representation of that theorem (cf. Remark 4.8).

In summary, if Theorem 4.6 is known to be valid under Hypotheses 2.1, 4.2, 4.5 and 4.7, then Theorem 1.3 can be proved under Hypotheses 2.1, 4.2, 4.5, 4.7 and 4.9. In §4.5 below, we briefly explain how to obtain Theorem 1.3 by removing all the hypotheses. Starting from §5, our focus will be how to tackle Theorem 4.6 under Hypotheses 2.1, 4.2, 4.5 and 4.7.

4.5 Removal of hypotheses

Suppose that Theorem 1.3 is shown under Hypotheses 2.1, 4.2, 4.5, 4.7 and 4.9. In other words, we assume that the desired Galois representations are constructed under these hypotheses. Using them as initial seeds, we can remove all the hypotheses except Hypothesis 4.9 by various tricks. Among key ingredients are Arthur-Clozel's base change for general linear groups ([AC89]) and the so-called patching lemma (due to Blasius-Ramakrishnan [BR89] and generalized by [Sor]) among others. Rather than delving into detail, we refer the reader to [CH13, 3.1] or the proof of Theorem VII.1.9 in [HT01] for this type of argument. (The corresponding part in [Shi11] appears in the proof of Proposition 7.4 and Theorem 7.5.)

Finally Hypothesis 4.9 needs to go away. Chenevier and Harris constructed Galois representations without Hypothesis 4.9 (but assuming the other hypotheses and then removing them again) by p-adic congruences. Namely they made use of p-adic families of Galois representations on eigenvarieties for definite unitary groups. They derived various expected properties for $\rho_{l,\iota_l}(\Pi)$, including l-adic Hodge theoretic properties, but established a weaker form of (1.1). Caraiani [Car12, Car14] obtained (1.1) in general (without Hypothesis 4.9). A main point in her work is to show that the Galois representations associated to Π in the cohomology of $U(1, n-1) \times U(1, n-1)$-Shimura varieties are pure in a spirit similar to [TY07].

5 Local geometry of Shimura varieties

As we commented at the end of §4.4, our aim in the rest of the article is to sketch the idea of proof of Theorem 1.3 under Hypotheses 2.1, 4.2, 4.5, 4.7 and 4.9 assuming $v \nmid l$.

In §5.1 we briefly discuss integral models for Shimura varieties. We omit the detail on the integral models with bad reduction (defined in terms of Drinfeld level structure at p) but note that these play a crucial role in establishing the first basic identity in §5.5. It is worth noting at the outset that the contents of §5.2–§5.5 are not needed in the analysis of good reduction (namely the methods (1) and (2) of §1.5). Indeed, in the case of good reduction modulo p, Kottwitz ([Kot92b]) derived a nice counting point formula for the whole special fiber of Shimura varieties (generalizing earlier work of Ihara and Langlands) rather than for an individual Newton stratum. Kottwitz's formula describes the Hecke action outside a prime p and the Frobenius action at p on the cohomology of Shimura varieties. However, the strategy should be modified quite a bit in the case of bad reduction and our aim is to introduce some of the new tools, which are largely due to Harris and Taylor in the setting of unitary Shimura varieties. In the case of GL_2 the tools were developed earlier by Deligne and Carayol.

5.1 Moduli definition of integral models

Keep the notation and hypotheses from §4.4. It is convenient to define a place $u := w|_E$ of E (as a restriction of w to E).

Let $U = U^p \times U_p$ be an open compact subgroup of $G(\mathbb{A}^\infty)$. We will assume that U_p is a maximal compact subgroup of $G(\mathbb{Q}_p)$. (It may not be hyperspecial as p may ramify in F.) Consider the following moduli problem which associates to a connected locally noetherian scheme S the set of equivalence classes $\{(A, \lambda, i, \bar{\eta})\}/\sim$ where

- A is an abelian scheme over S.
- $\lambda : A \to A^\vee$ is a prime-to-p polarization.
- $i : \mathcal{O}_F \hookrightarrow \mathrm{End}(A) \otimes_{\mathbb{Z}} \mathbb{Z}_{(p)}$ such that $\lambda \circ i(f) = i(f^c)^\vee \circ \lambda, \forall f \in \mathcal{O}_F$.
- $\bar{\eta}^p$ is a $\pi_1(S, s)$-invariant U^p-orbit of isomorphisms of $F \otimes_{\mathbb{Q}} \mathbb{A}^{\infty,p}$-modules η^p : $V \otimes_{\mathbb{Q}} \mathbb{A}^{\infty,p} \xrightarrow{\sim} V^p A_s$ which take the pairing lg $\cdot,\cdot\rangle$ to the λ-Weil pairing up to $(\mathbb{A}^{\infty,p})^\times$-multiples. (Here s is any geometric point of S. The map $\bar{\eta}^p$ for any two choices of s can be identified.)
- Lie A with the induced $\mathcal{O}_F$-action satisfies a "determinant" condition ([Kot92b, §5]).
- Two quadruples $(A_1, \lambda_1, i_1, \bar{\eta}_1^p)$ and $(A_2, \lambda_2, i_2, \bar{\eta}_2^p)$ are equivalent if there is a prime-to-p isogeny $A_1 \to A_2$ taking $\lambda_1, i_1, \bar{\eta}_1^p$ to $\gamma\lambda_2, i_2, \bar{\eta}_2^p$ for some $\gamma \in \mathbb{Z}_{(p)}^\times$.

If U^p is sufficiently small, the above functor is representable by a smooth projective $\mathcal{O}_{F_w}$-scheme, which is denoted Sh_{U^p}. (Recall that U_p is maximal, which means we are in prime-to-p level.)

If U_p is not maximal but a certain congruence subgroup of $G(\mathbb{Q}_p)$, the integral model for Sh_U can be constructed by adding Drinfeld level structure at p to the moduli problem

([HT01, Ch II.2, III.4]). In general Sh_U has bad reduction mod p, although Sh_U is smooth over $\mathcal{O}_{F_w}$ if U_p is maximal compact. Although we will not discuss this further, the integral model with bad reduction plays a crucial role in the proof of Theorem 5.11.

5.2 Newton stratification

The prototype for Newton stratification is seen on the mod p fiber $\overline{Y}$ of an affine elliptic modular curve Y. Let k be a field of characteristic p. A k-point on $\overline{Y}$ corresponds to an elliptic curve over k (with level structure). Thus there is a set-theoretic decomposition

$$\overline{Y} = \overline{Y}^{\mathrm{ord}} \coprod \overline{Y}^{\mathrm{ss}}$$

such that the points of $\overline{Y}^{\mathrm{ord}}$ (resp. $\overline{Y}^{\mathrm{ss}}$) correspond to ordinary (resp. supersingular) elliptic curves. Note that $\overline{Y}^{\mathrm{ord}}$ (resp. $\overline{Y}^{\mathrm{ss}}$) is Zariski open (resp. closed) in $\overline{Y}$.

There is an analogous construction for the mod w fiber $\overline{\mathrm{Sh}}_{U^p}$ of Sh_{U^p} (and also for the mod w fiber of integral models with Drinfeld level structure at p). To do this, we need a little preparation. Let $(\mathscr{A}^{\mathrm{univ}}, \lambda^{\mathrm{univ}}, \imath^{\mathrm{univ}})$ denote the universal abelian scheme over $\overline{\mathrm{Sh}}_{U^p}$ with polarization and endomorphism structure. Let s be an $\overline{\mathbb{F}}_p$-point of $\overline{\mathrm{Sh}}_{U^p}$. Then $\mathscr{A}_s^{\mathrm{univ}}$, denoting the fiber of $\mathscr{A}^{\mathrm{univ}}$ at s, is an abelian variety over $\overline{\mathbb{F}}_p$. Since its p-divisible group $\mathscr{A}_s^{\mathrm{univ}}[p^\infty]$ is equipped with an action of $\mathcal{O}_F \otimes_{\mathbb{Z}} \mathbb{Z}_p \simeq \prod_{x|p} \mathcal{O}_{F_x}$ (and a polarization), it is decomposed as

$$\mathscr{A}_s^{\mathrm{univ}}[p^\infty] = \bigoplus_{x|p} \mathscr{A}_s^{\mathrm{univ}}[x^\infty]$$

with respect to that action. The $\mathcal{O}_{F_w}$-height of each $\mathscr{A}_s^{\mathrm{univ}}[x^\infty]$ is n. The determinant condition in the moduli problem implies that the dimension of $\mathscr{A}_s^{\mathrm{univ}}[x^\infty]$, which is the same as $\dim_{\overline{\mathbb{F}}_p} \mathrm{Lie}\, \mathscr{A}_s^{\mathrm{univ}}[x^\infty]$, equals 1 if $x = w$ and 0 if $x \neq w$ and $x|_E = u$. In other words, in the latter case, $\mathscr{A}_s^{\mathrm{univ}}[x^\infty]$ is an étale p-divisible group. (If $x|_E \neq w|_E$ then $x^c|_E = w|_E$ and $\mathscr{A}_s^{\mathrm{univ}}[x^\infty]$ is isomorphic to $\mathscr{A}_s^{\mathrm{univ}}[(x^c)^\infty]^\vee$ via the prime-to-p polarization induced by λ^{univ}.) The upshot is that there is a stratification of $\overline{\mathrm{Sh}}_{U^p}$ into locally closed subsets

$$\overline{\mathrm{Sh}}_{U^p} = \coprod_{0 \le h \le n-1} \overline{\mathrm{Sh}}_{U^p}^{(h)}$$

such that $s \in \overline{\mathrm{Sh}}_{U^p}(\overline{\mathbb{F}}_p)$ lands in $\overline{\mathrm{Sh}}_{U^p}^{(h)}$ if and only if the maximal étale quotient of $\mathscr{A}_s^{\mathrm{univ}}[w^\infty]$ has $\mathcal{O}_{F_w}$-height h. For any $0 \le h' \le n-1$, the union of $\overline{\mathrm{Sh}}_{U^p}^{(h)}$ for all $0 \le h \le h'$ is Zariski closed in $\overline{\mathrm{Sh}}_{U^p}$. The last fact reflects the principle that the Newton polygon of a p-divisible group can only go up under specialization on the base scheme. Each locally closed subset $\overline{\mathrm{Sh}}_{U^p}^{(h)}$ is viewed as a scheme with the reduced subscheme structure. It turns out that each $\overline{\mathrm{Sh}}_{U^p}^{(h)}$ has dimension h (but it is a nontrivial fact that every stratum is nonempty).

Remark 5.1. Recall that Newton polygons may be used to classify isogeny classes of p-divisible groups over $\overline{\mathbb{F}}_p$ in the context of Dieudonné theory. In general the Newton stratification for Shimura varieties is defined according to isogeny classes of p-divisible groups with polarization and endomorphism structure. In our case it can be shown that the latter isogeny classes are classified by the integer h introduced above (which amounts to the Newton polygon of height n, dimension 1 and "étale height" h with nonnegative slopes).

Remark 5.2. Our Shimura varieties are nice in that the study of the universal p-divisible group $\mathscr{A}^{\mathrm{univ}}[p^\infty]$ over $\overline{\mathrm{Sh}}_{U^p}$ (with additional structure) is essentially reduced to the study of p-divisible groups of dimension 1. This dimension is linked to the signature of $G^1(\mathbb{R})$ in §4.1 via the determinant condition in the moduli problem. (For a different choice of signatures, the dimension is higher than 1 in general.) In the dimension 1 case, the deformation theory of p-divisible groups works nicely and the Drinfeld level structure for the integral models behaves well. This gives another reason why our choice of G is favorable (cf. §4.1, Remark 4.3).

5.3 Igusa varieties

Igusa varieties in the case of elliptic modular curves were studied in [Igu59] and [KM85]. They also appear in the context of p-adic automorphic forms as in Hida's work. (See [Hid04] for instance.) In work of Harris and Taylor, Igusa varieties and Rapoport-Zink spaces are used to study the bad reduction of Shimura varieties. We will use the same strategy here. In this subsection we give basic definitions for Igusa varieties.

Temporarily fix $0 \le h \le n-1$. Let $\Sigma^{(h)} := \prod_{x|p} \Sigma_x$ and $\Sigma_w = \Sigma_w^0 \times \Sigma_w^{\mathrm{et}}$ where each Σ_x denotes the p-divisible group over $\overline{\mathbb{F}}_p$ with an $\mathcal{O}_{F_x}$-action and

- Σ_w^0 is connected of $\mathcal{O}_{F_w}$-height $n-h$ and dimension 1,
- Σ_w^{et} is étale of $\mathcal{O}_{F_w}$-height h,
- Σ_x is étale of $\mathcal{O}_{F_x}$-height n for $x \ne w$, $x|u$,
- $\Sigma_{x^c} = \Sigma_x^\vee$ for $x|u$.

The Newton stratification is defined so that the geometric fibers of $\mathscr{A}^{\mathrm{univ}}[x^\infty]$ are isogenous to Σ_x as p-divisible groups with $\mathcal{O}_{F_x}$-action on $\overline{\mathrm{Sh}}_{U^p}^{(h)}$. (In fact, a little more is true in our particular case. Namely, $\mathscr{A}^{\mathrm{univ}}[p^\infty]$ is fiberwise isogenous to $\Sigma^{(h)}$ with $\mathcal{O}_F \otimes \mathbb{Z}_p$-action and polarization.)

Define

$$J_w^{(h)}(\mathbb{Q}_p) := \mathrm{Aut}_{\mathcal{O}_{F_w}}(\Sigma_w) \simeq D_{n-h,F_w}^\times \times GL_h(F_w)$$

and

$$J^{(h)}(\mathbb{Q}_p) := \mathbb{Q}_p^\times \times J_w^{(h)}(\mathbb{Q}_p) \times \prod_{x|u,\ x\ne w} GL_n(F_x)$$

where D_{n-h} is a central division algebra over $\mathcal{O}_{F_w}$ with Hasse invariant $\frac{1}{n-h}$.

Remark 5.3. The group $J^{(h)}(\mathbb{Q}_p)$ may be naturally identified with the automorphism group of $\Sigma^{(h)}$ with $\mathcal{O}_F \otimes_{\mathbb{Z}} \mathbb{Z}_p$-action and polarization (the latter preserved up to $\mathbb{Q}_p^\times$-multiples) in the isogeny category. We will often ignore the $\mathbb{Q}_p^\times$-part to simplify exposition, and this allows us to view $J^{(h)}(\mathbb{Q}_p)$ loosely as the automorphism group of $\prod_{x|u}\Sigma_x$ with $\prod_{x|u}\mathcal{O}_{F_x}$-action in the isogeny category.

Note that

$$G(\mathbb{Q}_p) \simeq \mathbb{Q}_p^\times \times \prod_{x|u} GL_n(F_x). \tag{5.1}$$

Thus $J^{(h)}$ is an inner form of a Levi subgroup of (a parabolic subgroup of) G over $\mathbb{Q}_p$.

Example 5.4. In the case of elliptic modular curves, the analogue of $J^{(h)}(\mathbb{Q}_p)$ is $\mathbb{Q}_p^\times \times \mathbb{Q}_p^\times$ (resp. $D^\times$) for the ordinary (resp. supersingular) stratum, where D is a central division algebra over $\mathbb{Q}_p$ of degree 4.

By abuse of notation, the pullback of $\mathscr{A}^{\mathrm{univ}}$ from $\overline{\mathrm{Sh}}_{U^p}$ to $\overline{\mathrm{Sh}}^{(h)}_{U^p}$ will still be denoted by $\mathscr{A}^{\mathrm{univ}}$. Let $\mathscr{G}^0$ (resp. $\mathscr{G}^{\mathrm{et}}$) be the maximal connected sub p-divisible group (resp. maximal quotient p-divisible group) of $\mathscr{G} := \mathscr{A}^{\mathrm{univ}}[p^\infty]$.

We will introduce Igusa varieties and Rapoport-Zink spaces which are closely related to the stratum $\overline{\mathrm{Sh}}^{(h)}_{U^p}$. Let $m \geq 1$. The first object $\mathrm{Ig}^{(h)}_{U^p,m}$ is the moduli space over $\overline{\mathrm{Sh}}^{(h)}_{U^p}$ which associates to a scheme S over $\overline{\mathrm{Sh}}^{(h)}_{U^p}$ the quadruple of isomorphisms $j = (j_{p,0}, j^0_w, j^{\mathrm{et}}_w, \{j_x\})$ where

- $j_{p,0} : \mathbb{Z}_p^\times \xrightarrow{\sim} \mathbb{Z}_p(1)^\times$.
- $j^0_w : \Sigma^0_w[w^m] \times_{\overline{\mathbb{F}}_p} S \xrightarrow{\sim} \mathscr{G}^0[w^m] \times_{\overline{\mathrm{Sh}}^{(h)}_{U^p}} S$ which is compatible with $\mathcal{O}_{F_w}$-actions,
- $j^{\mathrm{et}}_w : \Sigma^{\mathrm{et}}_w[w^m] \times_{\overline{\mathbb{F}}_p} S \xrightarrow{\sim} \mathscr{G}^{\mathrm{et}}[w^m] \times_{\overline{\mathrm{Sh}}^{(h)}_{U^p}} S$ which is compatible with $\mathcal{O}_{F_w}$-actions,
- $j_x : \Sigma_x[x^m] \times_{\overline{\mathbb{F}}_p} S \xrightarrow{\sim} \mathscr{G}[x^m] \times_{\overline{\mathrm{Sh}}^{(h)}_{U^p}} S$ for $x|u$ and $x \neq w$ which is compatible with $\mathcal{O}_{F_x}$-actions and

To be precise, there is a technical condition imposed on the data that $(j_{p,0}, j^0_w, j^{\mathrm{et}}_w, \{j_x\})$ should be liftable to the level of p-divisible groups.

Remark 5.5. An equivalent formulation is that $\mathrm{Ig}^{(h)}_{U^p,m}$ parametrizes graded isomorphisms $j : \Sigma^{(h)}[p^m] \xrightarrow{\sim} \mathrm{gr}(\mathscr{G})[p^m]$ compatible with $\mathcal{O}_F \otimes_{\mathbb{Z}} \mathbb{Z}_p$-actions and polarizations, the latter up to $(\mathbb{Z}/p^m\mathbb{Z})^\times$-multiple, where the grading is given by slope filtration. (See [Man05, Def 3].) We have chosen to give a more down-to-earth moduli problem above.

It turns out that each $\mathrm{Ig}^{(h)}_{U^p,m}$ is a smooth variety over $\overline{\mathbb{F}}_p$ and finite Galois over $\overline{\mathrm{Sh}}^{(h)}_{U^p}$. By abuse of notation, $\mathscr{L}_\xi$ will also denote its pullback from $\overline{\mathrm{Sh}}^{(h)}_{U^p}$ to each $\mathrm{Ig}^{(h)}_{U^p,m}$. Define

$$H_c(\mathrm{Ig}^{(h)}, \mathscr{L}_\xi) := \sum_{k\geq 0} (-1)^k \varinjlim_{U^p,m} H^k_c(\mathrm{Ig}^{(h)}_{U^p,m}, \mathscr{L}_\xi),$$

which is naturally a virtual representation of $G(\mathbb{A}^{\infty,p}) \times J^{(h)}(\mathbb{Q}_p)$. Indeed, the action of $G(\mathbb{A}^{\infty,p})$ is inherited from the Hecke action on the tower $\overline{\mathrm{Sh}}^{(h)}_{U^p}$. (Note that the tower of $\overline{\mathrm{Sh}}^{(h)}_{U^p}$ for varying U^p is invariant under the Hecke action of $G(\mathbb{A}^{\infty,p})$.) The action of $J^{(h)}(\mathbb{Q}_p)$ is defined by extending the natural action of

$$\mathbb{Z}_p^\times \times (\mathcal{O}_{D_{n-h,F_w}} \times GL_h(\mathcal{O}_{F_w})) \times \prod_{x|u, x\neq w} GL_n(\mathcal{O}_{F_x})$$

on the quadruple $(j_{p,0}, j_w^0, j_w^{\mathrm{et}}, \{j_x\})$. The computation of $H_c(\mathrm{Ig}^{(h)}, \mathcal{L}_\xi)$, which we will focus on in §6, is the most innovative part of [Shi11] and a vital input for the computation of $H(\mathrm{Sh}, \mathcal{L}_\xi)$.

5.4 Rapoport-Zink spaces and the map Mant$^{(h)}$

The Rapoport-Zink spaces are local analogues of PEL Shimura varieties. Indeed, the former are moduli spaces of p-divisible groups with additional structure whereas the latter are moduli spaces of abelian schemes with additional (PEL) structure. Just as a Shimura variety is constructed from a reductive group over $\mathbb{Q}$ and other data, a Rapoport-Zink space is associated with a reductive group over $\mathbb{Q}_p$ and some other data. Recall that our main strategy is based on the philosophy that the cohomology of Shimura varieties is closely related to the global Langlands correspondence. Similarly it is believed that the cohomology of Rapoport-Zink spaces has a lot to do with the local Langlands correspondence. Indeed, this was one of the main motivations to study these spaces. For some precise conjectures, see Remark 5.10 below.

Rapoport-Zink spaces were introduced by Rapoport and Zink in an attempt to generalize the non-abelian Lubin-Tate spaces and the Drinfeld spaces. The latter two spaces are associated with general linear groups and the unit groups of division algebras, and had been studied the most in connection with the local Langlands correspondence for GL_n (and in fact the Jacquet-Langlands correspondence as well). When $n = 1$, this reduces to the well-known relationship between the classical Lubin-Tate theory and local class field theory.

In our setting, the relevant Rapoport-Zink spaces are associated with $G(\mathbb{Q}_p)$ and isomorphic to (products of) the so-called non-abelian Lubin-Tate spaces. To be more concrete, the Rapoport-Zink space with no level structure, denoted $\mathrm{RZ}_w^{(h)}$, represents (as a formal scheme) the moduli problem

$$RZ_w^{(h)}(S) = \{(H, i, \beta)\}/\simeq$$

from the category of $\mathcal{O}_{F_w^{\mathrm{ur}}}$-schemes in which p is locally nilpotent to the category of sets, where

- H is a p-divisible group over S,
- $i : \mathcal{O}_{F_w} \hookrightarrow \mathrm{End}(H)$ is a $\mathbb{Z}_p$-algebra morphism,
- $\beta : \Sigma_w \times_{\overline{\mathbb{F}}_p} \overline{S} \to H \times_S \overline{S}$ is a quasi-isogeny compatible with $\mathcal{O}_{F_w}$-actions, where $\overline{S}$ is the closed subscheme of S defined by the ideal sheaf $p\mathcal{O}_S$, such that
- the determinant condition as in [RZ96, 3.23.(a)] is satisfied.

Then $\mathrm{RZ}_w^{(h)}$ is a Rapoport-Zink space associated to $R_{F_w/\mathbb{Q}_p}GL_n$. The Rapoport-Zink space $\mathrm{RZ}_0^{(h)}$ (without level structure) associated to G can be defined as the product of $\mathrm{RZ}_w^{(h)}$ and the zero-dimensional Rapoport-Zink spaces accounting for $\mathbb{Q}_p^\times$ and $\prod_{x|u,x\neq w} GL_n$ in the decomposition 5.1.

The formal scheme $\mathrm{RZ}_0^{(h)}$ gives rise to a rigid analytic space. With this space at the bottom, one can throw in level structure to construct a projective system of rigid analytic spaces $RZ^{(h)} = \{RZ_{U_p}^{(h)}\}$ indexed by open compact subgroups U_p of $G(\mathbb{Q}_p)$. They are (not of finite type but) locally of finite type over $\hat{F}_w^{\mathrm{ur}}$.

We will not give detail, but the l-adic cohomology of $\mathrm{RZ}^{(h)}$ comes equipped with a natural commuting action of $J^{(h)}(\mathbb{Q}_p)$, $G(\mathbb{Q}_p)$ and W_{F_w}. (The group $J^{(h)}(\mathbb{Q}_p)$ acts on the deformation datum for each fixed level U_p. On the other hand, $G(\mathbb{Q}_p)$ acts on the tower $\mathrm{RZ}^{(h)}$ in the style of Hecke correspondences.) To study its cohomology effectively, especially in connection with the cohomology of Shimura varieties, it is useful to define the following map

$$\mathrm{Mant}_n^{(h)} : \mathrm{Groth}(J^{(h)}(\mathbb{Q}_p)) \to \mathrm{Groth}(G(\mathbb{Q}_p) \times W_{F_w})$$

as

$$\mathrm{Mant}_n^{(h)}(\rho) := \sum_{i,j\geq 0} (-1)^{i+j} \varinjlim_{U_p} \mathrm{Ext}^i_{J^{(h)}(\mathbb{Q}_p)}(H_c^j(\mathrm{RZ}_{U_p}^{(h)}, \overline{\mathbb{Q}}_l), \rho).$$

In the case under consideration, a complete description of $\mathrm{Mant}^{(h)}$ was given by Harris-Taylor. (See [Shi11, Prop 2.2] for a summary of results.) The case $h = 0$ turns out to be the most interesting. To show the flavor we state a result in the supercuspidal case, which was proved by Carayol ([Car86]) for $n = 2$ and Harris-Taylor for any n. (The analogue of the Mant map for Drinfeld spaces was computed in [Car90] for $n = 2$ and in [Dat07] for any n.) When $n = 1$, the formula essentially follows from the classical Lubin-Tate theory.

Theorem 5.6 ([HT01, Thm VII.1.3]). *Let $\rho \in \mathrm{Irr}(J^{(h)}(\mathbb{Q}_p))$ be such that $JL(\rho)$ is supercuspidal. Then*

$$\mathrm{Mant}_n^{(0)}(\rho) = (-1)^{n-1} \cdot JL(\rho) \otimes \mathscr{L}_{F_w}(JL(\rho)).$$

The known proofs of the theorem are global in nature in that a key input comes from the cohomology of Shimura varieties and its interaction with $\mathrm{Mant}^{(h)}$ (as in Theorem 5.11 below). The proof of the local Langlands correspondence for GL_n over p-adic fields, either by Henniart or by Harris-Taylor, is also global as it relies on a result on the cohomology of Shimura varieties.

For $h > 0$ we have an induction formula, which is implicit in [HT01].

Theorem 5.7. $\mathrm{Mant}_n^{(h)}(\rho_1 \otimes \rho_2) = \mathrm{Ind}_{GL_{n-h,h}}^{GL_n}(\mathrm{Mant}_{n-h}^{(0)}(\rho_1) \otimes \rho_2).$

As the notation suggests,

$$\mathrm{Mant}_{n-h}^{(0)} : \mathrm{Groth}(\mathbb{Q}_p^\times \times D_{n-h}^\times) \to \mathrm{Groth}(\mathbb{Q}_p^\times \times GL_{n-h}(F_w) \times W_{F_w})$$

is built from the Rapoport-Zink spaces for GL_{n-h}, corresponding to the Newton polygon height $n-h$ and dimension 1. As we already remarked, $\mathrm{Mant}^{(0)}_{n-h}$ can be explicitly described thanks to Harris and Taylor. Therefore Theorem 5.7 enables us to compute $\mathrm{Mant}^{(h)}_n$ for all $1 \le h \le n-1$ as well.

Remark 5.8. It would natural and interesting to figure out $H^j_c(RZ^{(h)}_{U_p}, \overline{\mathbb{Q}}_l)$ directly, without taking the alternating sum over j. In the case at hand, the result was obtained by Boyer ([Boy09]).

Remark 5.9. In the history of class field theory, local class field theory was first proved by using its global counterpart, but later established by purely local methods. Thus it would be desirable to find a purely local proof of Theorem 5.6 and the local Langlands correspondence. Let us mention some partial results (which are not meant to be exhaustive by any means). For the first problem, see [Str05]. As for the second problem, Bushnell and Henniart ([BH05a], [BH05b]) explicitly constructed the Langlands correspondence in the "essentially tame" case.

Remark 5.10. Theorem 5.6 is concerned with the Rapoport-Zink spaces for GL_n corresponding to the "basic" Newton polygon of pure slope $1/n$. It is basic in the sense that it lies above the other Newton polygons with the same end points. A natural generalization of Theorem 5.6 would be a description of the analogue of $\mathrm{Mant}^{(h)}$ for other Rapoport-Zink spaces. In this direction of research, the most prominent conjectures seem to be the following, which are very precise but stated here only loosely.

- Kottwitz's conjecture ([Rap95, Conj 5.1], cf. [Har01, Conj 5.3, 5.4]) - the Mant map for a basic Newton polygon is described in terms of discrete L-parameters.
- Harris's conjecture ([Har01, Conj 5.2]) - the Mant map for a non-basic Newton polygon is obtained by an induction formula.

In fact Theorem 5.7 is a special case of Harris's conjecture, but the general case is a wide open question. For a progress toward the first (resp. second) conjecture, see [Far04] (resp. [Man08]). The paper [Shi12] provides a little extra information in the case of Rapoport-Zink spaces for GL_n.

5.5 The first basic identity

The cohomology of each Newton stratum in the special fiber of Shimura varieties can be related to the cohomology of Igusa varieties and Rapoport-Zink spaces via a very neat formula. For our Shimura varieties, the result is due to Harris and Taylor. (Although their Shimura varieties are attached to an inner form of our G, their proof carries over to our case.) The following is a reformulation in the style of Mantovan's theorem. See the remark below Theorem 5.11.

Theorem 5.11 (Harris-Taylor). *The following holds in* $\mathrm{Groth}(G(\mathbb{A}^\infty)\times W_{F_w})$. *(We regard* $\mathrm{Mant}^{(h)}$ *as the identity map on the space of* $G(\mathbb{A}^{\infty,p})$*-representations.)*

$$H(\mathrm{Sh}, \mathscr{L}_\xi) = \sum_{0\le h\le n-1} \mathrm{Mant}^{(h)}(H_c(\mathrm{Ig}^{(h)}, \mathscr{L}_\xi))$$

Sketch of proof. For simplicity, we assume $\mathscr{L}_\xi = \overline{\mathbb{Q}}_l$. Let $R\Psi_{\mathrm{Sh}}$ denote the nearby cycle complex on $\overline{\mathrm{Sh}}$ associated with $\overline{\mathbb{Q}}_l$. Then

$$H(\mathrm{Sh}, \overline{\mathbb{Q}}_l) = H(\overline{\mathrm{Sh}}, R\Psi_{\mathrm{Sh}}) = \sum_{0 \le h \le n-1} H(\overline{\mathrm{Sh}}^{(h)}, R\Psi_{\mathrm{Sh}})$$

in $\mathrm{Groth}(G(\mathbb{A}^\infty) \times W_{F_w})$, where $R\Psi_{\mathrm{Sh}}$ also denotes the sheaves on $\overline{\mathrm{Sh}}^{(h)}$ by abuse of notation.

Let us pretend that $m : \overline{\mathrm{RZ}}^{(h)} \times_{\overline{\mathbb{F}}_p} \mathrm{Ig}^{(h)} \to \overline{\mathrm{Sh}}^{(h)}$ is a Galois covering with Galois group $J^{(h)}(\mathbb{Q}_p)$. Although this is not literally true, we hope that this helps the reader to grasp some core ideas more easily.

Let $p_1 : \overline{\mathrm{RZ}}^{(h)} \times_{\overline{\mathbb{F}}_p} \mathrm{Ig}^{(h)} \to \overline{\mathrm{RZ}}^{(h)}$ denote the projection map. Berkovich's theory provides $R\Psi_{\mathrm{RZ}}$ on $\overline{\mathrm{RZ}}^{(h)}$, which is the analogue for formal schemes of nearby cycle complexes. Harris and Taylor proved (and Mantovan generalized) a deep fact that

$$m^* R\Psi^{(h)}_{\mathrm{Sh}} \simeq p_1^* R\Psi_{\mathrm{RZ}},$$

which roughly asserts that Shimura varieties and Rapoport-Zink spaces present the same kind of singularities along the Newton stratum for h. Then the following is morally true. The second last equality uses the fact that $H_c(\overline{\mathrm{RZ}}^{(h)}, R\Psi_{\mathrm{RZ}})$ is dual to $H_c(\mathrm{RZ}^{(h)}, \overline{\mathbb{Q}}_l)$ via Berkovich's theory.

$$\begin{aligned} H(\overline{\mathrm{Sh}}^{(h)}, R\Psi_{\mathrm{Sh}}) &= H_*(J^{(h)}(\mathbb{Q}_p), H_c(\overline{\mathrm{RZ}}^{(h)} \times \mathrm{Ig}^{(h)}, m^* R\Psi_{\mathrm{Sh}}) \\ &= H_*(J^{(h)}(\mathbb{Q}_p), H_c(\overline{\mathrm{RZ}}^{(h)} \times \mathrm{Ig}^{(h)}, p_1^* R\Psi_{\mathrm{RZ}}) \\ &= \mathrm{Tor}_{C_c^\infty(J^{(h)}(\mathbb{Q}_p))}(H_c(\overline{\mathrm{RZ}}^{(h)}, R\Psi_{\mathrm{RZ}}), H_c(\mathrm{Ig}^{(h)}, \overline{\mathbb{Q}}_l)) \\ &= \mathrm{Ext}_{J^{(h)}(\mathbb{Q}_p)}(H_c(\mathrm{RZ}^{(h)}, \overline{\mathbb{Q}}_l), H_c(\mathrm{Ig}^{(h)}, \overline{\mathbb{Q}}_l)) \\ &= \mathrm{Mant}^{(h)}(H_c(\mathrm{Ig}^{(h)}, \overline{\mathbb{Q}}_l)). \end{aligned}$$

□

Remark 5.12. Theorem 5.11 was extended by Mantovan ([Man05], [Man11]) to PEL-type Shimura varieties of unitary or symplectic type when the PEL datum is "unramified" at p (which amounts to the running assumption in [Kot92b]) and Kottwitz's integral model of Sh (with good reduction) is proper over $\mathcal{O}_{F_w}$. Fargues obtained a similar formula ([Far04, Cor 4.6.3]) when restricted to the basic (cf. Remark 5.10) stratum.

Remark 5.13. In the problem of understanding $H(\mathrm{Sh}, \mathscr{L}_\xi)$ as a virtual representation of $G(\mathbb{A}^\infty) \times W_{F_w}$, two sources of difficulty are bad reduction (or ramified Galois action) at w and global endoscopy for G. Theorem 5.11 enables us to separate the two kinds of difficulty. Namely, the information of bad reduction is mostly contained in $\mathrm{Mant}^{(h)}$ (which arises from a purely local geometric object) while the global endoscopy is captured by $H_c(\mathrm{Ig}^{(h)}, \mathscr{L}_\xi)$. It is worth noting that $\mathrm{Ig}^{(h)}$ is global in nature whereas $\mathrm{RZ}^{(h)}$ is a purely local object which can be defined independently of Shimura varieties.

Write $H_c(\mathrm{Ig}^{(h)}, \mathcal{L}_\xi) = \sum_{i\in I} n_i[\pi_i^{\infty,p}][\rho_i]$ where I is an index set, $\pi^{\infty,p} \in \mathrm{Irr}(G(\mathbb{A}^{\infty,p}))$ and $\rho_p \in \mathrm{Irr}(J^{(h)}(\mathbb{Q}_p))$. Define the "$\Pi^{\infty,p}$-part" as

$$H_c(\mathrm{Ig}^{(h)}, \mathcal{L}_\xi)\{\Pi^{\infty,p}\} := \sum_{\mathrm{WBC}(\pi_i^{\infty,p})=\Pi^{\infty,p}} n_i[\rho_i] \in \mathrm{Groth}(J^{(h)}(\mathbb{Q}_p)). \tag{5.2}$$

This is independent of the expansion of $H_c(\mathrm{Ig}^{(h)}, \mathcal{L}_\xi)$. Now let $\pi_p \in \mathrm{Irr}(G(\mathbb{Q}_p))$ be the unique representation whose base change is Π_p. (The base change in this case is elementary as p splits in E, cf. §3.3 and §3.4.) We can prove

Corollary 5.14. *In* $\mathrm{Groth}(G(\mathbb{Q}_p) \times W_{F_w})$,

$$\pi_p \otimes R_l(\Pi) = \sum_{0\le h\le n-1} \mathrm{Mant}^{(h)}(H_c(\mathrm{Ig}^{(h)}, \mathcal{L}_\xi)\{\Pi^{\infty,p}\}).$$

Proof. The corollary is basically obtained by taking the $\{\Pi^{\infty,p}\}$-part of Theorem 5.11. □

Since we know how $\mathrm{Mant}^{(h)}$ works, the proof of Theorem 4.6 is reduced to the problem of understanding $H_c(\mathrm{Ig}^{(h)}, \mathcal{L}_\xi)\{\Pi^{\infty,p}\}$ as a virtual representation of $J^{(h)}(\mathbb{Q}_p)$. This brings us to the next section.

6 Cohomology of Igusa varieties

6.1 Counting point formula

The action of $G(\mathbb{A}^{\infty,p}) \times J^{(h)}(\mathbb{Q}_p)$ on $H_c(\mathrm{Ig}^{(h)}, \mathcal{L}_\xi)\{\Pi^{\infty,p}\}$ is basically given by "Hecke correspondences". (The $G(\mathbb{A}^{\infty,p})$-action is indeed compatible with the Hecke action on Shimura varieties but the $J^{(h)}(\mathbb{Q}_p)$-action is more subtle. We will ignore the subtlety here, though.) The trace of the Hecke action on $H_c(\mathrm{Ig}^{(h)}, \mathcal{L}_\xi)$ can be computed in terms of fixed points of $\mathrm{Ig}^{(h)}(\overline{\mathbb{F}}_p)$ under algebraic correspondences thanks to Fujiwara and Varshavsky, who proved Deligne's conjecture ([Fuj97, Cor 5.4.5], [Var07, Thm 2.3.2]). The latter is a version of the Grothendieck-Lefschetz trace formula and needed here as $\mathrm{Ig}^{(h)}$ is usually non-proper over $\overline{\mathbb{F}}_p$. Roughly speaking, Deligne's conjecture says that the Grothendieck-Lefschetz trace formula holds for an algebraic correspondence on a non-proper variety if that correspondence is twisted by a large enough power of Frobenius.

Since $\mathrm{Ig}^{(h)}$ is a moduli space for $(A, \lambda, i, \bar{\eta}^p)$ as well as certain isomorphisms of p-divisible groups, the fixed points under a correspondence on $\mathrm{Ig}^{(h)}$ are naturally described in terms of the moduli data. Our hope is to extract some automorphic information from the fixed point formula. The best way might be to relate the fixed point formula to an analogous[3] formula in automorphic representation theory, such as the Arthur-Selberg trace formula. But the latter formula has obviously no reference to abelian varieties or their structures. So the main problem is to massage the fixed-point formula for $\mathrm{Ig}^{(h)}$ to obtain a trace formula for the Hecke action on $\mathrm{Ig}^{(h)}$ which resembles the geometric side of the trace formula for G. (So to speak, it is about the passage from (6.2) to the statement of Theorem 6.3 below.) In the

[3]Only remotely analogous, a priori.

context of unitary Shimura varieties, this was carried out by Harris and Taylor for a certain $U(1, n-1)$-type unitary group with no endoscopy. A trace formula for Igusa varieties was proved ([Shi10]) for any PEL-type Shimura varieties associated to unitary or symplectic groups (possibly with nontrivial endoscopy), in the spirit of Langlands-Kottwitz's formula ([Kot92b]) for Shimura varieties with good reduction.

Before stating the result, we define the notion of Kottwitz triples in our context (which are somewhat different from those for Shimura varieties with good reduction as in [Kot90, §2]).

Definition 6.1. By an **effective Kottwitz triple** (of type $0 \le h \le n-1$), we mean a triple $(\gamma_0; \gamma, \delta)$ where

- $\gamma_0 \in G(\mathbb{Q})$ is semisimple, and elliptic in $G(\mathbb{R})$
- $\gamma \in G(\mathbb{A}^{\infty,p})$ and $\gamma_0 \sim_{\overline{\mathbb{A}}^{\infty,p}} \gamma$.
- $\delta \in J^{(h)}(\mathbb{Q}_p)$ is acceptable (to be explained) and $\gamma_0 \sim_{\overline{\mathbb{Q}}_p} \delta$ in $G(\overline{\mathbb{Q}}_p)$ via a natural embedding $J^{(h)}(\overline{\mathbb{Q}}_p) \hookrightarrow G(\overline{\mathbb{Q}}_p)$ (natural up to $G(\overline{\mathbb{Q}}_p)$-conjugacy).
- a certain Galois cohomology invariant $\alpha(\gamma_0; \gamma, \delta)$ vanishes.

Two Kottwitz triples $(\gamma_0; \gamma, \delta) \sim (\gamma_0'; \gamma', \delta')$ are said to be equivalent if $\gamma_0 \sim_{st} \gamma_0'$, $\gamma \sim_{\mathbb{A}^{\infty,p}} \gamma'$, and $\delta \sim \delta'$. For each $0 \le h \le n-1$, we define $KT^{(h),\mathrm{eff}}$ to be the set of equivalence classes of all effective Kottwitz triples of type h.

Remark 6.2. The word "effective" refers to the last condition, which is closely tied with the phenomenon of endoscopy. The analogous fact is that an element $\gamma \in G(\mathbb{A})$ is not always $G(\overline{\mathbb{A}})$-conjugate to an element of $G(\mathbb{Q})$. The failure is detected by the nonvanishing of a similar Galois cohomology invariant. When we say that a group G over $\mathbb{Q}$ has "no endoscopy", it indicates that this failure does not occur. (In that case no endoscopic groups other than the quasi-split inner form of G will contribute to the stable trace formula for G.)

The counting point formula is stated below. The terminology "acceptable" will not be defined but morally means that "twisted by enough power of Frobenius" in a suitable sense.

Theorem 6.3 ([Shi09, Thm 13.1]). *If $\varphi \in C_c^\infty(G(\mathbb{A}^{\infty,p}) \times J^{(h)}(\mathbb{Q}_p))$ is acceptable, then*

$$\mathrm{tr}(\varphi | H_c(\mathrm{Ig}^{(h)}, \mathscr{L}_\xi)) = \sum_{(\gamma_0;\gamma,\delta)\in KT^{(h),\mathrm{eff}}} \mathrm{vol}(I_\infty(\mathbb{R})^1)^{-1} |A(I_0)| \,\mathrm{tr}\xi(\gamma_0) \cdot O_{(\gamma,\delta)}^{G(\mathbb{A}^{\infty,p})\times J^{(h)}(\mathbb{Q}_p)}(\varphi)$$

Sketch of proof. For simplicity, assume ξ is trivial so that $\mathscr{L}_\xi = \overline{\mathbb{Q}}_l$. Let $U_p(m) \subset J^{(h)}(\mathbb{Q}_p)$ denote the kernel of $\mathrm{Aut}(\Sigma^{(h)}) \to \mathrm{Aut}(\Sigma^{(h)}[p^m])$. It is enough to treat the case where $\varphi = \varphi^{\infty,p}\varphi_p'$ with $\varphi^{\infty,p} = \mathrm{char}_{U^p g^p U^p}$ and $\varphi_p' = \mathrm{char}_{U_p(m) g_p U_p(m)}$, as the general case is obtained by taking linear combinations. Then the left hand side is identified with

$$\mathrm{tr}([U^p g^p U^p] \times [U_p(m) g_p U_p(m)] | H_c(\mathrm{Ig}^{(h)}_{U^p,m}, \overline{\mathbb{Q}}_l)) \tag{6.1}$$

where $[U^p g^p U^p]$ and $[U_p(m) g_p U_p(m)]$ are Hecke correspondences for $\mathrm{Ig}^{(h)}_{U^p,m}$. The solution of Deligne's conjecture allows us to evaluate (6.1) as the number of fixed points on $\mathrm{Ig}^{(h)}(\overline{\mathbb{F}}_p)$ under the product correspondence. Recall from §5.3 that

$$\mathrm{Ig}^{(h)}(\overline{\mathbb{F}}_p) = \{(A, \lambda, i, \overline{\eta}^p, j)\}/\simeq$$

where A is an abelian variety over $\overline{\mathbb{F}}_p$ equipped with additional structure. It is not difficult to show that the number of fixed points corresponding to a given (A, λ, i) equals a sum of orbital integral of φ. To summarize the situation more precisely, we have

$$\mathrm{tr}(\varphi | H_c(\mathrm{Ig}^{(h)}, \overline{\mathbb{Q}}_l)) = \sum_{\{(A,\lambda,i)\}/\simeq} \left(\sum_{a \in \mathrm{Aut}^0(A,\lambda,i)} (\mathrm{const.}) \cdot O_a^{G(\mathbb{A}^{\infty,p}) \times J^{(h)}(\mathbb{Q}_p)}(\varphi) \right) \qquad (6.2)$$

where Aut^0 is the automorphism in the isogeny category (with additional structure). Then the proof is essentially completed by proving a natural bijection between $KT^{(h),\mathrm{eff}}$ and the set of A, λ, i, a in the sum. This is the core of the argument and involves a refinement of Honda-Tate theory, CM-lifting of abelian varieties from characteristic p to characteristic 0, Galois cohomology computations, theory of isocrystals over $\overline{\mathbb{F}}_p$ and others. □

We do not lose generality by restricting ourselves to acceptable functions. More precisely,

Lemma 6.4 ([Shi09, Lem 6.4]). *For $\Pi_1, \Pi_2 \in \mathrm{Groth}(G(\mathbb{A}^{\infty,p}) \times J^{(h)}(\mathbb{Q}_p))$,*

$$\mathrm{tr}\Pi_1(\varphi) = \mathrm{tr}\Pi_2(\varphi)$$

for all acceptable functions $\varphi \in C_c^\infty(G(\mathbb{A}^{\infty,p}) \times J^{(h)}(\mathbb{Q}_p))$ if and only if $\Pi_1 = \Pi_2$ in the Grothendieck group.

Remark 6.5. The lemma was used in [HT01] without stating it as a lemma.

6.2 Stabilization

We intend to use Theorem 6.3 to understand $H_c(\mathrm{Ig}_b, \mathscr{L}_\xi)$ in terms of automorphic representations of G. It is very common (e.g. in the trace formula approach to the Langlands functoriality) that the trace formula should be stabilized to have interesting applications. Thus it is natural to attempt to stabilize the right hand side of Theorem 6.3. In other words, we want to rewrite the sum of orbital integrals as a sum of stable orbital integrals on (G and its endoscopic groups). As far as elliptic conjugacy classes are concerned, the stabilization process has been well-known thanks to Langlands and Kottwitz. In fact it has been conditional on the fundamental lemma, but the latter is recently established by Laumon, Ngô, Waldspurger and others.

However, there is an immediate obstacle due to the peculiarity of our trace formula. First of all, $G(\mathbb{A}^{\infty,p}) \times J^{(h)}(\mathbb{Q}_p)$ is a strange topological group in that it is not the set of $\mathbb{A}^\infty$-points of any reductive group over $\mathbb{Q}$. So it is not a priori clear how to adapt the stabilization process and make sense of the Langlands-Shelstad transfer at p. This problem is successfully solved in [Shi10] for PEL-Shimura varieties of unitary or symplectic type. The result has the following form.

Theorem 6.6 ([Shi10, Thm 7.2]). *Let $\varphi \in C_c^\infty(G(\mathbb{A}^{\infty,p}) \times J^{(h)}(\mathbb{Q}_p))$ be an acceptable function. For each $G_{\vec{n}} \in \mathscr{E}^{\mathrm{ell}}(G)$, one can construct a function $\phi^{\vec{n}}$ from φ such that*

$$\mathrm{tr}(\varphi | H_c(\mathrm{Ig}^{(h)}, \mathscr{L}_\xi)) = |\ker^1(\mathbb{Q}, G)| \sum_{G_{\vec{n}} \in \mathscr{E}^{\mathrm{ell}}(G)} \iota(G, G_{\vec{n}}) ST_e^{G_{\vec{n}}}(\phi^{\vec{n}}).$$

(See the end of §6.2 for nice properties enjoyed by $\phi^{\vec{n}}$.)

Sketch of idea. It suffices to handle the case $\varphi = (\prod_{v \neq p,\infty} \varphi_v) \times \varphi'_p$. Away from p and ∞, the function $\phi^{\vec{n}}_v$ is the Langlands-Shelstad transfer of φ_v via $\widetilde{\eta}_{\vec{n}}$. At $v = \infty$ one has an explicit construction using Shelstad's real endoscopy and Clozel-Delorme's pseudo-coefficients for discrete series. The most interesting and important for applications is the case of $v = p$. Here the construction of $\phi^{\vec{n}}_p$ from φ'_p has no analogue in the usual trace formula business, as we must find a natural transfer from $J^{(h)}(\mathbb{Q}_p)$ (not $G(\mathbb{Q}_p)$) to $G_{\vec{n}}(\mathbb{Q}_p)$ which is an endoscopic group of $G(\mathbb{Q}_p)$ (but typically not of $J^{(h)}(\mathbb{Q}_p)$).

The idea is that there is a natural finite set of groups $\{M_H\}$ where each M_H is simultaneously an endoscopic group of $J^{(h)}$ and a Levi subgroup of $G_{\vec{n}}$. For each M_H, φ'_p transfers to M_H by the Langlands-Shelstad transfer and then to $G_{\vec{n}}$ by a certain non-standard transfer (which makes sense if φ' is acceptable in the same sense as in Theorem 6.3). Then $\varphi^{\vec{n}}_p$ is constructed as the signed sum of these transfers over the set of all M_H which intervene. □

After all the stable trace formula above will be used to extract spectral information. Thanks to a concrete description of $\phi^{\vec{n}}$ as sketched above (via the Langlands-Shelstad transfer or other means), the following is known (when φ and $\phi^{\vec{n}}$ admit product decompositions). The notation $\mathrm{Red}^{(h)}_{\vec{n}}$ will be defined in the next subsection when $\vec{n} = (n)$ and $\vec{n} = (n-1, 1)$. These are the only cases which concern us.

Proposition 6.7. (1) *When $v \neq p, \infty$, the identity in Proposition 3.1 holds (whenever case (i) or (ii) applies).*

(2) ([Shi11, Lem 5.10]) *When $v = p$, we have for all $\pi_{H,v} \in \mathrm{Irr}(G_{\vec{n}}(\mathbb{Q}_p))$,*[4]

$$\mathrm{tr}\left(\mathrm{Red}^{(h)}_{\vec{n}}(\pi_{H,v})\right)(\varphi'_p) = \mathrm{tr}\pi_{H,v}(\phi^{\vec{n}}_p).$$

(3) *When $v = \infty$, the trace of $\phi^{\vec{n}}_\infty$ on any discrete series of $G_{\vec{n}}(\mathbb{R})$ is given explicitly.*

[4]Here H is just another name for the endoscopic group $G_{\vec{n}}$; in fact it was already used in Proposition 3.1. This notation is convenient in the next subsection where $\vec{n} = (n-1, 1)$ is considered.

6.3 The maps $\mathrm{Red}_n^{(h)}$ and $\mathrm{Red}_{n-1,1}^{(h)}$

The main reference for this subsection is [Shi11, §5.5]. Recall that

$$J^{(h)}(\mathbb{Q}_p) \simeq \mathbb{Q}_p^\times \times (D_{n-h,F_w}^\times \times GL_h(F_w)) \times \prod_{x|u,\ x\neq w} GL_n(F_x),$$

$$G(\mathbb{Q}_p) \simeq \mathbb{Q}_p^\times \times \prod_{x|u} GL_n(F_x).$$

Denote by

$$\mathrm{LJ}_{n-h,h} : \mathrm{Groth}(GL_{n-h}(F_w) \times GL_h(F_w)) \to \mathrm{Groth}(D_{n-h,F_w}^\times \times GL_h(F_w))$$

Badulescu's Jacquet-Langlands map ([Bad07]) on the first factor and the identity map on the second. For $\pi_w \in \mathrm{Irr}(GL_n(F_w))$, set

$$\mathrm{Red}_{n;w}^{(h)}(\pi_w) := \mathrm{LJ}_{n-h,h}(\mathrm{Jac}_{n-h,h}(\pi_w))$$

where $\mathrm{Jac}_{n-h,h}$ is the Jacquet module from GL_n to $GL_{n-h} \times GL_h$. Define

$$\mathrm{Red}_n^{(h)} : \mathrm{Groth}(G(\mathbb{Q}_p)) \to \mathrm{Groth}(J^{(h)}(\mathbb{Q}_p))$$

so that for irreducible $\pi_p = \pi_{p,0} \otimes (\otimes_{x|u}\pi_x)$,

$$\mathrm{Red}_n^{(h)}(\pi_p) := \pi_{p,0} \otimes \mathrm{Red}_{n;w}^{(h)}(\pi_w) \otimes \left(\bigotimes_{x|u, x\neq w} \pi_x \right).$$

Remark 6.8. We are not being precise about normalization, for instance that of the Jacquet modules and parabolic inductions in this subsection (and other places). Also we dropped the sign appearing in [Shi11, §5.5] as the final result will be stated up to sign. See [Shi11, §5.5] for precise normalizations and signs.

The definition of $\mathrm{Red}_{n-1,1}^{(h)}$ is more technical as it is supposed to account for endoscopic terms. Recall

$$G_{n-1,1}(\mathbb{Q}_p) \simeq \mathbb{Q}_p^\times \times \prod_{x|u} (GL_{n-1}(F_x) \times GL_1(F_x)).$$

Let us define its w-part

$$\mathrm{Red}_{n-1,1;w}^{(h)} : \mathrm{Groth}(GL_{n-1}(F_w) \times GL_1(F_w)) \to \mathrm{Groth}(D_{n-h,F_w}^\times \times GL_h(F_w))$$

$$\pi_{w,1} \otimes \pi_{w,2}$$

$$\mapsto \begin{cases} 0, & \text{if } h = 0, \\ \mathrm{Ind}_{GL_{n-h,h-1,1}}^{GL_{n-h,h}} \left(\mathrm{LJ}_{n-h,h-1}(\mathrm{Jac}_{n-h,h-1}(\pi_{w,1})) \otimes \pi_{w,2}\right), & \text{if } 0 < h < n-1, \\ \mathrm{Ind}_{GL_{n-h,h-1,1}}^{GL_{n-h,h}} \left(\mathrm{LJ}_{n-h,h-1}(\mathrm{Jac}_{n-h,h-1}(\pi_{w,1})) \otimes \pi_{w,2}\right) - \pi_{w,2} \otimes \pi_{w,1}, & \text{if } h = n-1. \end{cases}$$

The notation $\mathrm{Ind}^{GL_{n-h,h}}_{GL_{n-h,h-1,1}}$ means the obvious parabolic induction from $GL_{n-h} \times GL_{h-1} \times GL_1$ to $GL_{n-h} \times GL_h$. Again we avoid the issue of precise sign and normalization. Finally, extend $\mathrm{Red}^{(h)}_{n-1,1;w}$ by the identity map outside w to define

$$\mathrm{Red}^{(h)}_{n-1,1} : \mathrm{Groth}(G_{n-1,1}(\mathbb{Q}_p)) \to \mathrm{Groth}(J^{(h)}(\mathbb{Q}_p))$$

for irreducible $\pi_{H,p} = \pi_{H,p,0} \otimes \pi_{H,w} \otimes (\otimes_{x|u,x\neq w} \pi_{H,x,1} \otimes \pi_{H,x,2})$ by

$$\mathrm{Red}^{(h)}_{n-1,1}(\pi_{H,p}) := \pi_{H,p,0} \otimes \mathrm{Red}^{(h)}_{n-1,1;w}(\pi_{H,w}) \otimes \bigotimes_{x|u,x\neq w} \mathrm{Ind}^{GL_n}_{GL_{n-1,1}}(\pi_{H,x,1} \otimes \pi_{H,x,2}). \tag{6.3}$$

Example 6.9. Let $\pi_{w,1} \in \mathrm{Irr}(GL_{n-1}(F_w))$, $\pi_{w,2} \in \mathrm{Irr}(GL_1(F_w))$ be supercuspidal representations. Let $\pi_w := \pi_{w,1} \boxplus \pi_{w,2}$. (The induction is always irreducible as $n \geq 3$.) Then the above formulas yield

$$\mathrm{Red}^{(h)}_{n-1,1;w}(\pi_w) = \begin{cases} 0, & \text{if } h \neq 1, n-1, \\ \mathrm{LJ}(\pi_{w,1}) \otimes \pi_{w,2}, & \text{if } h = 1, \\ \pi_{w,2} \otimes \pi_{w,1}, & \text{if } h = n-1. \end{cases} \tag{6.4}$$

$$\mathrm{Red}^{(h)}_{n-1,1;w}(\pi_{w,1} \otimes \pi_{w,2}) = \begin{cases} 0, & \text{if } h \neq 1, n-1, \\ \mathrm{LJ}(\pi_{w,1}) \otimes \pi_{w,2}, & \text{if } h = 1, \\ -\pi_{w,2} \otimes \pi_{w,1}, & \text{if } h = n-1. \end{cases} \tag{6.5}$$

Remark 6.10. It may appear that $\mathrm{Red}^{(h)}_{n-1,1}$ is a very unnatural map, but it is not. It is the signed sum of two functorial transfers from $G_{n-1,1}$ to $J^{(h)}$ represented by the L-morphisms which occur naturally in the stabilization problem of §6.2. Since only inner forms of general linear groups are involved, the transfers may be made explicit, and thereby $\mathrm{Red}^{(h)}_{n-1,1}$ was obtained in [Shi11].

6.4 Application of the twisted trace formula

Denote by $\pi_p \in \mathrm{Irr}(G(\mathbb{Q}_p))$ a representation such that $\mathrm{BC}(\pi_p) \simeq \Pi_p$. (Such a π_p is unique up to isomorphism as p splits in E.) When m is even, define $\pi_{H,p} \in \mathrm{Irr}(G_{n-1,1}(\mathbb{Q}_p))$ such that $\mathrm{BC}(\pi_{H,p}) \simeq \psi_p \otimes \Pi_{1,p} \otimes \Pi_{2,p}$ in the notation of §4.3. In the latter case we recall (4.1), which says in particular

$$\Pi^1_w = \mathrm{Ind}(\Pi_{1,w} \otimes \Pi_{2,w}). \tag{6.6}$$

(The parabolic induction can be shown to be irreducible.) Recall that $H_c(\mathrm{Ig}^{(h)}, \mathscr{L}_\xi)\{\Pi^{\infty,p}\}$ was defined in (5.2).

Theorem 6.11. *For each $0 \leq h \leq n-1$, the following equalities hold in $\mathrm{Groth}(J^{(h)}(\mathbb{Q}_p))$, where the sign e_2 is independent of h. The constants in the formulas are some explicit positive integers.*

(1) *(m odd)*

$$H_c(\mathrm{Ig}^{(h)}, \mathscr{L}_\xi))\{\Pi^{\infty,p}\} = (\mathrm{const.}) \cdot [\mathrm{Red}^{(h)}_n(\pi_p)]. \tag{6.7}$$

(2) *(m even)*

$$H_c(\mathrm{Ig}^{(h)}, \mathscr{L}_\xi)\{\Pi^{\infty,p}\} = (\mathrm{const.})\cdot\frac{1}{2}\left[(\mathrm{Red}_n^{(h)}(\pi_p) + e_2\mathrm{Red}_{n-1,1}^{(h)}(\pi_{H,p}))\right]. \quad (6.8)$$

Sketch of proof. Suppose that φ admits a product decomposition and that each $\phi^{\vec{n}}$ in Theorem 6.6 is a transfer of $f^{\vec{n}}$ in base change (§2.3). Then the following identity is essentially due to Labesse:

$$ST_e^{G_{\vec{n}}}(\phi^{\vec{n}}) = \widetilde{T}^{\mathbb{G}_{\vec{n}}}(f^{\vec{n}})$$

where the right hand side is the twisted trace formula for $\mathbb{G}_{\vec{n}}$ (with respect to θ in §2.2). Thus Theorem 6.6 implies that

$$\begin{aligned}\mathrm{tr}(\varphi|H_c(\mathrm{Ig}^{(h)}, \mathscr{L}_\xi)) &= |\ker^1(\mathbb{Q}, G)|\sum_{\vec{n}} \iota(G, G_{\vec{n}})\widetilde{T}^{\mathbb{G}_{\vec{n}}}(f^{\vec{n}}) \quad (6.9)\\ &\sim \widetilde{T}^{\mathbb{G}_n}(f^n) + \frac{1}{2}\widetilde{T}^{\mathbb{G}_{n-1,1}}(f^{n-1,1}) + (\text{other terms}).\end{aligned}$$

The notation $\sim$ indicates that $|\ker^1(\mathbb{Q}, G)|$ is ignored. The spectral expansion of the twisted trace formula looks like

$$\widetilde{T}^{\mathbb{G}_n}(f^n) = \sum_{\Pi'} \widetilde{\mathrm{tr}}\,\Pi'(f^n) + \frac{1}{2}\sum_{\Pi'_M} \widetilde{\mathrm{tr}}(\mathrm{Ind}(\Pi'_M))(f^n) + (\text{other terms}).$$

The first (resp. second) sum runs over θ-stable automorphic representations of Π' of $GL_n(\mathbb{A}_F)$ (resp. Π'_M of $(GL_{n-1}\times GL_1)(\mathbb{A}_F)$). The twisted trace with respect to θ is denoted by $\widetilde{\mathrm{tr}}$. When $\vec{n} = (n-1, 1)$,

$$\widetilde{T}^{\mathbb{G}_{n-1,1}}(f^{n-1,1}) = \sum_{\Pi'_H} \widetilde{\mathrm{tr}}\,\Pi'_H(f^{n-1,1}) + (\text{other terms}).$$

What (3.4) means for us is essentially (ignoring the character twist there)

$$\widetilde{\mathrm{tr}}(\Pi'_H)^{\infty,p}((f^{n-1,1})^{\infty,p}) = \widetilde{\mathrm{tr}}\,\mathrm{Ind}((\Pi'_H)^{\infty,p})((f^n)^{\infty,p}).$$

The identity holds outside p and ∞ because everything is the usual transfer along the way outside p, ∞ but there are some deviations from the usual transfer at p and ∞, as we have seen in the stabilization process.

On the other hand, the base change identities in Proposition 2.2 allows to rewrite the left hand side of (6.9) as

$$\mathrm{tr}(\varphi|H_c(\mathrm{Ig}^{(h)}, \mathscr{L}_\xi)) = \mathrm{tr}((f^n)^{\infty,p}\varphi'_p|BC^p(H_c(\mathrm{Ig}^{(h)}, \mathscr{L}_\xi))) \quad (6.10)$$

where BC^p is the map applying local base change away from p and ∞.

Now we look back at the situation of §4.3 and suppose that m is odd. We can separate the $\Pi^{\infty,p}$-part from the two sides of (6.9) (with (6.10) applied to the left hand side) by varying test functions outside p and ∞. Then only the $\Pi' = \Pi$ term survives on the right hand side. (For this we appeal to the strong multiplicity one theorem of Jacquet and Shalika.) Hence

$$\mathrm{tr}(\varphi'_p|H_c(\mathrm{Ig}^{(h)}, \mathscr{L}_\xi)\{\Pi^{\infty,p}\}) \sim \widetilde{\mathrm{tr}}\Pi_p(f_p^n)\cdot\widetilde{\mathrm{tr}}\,\Pi_\infty(f_\infty^n).$$

Propositions 2.2.(2) and 6.7.(2) tell us that

$$\widetilde{\mathrm{tr}}\Pi_p(f_p^n) = \mathrm{tr}\pi_p(\phi_p^n) = \mathrm{tr}\mathrm{Red}_n^{(h)}(\pi_p)(\varphi'_p) \tag{6.11}$$

and $\widetilde{\mathrm{tr}}\Pi_\infty(f_\infty^n)$ turns out to be a constant (depending only on ξ). Hence we obtain

$$\mathrm{tr}(\varphi'_p|H_c(\mathrm{Ig}^{(h)}, \mathscr{L}_\xi)\{\Pi^{\infty,p}\}) \sim \mathrm{tr}\mathrm{Red}_n^{(h)}(\pi_p)(\varphi'_p).$$

Since φ'_p can be chosen to be an arbitrary acceptable function, Lemma 6.4 concludes the proof.

It remains to treat the case when m is even. Again we can separate the $\Pi^{\infty,p}$-part from the two sides of (6.9) and notice that only the terms for $\Pi'_M = \Pi_M$ and $\Pi'_H = \Pi_M$ survive. Hence

$$\begin{aligned}\mathrm{tr}(\varphi'_p|H_c(\mathrm{Ig}^{(h)}, \mathscr{L}_\xi)\{\Pi^{\infty,p}\}) \sim{}& \frac{1}{2}\widetilde{\mathrm{tr}}\mathrm{Ind}(\Pi_{M,p})(f_p^n)\cdot\widetilde{\mathrm{tr}}\mathrm{Ind}(\Pi_{M,\infty})(f_\infty^n)\\ &+\frac{1}{2}\widetilde{\mathrm{tr}}\Pi_{M,p}(f_p^{n-1,1})\cdot\widetilde{\mathrm{tr}}\Pi_{M,\infty}(f_\infty^{n-1,1}).\end{aligned}$$

By applying Propositions 2.2.(2) and 6.7.(2) to the second term at p, we obtain

$$\widetilde{\mathrm{tr}}\Pi_{M,p}(f_p^{n-1,1}) = \mathrm{tr}\pi_{H,p}(\phi_p^{n-1,1}) = \mathrm{tr}\mathrm{Red}_{n-1,1}^{(h)}(\pi_{H,p})(\varphi'_p).$$

One can compute that $\widetilde{\mathrm{tr}}\mathrm{Ind}(\Pi_{M,\infty})(f_\infty^n)$ and $\widetilde{\mathrm{tr}}\Pi_{M,\infty}(f_\infty^{n-1,1})$ are constants which coincide up to ±1. Assign $e_2 = 1$ if they are the same and $e_2 = -1$ otherwise. By the above identity and (6.11),

$$\mathrm{tr}(\varphi'_p|H_c(\mathrm{Ig}^{(h)}, \mathscr{L}_\xi)\{\Pi^{\infty,p}\}) \sim \frac{1}{2}\mathrm{tr}\mathrm{Red}_n^{(h)}(\pi_p)(\varphi'_p) + \frac{1}{2}e_2\cdot\mathrm{tr}\mathrm{Red}_{n-1,1}^{(h)}(\pi_{H,p})(\varphi'_p).$$

The proof is finished by Lemma 6.4 as in the case when m is odd. □

Remark 6.12. In work of Harris-Taylor, where no endoscopy arises, Theorem [HT01, Thm V.5.4] corresponds to part (1) of Theorem 6.11. In fact their theorem takes the form

$$H_c(\mathrm{Ig}^{(h)}, \mathscr{L}_\xi) = (\mathrm{const.})\cdot\mathrm{Red}_n^{(h)}H(\mathrm{Sh}, \mathscr{L}_\xi) \tag{6.12}$$

and is justified by the comparison of the trace formulas for $H_c(\mathrm{Ig}^{(h)}, \mathscr{L}_\xi)$ and $H(\mathrm{Sh}, \mathscr{L}_\xi)$. (For the latter, the Arthur-Selberg trace formula suffices as the Galois action is to be forgotten in the identity.) As long as no endoscopy occurs, (6.12) generalizes ([Shi12, Thm 6.7]). In the presence of endoscopy, a simple identity like (6.12) is not expected and the type of argument in the above proof seems to be more effective.

7 Proof of Theorem 4.6 under hypotheses

In the remainder we finish the proof of Theorem 4.6 under running hypotheses, namely Hypotheses 2.1, 4.2, 4.5, 4.7 and 4.9. This achieves the goal of the article: it was already explained how Theorem 4.6 implies Theorem 1.3 under the same running hypotheses, and §4.5 gave a short outline and references for a strategy to get rid of those assumptions.

The main ingredients of this section are the computation of $\mathrm{Mant}_n^{(h)}$ (§5.4), Corollary 5.14 and Theorem 6.11. As in the previous sections the argument will be sketched while overlooking some delicate points. A precise treatment can be found in section 6.2 of [Shi11], especially the proof of Theorem 6.4 there.

7.1 In case m is odd

Lemma 7.1 ([Shi11, Prop 2.3]). *For every $\pi_p \in \mathrm{Irr}(G(\mathbb{Q}_p))$, the following holds in* $\mathrm{Groth}(G(\mathbb{Q}_p) \times W_{F_w})$.

$$\sum_{h=0}^{n-1} \mathrm{Mant}_n^{(h)}(\mathrm{Red}_n^{(h)}(\pi_p)) = \pi_p \otimes \mathscr{L}_{F_w}(\Pi_w^1) \tag{7.1}$$

Idea of proof. By taking the explicit description of $\mathrm{Mant}_n^{(h)}$ and $\mathrm{Red}_n^{(h)}$ (§5.4, §6.3) as inputs, one proves the lemma by computations with representations of p-adic general linear groups. We omit the detail, but see Example 7.2 below. □

Corollary 5.14 and Theorem 6.11 show that

$$\pi_p \otimes R(\Pi)|_{W_{F_w}} \sim \sum_{h=0}^{n-1} \mathrm{Mant}_n^{(h)}(\mathrm{Red}_n^{(h)}(\Pi_w^1)).$$

The above lemma tells us that the latter is equal to $\pi_p \otimes \mathscr{L}_{F_w}(\Pi_w^1)$. Hence the first part of Theorem 4.6 holds.

Example 7.2. We illustrate the proof of Lemma 7.1 in a particular case. Let

$$\pi_p = \pi_{p,0} \otimes (\otimes_{x|u}\pi_x) \tag{7.2}$$

where $\pi_w = \pi_{w,1} \boxplus \pi_{w,2}$ is as in Example 6.9. According to Theorems 5.6 and 5.7,

$$\begin{aligned} \mathrm{Mant}_n^{(1)}(\mathrm{LJ}(\pi_{w,1}) \otimes \pi_{w,2}) &= \mathrm{Ind}_{GL_{n-1,1}}^{GL_n}(\mathrm{Mant}_{n-1}^{(0)}(\pi_{w,1}) \otimes \pi_{w,2}) \\ &= (\pi_{w,1} \boxplus \pi_{w,2}) \otimes \mathscr{L}_{F_w}(\pi_{w,1}). \end{aligned} \tag{7.3}$$

Similarly

$$\mathrm{Mant}_n^{(n-1)}(\pi_{w,2} \otimes \pi_{w,1}) = (\pi_{w,1} \boxplus \pi_{w,2}) \otimes \mathscr{L}_{F_w}(\pi_{w,2}). \tag{7.4}$$

In light of (6.4), the left hand side of (7.1) is computed as

$$\begin{aligned} &\pi_{p,0} \otimes \Big(\mathrm{Mant}_n^{(1)}(\mathrm{LJ}(\pi_{w,1}) \otimes \pi_{w,2}) + \mathrm{Mant}_n^{(n-1)}(\pi_{w,2} \otimes \pi_{w,1})\Big) \otimes (\otimes_{x|u}\pi_x) \\ &\quad = \pi_p \otimes (\mathscr{L}_{F_w}(\pi_{w,1}) + \mathscr{L}_{F_w}(\pi_{w,2})) = \pi_p \otimes \mathscr{L}_{F_w}(\pi_{w,1} \boxplus \pi_{w,2}) \\ &\quad = \pi_p \otimes \mathscr{L}_{F_w}(\pi_w) = \pi_p \otimes \mathscr{L}_{F_w}(\Pi_w^1). \end{aligned}$$

The last identity uses $\Pi_w^1 \simeq \pi_w$, which follows from the fact that $\Pi_p = BC(\pi_p)$ (cf. §2.3.(ii)).

7.2 In case m is even

Lemma 7.3. *For every $\pi_{H,p} \in \mathrm{Irr}(G_{n-1,1}(\mathbb{Q}_p))$, the following holds in* $\mathrm{Groth}(G(\mathbb{Q}_p) \times W_{F_w})$.

$$\sum_{h=0}^{n-1} \mathrm{Mant}_n^{(h)}(\mathrm{Red}_{n-1,1}^{(h)}(\pi_{H,p})) = \pi_p \otimes (\mathscr{L}_{F_w}(\Pi_{1,w}) - \mathscr{L}_{F_w}(\Pi_{2,w})) \tag{7.5}$$

Proof. The proof is contained in the proof of [Shi11, Thm 6.4.(ii)]. Also see Example 7.5 below. □

Remark 7.4. This is an amazing identity. Lemmas 7.1 and 7.3 demonstrate how the representations in different Newton polygon strata add up to the expected Galois representation, even in the endoscopic case.

Again by Corollary 5.14 and Theorem 6.11,

$$\pi_p \otimes R(\Pi)|_{W_{F_w}} \sim \frac{1}{2} \sum_{h=0}^{n-1} \mathrm{Mant}_n^{(h)} \left(\mathrm{Red}_n^{(h)}(\pi_p) \pm \mathrm{Red}_{n-1,1}^{(h)}(\pi_{H,p}) \right),$$

where the sign depends on e_1 and e_2 in the cited theorem. The equality (6.6) and Lemmas 7.1 and 7.3 identify the right hand side with

$$\begin{aligned} &\pi_p \otimes \left(\frac{1}{2} (\mathscr{L}_{F_w}(\Pi_{1,w} \boxplus \Pi_{2,w}) \pm (\mathscr{L}_{F_w}(\Pi_{1,w}) - \mathscr{L}_{F_w}(\Pi_{2,w}))) \right) \\ &\quad = \pi_p \otimes \left(\frac{1}{2} (\mathscr{L}_{F_w}(\Pi_{1,w}) + \mathscr{L}_{F_w}(\Pi_{2,w}) \pm (\mathscr{L}_{F_w}(\Pi_{1,w}) - \mathscr{L}_{F_w}(\Pi_{2,w}))) \right) \\ &\quad = \pi_p \otimes \mathscr{L}_{F_w}(\Pi_{1,w}) \quad \text{or} \quad \pi_p \otimes \mathscr{L}_{F_w}(\Pi_{2,w}) \end{aligned}$$

depending on the sign. This concludes the proof of Theorem 4.6 in the second case.

Example 7.5. Lemma 7.3 can be shown without pain in a particular case as follows. Take π_p as in (7.2) where π_w is as in the setting of Example 6.9. Recall that $\pi_{H,p}$ was given at the start of §6.4. If we write

$$\pi_{H,p} = \pi_{H,p,0} \otimes \pi_{H,w} \otimes (\otimes_{x|u, x \neq w} \pi_{H,x,1} \otimes \pi_{H,x,2})$$

then

$$\pi_x = \pi_{H,x,1} \boxplus \pi_{H,x,2}, \quad \forall x|u.$$

Now the left hand side of (7.5) is identified with the following, with help of (6.3), (6.5), (7.3) and (7.4).

$$\begin{aligned} &\pi_{p,0} \otimes \left(\mathrm{Mant}_n^{(1)}(\mathrm{LJ}(\pi_{w,1}) \otimes \pi_{w,2}) - \mathrm{Mant}_n^{(n-1)}(\pi_{w,2} \otimes \pi_{w,1}) \right) \otimes (\otimes_{x|u} \pi_x) \\ &\quad = \pi_p \otimes (\mathscr{L}_{F_w}(\pi_{w,1}) - \mathscr{L}_{F_w}(\pi_{w,2})) . \end{aligned}$$

References

[AC89] James Arthur and Laurent Clozel, *Simple algebras, base change, and the advanced theory of the trace formula*, Annals of Mathematics Studies, vol. 120, Princeton University Press, Princeton, NJ, 1989. MR 1007299

[Art05] James Arthur, *An introduction to the trace formula*, Harmonic analysis, the trace formula, and Shimura varieties, Clay Math. Proc., vol. 4, Amer. Math. Soc., Providence, RI, 2005, pp. 1–263. MR 2192011

[Bad07] Alexandru Ioan Badulescu, *Jacquet-Langlands et unitarisabilité*, J. Inst. Math. Jussieu **6** (2007), no. 3, 349–379. MR 2329758

[BC11] Joël Bellaïche and Gaëtan Chenevier, *The sign of Galois representations attached to automorphic forms for unitary groups*, Compos. Math. **147** (2011), no. 5, 1337–1352. MR 2834723

[BG14] Kevin Buzzard and Toby Gee, *The conjectural connections between automorphic representations and Galois representations*, Automorphic forms and Galois representations. Vol. 1, London Math. Soc. Lecture Note Ser., vol. 414, Cambridge Univ. Press, Cambridge, 2014, pp. 135–187. MR 3444225

[BH05a] Colin J. Bushnell and Guy Henniart, *The essentially tame local Langlands correspondence. I*, J. Amer. Math. Soc. **18** (2005), no. 3, 685–710. MR 2138141

[BH05b] ———, *The essentially tame local Langlands correspondence. II. Totally ramified representations*, Compos. Math. **141** (2005), no. 4, 979–1011. MR 2148193

[BLGGT14] Thomas Barnet-Lamb, Toby Gee, David Geraghty, and Richard Taylor, *Potential automorphy and change of weight*, Ann. of Math. (2) **179** (2014), no. 2, 501–609. MR 3152941

[BLGHT11] T. Barnet-Lamb, D. Geraghty, M. Harris, and R. Taylor, *A family of Calabi-Yau varieties and potential automorphy II*, P.R.I.M.S. **98** (2011), 29–98.

[Bor79] A. Borel, *Automorphic L-functions*, Automorphic forms, representations and L-functions (Proc. Sympos. Pure Math., Oregon State Univ., Corvallis, Ore., 1977), Part 2, Proc. Sympos. Pure Math., XXXIII, Amer. Math. Soc., Providence, R.I., 1979, pp. 27–61. MR 546608 (81m:10056)

[Boy09] Pascal Boyer, *Monodromie du faisceau pervers des cycles évanescents de quelques variétés de Shimura simples*, Invent. Math. **177** (2009), no. 2, 239–280. MR 2511742

[BR89] Don Blasius and Dinakar Ramakrishnan, *Maass forms and Galois representations*, Galois groups over $\mathbf{Q}$ (Berkeley, CA, 1987), Math. Sci. Res. Inst. Publ., vol. 16, Springer, New York, 1989, pp. 33–77. MR 1012167

[BR94] Don Blasius and Jonathan D. Rogawski, *Zeta functions of Shimura varieties*, Motives (Seattle, WA, 1991), Proc. Sympos. Pure Math., vol. 55, Amer. Math. Soc., Providence, RI, 1994, pp. 525–571. MR 1265563 (95e:11051)

[Car] Ana Caraiani, *Lecture notes on perfectoid Shimura varieties*, `http://swc.math.arizona.edu/aws/2017/2017CaraianiNotes.pdf`.

[Car86] Henri Carayol, *Sur les représentations l-adiques associées aux formes modulaires de Hilbert*, Ann. Sci. École Norm. Sup. (4) **19** (1986), no. 3, 409–468. MR 870690 (89c:11083)

[Car90] H. Carayol, *Nonabelian Lubin-Tate theory*, Automorphic forms, Shimura varieties, and L-functions, Vol. II (Ann Arbor, MI, 1988), Perspect. Math., vol. 11, Academic Press, Boston, MA, 1990, pp. 15–39. MR 1044827

[Car12] Ana Caraiani, *Local-global compatibility and the action of monodromy on nearby cycles*, Duke Math. J. **161** (2012), no. 12, 2311–2413. MR 2972460

[Car14] ———, *Monodromy and local-global compatibility for $l = p$*, Algebra Number Theory **8** (2014), no. 7, 1597–1646. MR 3272276

[CH13] Gaëtan Chenevier and Michael Harris, *Construction of automorphic Galois representations, II*, Camb. J. Math. **1** (2013), no. 1, 53–73. MR 3272052

[CHL11] Laurent Clozel, Michael Harris, and Jean-Pierre Labesse, *Construction of automorphic Galois representations, I*, On the stabilization of the trace formula, Stab. Trace Formula Shimura Var. Arith. Appl., vol. 1, Int. Press, Somerville, MA, 2011, pp. 497–527. MR 2856383

[CHLN11] Laurent Clozel, Michael Harris, Jean-Pierre Labesse, and Bao-Châu Ngô (eds.), *On the stabilization of the trace formula*, Stabilization of the Trace Formula, Shimura Varieties, and Arithmetic Applications, vol. 1, International Press, Somerville, MA, 2011. MR 2742611

[CLH16] Ana Caraiani and Bao V. Le Hung, *On the image of complex conjugation in certain Galois representations*, Compos. Math. **152** (2016), no. 7, 1476–1488. MR 3530448

[Clo90] Laurent Clozel, *Motifs et formes automorphes: applications du principe de fonctorialité*, Automorphic forms, Shimura varieties, and L-functions, Vol. I (Ann Arbor, MI, 1988), Perspect. Math., vol. 10, Academic Press, Boston, MA, 1990, pp. 77–159. MR 1044819 (91k:11042)

[Clo91] ———, *Représentations galoisiennes associées aux représentations automorphes autoduales de* $\mathrm{GL}(n)$, Inst. Hautes Études Sci. Publ. Math. (1991), no. 73, 97–145. MR 1114211 (92i:11055)

[Clo13] ———, *Purity reigns supreme*, Int. Math. Res. Not. IMRN (2013), no. 2, 328–346. MR 3010691

[Dat07] J.-F. Dat, *Théorie de Lubin-Tate non-abélienne et représentations elliptiques*, Invent. Math. **169** (2007), no. 1, 75–152. MR 2308851 (2008g:22022)

[Far04] Laurent Fargues, *Cohomologie des espaces de modules de groupes p-divisibles et correspondances de Langlands locales*, Astérisque (2004), no. 291, 1–199, Variétés de Shimura, espaces de Rapoport-Zink et correspondances de Langlands locales. MR 2074714

[Fuj97] Kazuhiro Fujiwara, *Rigid geometry, Lefschetz-Verdier trace formula and Deligne's conjecture*, Invent. Math. **127** (1997), no. 3, 489–533. MR 1431137

[Hai05] Thomas J. Haines, *Introduction to Shimura varieties with bad reduction of parahoric type*, Harmonic analysis, the trace formula, and Shimura varieties, Clay Math. Proc., vol. 4, Amer. Math. Soc., Providence, RI, 2005, pp. 583–642. MR 2192017

[Har01] Michael Harris, *Local Langlands correspondences and vanishing cycles on Shimura varieties*, European Congress of Mathematics, Vol. I (Barcelona, 2000), Progr. Math., vol. 201, Birkhäuser, Basel, 2001, pp. 407–427. MR 1905332

[Hen00] Guy Henniart, *Une preuve simple des conjectures de Langlands pour* $\mathrm{GL}(n)$ *sur un corps p-adique*, Invent. Math. **139** (2000), no. 2, 439–455. MR 1738446

[Hid04] Haruzo Hida, *p-adic automorphic forms on Shimura varieties*, Springer Monographs in Mathematics, Springer-Verlag, New York, 2004. MR 2055355

[HLTT16] Michael Harris, Kai-Wen Lan, Richard Taylor, and Jack Thorne, *On the rigid cohomology of certain Shimura varieties*, Res. Math. Sci. **3** (2016), Paper No. 37, 308. MR 3565594

[HR12] Thomas J. Haines and Michael Rapoport, *Shimura varieties with $\Gamma_1(p)$-level via Hecke algebra isomorphisms: the Drinfeld case*, Ann. Sci. Éc. Norm. Supér. (4) **45** (2012), no. 5, 719–785 (2013). MR 3053008

[HT01] Michael Harris and Richard Taylor, *The geometry and cohomology of some simple Shimura varieties*, Annals of Mathematics Studies, vol. 151, Princeton University Press, Princeton, NJ, 2001, With an appendix by Vladimir G. Berkovich. MR 1876802 (2002m:11050)

[Igu59] Jun-ichi Igusa, *Kroneckerian model of fields of elliptic modular functions*, Amer. J. Math. **81** (1959), 561–577. MR 0108498

[KM85] Nicholas M. Katz and Barry Mazur, *Arithmetic moduli of elliptic curves*, Annals of Mathematics Studies, vol. 108, Princeton University Press, Princeton, NJ, 1985. MR 772569

[Kna94] A. W. Knapp, *Local Langlands correspondence: the Archimedean case*, Motives (Seattle, WA, 1991), Proc. Sympos. Pure Math., vol. 55, Amer. Math. Soc., Providence, RI, 1994, pp. 393–410. MR 1265560

[Kot86] Robert E. Kottwitz, *Stable trace formula: elliptic singular terms*, Math. Ann. **275** (1986), no. 3, 365–399. MR 858284 (88d:22027)

[Kot90] ——, *Shimura varieties and λ-adic representations*, Automorphic forms, Shimura varieties, and L-functions, Vol. I (Ann Arbor, MI, 1988), Perspect. Math., vol. 10, Academic Press, Boston, MA, 1990, pp. 161–209. MR 1044820

[Kot92a] ——, *On the λ-adic representations associated to some simple Shimura varieties*, Invent. Math. **108** (1992), no. 3, 653–665. MR 1163241

[Kot92b] ——, *Points on some Shimura varieties over finite fields*, J. Amer. Math. Soc. **5** (1992), no. 2, 373–444. MR 1124982

[Kud94] Stephen S. Kudla, *The local Langlands correspondence: the non-Archimedean case*, Motives (Seattle, WA, 1991), Proc. Sympos. Pure Math., vol. 55, Amer. Math. Soc., Providence, RI, 1994, pp. 365–391. MR 1265559

[Lab11] J.-P. Labesse, *Changement de base CM et séries discrètes*, On the stabilization of the trace formula, Stab. Trace Formula Shimura Var. Arith. Appl., vol. 1, Int. Press, Somerville, MA, 2011, pp. 429–470. MR 2856380

[Lan77] R. P. Langlands, *Shimura varieties and the Selberg trace formula*, Canad. J. Math. **29** (1977), no. 6, 1292–1299. MR 0498400

[Lan79] ——, *On the zeta functions of some simple Shimura varieties*, Canad. J. Math. **31** (1979), no. 6, 1121–1216. MR 553157

[Lan89] ——, *On the classification of irreducible representations of real algebraic groups*, Representation theory and harmonic analysis on semisimple Lie groups, Math. Surveys Monogr., vol. 31, Amer. Math. Soc., Providence, RI, 1989, pp. 101–170. MR 1011897

[Lan13] Kai-Wen Lan, *Arithmetic compactifications of PEL-type Shimura varieties*, London Mathematical Society Monographs Series, vol. 36, Princeton University Press, Princeton, NJ, 2013. MR 3186092

[LR92] Robert P. Langlands and Dinakar Ramakrishnan (eds.), *The zeta functions of Picard modular surfaces*, Université de Montréal, Centre de Recherches Mathématiques, Montreal, QC, 1992. MR 1155223

[Man] Elena Mantovan, *The Newton stratification*, in this volume.

[Man05] ——, *On the cohomology of certain PEL-type Shimura varieties*, Duke Math. J. **129** (2005), no. 3, 573–610. MR 2169874

[Man08] ——, *On non-basic Rapoport-Zink spaces*, Ann. Sci. Éc. Norm. Supér. (4) **41** (2008), no. 5, 671–716. MR 2504431

[Man11] ——, *l-adic étale cohomology of PEL type Shimura varieties with non-trivial coefficients*, WIN—women in numbers, Fields Inst. Commun., vol. 60, Amer. Math. Soc., Providence, RI, 2011, pp. 61–83. MR 2777800

[Mil05] J. S. Milne, *Introduction to Shimura varieties*, Harmonic analysis, the trace formula, and Shimura varieties, Clay Math. Proc., vol. 4, Amer. Math. Soc., Providence, RI, 2005, pp. 265–378. MR 2192012

[Min11] Alberto Minguez, *Unramified representations of unitary groups*, On the stabilization of the trace formula, Stab. Trace Formula Shimura Var. Arith. Appl., vol. 1, Int. Press, Somerville, MA, 2011, pp. 389–410. MR 2856377

[Mok15] Chung Pang Mok, *Endoscopic classification of representations of quasi-split unitary groups*, Mem. Amer. Math. Soc. **235** (2015), no. 1108, vi+248. MR 3338302

[Mor10] Sophie Morel, *On the cohomology of certain noncompact Shimura varieties*, Annals of Mathematics Studies, vol. 173, Princeton University Press, Princeton, NJ, 2010, With an appendix by Robert Kottwitz. MR 2567740 (2011b:11073)

[Mor16] ——, *Construction de représentations galoisiennes de torsion [d'après Peter Scholze]*, Astérisque (2016), no. 380, Séminaire Bourbaki. Vol. 2014/2015, Exp. No. 1102, 449–473. MR 3522182

[PT15] Stefan Patrikis and Richard Taylor, *Automorphy and irreducibility of some l-adic representations*, Compos. Math. **151** (2015), no. 2, 207–229. MR 3314824

[Rap95] Michael Rapoport, *Non-Archimedean period domains*, Proceedings of the International Congress of Mathematicians, Vol. 1, 2 (Zürich, 1994), Birkhäuser, Basel, 1995, pp. 423–434. MR 1403942

[Rap05] ——, *A guide to the reduction modulo p of Shimura varieties*, 2005, Automorphic forms. I, pp. 271–318. MR 2141705 (2006c:11071)

[Rog92] Jonathan D. Rogawski, *Analytic expression for the number of points mod p*, The zeta functions of Picard modular surfaces, Univ. Montréal, Montreal, QC, 1992, pp. 65–109. MR 1155227

[RZ96] M. Rapoport and Th. Zink, *Period spaces for p-divisible groups*, Annals of Mathematics Studies, vol. 141, Princeton University Press, Princeton, NJ, 1996. MR 1393439

[Sch] Peter Scholze, *The local Langlands correspondence for* GL_n *over p-adic fields, and the cohomology of compact unitary Shimura varieties*, in this volume.

[Sch13a] ——, *The Langlands-Kottwitz approach for some simple Shimura varieties*, Invent. Math. **192** (2013), no. 3, 627–661. MR 3049931

[Sch13b] ——, *The local Langlands correspondence for* GL_n *over p-adic fields*, Invent. Math. **192** (2013), no. 3, 663–715. MR 3049932

[Sch14] ——, *Perfectoid spaces and their applications*, Proceedings of the International Congress of Mathematicians—Seoul 2014. Vol. II, Kyung Moon Sa, Seoul, 2014, pp. 461–486. MR 3728623

[Sch15] ——, *On torsion in the cohomology of locally symmetric varieties*, Ann. of Math. (2) **182** (2015), no. 3, 945–1066. MR 3418533

[Sch16] ——, *Perfectoid Shimura varieties*, Jpn. J. Math. **11** (2016), no. 1, 15–32. MR 3510678

[Shi09] Sug Woo Shin, *Counting points on Igusa varieties*, Duke Math. J. **146** (2009), no. 3, 509–568. MR 2484281

[Shi10] ——, *A stable trace formula for Igusa varieties*, J. Inst. Math. Jussieu **9** (2010), no. 4, 847–895. MR 2684263

[Shi11] ——, *Galois representations arising from some compact Shimura varieties*, Ann. of Math. (2) **173** (2011), no. 3, 1645–1741. MR 2800722

[Shi12] ——, *On the cohomology of Rapoport-Zink spaces of EL-type*, Amer. J. Math. **134** (2012), no. 2, 407–452. MR 2905002

[Sor] Claus Sorensen, *A patching lemma*, in this volume.

[SS13] Peter Scholze and Sug Woo Shin, *On the cohomology of compact unitary group Shimura varieties at ramified split places*, J. Amer. Math. Soc. **26** (2013), no. 1, 261–294. MR 2983012

[Str05] Matthias Strauch, *On the Jacquet-Langlands correspondence in the cohomology of the Lubin-Tate deformation tower*, Astérisque (2005), no. 298, 391–410, Automorphic forms. I. MR 2141708

[Tat79] J. Tate, *Number theoretic background*, Automorphic forms, representations and L-functions (Proc. Sympos. Pure Math., Oregon State Univ., Corvallis, Ore., 1977),

Part 2, Proc. Sympos. Pure Math., XXXIII, Amer. Math. Soc., Providence, R.I., 1979, pp. 3–26. MR 546607 (80m:12009)

[Tay04] Richard Taylor, *Galois representations*, Ann. Fac. Sci. Toulouse Math. (6) **13** (2004), no. 1, 73–119. MR 2060030

[TY07] Richard Taylor and Teruyoshi Yoshida, *Compatibility of local and global Langlands correspondences*, J. Amer. Math. Soc. **20** (2007), no. 2, 467–493. MR 2276777 (2007k:11193)

[Var07] Yakov Varshavsky, *Lefschetz-Verdier trace formula and a generalization of a theorem of Fujiwara*, Geom. Funct. Anal. **17** (2007), no. 1, 271–319. MR 2306659

[Wei16] Jared Weinstein, *Reciprocity laws and Galois representations: recent breakthroughs*, Bull. Amer. Math. Soc. (N.S.) **53** (2016), no. 1, 1–39. MR 3403079

THE LOCAL LANGLANDS CORRESPONDENCE FOR GL_n OVER p-ADIC FIELDS, AND THE COHOMOLOGY OF COMPACT UNITARY SHIMURA VARIETIES

PETER SCHOLZE

Mathematisches Institut der Universität Bonn, Endenicher Allee 60, 53115 Bonn, Germany
E-mail address: scholze@math.uni-bonn.de

Abstract

We explain the key ideas behind our papers on the local Langlands correspondence and the cohomology of some Shimura varieties.

1 The Local Langlands Correspondence

Fix a p-adic field F, i.e., a finite extension of $\mathbb{Q}_p$, with ring of integers $\mathcal{O}$. Recall the statement of the Local Langlands Correspondence, which is now a theorem due to Harris-Taylor, [11], and Henniart, [14].

Theorem 1.1. *There is a natural bijection*

$$\begin{aligned} &\{\text{irreducible smooth representations of } GL_n(F)\}/\cong \\ &\xrightarrow{\cong} \{n\text{-dimensional Frobenius} - \text{semisimple} \\ &\quad \text{Weil} - \text{Deligne representations}\}/\cong \\ &\pi \mapsto (\sigma(\pi), N(\pi)) \ . \end{aligned}$$

Here, recall that an n-dimensional Frobenius-semisimple Weil-Deligne representation can be described as a pair (σ, N), where σ is a semisimple n-dimensional representation of the Weil group W_F of F, and $N : \sigma \to \sigma(-1)$ is a W_F-equivariant map, where $\sigma(-1) = \sigma \otimes \chi_{\mathrm{cycl}}^{-1}$ is the Tate twist of σ; we recall that N is necessarily nilpotent.

An important part of the theorem is to pin down what is meant by the term 'natural bijection'. Usually, this is done by requiring the matching of certain numerical invariants attached to both sides of the correspondence, the L- and ϵ-factors of pairs. With this requirement, the correspondence is unique by a theorem of Henniart, [13]. This characterization is useful when working with L-functions, but has the drawback of being extremely inexplicit: In order to check whether for a given π, one has constructed the right $(\sigma(\pi), N(\pi))$, one has to check matching of ϵ-factors of pairs for all π' coming from smaller GL_k's. In particular, one has to understand completely the local Langlands correspondence for smaller GL_k's.

We establish a new local characterization of the local Langlands correspondence, that gives at least in principle a formula for the character of $\sigma(\pi)$ in terms of the character of π.[1]

Date: February 6, 2017.

[1] We recall that $N(\pi)$ can be described through the Bernstein-Zelevinsky classification, [4], and hence we will ignore it in our discussion.

This will use the Bernstein center of $\mathrm{GL}_n(F)$. We recall that the Bernstein center, [5], can be described geometrically as conjugation-invariant distributions acting on $C_c^\infty(\mathrm{GL}_n(F))$. Spectrally, elements of the Bernstein center act by a scalar on any irreducible smooth representation, and one can describe explicitly which functions on the set of irreducible smooth representations come from elements of the Bernstein center.

Now, let $\tau \in W_F$ be any element. Using the local Langlands correspondence, one can describe an element f_τ of the Bernstein center of $\mathrm{GL}_n(F)$ by requiring that it acts as the scalar $\mathrm{tr}(\tau|\sigma(\pi))$ on any irreducible smooth representation π. In the following, we will describe a geometric construction of f_τ, considered as a conjugation-invariant distribution on $C_c^\infty(\mathrm{GL}_n(F))$. Concretely, this means that for any $h \in C_c^\infty(\mathrm{GL}_n(F))$, we have to construct the convolution $f_\tau * h$.

1.1 A geometric characterization of the Local Langlands Correspondence

In the following, we assume for ease of exposition that $F = \mathbb{Q}_p$. Recall that a conjecture of Carayol, now a theorem of Harris and Taylor, states that the local Langlands correspondence for supercuspidal π is realized in the cohomology of the Lubin-Tate tower. Here, the Lubin-Tate tower is the deformation space of the 1-dimensional formal group of height n over $\bar{\mathbb{F}}_p$.

We want to extend this conjecture to a conjecture describing the local Langlands correspondence for all π, not necessarily supercuspidal, and to make this possible, we have to consider the deformation spaces of all 1-dimensional p-divisible groups of height n. Recall that a 1-dimensional formal group of height n in characteristic p is the same as a connected 1-dimensional p-divisible group of height n. Moreover, we will consider such deformation spaces for groups defined over finite extensions $\mathbb{F}_{p^r}$ of $\mathbb{F}_p$, not just over the algebraic closure $\bar{\mathbb{F}}_p$.

Fix an integer $r \geq 1$, and let $\tau \in W_{\mathbb{Q}_p}$ be an element acting as the r-th power of geometric Frobenius on the residue field $\bar{\mathbb{F}}_p$. By Dieudonné theory, 1-dimensional p-divisible groups H of height n over $\mathbb{F}_{p^r}$ can be described as elements

$$\delta \in \mathrm{GL}_n(\mathbb{Z}_{p^r})\mathrm{diag}(p^{-1}, 1, \ldots, 1)\,\mathrm{GL}_n(\mathbb{Z}_{p^r}),$$

well-defined up to σ-conjugation by $\mathrm{GL}_n(\mathbb{Z}_{p^r})$. Here, $\mathbb{Z}_{p^r} = W(\mathbb{F}_{p^r})$, on which σ acts, lifting the p-th power map on $\mathbb{F}_{p^r}$. For this, we use covariant Dieudonné theory: The covariant Dieudonné module $M(H)$ of H can be trivialized as $\mathbb{Z}_{p^r}$-module to $M(H) \cong \mathbb{Z}_{p^r}^n$, and then the Frobenius operator F can be written as $F = p\delta\sigma$, for some δ in the given double coset. Changing the isomorphism $M(H) \cong \mathbb{Z}_{p^r}^n$ amounts to σ-conjugation $\delta \mapsto g^{-1}\delta g^\sigma$, for some $g \in \mathrm{GL}_n(\mathbb{Z}_{p^r})$.[2]

Fix some δ, giving rise to a p-divisible group H_δ over $\mathbb{F}_{p^r}$. Using the results of Illusie, [17], and Drinfeld, [8], one sees that the universal deformation ring R_δ of H_δ is isomorphic to $\mathbb{Z}_{p^r}[[T_1, \ldots, T_{n-1}]]$, and that for any $m \geq 1$, there is a finite covering $R_{\delta,m}$ of R_δ by a regular semilocal ring $R_{\delta,m}$, parametrizing Drinfeld-level-p^m-structures. Let H_δ^{univ} denote

[2] Here, we use the normalization of δ used in the paper [25], which is slightly different from the one used in [26].

the universal deformation of H_δ to R_δ; then Drinfeld-level-p^m-structures are given by sections $X_1, \ldots, X_n \in H_\delta^{\mathrm{univ}}[p^m]$ such that

$$[H_\delta^{\mathrm{univ}}[p^m]] = \sum_{i_1,\ldots,i_n \in \mathbb{Z}/p^m\mathbb{Z}} [i_1X_1 + \ldots + i_nX_n]$$

as relative Cartier divisors on $H_\delta^{\mathrm{univ}}/R_\delta$. After inverting p, the cover $R_{\delta,m}/R_\delta$ is finite étale and Galois with Galois group $\mathrm{GL}_n(\mathbb{Z}/p^m\mathbb{Z})$.

From theorems of Illusie, [17], (generalized by Faltings, [9], to general F) and Artin's algebraization theorem, one deduces the following important algebraization result. The key input is that R_δ is also the versal deformation space of $H_\delta[p^m]$ for $m \geq 1$.

Theorem 1.2. *There is a regular flat $\mathbb{Z}_{p^r}$-algebra $\mathcal{R}_{\delta,m}$ of finite type with an action of $\mathrm{GL}_n(\mathbb{Z}/p^m\mathbb{Z})$, an m-truncated Barsotti-Tate group $\mathcal{H}_{\delta,m}/\mathcal{R}_{\delta,m}$ and sections $\mathcal{X}_1, \ldots, \mathcal{X}_n \in \mathcal{H}_{\delta,m}$, and a finite closed subscheme*

$$Z \subset \mathrm{Spec}\ (\mathcal{R}_{\delta,m} \otimes \mathbb{F}_{p^r})$$

such that the completion of $(\mathcal{R}_{\delta,m}, \mathcal{H}_{\delta,m}, \mathcal{X}_1, \ldots, \mathcal{X}_n)$ along Z is $\mathrm{GL}_n(\mathbb{Z}/p^m\mathbb{Z})$-equivariantly isomorphic to $(R_{\delta,m}, H_\delta^{\mathrm{univ}}[p^m], X_1, \ldots, X_n)$. Moreover, the same algebraization can be chosen for all δ' in a small neighborhood of δ, more precisely the algebraization depends only on $H_\delta[p^m]$.

Let $X_{\delta,m}$ be the generic fibre of Spf $R_{\delta,m}$, considered as a rigid-analytic variety over the fraction field $\mathbb{Q}_{p^r}$ of $\mathbb{Z}_{p^r}$. Consider the étale cohomology groups (cf. e.g. Huber, [15]), where we fix $\ell \neq p$,

$$H_\delta^i = \varinjlim_m H_{\mathrm{\acute{e}t}}^i(X_{\delta,m} \hat{\otimes}_{\mathbb{Q}_{p^r}} \mathbb{C}_p, \bar{\mathbb{Q}}_\ell) .$$

They carry an action of $W_{\mathbb{Q}_{p^r}} \times \mathrm{GL}_n(\mathbb{Z}_p)$. From the algebraization theorem and [16], one deduces that the action of $W_{\mathbb{Q}_{p^r}}$ is continuous, and that the action of $\mathrm{GL}_n(\mathbb{Z}_p)$ is admissible.

Up to this point, we have adapted the definition of the Lubin-Tate tower to general 1-dimensional p-divisible groups, and have defined the corresponding cohomology groups. Now basically the function f_τ will be defined by means of Grothendieck's sheaf-function correspondence.

Definition 1.3. Recall the element $\tau \in W_{\mathbb{Q}_p}$ acting as $x \mapsto x^{p^{-r}}$ on the residue field, and fix some $h \in C_c^\infty(\mathrm{GL}_n(\mathbb{Z}_p))$ with values in $\mathbb{Q}$. Then define a function $\phi_{\tau,h}$ on $\mathrm{GL}_n(\mathbb{Q}_{p^r})$ with values in $\bar{\mathbb{Q}}_\ell$ by setting

$$\phi_{\tau,h}(\delta) = \mathrm{tr}(\tau \times h | H_\delta^*)$$

for δ as above, and 0 else.

From the algebraization theorem, and a result of Mieda on independence of ℓ, [21], one finds the following result.

Theorem 1.4. *The function $\phi_{\tau,h}$ is locally constant with compact support, with values in $\mathbb{Q}$, independent of ℓ, thus defining a function $\phi_{\tau,h} \in C_c^\infty(\mathrm{GL}_n(\mathbb{Q}_{p^r}))$.*

Let $f_{\tau,h} \in C_c^\infty(\mathrm{GL}_n(\mathbb{Q}_p))$ be a base-change transfer of $\phi_{\tau,h}$, cf. [1]. Roughly, this means that the orbital integrals of $f_{\tau,h}$ match up with the twisted orbital integrals of $\phi_{\tau,h}$. We note that the existence of a base-change transfer is proved by local means in the case of GL_n. We note that $f_{\tau,h}$ is not well-defined, but its orbital integrals are. This implies that the trace of $f_{\tau,h}$ on any representation is well-defined as well, by Weyl's integration formula.

Now our main theorem says that up to shifts and duals, $f_{\tau,h}$ is the same as $f_\tau * h$, in the sense that their orbital integrals agree. In fact, with suitable modification of the previous discussion, everything applies to general F, so we formulate the result for general F.

Theorem 1.5 ([26]). (a) *For any irreducible smooth representation π of $\mathrm{GL}_n(F)$ there is a unique semisimple n-dimensional representation $\mathrm{rec}(\pi)$ (up to isomorphism) of W_F such that for all $\tau \in W_F$ and $h \in C_c^\infty(\mathrm{GL}_n(\mathcal{O}))$ as above,*

$$\mathrm{tr}(f_{\tau,h}|\pi) = \mathrm{tr}(\tau|\mathrm{rec}(\pi))\mathrm{tr}(h|\pi)\ .$$

Write $\sigma(\pi) = \mathrm{rec}(\pi)^\vee(\frac{n-1}{2})$.

(b) *If π is a subquotient of the normalized parabolic induction of the irreducible representation $\pi_1 \otimes \cdots \otimes \pi_t$ of $\mathrm{GL}_{n_1}(F) \times \cdots \times \mathrm{GL}_{n_t}(F)$, then $\sigma(\pi) = \sigma(\pi_1) \oplus \ldots \oplus \sigma(\pi_t)$.*

(c) *The map $\pi \longmapsto \sigma(\pi)$ induces a bijection between the set of isomorphism classes of supercuspidal irreducible smooth representations of $\mathrm{GL}_n(F)$ and the set of isomorphism classes of irreducible n-dimensional representations of W_F.*

(d) *The bijection defined in* (c) *is compatible with twists, central characters, duals, and L- and ϵ-factors of pairs, hence is the standard correspondence.*

We remark that the normalizations here are slightly different from the one in [26], and are now made compatible with those in [25].

The uniqueness assertion in (a) is clear, as the condition determines $\mathrm{tr}(\tau|\mathrm{rec}(\pi))$ if τ projects to a positive power of geometric Frobenius, and it is well-known that this determines the representation up to semisimplification. Hence this theorem gives a new local characterization of the Local Langlands Correspondence, which can be informally summarized by saying that the Local Langlands Correspondence is realized in the cohomology of the moduli space of one-dimensional p-divisible groups of height n, for all irreducible smooth representations. Also note that the extraneous division algebra acting in the Lubin-Tate setting disappears in our formulation.

1.2 Local-global compatibility

Using our characterization of the local Langlands correspondence, it is not hard to show that it is compatible with global correspondences. This was done previously by Harris and Taylor, [11], using Igusa varieties and counting points for them. In our approach, all we need is Kottwitz's paper [20] describing the points of the special fibre in the case of good reduction.

Let Sh_K, $K \subset G(\mathbb{A}_f)$, be a Harris-Taylor type Shimura variety associated to $G = GU(1, n-1)$, for an imaginary-quadratic extension $E/\mathbb{Q}$, and a division algebra D over E.

In particular, Sh_K is a projective variety over E. We assume that p splits in E, fix a place $\mathfrak{p}$ of E above p, and we assume that D is split at $\mathfrak{p}$. In particular, $G_{\mathbb{Q}_p} \cong \mathrm{GL}_n \times \mathbb{G}_m$. In the following, we will often ignore the $\mathbb{G}_m$-factor, which causes little trouble. Let

$$H^*_{\mathrm{Sh}} = \varinjlim_K H^*(\mathrm{Sh}_K \otimes \bar{\mathbb{Q}}, \bar{\mathbb{Q}}_\ell)$$

denote the cohomology of the Shimura variety, where we fix some $\ell \neq p$.

Fix the hyperspecial maximal compact subgroup $K_p^0 = \mathrm{GL}_n(\mathbb{Z}_p) \times \mathbb{Z}_p^\times \subset G(\mathbb{Q}_p)$, and some sufficiently small level K^p away from p. Then, cf. e.g. [20], there is proper smooth model $\mathcal{M}_{K^p}$ of $\mathrm{Sh}_{K_p^0 K^p}$ over $\mathcal{O}_{E,\mathfrak{p}} \cong \mathbb{Z}_p$. By the theorem of Serre and Tate and the choice of the Shimura variety (specifically, signature $(1, n-1)$), for any point $x \in \mathcal{M}_{K^p}(\mathbb{F}_{p^r})$, the completion $\widehat{\mathcal{M}_{K^p,x}}$ of $\mathcal{M}_{K^p}$ at x is isomorphic to Spf R_δ, where δ corresponds to (a part of) the p-divisible group of the universal abelian variety at x.

Let τ and h be as above, and $f^p \in C_c^\infty(G(\mathbb{A}_f^p))$. Fix some open normal subgroup $K_p \subset K_p^0$ such that h is biinvariant under K_p, and let $\pi : \mathrm{Sh}_{K_p K^p} \to \mathrm{Sh}_{K_p^0 K^p}$ be the natural projection. Then the proper base-change theorem implies that

$$\mathrm{tr}(\tau \times h f^p | H^*_{\mathrm{Sh}}) = \mathrm{tr}(c | H^*(\mathcal{M}_{K^p} \otimes \bar{\mathbb{F}}_p, R\psi\pi_*\bar{\mathbb{Q}}_\ell)) ,$$

where c is some correspondence, and $R\psi$ denotes the nearby cycle sheaves, associated to the local system $\pi_*\bar{\mathbb{Q}}_\ell$ in this case. Essentially, we just used that

$$H^*(\mathrm{Sh}_{K_p K^p} \otimes \bar{\mathbb{Q}}, \bar{\mathbb{Q}}_\ell) \cong H^*(\mathrm{Sh}_{K_p^0 K^p}, \pi_*\bar{\mathbb{Q}}_\ell) \cong H^*(\mathcal{M}_{K^p} \otimes \bar{\mathbb{F}}_p, R\psi\pi_*\bar{\mathbb{Q}}_\ell) .$$

We note that if $K_p = (1 + p^m M_n(\mathbb{Z}_p)) \times \mathbb{Z}_p^\times$, $m \geq 1$, then $\mathrm{Sh}_{K_p K^p} \to \mathrm{Sh}_{K_p^0 K^p}$ looks locally like $X_{\delta,m} \to X_\delta$, where X_δ is the generic fibre of Spf R_δ. In particular, for $x \in \mathcal{M}_{K^p}(\mathbb{F}_{p^r})$ and a geometric point $\bar{x}$ above x, we have an identification

$$(R\psi\pi_*\bar{\mathbb{Q}}_\ell)_{\bar{x}} \cong H^*(X_{\delta,m}, \bar{\mathbb{Q}}_\ell) .$$

Now we apply the Lefschetz trace formula (as given in [30]) to find that

$$\mathrm{tr}(c | H^*(\mathcal{M}_{K^p} \otimes \bar{\mathbb{F}}_p, R\psi\pi_*\bar{\mathbb{Q}}_\ell)) = \sum_{x \in \mathrm{Fix}(c | \mathcal{M}_{K^p}(\bar{\mathbb{F}}_p))} \mathrm{tr}(c_x | (R\psi\pi_*\bar{\mathbb{Q}}_\ell)_x) .$$

Here, c_x denotes the correspondence, localized at a fixed point. We stress that the fixed point set $\mathrm{Fix}(c|\mathcal{M}_{K^p}(\bar{\mathbb{F}}_p))$ is the set of fixed points on a model of the Shimura variety with no level at p, and these were described by Kottwitz, [20]. Roughly, isogeny classes of fixed points correspond to so-called Kottwitz triples $(\gamma_0; \gamma, \delta)$, where $\gamma_0 \in G(\mathbb{Q})$ is a stable conjugacy class, elliptic in $G(\mathbb{R})$, $\gamma \in G(\mathbb{A}_f^p)$ is a conjugacy class, stably conjugate to γ_0, and $\delta \in G(\mathbb{Q}_{p^r})$ is a σ-conjugacy class such that $N\delta$ is stably conjugate to γ_0, where $N\delta = \delta\delta^\sigma \cdots \delta^{\sigma^{r-1}}$ denotes the norm of δ.[3]

[3] The compatibility condition $\alpha(\gamma_0; \gamma, \delta) = 1$ that occurs for general Shimura varieties is vacuous in this case, as we used a division algebra to define the Shimura variety, cf. [19]. This means that one has 'trivial endoscopy', and that later on one will not need to stabilize the trace formula. In particular, the endoscopic fundamental lemma of Ngô is not needed.

On the other hand, using our identifications, the local term

$$\mathrm{tr}(c_x|(R\psi\pi_*\bar{\mathbb{Q}}_\ell)_x) = \mathrm{tr}(\tau \times h|H^*_\delta) = \phi_{\tau,h}(\delta)$$

is given by the function $\phi_{\tau,h}$ defined earlier.

Now the manipulations made in [20] show that one can rewrite the Lefschetz trace formula as

$$\mathrm{tr}(\tau \times hf^p|H^*_{\mathrm{Sh}}) = \sum_{(\gamma_0;\gamma,\delta)} c(\gamma_0;\gamma,\delta)O_\gamma(f^p)TO_{\delta\sigma}(\phi_{\tau,h})\,,$$

where $c(\gamma_0;\gamma,\delta)$ is a certain volume factor, and $O_\gamma(f^p)$ and $TO_{\delta\sigma}(\phi_{\tau,h})$ denote orbital and twisted orbital integrals, respectively. As endoscopy is trivial in our case, one can use the so-called pseudo-stabilization to rewrite this as the geometric side of the Arthur-Selberg trace formula for G. The spectral side of the Arthur-Selberg trace formula can be rewritten as the trace on H^*_{Sh}, by using Matsushima's formula. Summing up, one gets the formula

$$n\mathrm{tr}(\tau \times hf^p|H^*_{\mathrm{Sh}}) = \mathrm{tr}(f_{\tau,h}f^p|H^*_{\mathrm{Sh}})\,,$$

cf. [19] for these manipulations.

The key observation is that the right-hand side of this formula depends only on H^*_{Sh} as a $G(\mathbb{A}_f)$-representation, whereas the left-hand side sees the Galois action. This means that we have given a formula for the Galois action in terms of the automorphic action. Let us make more precise what we get.

Choose some irreducible smooth representation π_f of $G(\mathbb{A}_f)$ that occurs in H^*_{Sh}, and let

$$H^*_{\mathrm{Sh}}(\pi_f) = \mathrm{Hom}_{G(\mathbb{A}_f)}(\pi_f, H^*_{\mathrm{Sh}})$$

be the corresponding representation of $\mathrm{Gal}(\bar{\mathbb{Q}}/E)$; recall that H^*_{Sh} is semisimple as $G(\mathbb{A}_f)$-representation by Matsushima's formula. An important fact is that if π'_f is another irreducible smooth representation of $G(\mathbb{A}_f)$ that occurs in H^*_{Sh}, and such that the representations away from p are isomorphic, i.e. $\pi_f^p \cong \pi_f'^p$, then in fact $\pi_f \cong \pi'_f$. This is proved in [11] by reduction to the analogous result for GL_n, i.e. strong multiplicity 1, using stable base-change from G to GL_n/E, which was established under the given special circumstances by Clozel.

It follows that one can choose $f^p \in C_c^\infty(G(\mathbb{A}_f^p))$ to cut out π_f. The result is that for all τ and h,

$$\begin{aligned} n\mathrm{tr}(\tau|H^*_{\mathrm{Sh}}(\pi_f))\mathrm{tr}(h|\pi_p) &= n\mathrm{tr}(\tau \times h|H^*_{\mathrm{Sh}}(\pi_f) \otimes \pi_p) \\ &= \mathrm{tr}(f_{\tau,h}|H^*_{\mathrm{Sh}}(\pi_f) \otimes \pi_p) = \dim H^*_{\mathrm{Sh}}(\pi_f)\mathrm{tr}(f_{\tau,h}|\pi_p)\,. \end{aligned}$$

Here, everything depends only on the local component π_p, except the first factor $\mathrm{tr}(\tau|H^*_{\mathrm{Sh}}(\pi_f))$ (and the dimension of $H^*_{\mathrm{Sh}}(\pi_f)$, which involves some automorphic multiplicity). This implies that the virtual representation

$$\frac{n}{\dim H^*_{\mathrm{Sh}}(\pi_f)} H^*_{\mathrm{Sh}}(\pi_f)|_{W_{\mathbb{Q}_p}}$$

of $W_{\mathbb{Q}_p}$ depends only on π_p, and in fact our characterization of local Langlands says that it is equal to $\mathrm{rec}(\pi_p)$. This gives exactly the desired local-global-compatibility result. Moreover, we use this argument to establish the existence of $\mathrm{rec}(\pi_p)$ for many π_p, by embedding the local situation into a suitable global picture.

1.3 Bijectivity of the correspondence

Our main innovation is a simplified proof of the bijectivity of the local Langlands correspondence. In former proofs, this was done by appealing to Henniart's numerical correspondence, [12], which says that after putting suitable ramification conditions on both of the correspondence, the number of representations on each side of the correspondence is finite and the same.

The key input needed to prove the bijectivity is the following purely representation-theoretic result. Recall that for any cyclic extension F'/F of prime degree of p-adic fields, the work of Arthur and Clozel, [1], defines a base-change map $\mathrm{BC}_F^{F'}$ from irreducible smooth representation of $\mathrm{GL}_n(F)$ to irreducible smooth representations of $\mathrm{GL}_n(F')$. This mirrors the restriction functor from W_F to $W_{F'}$ on the Galois side. On the Galois side, every irreducible representation becomes unramified after a finite series of such base-change operations. If the local Langlands correspondence is true, then the same should be true on the automorphic side: For any supercuspidal representation π of $\mathrm{GL}_n(F)$, there is a series of cyclic extensions of prime degree $F_0 = F \subset F_1 \subset \ldots \subset F_m$ such that

$$\mathrm{BC}_{F_{m-1}}^{F_m} \mathrm{BC}_{F_{m-2}}^{F_{m-1}} \cdots \mathrm{BC}_{F_0}^{F_1} \pi$$

is an unramified representation of $\mathrm{GL}_n(F_m)$. Note that, by induction, it is enough to ensure that π does not stay supercuspidal after any series of base-changes. We give a direct proof of this result.

In fact, from the local-global-compatibility, one deduces that the correspondence $\pi \mapsto \sigma(\pi)$ is compatible with base-change, i.e.

$$\sigma(\mathrm{BC}_F^{F'} \pi) = \sigma(\pi)|_{W_{F'}} .$$

In particular, after a series of base-changes, we can assume that $\sigma(\pi)$ has invariants under the inertia group $I_F \subset W_F$.[4] Hence, it is enough to prove the following result.

Theorem 1.6. *Let π be a supercuspidal representation of $\mathrm{GL}_n(F)$, $n \geq 2$. Then $\sigma(\pi)^{I_F} = 0$.*

This theorem follows from the following two geometric theorems.

Theorem 1.7 (cf. [24, Theorem 3.4]). *Let $H/\mathbb{F}_{p^r}$ be a connected 1-dimensional p-divisible group of height n, corresponding to some δ. Let*

$$(\mathcal{R}_{\delta,m}, \mathcal{H}_{\delta,m}, \mathcal{X}_1, \ldots, \mathcal{X}_n)$$

be an algebraization of its deformation ring with Drinfeld-level-p^m-structure, as in Theorem 1.2. Then, after replacing $\mathcal{R}_{\delta,m}$ by a suitable localization, $\mathcal{R}_{\delta,m}$ is a connected

[4] We might even assume that it is unramified, but this will not help much.

regular ring, whose special fibre Spec $(\mathcal{R}_{\delta,m} \otimes \mathbb{F}_{p^r})$ *has a stratification into locally closed strata* Z_i, $i \in I$, *such that all* $\overline{Z}_i$ *are regular. In fact, one may identify*

$$I = \{0 \neq L \subset (\mathbb{Z}/p^m\mathbb{Z})^n \text{ direct summand}\},$$

and $\overline{Z}_L$ *as the closed subscheme of* Spec $(\mathcal{R}_{\delta,m} \otimes \mathbb{F}_{p^r})$ *where for all* $(i_1, \ldots, i_n) \in L$, $i_1 X_1 + \ldots + i_n X_n = e$ *is the identity section of* $\mathcal{H}_{\delta,m}$.

We remark that this looks formally a little bit like the case of semistable reduction, but we caution the reader that the irreducible components of the special fibre have usually a very large multiciplicity, divisible by large powers of p, and that the intersection between different components is highly nontransversal, with the intersection multiplicities again being divisible by large powers of p. The proof of this theorem is not hard. In fact, it is enough to check these statements for the $R_{\delta,m}$ completion of $\mathcal{R}_{\delta,m}$. In that case, the computations of Drinfeld, [8], do the job, which say that $R_{\delta,m}$ is a regular ring with $(X_1, \ldots, X_n)$ as a system of regular parameters.

Theorem 1.8 (cf. [24, Theorem 5.3]). *Let* R *be a regular flat* $\mathbb{Z}_{p^r}$*-algebra whose special fibre has a stratification into locally closed strata* Z_i, $i \in I$, *such that all* $\overline{Z}_i$ *are regular. Moreover, assume a certain combinatorial condition on the combinatorics of the strata. Then there is an explicit description of the derived inertia invariants* $(R_{I_{\mathbb{Q}_p}} R\psi\bar{\mathbb{Q}}_\ell)_{\bar{x}}$ *of* $(R\psi\bar{\mathbb{Q}}_\ell)_{\bar{x}}$, $x \in (\mathrm{Spec}\, R)(\mathbb{F}_{p^r})$.

The combinatorial condition can be verified in the case at hand; it reduces to the combinatorics of the unipotent representations of $\mathrm{GL}_n(F)$. The key input in the second theorem is Grothendieck's purity conjecture, proved by Thomason, [29], in the form with rational coefficients that we need, and by Gabber in general.

Example 1.9. Let $n = 2$, $F = \mathbb{Q}_p$. In that case, the irreducible components of the special fibre of $\mathcal{R}_{\delta,m}$ are parametrized by $\mathbb{P}^1(\mathbb{Z}/p^m\mathbb{Z})$, and there is one point x in the special fibre such that any two irreducible components intersect precisely at x. One gets identifications

$$(R^k_{I_{\mathbb{Q}_p}} R\psi\bar{\mathbb{Q}}_\ell)_{\bar{x}} = \begin{cases} \bar{\mathbb{Q}}_\ell & k = 0 \\ \bar{\mathbb{Q}}_\ell^{\#\mathbb{P}^1(\mathbb{Z}/p^m\mathbb{Z})}(-1) & k = 1 \\ \bar{\mathbb{Q}}_\ell^{\#\mathbb{P}^1(\mathbb{Z}/p^m\mathbb{Z})-1}(-2) & k = 2\,. \end{cases}$$

In fact, in the situation of Theorem 1.8, it is true in general that

$$(R^k_{I_{\mathbb{Q}_p}} R\psi\bar{\mathbb{Q}}_\ell)_{\bar{x}} \cong V(-k)\,,$$

for some finite-dimensional $\bar{\mathbb{Q}}_\ell$-vector space V.

These theorems solve the problem of determining $\sigma(\pi)^{I_F}$ for supercuspidal π: Informally, by our characterization, $\sigma(\pi)$ is realized in the nearby cycle sheaves, and hence $\sigma(\pi)^{I_F}$ is realized in the inertia invariants of $R\psi\bar{\mathbb{Q}}_\ell$. On the other hand, these inertia invariants are computed explicitly in terms of unipotent representations, so do not contain supercuspidal contributions. The outcome is that $\sigma(\pi)^{I_F} = 0$, as desired.

2 The cohomology of compact unitary group Shimura varieties at ramified split places

The methods used to prove the local Langlands correspondence apply to more general Shimura varieties. In joint work with Sug Woo Shin, [27], we used this to prove new results about the cohomology of compact unitary group Shimura varieties.

2.1 Statement of results

Let $G/\mathbb{Q}$ be a unitary similitude group associated to a CM field F, with totally real subfield $F^+ \subset F$. We assume that G is anisotropic modulo center. This will be achieved by using either a division algebra to define G, or by using signature $(0, n)$ at one infinite place. Let Sh_K, $K \subset G(\mathbb{A}_f)$, be the corresponding Shimura variety, which is of PEL type. It is defined over the reflex field $E \subset \mathbb{C}$, a finite extension of $\mathbb{Q}$. Moreover, we fix a prime p such that $G_{\mathbb{Q}_p}$ is a product of Weil restrictions of GL_k's. Concretely, this means that we assume that all places of F^+ above p split in F, and that G is quasisplit at p (i.e., in case G is defined by a division algebra, that the division algebra splits at all places above p). In particular, it follows that the local Langlands correspondence for $G_{\mathbb{Q}_p}$ is known, by the result for GL_k over p-adic fields. The assumption also implies that there is a unique conjugacy class of maximal compact subgroups of $G(\mathbb{Q}_p)$. In fact, there is an extension of G to a connected algebraic group over $\mathbb{Z}_{(p)}$ whose $\mathbb{Z}_p$-valued are such a maximal compact subgroup; we fix one such an extension. Also fix some place $v|p$ of E.

Again, we fix some $\ell \neq p$ and consider the cohomology of the Shimura variety,

$$H^*_{\mathrm{Sh}} = \varinjlim_K H^*(\mathrm{Sh}_K \otimes \bar{\mathbb{Q}}, \bar{\mathbb{Q}}_\ell)\,.^5$$

The data defining the Shimura variety also give rise to an algebraic representation r of the L-group ${}^LG$ of G. Restricted to the dual group $\hat{G}$, it is given by the highest weight representation of $\widehat{G}$ corresponding to the cocharacter $-\mu$ of G under duality, where μ is the usual minuscule cocharacter from the definition of a Shimura variety.

Theorem 2.1. *Assume that G is associated to a division algebra over F as in [19]. Then there is an equality of virtual $W_{E_v} \times G(\mathbb{Z}_p) \times G(\mathbb{A}_f^p)$-representations,*

$$H^*_{\mathrm{Sh}} = \sum_{\pi_f} a(\pi_f)\pi_f \otimes (r \circ \varphi_{\pi_p}|_{W_{E_v}})| \cdot |^{-\dim \mathrm{Sh}/2}\,.$$

Here π_f runs through irreducible admissible representations of $G(\mathbb{A}_f)$, the integer $a(\pi_f)$ is as in [19], p. 657, and φ_{π_p} is the local Langlands parameter associated to π_p.

We remark that from the assumption that G is associated to a division algebra, it follows that endoscopy is trivial in this situation, cf. [19]. We stress that the theorem works at places where F^+ is ramified, and in particular where the Shimura variety does not have a smooth

[5] In [27], we also consider cohomology with values in local systems associated to algebraic representations ξ of G; everything goes through in that generality.

integral model, not even for maximal compact level at p. Under such situations, very little was known before.

Corollary 2.2. *In the situation of the theorem, let $K \subset G(\mathbb{A}_f)$ be any sufficiently small compact open subgroup. Then the semisimple local Hasse-Weil zeta function of Sh_K at the place $v|p$ of E is given by*

$$\zeta_v^{\mathrm{ss}}(\mathrm{Sh}_K, s) = \prod_{\pi_f} L^{\mathrm{ss}}(s - \dim \mathrm{Sh}/2, \pi_p, r)^{a(\pi_f)\dim \pi_f^K}\ .$$

In particular because the weight-monodromy conjecture is not known, one can currently not get results about the local zeta function itself, and we have to resort to its semisimple version.

Our second main theorem reads roughly as follows. For precise statements, we refer to the paper [27].

Theorem 2.3. *Assume that G is quasisplit at all finite places. Then for both stable and endoscopic π_f, we give an explicit description of the π_f-isotypic component of H^*_{Sh} as a W_{E_v}-representation, and verify the expected local-global compatibility result.*

Using this result, we can reprove a result of Shin[6] on the existence of Galois representations attached to regular algebraic conjugate self-dual automorphic forms.

Theorem 2.4 ([28, Thm 1.2]). *Let F be any CM field. Let Π be a cuspidal automorphic representation of $\mathrm{GL}_n(\mathbb{A}_F)$ such that*

(i) $\Pi^\vee \simeq \Pi \circ c$, *where $c : F \to F$ denotes the complex conjugation,*

(ii) Π_∞ *is regular algebraic.*

(iii) Π_∞ *is slightly regular, if n is even.*

Then for each prime ℓ and an isomorphism $\iota_\ell : \bar{\mathbb{Q}}_\ell \simeq \mathbb{C}$, there exists a continuous semisimple representation

$$R_{\ell,\iota_\ell}(\Pi) : \mathrm{Gal}(\bar{\mathbb{Q}}/F) \to \mathrm{GL}_n(\bar{\mathbb{Q}}_\ell)$$

such that for any finite place v of F not dividing ℓ, the Frobenius semisimple Weil-Deligne representation associated to $R_{\ell,\iota_\ell}(\Pi)|_{W_{F_v}}$ corresponds to $\iota_\ell^{-1}\Pi_v$ via (a suitably normalized) local Langlands correspondence, except possibly for the monodromy operator N.

We note that under the assumption that Π_v is square-integrable for one finite place v, this was proved (without assumption (iii)) by Harris and Taylor, [11]. Also, there are now significant improvements, cf. [7], [6], [3], [2], for instance (iii) has become unnecessary, a similar conclusion holds at y dividing ℓ, and also the monodromy operator N is known to match.

---

[6] building on the work and ideas of many others, including Kottwitz, Clozel, Blasius, Rogawski, Harris, Labesse, Taylor, …, with one key new input in general being the fundamental lemma, settled by work of Ngô, Waldspurger, ….

In the following, we describe the definition of generalizations of the functions $\phi_{\tau,h}$ introduced in the first talk in the context of general PEL data, and their conjectural spectral properties. These results form the key geometric input in the proofs of the theorems above.

2.2 Definition of test functions

For PEL data of type A and C, these test functions are defined in [25]. Let us explain just the case necessary for our global applications. Fix a finite extension F of $\mathbb{Q}_p$ with ring of integers $\mathcal{O}_F$, an integer $n \geq 1$, and let $G = \mathrm{Res}_{\mathcal{O}_F/\mathbb{Z}_p} \mathrm{GL}_n$. Let

$$\mu : \mathbb{G}_m \to G_{\bar{\mathbb{Q}}_p}$$

be a cocharacter such that for any embedding $\alpha : F \hookrightarrow \bar{\mathbb{Q}}_p$, the corresponding composite map

$$\mathbb{G}_m \xrightarrow{\mu} G_{\bar{\mathbb{Q}}_p} \to \mathrm{GL}_n$$

corresponds to an action of $\mathbb{G}_m$ on $\bar{\mathbb{Q}}_p^n$ with weights only 0 and 1. Let $m(\alpha)$ denote the multiplicity of weight 1 in this decomposition.[7] The conjugacy class $\overline{\mu}$ of μ is defined over a finite extension E of $\mathbb{Q}_p$. We note that $\mathrm{Gal}(\bar{\mathbb{Q}}_p/E)$ is exactly the stabilizer of the function $\alpha \mapsto m(\alpha)$. Let $\mathcal{O}_E \subset E$ be the ring of integers.

Let $\mathcal{D} = (F, n, \overline{\mu})$ denote the given EL data.

Definition 2.5. Let R be an $\mathcal{O}_E$-algebra which is p-adically complete. A p-divisible group with $\mathcal{D}$-structure over R is a pair (H, ι), where H/R is p-divisible group, $\iota : \mathcal{O}_F \to \mathrm{End}(H)$ is an action of $\mathcal{O}_F$, such that the Lie algebra of the universal vector extension of H is locally on Spec R a free $R \otimes_{\mathbb{Z}_p} \mathcal{O}_F$-module of rank n, and the Kottwitz signature condition is satisfied.

We refer the reader to [23, 3.23] for a detailed discussion of the last condition; it depends on the cocharacter μ. We note that if R is flat over $\mathcal{O}_E$, then the Lie algebra $\mathrm{Lie} H$ decomposes under the action of $\mathcal{O}_F$ as

$$\mathrm{Lie} H \otimes_{\mathcal{O}_E} \bar{\mathbb{Q}}_p = \prod_{\alpha : F \hookrightarrow \bar{\mathbb{Q}}_p} (\mathrm{Lie} H \otimes_{\mathcal{O}_E} \bar{\mathbb{Q}}_p)_\alpha ,$$

where F acts on $(\mathrm{Lie} H \otimes_{\mathcal{O}_E} \bar{\mathbb{Q}}_p)_\alpha$ via the embedding $\alpha : F \hookrightarrow \bar{\mathbb{Q}}_p$. In that case, the signature condition is equivalent to the condition that the locally free $R \otimes_{\mathcal{O}_E} \bar{\mathbb{Q}}_p$-module $(\mathrm{Lie} H \otimes_{\mathcal{O}_E} \bar{\mathbb{Q}}_p)_\alpha$ is of rank $m(\alpha)$.

For a p-divisible group H with $\mathcal{D}$-structure over a perfect field k, we let R_H be its universal deformation ring, with universal deformation H^{univ}. It is a complete noetherian local $W(k)$-algebra. Let us recall the following result of Wedhorn, [32].

Theorem 2.6. *If F is an unramified extension of $\mathbb{Q}_p$, then R_H is formally smooth, and R_H is the versal deformation ring of $H[p^m]$, for any $m \geq 1$.*

---

[7] We follow the convention on μ used in [20] and [27], which is opposite to the one in [25] and [23].

Both statements fail in general if F is ramified over $\mathbb{Q}_p$. If $m(\alpha) = 0$ for all but one embedding $F \hookrightarrow \bar{\mathbb{Q}}_p$, then Faltings's theory of strict $\mathcal{O}$-group schemes, [9], shows that the same conclusions hold if one uses a slightly different notion of truncation, which makes sense only in this very special situation. We note that the assumption on $m(\alpha)$ is satisfied in the first talk.

Definition 2.7. A p-divisible group with $\mathcal{D}$-structure H over a perfect field k is said to be liftable if R_H is not p-torsion.

We note that if H is liftable, then there exists a mixed characteristic discrete valuation ring R whose residue field is a finite extension of k, and a p-divisible group with $\mathcal{D}$-structure over R, whose special fibre is given by H.

Proposition 2.8. *If H is liftable, then the covariant Dieudonné module $M(H)$ is isomorphic to $(\mathcal{O}_F \otimes_{\mathbb{Z}_p} W(k))^n$ as $\mathcal{O}_F \otimes_{\mathbb{Z}_p} W(k)$-module.*

In particular, one gets an injection from the set of liftable p-divisible groups with $\mathcal{D}$-structure over k into the set of $G(W(k))$-σ-conjugacy classes in $G(W(k)[p^{-1}])$.

Let X_H be the rigid-analytic generic fibre of Spf R_H. For an open subgroup $K \subset G(\mathbb{Z}_p)$, define $X_{H,K}/X_H$ as the finite étale cover parametrizing level-K-structures. We set

$$H_H^* = \varinjlim_K H^*(X_{H,K} \hat{\otimes}_{W(k)[p^{-1}]} C, \bar{\mathbb{Q}}_\ell)\,,$$

where C is the completion of an algebraic closure of $W(k)[p^{-1}]$. These groups carry a continuous Galois action, and an action of $G(\mathbb{Z}_p)$.

Theorem 2.9. *The action of $G(\mathbb{Z}_p)$ is admissible.*

If F is unramified, resp. $m(\alpha) = 0$ for all but one α, this follows as in the first talk from an algebraization result for R_H and its covers, which in turn follows from Artin's algebraization results and the theorem of Wedhorn, resp. Faltings. In general, we know no direct proof of this result. However, in [27], we show that any p-divisible group with $\mathcal{D}$-structure occurs in a suitable Shimura variety, which provides the desired algebraization.[8]

Now we can define the test functions.

Definition 2.10. For $\tau \in W_E$ projecting to r-th power of geometric Frobenius, $r \geq 1$, and $h \in C_c^\infty(G(\mathbb{Z}_p))$ with values in $\mathbb{Q}$, define $\phi_{\tau,h} : G(\mathbb{Q}_{p^r}) \to \bar{\mathbb{Q}}_\ell$ by

$$\phi_{\tau,h}(\delta) = \mathrm{tr}(\tau \times h | H_H^*)$$

if δ corresponds to some H, and $\phi_{\tau,h}(\delta) = 0$ otherwise.

The following theorem is proved similarly to Theorem 1.4.

Theorem 2.11. *The function $\phi_{\tau,h} \in C_c^\infty(G(\mathbb{Q}_{p^r}))$ is locally constant with compact support, and with values in $\mathbb{Q}$ independent of ℓ.*

[8] In particular, this theorem is only proved at the very end, and one has to argue more carefully to not end up with a circular argument. This is taken care of in [25], but we will ignore these extra difficulties.

2.3 Spectral properties of test functions

In [27], we define these test functions in greater generality, where interesting local endoscopic groups for G exist, and are necessary to give conjectures that pin down the twisted orbital integrals of these test functions. In the case at hand, all of them are Levi subgroups, and the endoscopic group $H = G$ is enough to pin down the twisted orbital integrals of $\phi_{\tau,h}$. Therefore, we will not consider endoscopic groups in the following.

Let $f_{\tau,h} \in C_c^\infty(G(\mathbb{Q}_p))$ be a base-change transfer of $\phi_{\tau,h}$. Its existence follows as before from [1]. Recall that μ gives rise to a representation r of the L-group ${}^L(G_E) = \hat{G} \rtimes W_E$; restricted to the dual group

$$\hat{G} = \prod_{\alpha: F \hookrightarrow \bar{\mathbb{Q}}_p} \mathrm{GL}_n(\mathbb{C}),$$

the representation r is given by

$$\bigotimes_\alpha \bigwedge^{m(\alpha)} V_\alpha^\vee,$$

where $V_\alpha \cong \mathbb{C}^n$ denotes the tautological representation of $\mathrm{GL}_n(\mathbb{C})$.

For any irreducible smooth representation π of $G(\mathbb{Q}_p) = \mathrm{GL}_n(F)$, let $\varphi_\pi : W_{\mathbb{Q}_p} \to {}^L G$ be the associated (semisimple) L-parameter (i.e., we ignore monodromy again).[9] We define

$$\mathrm{rec}_\mu(\pi) = (r \circ \varphi_\pi|_{W_E})|\cdot|^{-\langle\rho,\mu\rangle}$$

as a representation of W_E, where ρ denotes the half-sum of positive roots. We note that $2\langle\rho,\mu\rangle$ is the dimension of X_H, whenever X_H is nonempty.

Theorem 2.12. *For any irreducible smooth representation π of $G(\mathbb{Q}_p) = \mathrm{GL}_n(F)$,*

$$\mathrm{tr}(f_{\tau,h}|\pi) = \mathrm{tr}(\tau|\mathrm{rec}_\mu(\pi))\mathrm{tr}(h|\pi).$$

This is a direct generalization of the corresponding identity in the first talk. For the proof, we embed the local situation into a global situation without endoscopy. In that case, the discussion of local-global compatibility given in the first talk goes through word-by-word, and one has to identify the π_f-isotypic $H^*_{\mathrm{Sh}}(\pi_f)$ as a representation of the local Weil group with $\mathrm{rec}_\mu(\pi)$, where $\pi = \pi_p$. For almost all p (where π_f is unramified), this is known by classical work of Kottwitz, [19]. On the other hand, by a suitable form of stable base change, one can use the Harris – Taylor case of Theorem 2.4 (which can also be proved by the methods of the first talk) to show that there is a Galois representation $R(\pi_f)$ that verifies local-global compatibility everywhere, and such that $r \circ R(\pi_f)$ agrees with $H^*_{\mathrm{Sh}}(\pi_f)$ at almost all finite places (by Kottwitz's result). By Chebotarev, they agree everywhere, which finally identifies $H^*_{\mathrm{Sh}}(\pi_f)$ as a representation of the local Weil group. A similar argument occurred earlier in work of Fargues, [10].

[9] Note that by Shapiro's lemma, giving a section $W_{\mathbb{Q}_p} \to {}^L G$ of the projection ${}^L G \to W_{\mathbb{Q}_p}$ is equivalent to giving a homomorphism $W_F \to \mathrm{GL}_n(\mathbb{C})$, so one gets φ_π from the local Langlands correspondence for $\mathrm{GL}_n(F)$.

From here, one deduces the desired results by applications of suitable trace formulas, using work of Ngô, [22], Waldspurger, [31], and others, in the endoscopic case. These arguments are similar to the arguments in the section on local-global-compatibility above: One starts with a Lefschetz trace formula whose local terms are given by the functions $\phi_{\tau,h}$. Then one stabilizes the trace formula, and relates it to an Arthur–Selberge trace formula. However, the details in the endoscopic case are significantly more complicated, and we refer the reader to the original paper [27].

It is to be expected that our results work in greater generality, but for now such results will be conditional on Arthur's conjectures for the groups involved. Once these are settled, there is hope that one can push through our arguments in general. One can also hope to extend these methods beyond PEL cases, using the work of Kisin, [18], on integral models for Shimura varieties of abelian type.

Acknowledgements

It is a pleasure to thank the organizers of the 2012 Program on Galois Representations and Shimura Varieties at the Fields Institute for their invitation, and for giving me the opportunity to speak at the Workshop on the Cohomology of Shimura Varieties. These notes follow closely the two talks given there, and were written while the author was a Clay Research Fellow.

References

[1] J. Arthur and L. Clozel. *Simple algebras, base change, and the advanced theory of the trace formula*, volume 120 of *Annals of Mathematics Studies*. Princeton University Press, Princeton, NJ, 1989.

[2] T. Barnet-Lamb, T. Gee, D. Geraghty, and R. Taylor. Local-global compatibility for $l = p$, I. *Ann. Fac. Sci. Toulouse Math. (6)*, 21(1):57–92, 2012.

[3] T. Barnet-Lamb, T. Gee, D. Geraghty, and R. Taylor. Local-global compatibility for $l = p$, II. *Ann. Sci. Éc. Norm. Supér. (4)*, 47(1):165–179, 2014.

[4] I. N. Bernstein and A. V. Zelevinsky. Induced representations of reductive $\mathfrak{p}$-adic groups. I. *Ann. Sci. École Norm. Sup. (4)*, 10(4):441–472, 1977.

[5] J. N. Bernstein. Le "centre" de Bernstein. In *Representations of reductive groups over a local field*, Travaux en Cours, pages 1–32. Hermann, Paris, 1984. Edited by P. Deligne.

[6] A. Caraiani. Local-global compatibility and the action of monodromy on nearby cycles. *Duke Math. J.*, 161(12):2311–2413, 2012.

[7] G. Chenevier and M. Harris. Construction of automorphic Galois representations, II. *Camb. J. Math.*, 1(1):53–73, 2013.

[8] V. G. Drinfel'd. Elliptic modules. *Mat. Sb. (N.S.)*, 94(136):594–627, 656, 1974.

[9] G. Faltings. Group schemes with strict $\mathcal{O}$-action. *Mosc. Math. J.*, 2(2):249–279, 2002. Dedicated to Yuri I. Manin on the occasion of his 65th birthday.

[10] L. Fargues. Cohomologie des espaces de modules de groupes p-divisibles et correspondances de Langlands locales. *Astérisque*, 291:1–200, 2004.

[11] M. Harris and R. Taylor. *The geometry and cohomology of some simple Shimura varieties*, volume 151 of *Annals of Mathematics Studies*. Princeton University Press, Princeton, NJ, 2001. With an appendix by Vladimir G. Berkovich.

[12] G. Henniart. La conjecture de Langlands locale numérique pour $\mathrm{GL}(n)$. *Ann. Sci. École Norm. Sup. (4)*, 21(4):497–544, 1988.

[13] G. Henniart. Caractérisation de la correspondance de Langlands locale par les facteurs ϵ de paires. *Invent. Math.*, 113(2):339–350, 1993.

[14] G. Henniart. Une preuve simple des conjectures de Langlands pour $\mathrm{GL}(n)$ sur un corps p-adique. *Invent. Math.*, 139(2):439–455, 2000.

[15] R. Huber. *Étale cohomology of rigid analytic varieties and adic spaces*. Aspects of Mathematics, E30. Friedr. Vieweg & Sohn, Braunschweig, 1996.

[16] R. Huber. A finiteness result for direct image sheaves on the étale site of rigid analytic varieties. *J. Algebraic Geom.*, 7(2):359–403, 1998.

[17] L. Illusie. Déformations de groupes de Barsotti-Tate (d'après A. Grothendieck). *Astérisque*, (127):151–198, 1985. Seminar on arithmetic bundles: the Mordell conjecture (Paris, 1983/84).

[18] M. Kisin. Integral models for Shimura varieties of abelian type. *J. Amer. Math. Soc.*, 23(4): 967–1012, 2010.

[19] R. E. Kottwitz. On the λ-adic representations associated to some simple Shimura varieties. *Invent. Math.*, 108(3):653–665, 1992.

[20] R. E. Kottwitz. Points on some Shimura varieties over finite fields. *J. Amer. Math. Soc.*, 5(2):373–444, 1992.

[21] Y. Mieda. On l-independence for the étale cohomology of rigid spaces over local fields. *Compos. Math.*, 143(2):393–422, 2007.

[22] B. C. Ngô. Le lemme fondamental pour les algèbres de Lie. *Publ. Math. Inst. Hautes Études Sci.*, (111):1–169, 2010.

[23] M. Rapoport and T. Zink. *Period spaces for p-divisible groups*, volume 141 of *Annals of Mathematics Studies*. Princeton University Press, 1996.

[24] P. Scholze. The Langlands-Kottwitz approach for some simple Shimura varieties. *Invent. Math.*, 192(3):627–661, 2013.

[25] P. Scholze. The Langlands-Kottwitz method and deformation spaces of p-divisible groups. *J. Amer. Math. Soc.*, 26(1):227–259, 2013.

[26] P. Scholze. The local Langlands correspondence for GL_n over p-adic fields. *Invent. Math.*, 192(3):663–715, 2013.

[27] P. Scholze and S. W. Shin. On the cohomology of compact unitary group Shimura varieties at ramified split places. *J. Amer. Math. Soc.*, 26(1):261–294, 2013.

[28] S. W. Shin. Galois representations arising from some compact Shimura varieties. *Annals of Math.*, 173:1645–1741, 2011.

[29] R. Thomason. Absolute cohomological purity. *Bull. Soc. Math. France*, 112(3):397–406, 1984.

[30] Y. Varshavsky. Lefschetz-Verdier trace formula and a generalization of a theorem of Fujiwara. *Geom. Funct. Anal.*, 17(1):271–319, 2007.

[31] J.-L. Waldspurger. Le lemme fondamental implique le transfert. *Comp. Math.*, 105(2): 153–236, 1997.

[32] T. Wedhorn. The dimension of Oort strata of Shimura varieties of PEL-type. In *Moduli of abelian varieties (Texel Island, 1999)*, volume 195 of *Progr. Math.*, pages 441–471. Birkhäuser, Basel, 2001.

UNE APPLICATION DES VARIÉTÉS DE HECKE DES GROUPES UNITAIRES

GAËTAN CHENEVIER

Laboratoire de Mathématiques d'Orsay, Université Paris-Sud, Université Paris-Saclay, 91405 Orsay, France
E-mail address: gaetan.chenevier@math.cnrs.fr

Dans cette note[1], nous expliquons la construction de variétés de Hecke p-adiques associées aux groupes unitaires définis d'un corps de nombres CM et décrivons des propriétés de la famille de représentations galoisiennes portée par ces variétés. Nous donnons une application à la construction de certaines représentations galoisiennes qui joue un rôle important dans [ChH]. Les variétés de Hecke ci-dessus sont aussi utilisées dans [BCh2].

La construction des variétés de Hecke que nous donnons est une combinaison essentiellement triviale des méthodes de [Ch] (cas d'un corps quadratique imaginaire, revisité dans [BCh, §7]) et de [Bu] (cas d'une algèbre de quaternions sur un corps totalement réel, voir aussi [Y]). Elle pourrait aussi se déduire comme cas particulier des travaux d'Emerton [E], et elle est aussi contenue dans les travaux récents de Loeffler [Loe]. En ce qui concerne les propriétés galoisiennes, nous utilisons et étendons certains résultats de [BCh, §7].

1 Variétés de Hecke des groupes unitaires définis

1.1 Notations et choix

Fixons un corps de nombres totalement réel F, $\mathcal{K}$ une extension quadratique totalement imaginaire de F, $n \geq 1$ un entier et U un groupe unitaire à n variables attaché à $\mathcal{K}/F$ vu comme groupe algébrique sur F. On suppose que U est *défini*, ce qui signifie que $U(F_v)$ est un groupe unitaire compact pour chaque place réelle v de F. Cela entraîne que les représentations automorphes (complexes, irréductibles) Π de U sont toutes discrètes et cohomologiques en degré 0 :

$$L^2(U(F)\backslash U(\mathbb{A}_F))_{K\text{-fini}} \simeq \bigoplus_{\Pi} m(\Pi)\Pi,$$

$K \subset U(\mathbb{A}_F)$ étant un sous-groupe compact ouvert quelconque, et m(Π) étant la multiplicité (finie) de la représentation Π. De plus, les composantes archimédiennes Π_∞ sont de dimension finie et les composantes Π_f aux places finies sont définies sur $\overline{\mathbb{Q}}$.

Afin d'associer aux Π_f des $\overline{\mathbb{Q}}$-structures de manière univoque, il convient de fixer un plongement $\iota_\infty : \overline{\mathbb{Q}} \longrightarrow \mathbb{C}$. Comme nous allons nous intéresser aux congruences entre ces Π, il nous faut aussi fixer un nombre premier p ainsi qu'un plongement $\iota_p : \overline{\mathbb{Q}} \to \overline{\mathbb{Q}}_p$. Dans toute cette note, $\overline{\mathbb{Q}}$ et $\overline{\mathbb{Q}}_p$ sont des clôtures algébriques fixées de $\mathbb{Q}$ et $\mathbb{Q}_p$. On munit $\overline{\mathbb{Q}}_p$ de sa

Gaëtan Chenevier est financé par le C.N.R.S. et par le projet ANR-14-CE25 (PerCoLaTor).

[1]Le texte date du 15 Avril 2009, avec des modifications très mineures apportées en Juin 2017. Nous remercions le rapporteur pour sa relecture.

norme $\|.\|$ et de sa valuation p-adique $\mathbf{v}(.)$ telles que $\mathbf{v}(p) = 1$ et $\|p\| = 1/p$. Si v est une place de F divisant p, nous noterons $\Sigma(v) \subset \mathrm{Hom}(F, \overline{\mathbb{Q}}_p)$ l'ensemble des plongements continus v-adiquement (i.e. induisant v) et $\Sigma_\infty(v) := \iota_\infty \iota_p^{-1}\Sigma(v) \subset \mathrm{Hom}(F, \mathbb{R})$ les plongements réels correspondants.

Si v est une place de F décomposée dans $\mathcal{K}$, et si $U(F_v) \simeq \mathrm{GL}_n(F_v)$, la donnée d'une place w de $\mathcal{K}$ au dessus de v permet d'identifier $U(F_v)$ avec $\mathrm{GL}_n(\mathcal{K}_w) = \mathrm{GL}_n(F_v)$ de manière unique à conjugaison près, ce que l'on fera parfois sans plus de précision. En particulier, si Π est une représentation automorphe de U on notera Π_w, ou même par abus Π_v, la représentation de $\mathrm{GL}_n(F_v)$ associée à un tel choix.

Les variétés de Hecke de U que nous allons considérer dépendent de la donnée de S_p, W_∞, $\mathcal{H}$ et e que nous fixons dès maintenant :

(En p) S_p est un *sous-ensemble non vide* de l'ensemble des places finies de F divisant p tel que $\forall v \in S_p$, $U(F_v) \simeq \mathrm{GL}_n(F_v)$. En particulier, chaque place $v \in S_p$ est décomposée dans $\mathcal{K}$. Il sera important dans les applications à [ChH] d'autoriser que S_p ne contienne pas toutes les places de F divisant p.

(En ∞) W_∞ est une représentation irréductible fixée de $\prod_w U(F_w)$, le produit portant sur toutes les places réelles w de F qui ne sont pas dans $\cup_{v \in S_p} \Sigma_\infty(v)$.

(Hors S_p) Fixons de plus :

- un ensemble fini S de places finies de F tel que $S \cap S_p = \emptyset$, et contenant les places v telles que $U(F_v)$ est ramifié,
- $K^S = \prod_{v \notin S} K_v \subset U(\mathbb{A}^S_{F,f})$ un sous-groupe compact ouvert tel que K_v est maximal hyperspécial pour $v \notin S \cup S_p$, et K_v est un sous-groupe d'Iwahori[2] pour tout $v \in S_p$,
- $\mathcal{H}$ une sous-algèbre commutative de l'algèbre de Hecke globale de $U(\mathbb{A}^{S_p}_{F,f})$ contenant l'algèbre de Hecke sphérique $\mathcal{H}^S = \mathbb{Z}[U(\mathbb{A}^{S \cup S_p})//K^{S \cup S_p}]$ hors de $S \cup S_p$.
- pour chaque $v \in S$, un idempotent e_v de l'algèbre de Hecke de $U(F_v)$ qui commute avec $\mathcal{H}$.

On pose $e = (\otimes_{v \in S} e_v) \otimes \mathbb{1}_{K^{S \cup S_p}}$, c'est un idempotent de l'algèbre de Hecke de $U(\mathbb{A}^{S_p}_{F,f})$ qui commute avec $\mathcal{H}$. On supposera que $\mathcal{H}$ et les e_v sont à coefficients dans un corps de nombres galoisien $E \subset \overline{\mathbb{Q}}$ et on notera L la clôture galoisienne de $\iota_p(E)$ et $\iota_p(\mathcal{K})$ dans $\overline{\mathbb{Q}}_p$.

On notera $\mathcal{A}$ l'ensemble des représentations automorphes irréductibles Π de U avec $e(\Pi_f)^{K_{S_p}} \neq 0$ et[3] $\bigotimes_w \Pi_w \simeq W_\infty$, le produit tensoriel portant sur toutes les places réelles w de F qui ne sont pas dans $\cup_{v \in S_p} \Sigma_\infty(v)$.

---

[2]On rappelle que si F_v est une extension finie de $\mathbb{Q}_p$, un sous-groupe d'Iwahori de $\mathrm{GL}_n(F_v)$ est un sous-groupe de la forme gIg^{-1} où $g \in \mathrm{GL}_n(F_v)$, et où I désigne le sous-groupe des éléments $(g_{i,j}) \in \mathrm{GL}_n(\mathcal{O}_{F_v})$ avec $\mathbf{v}(g_{i,j}) \geq 1$ pour tout $j < i$ (I est le *sous-groupe d'Iwahori standard*).

[3]Notons que nous ne spécifions pas Π_w pour $w \in \Sigma_\infty(v)$ avec $v \in S_p$, de sorte qu'il existera en général une infinité dénombrable de telles représentations, du moins si $\mathcal{A} \neq \emptyset$.

1.2 Congruences entre représentations automorphes

Si $\Pi \in \mathcal{A}$, alors $e(\Pi_f)^{K_{S_p}}$ est un $\mathcal{H}$-module de dimension finie sur $\mathbb{C}$, que l'on peut donc trigonaliser, ce qui nous fournit après semi-simplification un ensemble fini de morphismes d'anneaux $\mathcal{H} \to \mathbb{C}$ (dont on négligera les multiplicités). Comme Π_f, e et $\mathcal{H}$ sont définis sur $\overline{\mathbb{Q}}$, ces morphismes prennent leurs valeurs dans $\overline{\mathbb{Q}}$ via ι_∞^{-1}, que nous pouvons alors envoyer dans $\overline{\mathbb{Q}}_p$ via ι_p, et nous les verrons toujours de cette manière comme des morphismes d'anneaux

$$\psi : \mathcal{H} \longrightarrow \overline{\mathbb{Q}}_p,$$

que l'on appellera *systèmes de valeurs propres associés à* Π. Notons que $(\Pi_f^{S \cup S_p})^{K^{S \cup S_p}}$ étant de dimension 1, les systèmes de valeurs propres associés à un même Π coïncident sur $\mathcal{H}^S$ et contiennent par la théorie de Satake une information identique à $\Pi_f^{S \cup S_p}$.

Donnons déjà une version primitive, mais non triviale, des résultats de cette note. Supposons jusqu'à la fin de ce §1.2 que $\mathcal{H} = \mathcal{H}^S$ pour simplifier. Dans ce cas, chaque Π a un unique système de valeurs propres $\psi = \psi_\Pi$ associé.[4] La notion de congruence entre représentations automorphes de U qui nous intéresse est la suivante : *si* Π *et* Π' *sont dans* $\mathcal{A}$*, on écrit* $\Pi \equiv \Pi' \bmod p^m$ *si*

$$\forall h \in \mathcal{H}^S, \quad \|\psi_\Pi(h) - \psi_{\Pi'}(h)\| \leq p^{-m}.$$

Pour tout $v \in S_p$, il sera commode de fixer un isomorphisme $U(F_v) \simeq \mathrm{GL}_n(F_v)$ comme au §1.1, c'est-à-dire une place de $\mathcal{K}$ divisant v. On pose $T = \prod_{v \in S_p} T_v$ et $B = \prod_{v \in S_p} B_v$, où pour $v \in S_p$ on a noté T_v le tore diagonal de $\mathrm{GL}_n(F_v)$ et B_v le sous-groupe de Borel constitué des éléments triangulaires supérieurs de $\mathrm{GL}_n(F_v)$. On note $X^*(T)$ le groupe des caractères algébriques de T qui sont définis sur $\overline{\mathbb{Q}}_p$. C'est un groupe abélien libre de rang $n \sum_{v \in S_p} [F_v : \mathbb{Q}_p]$ muni du cône naturel des poids dominants relativement à B. Un élément κ de $X^*(T)$ est entièrement déterminé par une famille d'entiers $k_{i,\sigma}$, où $i \in \{1, \ldots, n\}$ et $\sigma \in \mathrm{Hom}(F, \overline{\mathbb{Q}}_p)$ induit une place v_σ dans S_p, et la relation

$$\kappa(x_{1,v}, x_{2,v}, \cdots, x_{n,v}) = \prod_{(i,\sigma)} \sigma(x_{i,v_\sigma})^{k_{i,\sigma}},$$

pour tout $(x_{1,v}, x_{2,v}, \cdots, x_{n,v})_v \in T$. Sous cette écriture, κ est dominant si pour tout σ, on a $k_{1,\sigma} \geq k_{2,\sigma} \geq \cdots \geq k_{n,\sigma}$. On dira aussi que κ est *M-loin des murs* si on a $|\kappa_{i,\sigma} - \kappa_{i+1,\sigma}| \geq M$ pour tout $\sigma \in \mathrm{Hom}(F, \overline{\mathbb{Q}}_p)$ et tout entier $i = 1, \ldots, n-1$.

À toute représentation $\Pi \in \mathcal{A}$ est associé un unique poids dominant $\kappa(\Pi) \in X^*(T)$, à savoir le plus haut poids de $\otimes_w \Pi_w$ avec w parcourant $\cup_{v \in S_p} \Sigma_\infty(v)$, défini au moyen de ι_∞ et ι_p (voir le §1.4 pour les détails). Fixons q un entier tel que pour tout $v \in S_p$, $|k_v^\times|$ divise $q-1$ où k_v est le corps résiduel de F_v. Le résultat suivant est une conséquence immédiate du théorème 1.6.

[4]Cependant, noter que les fibres (finies) de $\Pi \mapsto \psi_\Pi$ ne sont pas nécéssairement des singletons car la multiplicité un forte n'est pas satisfaite pour U quand $n > 1$.

Théorèm 1.3. *Soient* $\Pi \in \mathcal{A}$ *et* $m \in \mathbb{N}$. *Il existe* $M, M' \in \mathbb{N}$ *tels que pour tout*

$$\kappa \in \kappa(\Pi) + p^M(q-1)X^*(T)$$

dominant et M'*-loin des murs, il existe* $\Pi' \in \mathcal{A}$ *vérifiant*

$$\Pi' \equiv \Pi \bmod p^m \text{ et } \kappa(\Pi') = \kappa.$$

En particulier, l'image de $\Pi \mapsto (\psi_\Pi, \kappa(\Pi))$, $\mathcal{A} \to \mathrm{Hom}_{\mathbb{Z}}(\mathcal{H}^S, \overline{\mathbb{Q}}_p) \times (X^*(T) \otimes \mathbb{Z}_p)$, est sans point isolé (le membre de droite étant muni de la distance ultramétrique évidente).[5]

1.4 Poids et raffinements : l'application ν

De même que les congruences de Kummer entre les nombres de Bernoulli ont une incarnation analytique p-adique en la fonction ζ p-adique de Kubota-Leopold et Iwasawa (une fonction analytique p-adique sur l'espace des caractères p-adiques de $\mathbb{Z}_p^*$), les congruences ci-dessus admettent une variante analytique plus subtile, qui est le contenu des variétés de Hecke.

Plus exactement, la variété de Hecke de U attachée à $(S_p, W_\infty, \mathcal{H}, e)$ est un espace analytique p-adique interpolant tous les couples (ψ, ν) où ν contient la donnée du *poids* de Π_∞ et d'un *p-raffinement* de Π_{S_p} où Π est la représentation ayant donné lieu à ψ. À cet effet il nous reste à définir ces ν précisément. Du point de vue du parallèle avec les congruences de Kummer, ce ν représente la modification du facteur eulérien en p nécessaire à effectuer pour obtenir les congruences, en dimension $n > 1$ elle n'est pas unique et chaque choix possible conduit à des congruences différentes en général.

Pour tout $v \in S_p$, on note T_v^0 le sous-groupe compact maximal de T_v. On pose $T^0 = \prod_{v \in S_p} T_v^0$ et

$$\mathcal{T} = \mathrm{Hom}(T, \mathbb{G}_m^{\mathrm{rig}}), \ \ \mathcal{W} = \mathrm{Hom}(T^0, \mathbb{G}_m^{\mathrm{rig}})$$

les espaces analytiques[6] en groupes sur $\mathbb{Q}_p$ paramétrant respectivement les caractères continus p-adiques de T et de T^0. Le groupe $X^*(T)$ des caractères algébriques de T est un sous-groupe de $\mathcal{T}(L)$.

Nous allons tout d'abord définir une application naturelle

$$\mathcal{A} \longrightarrow \mathcal{T}(\overline{\mathbb{Q}}_p), \ \ \Pi \mapsto \kappa(\Pi),$$

déjà mentionnée plus haut. Soient $v \in S_p$, $w \in \Sigma_\infty(v)$, et W une représentation irréductible continue de $U(F_w)$. Alors $W_{|U(F)}$ est canoniquement définie sur $\overline{\mathbb{Q}}$ via ι_∞, et ce $\overline{\mathbb{Q}}$-modèle induit une F_v-représentation algébrique de $U(F_v)$ à coefficients dans $\overline{\mathbb{Q}}_p$ en étendant les scalaires à ι_p. Le plus haut poids de cette représentation, c'est-à-dire l'unique sous-caractère

[5] J'ignore si cette image est d'adhérence compacte en général. Il s'agirait de savoir si les $\psi_\Pi(\mathcal{H}^S)$, $\Pi \in \mathcal{A}$, restent dans une extension *finie* de $\mathbb{Q}_p$.

[6] Dans cette note, les espaces rigides analytiques que nous considérons sont pris au sens de Tate. En tant qu'espace analytique, $\mathcal{W}$ est une réunion disjointe de boules ouvertes et $\mathcal{T}$ est produit direct (non canoniquement) de $\mathcal{W}$ par des copies de $\mathbb{G}_m^{\mathrm{rig}}$. Nous renvoyons par exemple à [Bu, §8] pour des généralités sur ces espaces.

de sa restriction à B_v, définit un caractère algébrique $\kappa_{W,w} : T_v \longrightarrow \overline{\mathbb{Q}}_p^*$, que l'on étend trivialement en un caractère de T. On pose alors pour $\Pi \in \mathcal{A}$,

$$\kappa(\Pi) = \kappa(\Pi_\infty) = \prod_{v \in S_p, w \in \Sigma_\infty(v)} \kappa_{\Pi_w, w}.$$

Par construction, on a $\kappa(\Pi) \in X^*(T) \subset \mathcal{T}(L)$.

Pour terminer la définition de ν nous devons rappeler le concept de *p-raffinement* d'une représentation de $U(F_v)$ ([M, §3],[Ch, §4.8],[BCh, §6.4]), qui est au coeur de la théorie. On suppose encore $v \in S_p$. Soit π_v une représentation complexe lisse et irréductible de $\mathrm{GL}_n(F_v)$ admettant un vecteur invariant non nul sous le sous-groupe d'Iwahori K_v. Un résultat de Borel et Matsumoto assure que π_v est une sous-représentation d'une série principale non ramifiée. Autrement dit, il existe un caractère

$$\chi : T_v/T_v^0 \longrightarrow \mathbb{C}^*$$

tel que π_v se plonge comme sous-représentation dans l'induite parabolique lisse et normalisée de χ à $U(F_v)$:

$$\begin{aligned}\mathrm{Ind}_{B_v}^{\mathrm{GL}_n(F_v)} \chi = \{f : \mathrm{GL}_n(F_v) \to \mathbb{C} \text{ lisse}, f(bg) \\ = (\chi\delta_{B_v}^{1/2})(b)f(g)\ \forall b \in B_v, \forall g \in \mathrm{GL}_n(F_v)\}.\end{aligned}$$

On dira qu'un tel caractère χ est un *raffinement accessible*[7] de π_v.

Concrètement, un raffinement accessible de π_v est simplement la donnée d'un ordre particulier sur les valeurs propres de la classe de conjugaison semi-simple du Frobenius géométrique dans le paramètre de Langlands de π_v. Précisément, cet ordre $\varphi_{1,v}, \cdots, \varphi_{n,v}$ est défini par $\varphi_{i,v} = \chi(t_{i,v})$ où

$$t_{i,v} = (1, \ldots, 1, \varpi_v, 1, \ldots, 1) \ \in \ T_v,$$

l'uniformisante ϖ_v de F_v étant en i-ème position dans le tore diagonal T_v.

Pour revenir aux représentations automorphes de U, si $\Pi \in \mathcal{A}$ nous appellerons *raffinement de* Π, ou de Π_{S_p}, la donnée d'un caractère

$$\chi = \bigotimes_{v \in S_p} \chi_v : T/T^0 \longrightarrow \overline{\mathbb{Q}}_p^*$$

où pour chaque v, χ_v est un raffinement accessible de $\Pi_v|\det|^{\frac{1-n}{2}}$ composé par $\iota_p\iota_\infty^{-1}$. Au final, *à chaque représentation automorphe raffinée* (Π, χ), $\Pi \in \mathcal{A}$, *on peut associer l'élément* (produit)

$$\nu(\Pi, \chi) := \kappa(\Pi) \cdot \chi \cdot (\delta_B^{-1/2}|\det|^{\frac{n-1}{2}}) \in \mathcal{T}(\overline{\mathbb{Q}}_p).$$

[7]Plus généralement, on appelle *raffinement de* π_v un caractère de T_v/T_v^0 tel que π_v soit un *sous-quotient* de $\mathrm{Ind}_{B_v}^{\mathrm{GL}_n(F_v)}\chi$. Si π_v est sphérique et tempérée (ou même simplement générique) alors tous ses raffinements sont accessibles; en général c'est inexact, par exemple la représentation de Steinberg ou la représentation triviale n'admettent qu'un seul raffinement accessible. En revanche, il existe toujours au moins un raffinement accessible.

Notons que $\nu(\Pi, \chi)$ détermine exactement $\kappa(\Pi)$ et χ. La normalisation par le facteur constant $\delta_B^{-1/2}|\det|^{\frac{n-1}{2}}$ apparaît pour des raisons d'algébricité.

1.5 Énoncé du théorème

Nous avons maintenant suffisamment de notations pour énoncer le théorème. Considérons le sous-ensemble

$$\mathcal{Z} \subset \mathrm{Hom}_{\mathrm{ann}}(\mathcal{H}, \overline{\mathbb{Q}}_p) \times \mathcal{T}(\overline{\mathbb{Q}}_p)$$

formé des couples (ψ, ν) tels qu'il existe une représentation automorphe $\Pi \in \mathcal{A}$ de U et un raffinement accessible χ de Π_{S_p} de sorte que ψ est un système de valeurs propres de $\mathcal{H}$ associé à Π et $\nu = \nu(\Pi, \chi)$.

Théorèm 1.6. *Il existe un unique quadruplet (X, ψ, ν, Z) où :*

- *X est un espace rigide analytique sur L qui est réduit,*
- *$\psi : \mathcal{H} \longrightarrow \mathcal{O}(X)$ est un morphisme d'anneaux,*
- *$\nu : X \longrightarrow \mathcal{T}$ est un morphisme analytique fini,*
- *$Z \subset X(\overline{\mathbb{Q}}_p)$ est un sous-ensemble Zariski-dense et d'accumulation,*

tel que les conditions (i) et (ii) suivantes soient satisfaites :

(i) *Pour tout ouvert affinoide $V \subset \mathcal{T}$, l'application naturelle*

$$\psi \otimes \nu^* : \mathcal{H} \otimes_{\mathbb{Z}} \mathcal{O}(V) \longrightarrow \mathcal{O}(\nu^{-1}(V))$$

est surjective.

(ii) *L'application naturelle d'évaluation $X(\overline{\mathbb{Q}}_p) \longrightarrow \mathrm{Hom}_{\mathrm{ann}}(\mathcal{H}, \overline{\mathbb{Q}}_p)$,*

$$x \mapsto \psi_x := (h \mapsto \psi(h)(x)),$$

induit une bijection $Z \xrightarrow{\sim} \mathcal{Z}$, $z \mapsto (\psi_z, \nu(z))$.

Il satisfait de plus :

(iii) *X est équidimensionnel de dimension $\dim(\mathcal{W}) = n(\sum_{v \in S_p} [F_v : \mathbb{Q}_p])$.*

(iv) *Plus précisément, soit*

$$\kappa : X \longrightarrow \mathcal{W}$$

la composée de ν par la projection canonique $\mathcal{T} \to \mathcal{W}$. L'espace X est admissiblement recouvert par les ouverts affinoides $\Omega \subset X$ tels que $\kappa(\Omega)$ est un ouvert affinoide de $\mathcal{W}$ et tels que $\kappa_{|\Omega} : \Omega \to \kappa(\Omega)$ soit finie, et surjective restreinte à chaque composante irréductible de Ω. Enfin, l'image par κ de chaque composante irréductible de X est un ouvert Zariski de $\mathcal{W}$.

(v) $\psi(\mathcal{H}^S) \subset \mathcal{O}(X)^{\leq 1}$.

(vi) *(Critère de classicité) Soit $x \in X(\overline{\mathbb{Q}}_p)$ tel que $\kappa(x)$ soit algébrique et dominant. Pour chaque $(i, \sigma) \in \{1, \ldots, n\} \times \mathrm{Hom}(F, \overline{\mathbb{Q}}_p)$, notons $k_{i,\sigma}$ l'entier associé à $\kappa(x) \in X^*(T)$ comme au §1.4, $v_\sigma \in S_p$ la place induite par σ, et $\varphi_{i,\sigma} = \delta(t_{i,v_\sigma})(x)$. Supposons que pour tout $(i, \sigma) \in \{1, \ldots, n-1\} \times \mathrm{Hom}(F, \overline{\mathbb{Q}}_p)$, on ait*

$$\frac{\mathbf{v}(\varphi_{1,\sigma}\varphi_{2,\sigma}\cdots\varphi_{i,\sigma})}{\mathbf{v}(\varpi_{v_\sigma})} < k_{i,\sigma} - k_{i+1,\sigma} + 1,$$

alors $x \in Z$.

Nous renvoyons à [BCh, Ch. 7] pour une discussion détaillée de ce type d'énoncé.

Rappels : Bornons-nous ci-dessous à rappeler les notions de géométrie rigide utilisées. Si Y est un espace rigide sur L, nous notons $\mathcal{O}(Y)$ l'algèbre de ses fonctions globales et $\mathcal{O}(Y)^{\leq 1}$ le sous-anneau des fonctions bornées par 1. De plus, $Y(\overline{\mathbb{Q}}_p)$ désigne la réunion des $Y(L')$ où $L' \subset \overline{\mathbb{Q}}_p$ est une extension finie de L.

Une partie Z de $Y(\overline{\mathbb{Q}}_p)$ est dite Zariski-dense (resp. d'accumulation) si l'ensemble $|Z| \subset Y$ des points fermés sous-jacent l'est.[8] Une partie $Z \subset Y$ est dite Zariski-dense si le seul fermé analytique réduit de Y contenant Z est la nilréduction Y_{red} de Y. Si Y est de Stein et réduit (ce qui est le cas de $\mathcal{W}$, $\mathcal{T}$ et X) alors il est équivalent de demander qu'une fonction analytique globale sur Y s'annulant sur Z est identiquement nulle.

On dit que Z s'accumule en une partie Z' de Y si pour tout $z \in Z'$ et tout voisinage ouvert U de Y contenant z il existe un ouvert affinoide $V \subset U$ contenant z tel que $Z \cap V$ est Zariski-dense dans V. On dit que Z est d'accumulation si il s'accumule en lui-même.

Par exemple, le sous-ensemble de $\mathcal{W}(\overline{\mathbb{Q}}_p)$ paramétrant les caractères algébriques dominants est Zariski-dense, et son sous-ensemble des caractères réguliers s'accumule en tous les caractères algébriques (lemme 2.7).

2 Preuve du théorème

La preuve que nous donnons ci-dessous du théorème 1.6 est similaire à celle donnée dans [BCh, §7] dans le cas $F = \mathbb{Q}$, et auquel nous nous réfèrerons parfois. La variété de Hecke X associée à $(S_p, W_\infty, \mathcal{H}, e)$ est construite par un procédé formel, dû à Coleman et Coleman-Mazur [CM] (voir aussi [Bu], [Ch]), à partir de l'action d'opérateurs de Hecke compacts agissant sur la famille plate des espaces de Banach de formes automorphes p-adiques pour U. Une partie importante de la démonstration consiste à définir ces espaces, une autre étant de vérifier un analogue du critère de "classicité" de Coleman [C1].

2.1 Notations

Dans les §2.2 à §2.8 suivants, de nature purement locale, F_p est une extension finie de $\mathbb{Q}_p$, d'uniformisante ϖ et d'anneau d'entiers $\mathcal{O}_p$. Nous désignerons ici[9] par G le groupe

[8] Noter que l'ensemble $\mathcal{Z}$ introduit plus haut est stable par l'action de $\mathrm{Gal}(\overline{\mathbb{Q}}_p/L)$.

[9] Nous nous excusons du fait que nous les notations utilisées ici pour T et B (et plus loin $\mathcal{W}$) sont en léger conflit avec les notations utilisées dans l'introduction.

$\mathrm{GL}_n(F_p)$, B son sous-groupe triangulaire supérieur, N le radical unipotent de B, et T son tore diagonal. Soient T^0 le sous-groupe compact maximal de T et I le sous-groupe d'Iwahori de G constitué des éléments de $\mathrm{GL}_n(\mathcal{O}_p)$ qui sont triangulaires supérieurs modulo ϖ. Enfin, on note $T^- \subset T$ le sous-monoïde des éléments de la forme

$$\mathrm{diag}(x_1, \cdots, x_n), \quad x_i \in F_p^*, \quad \mathbf{v}(x_i) \geq \mathbf{v}(x_{i+1}) \quad \forall 1 \leq i < n,$$

$M \subset G$ le sous-monoïde engendré par I et T^-, et $T^{--} \subset T^-$ le sous-monoïde des éléments $\mathrm{diag}(x_1, \cdots, x_n)$ satisfaisant $\mathbf{v}(x_i) \geq \mathbf{v}(x_{i+1})+1$ pour $1 \leq i < n$ (on rappelle que $\mathbf{v}(p) = 1$).

2.2 Algèbre de Hecke-Iwahori et raffinements accessibles

Commençons par rappeler le lien existant entre algèbre de Hecke-Iwahori et raffinements. Nous renvoyons à [BCh, §6.2] ou à [Ch, §4.8] pour une discussion détaillée.

Soit $\mathfrak{A}^- \subset \mathbb{Z}[G//I]$ le sous-anneau engendré par les fonctions caractéristiques $\mathbb{1}_{ItI}$ des doubles classes ItI pour $t \in T^-$. Ces fonctions sont inversibles dans $\mathbb{Z}[1/p][G//I]$, on peut donc considérer le sous-anneau $\mathfrak{A} \subset \mathbb{Z}[1/p][G//I]$ engendré par les $\mathbb{1}_{ItI}$ $(t \in T^-)$, et leurs inverses. On verra l'inclusion $T^-/T^0 \subset T/T^0$ comme une symétrisation du monoïde T^-/T^0. Le lemme suivant récapitule un ensemble de résultats dûs à Bernstein, Borel, Casselman et Matsumoto (voir par exemple [BCh, §6.2]).

Lemme 2.3. (i) *On a l'égalité* $M = \coprod_{t\in T^-/T^0} ItI$. *De plus, l'application*

$$\tau : M \longrightarrow T^-/T^0,$$

définie par la relation $m \in I\tau(m)I$, *est un homomorphisme multiplicatif.*

(ii) *L'application* $T^-/T^0 \to \mathfrak{A}^-$, $t \mapsto \mathbb{1}_{ItI}$, *est multiplicative et s'étend en des isomorphismes d'anneaux*[10] $\mathbb{Z}[T^-/T^0] \xrightarrow{\sim} \mathfrak{A}^-$ *et* $\mathbb{Z}[T/T^0] \xrightarrow{\sim} \mathfrak{A}$.

Soit π *une représentation complexe lisse irréductible de G telle que* $\pi^I \neq 0$. *Regardons* π^I *comme module sur* $\mathbb{C}[T/T^0]$ *par le dernier isomorphisme du (ii), sa semi-simplification* $(\pi^I)^{\mathrm{ss}}$ *est une somme de caractères de* T/T^0.

(iii) *Un caractère* $\chi : T/T^0 \to \mathbb{C}^*$ *apparaît dans* $(\pi^I)^{\mathrm{ss}}$ *si, et seulement si,* $\chi\delta_B^{1/2}$ *est un raffinement accessible de* π. *En particulier, il existe toujours au moins un raffinement accessible de* π.

2.4 Préliminaires sur les fonctions et les caractères r-analytiques

Soient Λ un $\mathbb{Z}_p$-module libre de rang fini, $r \geq 0$ un entier, et A une $\mathbb{Q}_p$-algèbre affinoide. Choisissons $e_1, \ldots, e_d$ une $\mathbb{Z}_p$-base de Λ, et notons $u_1, \cdots, u_d \in \mathrm{Hom}_{\mathbb{Z}_p}(\Lambda, \mathbb{Z}_p)$ sa base duale. L'algèbre de Tate $A\langle u_1, \cdots, u_d\rangle$ s'injecte de manière naturelle dans la A-algèbre A^Λ des fonctions de Λ dans A, et son image dans cette dernière ne dépend pas du choix de la

[10] Il faut prendre garde que pour $n > 1$ cet homomorphisme n'envoie pas $t \in T$ sur $\mathbb{1}_{ItI}$, mais plutôt sur $\mathbb{1}_{IaI}/\mathbb{1}_{IbI}$ si $a, b \in T^-$ et $t = a/b$.

base (e_i). On la notera $A\langle\Lambda\rangle$. Si A est muni d'une norme, on munit $A\langle\Lambda\rangle \simeq A\langle u_1, \cdots, u_d\rangle$ de la norme de Gauss associée.

Une fonction $f : \Lambda \to A$ est dite r-analytique si pour chaque $\lambda \in \Lambda$, l'application

$$f_\lambda : \Lambda \to A, \quad x \mapsto f(\lambda + p^r x),$$

appartient à $A\langle\Lambda\rangle$. Ces fonctions forment une sous-A-algèbre de A^Λ que l'on munit de la norme $\sup_{\lambda\in\Lambda}|f_\lambda|_{\mathrm{Gauss}}$ (noter que le sup porte sur un sous-ensemble fini). On rappelle que suivant Coleman, un A-module de Banach est dit *orthonormalisable* s'il est isométrique à $A\langle u\rangle$. Le résultat suivant est alors immédiat.

Lemme 2.5. *Le A-module des fonctions r-analytiques $\Lambda \to A$ est un A-module de Banach orthonormalisable.*

Des considérations similaires s'appliquent aux caractères de T^0 comme suit. Un caractère $c : \mathcal{O}_p^* \to A^*$ est dit r-analytique si la fonction qui s'en déduit

$$x \mapsto c(1 + \varpi x), \mathcal{O}_p \to A,$$

est r-analytique sur le $\mathbb{Z}_p$-module $\mathcal{O}_p$; par multiplicativité il suffit en fait de vérifier que $x \mapsto c(1 + p^r \varpi x)$ est dans $A\langle\mathcal{O}_p\rangle$. De même, un caractère $T^0 = (\mathcal{O}_p^*)^n \to A^*$ est dit r-analytique si sa restriction à chaque facteur $\mathcal{O}_p^*$ l'est.

Le lemme suivant est bien connu : voir par exemple [Bu, Lemme 8.2] pour le (i), le (ii) découle du fait qu'un caractère continu $\mathbb{Z}_p \to A^*$, A disons affinoide, est r-analytique pour r assez grand.

Lemme 2.6. (i) *Le foncteur des $\mathbb{Q}_p$-algèbres affinoides dans les groupes, associant à A le groupe des morphismes de groupes continus $T^0 \to A^*$, est représentable par un espace rigide analytique en groupes sur $\mathbb{Q}_p$ que l'on notera $\mathcal{W}$.*

On notera aussi $\chi : T^0 \to \mathcal{O}(\mathcal{W})^*$ le caractère universel, et pour tout morphisme $V \to \mathcal{W}$ avec V affinoide, on notera enfin $\chi_V : T^0 \longrightarrow \mathcal{O}(V)^*$ le caractère induit par restriction.

(ii) *Un tel V étant fixé, il existe un plus petit entier r_V tel que χ_V est r-analytique pour tout $r \geq r_V$.*

En tant qu'espace rigide, $\mathcal{W}$ est une réunion disjointe de $|\mu|^n$ boules unités ouvertes de dimension $n[F_p : \mathbb{Q}_p]$, $\mu \subset F_p^*$ étant le sous-groupe des racines de l'unité. Si L est une extension galoisienne de $\mathbb{Q}_p$ contenant F_p, la restriction à T^0 induit une inclusion $X^*(T) \subset \mathcal{W}(L)$. Soit $X^*(T)_{\mathrm{reg}} \subset X^*(T)$ le sous-monoïde des caractères B-dominants et *réguliers* (c'est-à-dire qui ne sont pas dans un mur, ou encore que les $(k_{\sigma,i})$ associés satisfont $k_{\sigma,i} > k_{\sigma,i+1}$ pour tout $\sigma \in \mathrm{Hom}(F_p, \overline{\mathbb{Q}}_p)$ et $1 \leq i < n$).

Lemme 2.7. *$X^*(T)$ est Zariski-dense dans $\mathcal{W}$ et $X^*(T)_{\mathrm{reg}}$ s'accumule en $X^*(T)$.*

Preuve — Pour le premier point on peut supposer $n = 1$, donc $T = \mathcal{O}_p^*$, auquel cas le résultat suit car μ est un groupe cyclique et donc l'un quelconque des plongements de F_p dans L engendre $\mathrm{Hom}(\mu, L^*)$.

Pour le second, on vérifie immédiatement que si $\kappa \in X^*(T)$ et si Ω est un voisinage ouvert affinoide de κ dans $\mathcal{W}$, disons même une boule, alors pour tout entier N assez grand on a (en notation additive)

$$\kappa + (q-1)p^N X^*(T)_{\mathrm{reg}} \subset X^*(T)_{\mathrm{reg}} \cap \Omega,$$

où $q := |\mathcal{O}_p/\varpi\mathcal{O}_p|$. Il est alors élémentaire de vérifier que le terme de gauche est Zariski-dense dans la boule Ω, par l'indépendance L-linéaire des plongements de F_p dans L. Le point est que si $v_1, \ldots, v_r \in \mathcal{O}_L^r$ est L-libre, et si $f \in L\langle z_1, \ldots, z_r\rangle$ s'annule en tous les éléments du monoïde $\sum_{i=1}^r \mathbb{N}v_i$, alors $f = 0$. □

2.8 La série principale localement analytique et T^--stable d'un sous-groupe d'Iwahori

Soit $\overline{N}_0$ le sous-groupe des éléments triangulaires *inférieurs* de I. Soit J l'ensemble des couples (i, j), avec $1 \leq i < j \leq n$, et soit

$$\phi : \mathcal{O}_p^J \to \overline{N}_0, \ (\phi(x))_{i,j} = \varpi x_{(i,j)},$$

la bijection évidente. Si A est une $\mathbb{Q}_p$-algèbre affinoide, une fonction $f : \overline{N}_0 \to A$ est dite r-analytique si la composée de ϕ par f l'est, cela ne dépend pas du choix de ϖ intervenant dans ϕ. Ces fonctions forment une A-algèbre de Banach par transport de structure.

Afin de prolonger les caractères de T^0 en des caractères de B, nous aurons besoin de choisir une section du morphisme $T \to T/T^0$. Pour fixer les idées, notons $T^{\varpi} \subset T$ le sous-groupe constitué des

$$\mathrm{diag}(\varpi^{a_1}, \cdots, \varpi^{a_n}), \quad a_i \in \mathbb{Z}, \quad \forall 1 \leq i \leq n.$$

On a $T = T^{\varpi} \times T^0$ et si ψ est un caractère de T^0, nous noterons $\widetilde{\psi}$ le caractère de B tel que $\widetilde{\psi}_{|T^0} = \psi$ et $\widetilde{\psi}(T^{\varpi}) = \widetilde{\psi}(N) = \{1\}$.

La série principale localement analytique et T^--stable de I peut être définie comme suit. Pour tout morphisme $V \to \mathcal{W}$ avec V affinoide, et pour tout $r \geq r_V$, on pose

$$\mathcal{C}(V, r) = \left\{ \begin{array}{l} f : IB \longrightarrow \mathcal{O}(V), \ f(xb) = \widetilde{\chi_V}(b) f(x) \ \ \forall\, x \in IB, \ b \in B, \\ \\ f_{|\overline{N}_0} \text{ est } r\text{-analytique.} \end{array} \right\}.$$

En particulier, tout caractère continu $\delta : T^0 \to L^*$ avec $L/\mathbb{Q}_p$ finie définit un unique L-point V de $\mathcal{W}$, on notera aussi simplement $\mathcal{C}(\delta, r)$ au lieu de $\mathcal{C}(V, r)$ dans ce cas.

En général, $\mathcal{C}(V, r)$ est un $\mathcal{O}(V)$-module de manière naturelle. On le munit de la norme $|f| := |f_{|\overline{N}_0}|$ héritée des fonctions r-analytiques sur $\overline{N}_0$, $\mathcal{O}(V)$ étant muni de sa norme sup. Le lemme suivant, essentiellement immédiat, résume les propriétés élémentaires de la famille $\{\mathcal{C}(V, r), V, r \geq r_V\}$ ainsi définie.

Lemme 2.9. (i) *L'application $f \mapsto f_{|\overline{N}_0}$ induit une isométrie $\mathcal{O}(V)$-linéaire de $\mathcal{C}(V, r)$ sur les fonctions r-analytiques $\mathcal{O}(V)$-valuées sur $\overline{N}_0$. En particulier, c'est un $\mathcal{O}(V)$-module orthonormalisable.*

(ii) *$\mathcal{C}(V,r)$ est munie d'une action $\mathcal{O}(V)$-linéaire de M par translations à gauche*

$$(m.f)(x) = f(m^{-1}x).$$

Les éléments de M agissent par des endomorphismes continus de norme ≤ 1, et pour tout $t \in T^{--}$, l'action de t sur $\mathcal{C}(V,r)$ est $\mathcal{O}(V)$-compacte.

(iii) *Pour tout morphisme $V' \to V$, la restriction $\mathcal{C}(V,r) \to \mathcal{C}(V',r)$ induit un isomorphisme M-équivariant*

$$\mathcal{C}(V,r)\widehat{\otimes}_{\mathcal{O}(V)}\mathcal{O}(V') \xrightarrow{\sim} \mathcal{C}(V',r).$$

Si $r' \geq r$ l'inclusion naturelle $\mathcal{C}(V,r) \longrightarrow \mathcal{C}(V,r')$ est continue et $\mathcal{O}(V)[M]$-équivariante, $\mathcal{O}(V)$-compacte si de plus $r' > r$. Enfin, si $r \geq 1$ et $t \in T^{--}$, alors l'action de t se factorise par l'inclusion compacte $\mathcal{C}(V,r-1) \longrightarrow \mathcal{C}(V,r)$ ci-dessus.

Preuve — Le premier point du (i) vient de ce que la multiplication dans G induit une bijection (décomposition d'Iwahori) $\overline{N}_0 \times B \xrightarrow{\sim} IB$, et le reste du lemme 2.5. Remarquons que si $t \in T^-$, on a $t^{-1}\overline{N}_0 t \subset \overline{N}_0$, de sorte que $M^{-1}IB \subset IB$. Soit $i \in I$. La multiplication $n \mapsto in$ par i dans G et la décomposition d'Iwahori induisent une application

$$\overline{N}_0 \to \overline{N}_0 \times T^0 \times (N \cap I),\ n \mapsto (\alpha(n), \beta(n), \gamma(n)),$$

qui n'est que les $\mathcal{O}_p$-points d'un morphisme analogue défini au niveau des $\mathcal{O}_p$-schémas formels associés de manière évidente à $\overline{N}_0$, T^0 et $N \cap I$. En particulier via ϕ, $n \mapsto \alpha(n)$ est un automorphisme du $\mathcal{O}_p$-espace affine formel de rang $|J|$, de sorte que son action sur les fonctions préserve la r-analyticité pour tout r. De même, l'application $n \mapsto \beta(n)$ a la propriété que si $c : T^0 \to A^*$ est r-analytique, alors $n \mapsto c(\beta(n))$ est r-analytique sur $\overline{N}_0$. (Pour un calcul explicite des $\alpha(n)$ et $\beta(n)$, voir [Ch, §3]). Cela démontre le premier point du (ii). Le second point est standard si l'on remarque que

$$\forall t \in T^{--},\quad \phi^{-1}(t^{-1}\overline{N}_0 t) \subset p\,\mathcal{O}_p^J.$$

Le (iii) est une conséquence immédiate du (i). □

2.10 Plongement des représentations algébriques de G dans la série principale

Soit L une extension finie de $\mathbb{Q}_p$ et $\delta : T \longrightarrow L^*$ un caractère continu, étendu à B par $\delta(N) = \{1\}$. On note r_δ le plus petit entier r tel que $\delta_{|T^0}$ soit r-analytique. Soit

$$i_B^{IB}(\delta, r)$$

le L-Banach des fonctions $f : IB \to L$ dont la restriction à $\overline{N}_0$ est r-analytique et qui satisfont $f(xb) = \delta(b)f(x)$ pour tout $x \in IB$ et $b \in B$. Cet espace est encore muni d'une action continue L-linéaire de M par translations à gauche. Il coïncide avec l'espace $\mathcal{C}(\delta_{|T^0}, r)$ défini au paragraphe précédent si $\delta_{|T^\omega}$ est trivial. En général, nous aurons besoin d'un sorite permettant de comparer $i_B^{IB}(\delta, r)$ et $\mathcal{C}(\delta_{|T^0}, r)$.

Soit $\delta' : T \longrightarrow L^*$ un caractère tel que $\delta'(T^0) = \{1\}$. Ce caractère se prolonge en un morphisme de monoïdes $\delta' : M \longrightarrow L^*$ par $\delta'(m) := \delta'(\tau(m))$ (lemme 2.3 (i)), dont on vérifie facilement qu'il s'étend de manière unique en une fonction $\delta' : M^{-1}B = IB \to L^*$ telle que $\delta'(m^{-1}tn) = \delta'(m)^{-1}\delta'(t)$ pour tout $m \in M, t \in T$ et $n \in N$. Si $r \geq r_\delta$, on dispose donc d'une bijection naturelle

$$i_B^{IB}(\delta, r) \longrightarrow i_B^{IB}(\delta\delta', r), \ f \mapsto (x \mapsto \delta'(x)f(x)),$$

dont on vérifie immédiatement qu'elle induit une isométrie M-équivariante

$$(i_B^{IB}(\delta, r)) \otimes \delta'^{-1} \longrightarrow i_B^{IB}(\delta\delta', r). \tag{2.1}$$

Supposons jusqu'à la fin de ce paragraphe que $\delta \in X^*(T)$ est un caractère algébrique dominant, auquel cas $r_\delta = 0$. Soit W_δ la L-représentation de G de plus haut poids δ, elle est irréductible et absolument simple. Le choix d'un vecteur $v \in W_\delta$ de plus haut poids fournit un morphisme $L[M]$-équivariant non nul, donc injectif,

$$W_\delta^* \longrightarrow i_B^{IB}(\delta, 0), \ \varphi \mapsto (x \mapsto \varphi(x(v))).$$

On déduit alors de (2.1) le lemme suivant.

Lemme 2.11. *Soit δ^ϖ le caractère de T trivial sur T^0 et coïncidant avec δ sur T^ϖ. Il existe un plongement $L[M]$-équivariant*

$$W_\delta^* \otimes \delta^\varpi \longrightarrow \mathcal{C}(\delta_{|T^0}, 0) = i_B^{IB}(\delta, 0) \otimes \delta^\varpi.$$

En fait, le sous-espace de $\mathcal{C}(\delta_{|T^0}, 0)$ ainsi défini est exactement le sous-espace des fonctions sur IB qui sont la restriction d'un fonction polynomiale sur G (vu comme $\mathbb{Q}_p$-groupe).

Nous allons terminer ce paragraphe en introduisant un sous-espace intermédiaire entre W_δ^* et $\mathcal{C}(\delta_{|T^0}, 0)$. Supposons que L contienne une clôture galoisienne de F_p. Pour chaque $j \in J$ et $\sigma \in \mathrm{Hom}(F_p, L)$, on dispose d'une fonction $\sigma_j : F_p^J \to L$, $(x_i) \mapsto \sigma(x_j)$. Ces éléments forment une base du L-espace vectoriel des applications $\mathbb{Q}_p$-linéaires de F_p^J dans L. Disons qu'une fonction $f : \mathcal{O}_p^J \to L$ est $\sharp$-analytique si c'est la restriction à $\mathcal{O}_p^J$ d'un élément de l'algèbre de Tate sur les σ_j :

$$L\langle \ \{\sigma_j\}_{(j,\sigma)\in J\times \mathrm{Hom}(F_p, L)} \ \rangle.$$

Par transport de structure via ϕ, cela nous fournit une notion de fonction $\sharp$-analytique sur $\overline{N}_0$, elles forment une L-algèbre de Banach pour la norme de Gauss de l'algèbre ci-dessus. On peut considérer l'espace

$$\mathcal{C}(\delta_{|T^0}, 0)^\sharp = \left\{ \begin{array}{l} f : IB \longrightarrow L, \ f(xb) = \widetilde{\delta_{|T^0}}(b) f(x) \ \ \forall \ x \in IB, \ b \in B, \\ \\ f_{|\overline{N}_0} \text{ est } \sharp\text{-analytique.} \end{array} \right\}.$$

C'est un L-Banach pour la norme introduite plus haut sur les fonctions $\sharp$-analytiques. Une fonction $\sharp$-analytique étant 0-analytique, et une fonction polynomiale sur G étant $\sharp$-analytique, on a des inclusions $W_\delta^* \otimes \delta^\varpi \subset \mathcal{C}(\delta_{|T^0}, 0)^\sharp \subset \mathcal{C}(\delta_{|T^0}, 0)$. Quand F_p est non ramifiée sur $\mathbb{Q}_p$ cette dernière inclusion est une égalité, mais pas en général.

Lemme 2.12. (i) $\mathcal{C}(\delta_{|T^0}, 0)^\sharp$ *est un sous-espace dense de* $\mathcal{C}(\delta_{|T^0}, 0)$ *stable par* M.

(ii) *Les éléments de* M *sont de norme* ≤ 1 *sur* $\mathcal{C}(\delta_{|T^0}, 0)^\sharp$ *et ceux de* T^{--} *sont compacts.*

(iii) *Il existe* $m \in \mathbb{N}$ *tel que pour tout* $t \in T^{--}$ *et* $f \in \mathcal{C}(\delta_{|T^0}, 0)$, *alors* $t^m(f) \in \mathcal{C}(\delta_{|T^0}, 0)^\sharp$.

Preuve — Les arguments donnés au (i) et (ii) du lemme 2.9 montrent que $\mathcal{C}(\delta_{|T^0}, 0)^\sharp$ est stable par M, ainsi que le (ii) ci-dessus. L'assertion de densité dans le (i) vient de ce que les fonctions $\mathbb{Z}_p$-polynomiales $\mathcal{O}_p^J \to L$ sont denses dans les fonctions 0-analytiques, et sont $\sharp$-analytiques. Le sous-$\mathcal{O}_L$-module

$$\sum_{\sigma \in \mathrm{Hom}(F_p, L)} \mathcal{O}_L \sigma \;\subset\; \mathrm{Hom}_{\mathbb{Z}_p}(\mathcal{O}_p, \mathcal{O}_L)$$

est un $\mathcal{O}_L$-réseau par indépendance linéaire des $[F_p : \mathbb{Q}_p]$ plongements. On en déduit que (iii) est satisfait pour tout entier m tel que l'indice de ce réseau divise p^m. □

L'intérêt de l'espace $\mathcal{C}(\delta_{|T^0}, 0)^\sharp$ est qu'il admet une présentation simple, la *présentation de Plücker*, permettant d'obtenir des estimées raisonnablement précises des normes des éléments $m \in M$ agissant sur $\mathcal{C}(\delta_{|T^0}, 0)^\sharp / W_\delta^*$. Cette présentation, classique pour les W_δ et introduite dans [Ch] pour la série principale analytique de GL_n, est la présentation obtenue en plongeant la variété des drapeaux complets de GL_n sur $\mathcal{O}_p$ dans le produit des espaces projectifs des $\Lambda^i(\mathcal{O}_p^n)$ (plongement de Plücker). Considérons à cet effet la décomposition de δ en poids fondamentaux :

$$\delta = \prod_{\sigma \in \mathrm{Hom}(F_p, L), 1 \leq i \leq n} \delta_{i,\sigma}$$

où $\delta_{i,\sigma}$ est le plus haut poids de la puissance symétrique $(k_{i,\sigma} - k_{i+1,\sigma})$-ième de la représentation fondamentale

$$\mathrm{GL}_n(F_p) \xrightarrow{\Lambda^i} \mathrm{GL}(\Lambda^i(F_p^n)) \xrightarrow{\sigma} \mathrm{GL}(\Lambda^i(L^n))$$

(on conviendra que $k_{n+1,\sigma} = 0$ pour tout σ).

Lemme 2.13. *Pour chaque* $\sigma \in \mathrm{Hom}(F_p, L)$ *et* $1 \leq i \leq n$ *il existe un* L*-Banach* $P_{i,\sigma}$ *muni d'une action linéaire continue de* M *par éléments de norme* ≤ 1, *ayant les propriétés suivantes :*

(i) *Il existe un diagramme commutatif de* $L[M]$*-modules de Banach,*

$$\begin{array}{ccc} \otimes_{i,\sigma}(W^*_{\delta_{i,\sigma}} \otimes \delta^{\varpi}_{i,\sigma}) & \longrightarrow & \widehat{\otimes}_{i,\sigma} P_{i,\sigma} \\ \big\downarrow & & \big\downarrow \\ W^*_\delta \otimes \delta^{\varpi} & \longrightarrow & \mathcal{C}(\delta_{|T^0}, 0)^\sharp \end{array} ,$$

les flèches verticales (resp. horizontales) étant surjectives (resp. injectives).

(ii) T^{--} *agit sur les* $P_{i,\sigma}$ *par des opérateurs compacts.*

(iii) *Si* $i < n$ *et*[11] $m \in It_1t_2\cdots t_iI \subset M$, *alors* $\frac{m}{\varpi^{k_{i,\sigma}-k_{i+1,\sigma}+1}}$ *est de norme* ≤ 1 *sur*

$$P_{i,\sigma}/(W^*_{i,\sigma} \otimes \delta^{\varpi}_{i,\sigma}).$$

Preuve — Définissons tout d'abord $P_{i,\sigma}$. Soit $1 \leq i \leq n$. Soient $G_i = \mathrm{GL}(\Lambda^i(F_p^n))$, $R_i = \Lambda^i(\mathcal{O}_p^n)$ et écrivons

$$R_i = \mathcal{O}_p e \oplus R'$$

une décomposition en $\mathcal{O}_p$-modules T^0-stables telle que e est un vecteur de plus haut poids de $R_i[1/p]$, vu comme F_p-représentation de $\mathrm{GL}_n(F_p)$ muni du Borel supérieur. Soient B_i le stabilisateur de $F_p e$ dans G_i, $\overline{B}_i$ le parabolique opposé pour la décomposition de R_i ci-dessus, I_i le sous-groupe parahorique des éléments $g \in G_i$ tels que $g(R_i) \subset R_i$ et $g(e) \in \mathcal{O}_p e + \varpi R_i$, et enfin $\overline{N}_i$ le sous-groupe des éléments unipotents de $\overline{B}_i \cap I_i$. L'application

$$g \mapsto \frac{g(e) - e}{\varpi}$$

identifie $\overline{N}_i$ à R'. On notera $\mathcal{O}(\overline{N}_i)$ l'algèbre de Tate des fonctions F_p-analytiques $\overline{N}_i \to F_p$, c'est-à-dire $F_p\langle u_1, \cdots, u_s\rangle$ où (u_i) est une $\mathcal{O}_p$-base du $\mathcal{O}_p$-dual de R'. Soient $\delta' : B_i \to F_p^*$ le caractère de B_i sur $F_p\, e$ et $m \in \mathbb{Z}$. On pose

$$P_i(m) = \left\{ \begin{array}{l} f : I_iB_i \longrightarrow F_p,\ f(xb) = \delta'(b)^m f(x) \ \ \forall\, x \in I_iB_i,\ b \in B_i, \\ \\ f_{|\overline{N}_i} \in \mathcal{O}(\overline{N}_i). \end{array} \right\}.$$

C'est un F_p-Banach isométrique à $\mathcal{O}(\overline{N}_i)$ (via $f \mapsto f_{|\overline{N}_i}$) qui est muni d'une action continue du sous-monoïde de G_i préservant I_iB_i par translations à gauche, et que l'on restreint via Λ^i en une représentation de M. On pose alors

$$P_{i,\sigma} := (P_i(k_{i,\sigma} - k_{i+1,\sigma}) \otimes_\sigma L) \otimes \delta^{\varpi}_{i,\sigma}.$$

Par construction, $P_{i,\sigma}$ contient $W^*_{\delta_{i,\sigma}} \otimes \delta_{i,\sigma}$ comme sous-$L[M]$-module des fonctions F_p-polynomiales. De plus, on dispose d'une application naturelle $L[M]$-équivariante

$$((f_{i,\sigma} \otimes_\sigma 1)_{i,\sigma}) \to (x \mapsto \prod_{i,\sigma} \sigma(f_{i,\sigma}(\Lambda^i(x)))),\ \widehat{\otimes}_{i,\sigma} P_{i,\sigma} \to \mathcal{C}(\delta_{|T^0}, 0). \qquad (2.2)$$

L'application $(\Lambda^i)_{i=1\ldots n} : \overline{N}_0 \to \prod_i \overline{N}_i$ étant induite par une immersion fermée de schémas formels sur $\mathcal{O}_p$, l'image de l'application (2.2) est exactement le sous-espace $\mathcal{C}(\delta_{|T^0}, 0)^\sharp$. De plus, l'application obtenue

$$\widehat{\otimes}_{i,\sigma} P_{i,\sigma} \to \mathcal{C}(\delta_{|T^0}, 0)^\sharp$$

[11] Comme au §1.4, on note $t_i \in T$ l'élément diagonal valant ϖ à la place i et 1 ailleurs.

est continue de norme ≤ 1. Nous avons donc construit la flèche verticale de droite du diagramme du (i). Cette flèche induit par construction la surjection de Plücker sur les sous-espaces de fonctions polynomiales, ce qui démontre le (i). L'assertion (ii) est immédiate.

L'assertion (iii) est conséquence du fait trivial suivant : si un endomorphisme continu θ de la F_p-algèbre de Tate $V := F_p\langle u_1, \cdots, u_s\rangle$ est tel que pour tout $i = 1 \dots s$,

$$\theta(u_i) \in \varpi\left(\sum_{j=1}^{s} \mathcal{O}_p u_j\right),$$

et si $V_n \subset V$ est le sous-espace des polynômes de degré total $\leq n$ en les u_i, alors la norme de θ/ϖ^{n+1} sur V/V_n est ≤ 1. □

2.14 Extension de ces constructions au cas du groupe $U(F_{S_p})$

Pour chaque place $v \in S_p$, choisissons une place $\tilde{v}$ de $\mathcal{K}$ divisant p, ce qui nous permet d'identifier une fois pour toutes F_v à $\mathcal{K}_{\tilde{v}}$ et $U(F_v)$ à $\mathrm{GL}_n(F_v)$ (cf. §1.1). Les constructions des paragraphes 2.2 à 2.10 s'appliquent et nous fournissent des objets que nous noterons dès à présent par les même symboles mais indicés par les $v \in S_p$. Pour les groupes ou monoïdes en jeu, le symbole $*$ sans indice sera désormais utilisé pour désigner le produit sur tous les $v \in S_p$ des $*_v$. Par exemple,

$$T = \prod_{v \in S_p} T_v, \quad I = \prod_{v \in S_p} I_v, \quad \text{etc} \dots$$

Cela nous fournit des G, B, N, T, I, M, ainsi que leurs décorations précédemment introduites. On désignera aussi par $\mathcal{W}$ le produit $\prod_{v \in S_p} \mathcal{W}_v$. Ces notations sont alors compatibles avec celle du §1. On pose de plus $\mathfrak{A}^- = \otimes_{v \in S_p} \mathfrak{A}_v^-$ et

$$\mathcal{C}\left(\prod_v V_v, r\right) := \widehat{\otimes}_v \mathcal{C}(V_v, r).$$

La définition de $\mathcal{C}(V, r)$ ci-dessus s'étend trivialement aux affinoides $V \to \mathcal{W}$ généraux (plutôt qu'aux produits de $V_v \to \mathcal{W}_v$). Tout ce qui a été dit au §2.8 sur les représentations $\mathcal{C}(V_v, r)$ des M_v s'étend trivialement aux $\mathcal{C}(V, r)$ vus comme $\mathcal{O}(V)[M]$-modules.

2.15 Formes automorphes p-adiques de U de type $(S_p, W_\infty, e, \mathcal{H})$

Le travail local précédemment effectué nous permet enfin de définir les formes automorphes p-adiques de U qui nous intéressent. On reprend les notations du § 1.1 concernant S_p, W_∞, S, K^S, $\mathcal{H}$ et e. On peut supposer que pour $v \in S_p$ le sous-groupe d'Iwahori K_v coïncide avec I_v.

La représentation W_∞ donnée peut être vue comme une représentation algébrique du $\mathbb{Q}_p$-groupe $\prod_{v \notin S_p} U(F_v)$ qui est définie sur L, via ι_p et ι_∞. On pose

$$\mathcal{H}^- := \mathfrak{A}^- \otimes \mathcal{H}.$$

Choisissons un sous-groupe compact ouvert de la forme $I \times K' \subset K$ suffisament petit de sorte que $e \in E[U(\mathbb{A}_{F,f}^{S_p})//K']$ et que K'_v soit sans torsion à au moins une place v.

Il sera commode d'introduire le foncteur $F : \mathrm{Mod}(L[M]) \longrightarrow \mathrm{Mod}(\mathfrak{A}^- \otimes E[U(\mathbb{A}_{F,f}^{S_p})//K'])$ défini par

$$F(C) := \left\{ \begin{array}{l} f : U(F)\backslash U(\mathbb{A}_{F,f}) \longrightarrow C, \\ \\ f(gk) = (\prod_{v \in S_p} k_v)^{-1} f(g), \ \forall g \in U(\mathbb{A}_{F,f}), \ \forall k \in I \times K'. \end{array} \right\}$$

L'ensemble $U(F)\backslash U(\mathbb{A}_{F,f})/(I \times K')$ est fini par finitude du nombre de classes [Bo], notons h son cardinal et choisissons une décomposition

$$U(\mathbb{A}_{F,f}) = \coprod_{i=1}^{h} U(F)x_i(I \times K').$$

Les groupes $x_i^{-1}U(F)x_i \cap (I \times K')$ sont finis (car $U(F_v)$ est compact si v est archimédienne) et sans torsion, donc triviaux. L'application $f \mapsto (f(x_1), \cdots, f(x_h))$ induit donc un isomorphisme L-linéaire

$$F(C) \tilde{\rightarrow} C^h. \tag{2.3}$$

En particulier $F(C)$, ainsi que son facteur direct $eF(C)$, héritent de la plupart des propriétés supplémentaires satisfaites par C. Par exemple, si C est muni d'une norme $|.|$ de L-espace vectoriel, il en va de même de $eF(C)$ en posant

$$|f| := \sup_{x \in U(\mathbb{A}_{F,f})} |f(x)| = \sup_{i=1\ldots h} |f(x_i)|.$$

Si M agit par éléments de norme ≤ 1 sur C, l'espace normé $eF(C)$ est un facteur direct topologique de $F(C) \tilde{\rightarrow} C^h$, et si de plus $t \in T^-$ est de norme $\leq s$ sur C, alors $\mathbb{1}_t$ est aussi de norme $\leq s$ sur $eF(C)$.

Soit $V \to \mathcal{W}$ avec V affinoide, et soit $r \geq r_V$. On définit un $\mathcal{H}^-$-module en posant

$$\mathcal{S}(V, r) := eF(\mathcal{C}(V, r)).$$

C'est la définition que nous adoptons pour l'espace des *formes automorphes p-adiques de U de type* (S_p, W_∞, e), *rayon de convergence r et de poids dans* V. C'est un $\mathcal{O}(V)$-module de Banach pour la norme qu'il hérite de $\mathcal{C}(V, r)$ comme plus haut, qui est facteur direct topologique d'un $\mathcal{O}(V)$-Banach orthonormalisable (c'est la condition (Pr) de [Bu, p. 72]). Il est muni d'une action $\mathcal{O}(V)$-linéaire de $\mathcal{H}^-$ telle que chaque $h \in \mathcal{H}^-$ est borné par 1 et chaque $t \in T^{--}$ est $\mathcal{O}(V)$-compact. La formule (2.3) assure que la collection des espaces $\{\mathcal{S}(V, r), V, r \geq r_V\}$ satisfait des compatibilités analogues à celles des $\{\mathcal{C}(V, r), \ V, \ r \geq r_V\}$ quand V et r varient (lemme 2.9).

2.16 Formes automorphes classiques et critère de classicité

Soit $\delta \in X^*(T) \subset \mathcal{W}(L)$ un caractère dominant relativement au $\mathbb{Q}_p$-sous-groupe de Borel B, en particulier $r_\delta = 0$. Il est bien connu, et trivial de vérifier, que la donnée de ι_p et

ι_∞ réalise le $\mathcal{H}^-$-module $eF(W_\delta^*)$ est comme une L-structure du $\mathcal{H}^-$module des formes automorphes complexes

$$\bigoplus_{\Pi\in\mathcal{A},\Pi_\infty\simeq_\iota W_\delta\otimes W_\infty} m(\Pi)e(\Pi_f)^{K_{S_p}}.$$

(voir par exemple [Ch, §4.2].) De plus, le lemme 2.11 fournit une inclusion naturelle $\mathcal{H}^-$-équivariante

$$eF(W_\delta^*)\otimes\delta^\varpi \hookrightarrow \mathcal{S}(\delta,0)=eF(\delta,0), \tag{2.4}$$

le caractère δ^ϖ de T/T^0 étant vu comme un caractère de M via le lemme 2.3 (ii), dont l'image est en général appelée *l'espace des formes automorphes p-adiques classiques* de poids δ.

Dans le reste de ce paragraphe nous allons établir le *critère de classicité*, qui est un critère assurant qu'un élément $f\in\mathcal{S}(\delta,r)$ est classique, *i.e.* qu'il appartient à $eF(W_\delta^*)$. Une condition nécessaire est qu'il soit *de pente finie* aux places v divisant S_p. Rappelons qu'un élément $f\in\mathcal{S}(\delta,r)$ est dit de pente finie si pour un $t\in T^{--}$ (vu comme élément de $\mathfrak{A}^-$) :

- le sous-espace $L[t].f\subset\mathcal{S}(\delta,r)$ est de dimension finie,
- et $t_{|L[t].f}$ est inversible.

Comme un tel t agit de manière compacte sur $\mathcal{S}(\delta,r)$, et comme $\mathfrak{A}^-$ est commutative, cela entraîne que le sous-espace $\mathfrak{A}^-.f$ est de dimension finie sur L, stable par t, et que la restriction de t y est inversible. On vérifie immédiatement que si f est de pente finie, alors pour tout $t\in T^{--}$, $L[t].f$ est de dimension finie et $t_{|L[t].f}$ est inversible. Par conséquent les éléments de pente finie forment un sous-L-espace vectoriel

$$\mathcal{S}(\delta,r)^{\mathrm{fs}}\subset\mathcal{S}(\delta,r),$$

sur lequel tous les éléments de T^- sont inversibles. En particulier, $\mathcal{S}(\delta,r)^{\mathrm{fs}}$ s'étend naturellement en un $\mathfrak{A}$-module et

$$eF(W_\delta^*)\otimes\delta^\varpi\subset\mathcal{S}(\underline{k},r)^{\mathrm{fs}}$$

en est un sous-$\mathfrak{A}$-module. Rappelons que l'on a défini au §1.4 des éléments $t_{i,v}\in T_v$ pour $v\in S_p$ et $1\le i\le n$, on les verra aussi comme des éléments de $\mathfrak{A}_v$ (lemme 2.3 (ii)). On a aussi défini loc. cit. des $k_{i,\sigma}\in\mathbb{Z}$ associés à δ.

Proposition 2.17. *Soit $f\in\mathcal{S}(\delta,r)^{\mathrm{fs}}\otimes_L\overline{\mathbb{Q}}_p$ une forme propre pour tous les éléments de $\mathfrak{A}$. Pour $v\in S_p$ et $i=1,\dots,n$ écrivons $t_{i,v}(f)=\varphi_{i,v}f$ avec $\varphi_{i,v}\in\overline{\mathbb{Q}}_p^*$. Supposons*

$$\frac{\mathbf{v}(\varphi_{1,v}\varphi_{2,v}\cdots\varphi_{i,v})}{\mathbf{v}(\varpi_v)}<k_{i,\sigma}-k_{i+1,\sigma}+1,$$

pour tout $v\in S_p$, $1\le i<n$ et $\sigma\in\mathrm{Hom}(F_v,\overline{\mathbb{Q}}_p)$, alors f est classique.

Plus précisément, le sous-espace caractéristique tout entier de f sous l'action de $\mathfrak{A}$ est inclus dans $eF(W_\delta^*)$.

Nous renvoyons à [BCh, §7.3.5] pour une discussion de ce résultat. La preuve que nous donnons est une légère variante de celle donnée loc. cit, elle-même issue de [Ch, Prop. 4.7.4], reposant sur les propriétés de la présentation de Plücker (lemme 2.13).

Preuve — En raisonnant comme dans [BCh, §7.3.5], on peut supposer que $e = \mathbb{1}_{K'}$, c'est à dire que $eF(E) = F(E)$, et aussi que $r = 0$ par un argument de série caractéristique dû à Coleman. D'après le lemme 2.12,

$$\mathcal{S}(\delta, 0)^{\mathrm{fs}} \subset \mathcal{S}(\delta, 0)^{\sharp} := F(\mathcal{C}(\delta, 0)^{\sharp}).$$

Le foncteur F étant exact, l'image par F du diagramme du lemme 2.13 (i) nous fournit un diagramme commutatif de $\mathcal{H}^- \otimes L$-modules

$$\begin{array}{ccc} F(A) & \longrightarrow & F(B) \\ \downarrow & & \downarrow \\ F(W_\delta^*) \otimes \delta^{\varpi} & \longrightarrow & \mathcal{S}(\delta, 0)^{\sharp} \end{array}$$

où $A = \otimes_{i,\sigma}(W^*_{\delta_{i,\sigma}} \otimes \delta^{\varpi}_{i,\sigma})$ et $B = \widehat{\otimes}_{i,\sigma} P_{i,\sigma}$, les flèches verticales (resp. horizontales) étant surjectives (resp. injectives).

Soit $\psi : \mathfrak{A}^- \to \overline{\mathbb{Q}}_p$ le système de valeurs propres associé à la forme f de l'énoncé ; si V est un $\mathfrak{A}^-$-module, on désignera par $V[\psi] \subset V$ le sous-espace caractéristique associé à ψ. Comme T^{--} agit par opérateurs compacts sur B, la surjection de droite ci-dessus induit encore une surjection $F(B)[\psi] \to \mathcal{S}(\delta, 0)^{\sharp}[\psi]$. On a une injection naturelle :

$$B/A \hookrightarrow \prod_{(i,\sigma)} C_{i,\sigma}, \quad \text{où} \quad C_{i,\sigma} := (\widehat{\otimes}_{(i',\sigma') \neq (i,\sigma)} P_{i',\sigma'}) \widehat{\otimes} (P_{i,\sigma}/(W^*_{\delta_{i,\sigma}} \otimes \delta^{\varpi}_{i,\sigma})),$$

et une injection similaire de $\mathcal{H}^-$-modules qui s'en déduit en appliquant F, de sorte qu'il suffit de montrer que $F(C_{i,\sigma})[\psi] = 0$ pour tout i et σ. C'est évident si $i = n$ car $C_{n,\sigma} = 0$. Si $i < n$ cela découle exactement de l'estimée de la norme de l'élément $t_{1,v} t_{2,v} \ldots t_{i,v} \in \mathfrak{A}^-$ donnée par le lemme 2.13 (iii), $v \in S_p$ étant la place induite par σ. □

2.18 Fin de la démonstration

La méthode de Coleman et Coleman-Mazur [CM] permet de construire formellement la variété de Hecke X à partir de l'action de $\mathcal{H}^-$ sur la famille des modules de Banach $\mathcal{S}(V, r)$, avec $V \subset \mathcal{W}$ et $r \geq r_V$. L'existence de (X, ψ, ν, Z) comme dans le théorème 1.6 satisfaisant (i) à (vi) découle de manière standard de cette construction, de la proposition 2.17 et du lemme 2.7 : on choisit n'importe quel élément dans T^{--} et on construit X au dessus de l'hypersurface de Fredholm de cet élément. On renvoie à [CM], [Ch, §6.3], [Bu] et [BCh, §7.3.6] pour plus de détails. L'assertion d'unicité est [Ch, prop. 7.2.8].

3 Famille de pseudo-caractères galoisiens sur les variétés de Hecke

Dans cette section, nous donnons une application galoisienne des constructions précédentes qui sera utilisée dans [ChH].

3.1 Hypothèses et énoncé du résultat

Plaçons nous dans le contexte simplifié où

- $[F : \mathbb{Q}]$ est pair,
- $\mathcal{K}/F$ est non ramifiée à toutes les places finies de $\mathcal{K}$,
- n est pair,
- au moins une place de F divisant p se décompose dans $\mathcal{K}$.

On choisit l'ensemble S_p non vide et quelconque comme au § 1.1 et on notera de plus S'_p le complémentaire de S_p dans l'ensemble des places de F divisant p. Par abus de langage, on verra parfois S_p et S'_p comme des ensembles de places de $\mathcal{K}$ (celles au dessus de S_p et S'_p respectivement).

Par le principe de Hasse pour les groupes unitaires, il existe un (unique) groupe unitaire U à n variables attaché à $\mathcal{K}/F$ qui est quasi-déployé à toutes les places finies de F et tel que $U(F_v)$ est compact pour toute place archimédienne v de F. Par exemple, si $N(.)$ désigne la norme de $\mathcal{K}$ sur F, le groupe unitaire associé à la forme hermitienne $\sum_{i=1}^n N(z_i)$ sur $\mathcal{K}^n$ convient quand n est multiple de 4.

Soit $\mathcal{A}_0$ l'ensemble des représentations automorphes Π de U qui sont non ramifiées aux places finies de F non décomposées dans $\mathcal{K}$ et telles que Π_v a des Iwahori-invariants non nuls pour chaque $v \in S_p$. Remarquons le changement de base local à GL_n est bien défini pour les composantes locales des éléments Π de $\mathcal{A}_0$, et on posera

$$\mathrm{BC}(\Pi) := \bigotimes_v{}' \mathrm{BC}(\Pi_v),$$

c'est une représentation irréductible de $\mathrm{GL}_n(\mathbb{A}_{\mathcal{K}})$. Soit $\mathcal{A}_0^{\mathrm{reg}} \subset \mathcal{A}_0$ le sous-ensemble des Π tels que *pour au moins une place archimédienne* v de F, le poids extrémal de la représentation de dimension finie Π_v de $U(F_v)$ n'est dans aucun mur d'une chambre de Weyl. D'après un résultat de Labesse [Lab, Cor. 5.4], pour toute représentation $\Pi \in \mathcal{A}_0^{\mathrm{reg}}$ alors $\mathrm{BC}(\Pi)$ est une représentation automorphe (non nécessairement discrète). Rappelons le théorème collectif suivant, combinant des énoncés démontrés dans les livres 1 et 2 et par S.W. Shin (voir [ChH, Thm. 1.4], [Shin]).

Si π_v est une représentation irréductible de $\mathrm{GL}_n(\mathcal{K}_v)$, nous désignons par $\mathcal{L}(\pi_v)$ la représentation de Weil-Deligne associée par la correspondance de Langlands locale (normalisée comme dans [ChH], et définie sur $\overline{\mathbb{Q}}$ si π_v l'est) et par $\mathcal{L}_W(\pi_v) : W_{\mathcal{K}_v} \to \mathrm{GL}_n(\mathbb{C})$ la représentation (semi-simple) du groupe de Weil $W_{\mathcal{K}_v}$ de $\mathcal{K}_v$ associée.

Théorèm 3.2. *Soit* $\Pi \in \mathcal{A}_0^{\mathrm{reg}}$, $\pi := \mathrm{BC}(\Pi)$. *Il existe une unique représentation*

$$\rho_\Pi : \Gamma_{\mathcal{K}} \longrightarrow \mathrm{GL}_n(\overline{\mathbb{Q}}_p),$$

semi-simple et continue, telle que :

(a) *Pour toute place finie* v *de* $\mathcal{K}$ *première à* p,

$$((\rho_\Pi)_{|\Gamma_{\mathcal{K}_v}})^{F-ss} \simeq \iota_p \iota_\infty^{-1} \mathcal{L}(\pi_v | \bullet |^{\frac{1-n}{2}}).$$

(b) *Pour toute place finie v de $\mathcal{K}$ divisant p, $(\rho_\Pi)_{|\Gamma_{\mathcal{K}_v}}$ est de de Rham et à poids de Hodge-Tate distincts déterminés par Π_∞ (voir [ChH, §1.5] pour la recette).*

(c) *Soit v une place finie de $\mathcal{K}$ divisant p telle que π_v ait des Iwahori-invariants. Alors $(\rho_\Pi)_{|\Gamma_{\mathcal{K}_v}}$ est semi-stable.*

(d) *Soit v une place finie de $\mathcal{K}$ divisant p telle que π_v soit non ramifiée. Alors $\rho_v := (\rho_\Pi)_{|\Gamma_{\mathcal{K}_v}}$ est crystalline, et si φ_v désigne la plus petite puissance linéaire*[12] *du Frobenius cristallin de $D_{\mathrm{cris}}(\rho_v)$, on a l'égalité dans $\overline{\mathbb{Q}}_p[T]$:*

$$\det(T - \varphi_v) = \iota_p \iota_\infty^{-1} \det(T - \mathcal{L}_W(\pi_v | \bullet |^{\frac{1-n}{2}})(\mathrm{Frob}_v)).$$

Le reste de cet article est consacré à la preuve du résultat suivant :

Théorèm 3.3. *Soit $\Pi \in \mathcal{A}_0$, $\pi = \mathrm{BC}(\Pi)$. Il existe une unique représentation*

$$\rho_\Pi : \Gamma_{\mathcal{K}} \longrightarrow \mathrm{GL}_n(\overline{\mathbb{Q}}_p),$$

semi-simple et continue, telle que :

(a') *Pour toute place finie v de $\mathcal{K}$ première à p,*

$$((\rho_\Pi)_{|\Gamma_{\mathcal{K}_v}})^{ss} \simeq \iota_p {\iota_\infty}^{-1} \mathcal{L}_W(\pi_v \otimes | \bullet |^{\frac{1-n}{2}}).$$

De plus, $(\rho_\Pi)_{|\Gamma_{\mathcal{K}_v}}$ est non ramifiée si π_v l'est.

(b) *Pour toute place finie $v \in S'_p$, $(\rho_\Pi)_{|\Gamma_{\mathcal{K}_v}}$ est de de Rham et à poids de Hodge-Tate distincts déterminés par les Π_w avec $w \in \Sigma_\infty(v)$ (même recette).*

(c) *Soit $v \in S'_p$ telle que π_v ait des Iwahori-invariants. Alors $(\rho_\Pi)_{|\Gamma_{\mathcal{K}_v}}$ est semi-stable.*

(d) *Soit $v \in S'_p$ telle que π_v soit non ramifiée. Alors $\rho_v := (\rho_\Pi)_{|\Gamma_{\mathcal{K}_v}}$ est crystalline, et si φ_v désigne la plus petite puissance linéaire*[13] *du Frobenius cristallin de $D_{\mathrm{cris}}(\rho_v)$, on a l'égalité dans $\overline{\mathbb{Q}}_p[T]$:*

$$\det(T - \varphi_v) = \iota_p {\iota_\infty}^{-1} \det(T - \mathcal{L}_W(\pi_v | \bullet |^{\frac{1-n}{2}})(\mathrm{Frob}_v)).$$

(e) *Si $v \in S_p$, alors $(\rho_\Pi)_{|\Gamma_{\mathcal{K}_v}}$ est de Hodge-Tate, à poids distincts et déterminés par les Π_w avec $w \in \Sigma_\infty(v)$ (même recette).*

Nous allons en fait démontrer un raffinement du théorème 3.3 (a'). Soit $\mathcal{I}$ l'ensemble des représentations irréductibles continues de $W_{\mathcal{K}_v}$, prises à isomorphisme et torsion par un caractère non ramifié près. Si ρ est une représentation de Weil-Deligne de $\mathcal{K}_v$,

$$\rho = \bigoplus_{\sigma \in \mathcal{I}} \rho[\sigma]$$

où $\rho[\sigma]$ est la plus grande sous-représentation de ρ dont les facteurs de Jordan-Hölder sont des torsions non ramifiées de σ. De plus, pour tout $\sigma \in \mathcal{I}$, il existe une unique suite

[12] Autrement dit, $\varphi_v = F^r$ où F est le Frobenius cristallin et r le degré du corps résiduel de F_v.
[13] Autrement dit, $\varphi_v = F^r$ où F est le Frobenius cristallin et r le degré du corps résiduel de F_v.

décroissante d'entiers positifs ou nuls $m_1(\sigma) \geq m_2(\sigma) \geq \cdots$, et des caractères non ramifiés χ_i, tels que

$$\rho[\sigma] = \bigoplus_{i=1} \sigma_i(m_i(\sigma)), \quad \sigma_i := \sigma \otimes \chi_i,$$

où la notation $\sigma(m)$ désigne la représentation de Weil-Deligne indécomposable de longueur m et de sous-représentation simple σ. On verra cette suite $(m_i(\sigma))$ comme une partition $p(\rho, \sigma)$ de l'entier $\dim(\rho[\sigma])/\dim(\sigma)$. On note $\prec$ la relation de dominance usuelle sur les partitions ([Mac, §1]).

Définition 3.4. *Si ρ et ρ' sont des représentations de Weil-Deligne de $\mathcal{K}_v$, on note $\rho \prec \rho'$ si ρ et ρ' ont même semi-simplifiée et si de plus, pour tout $\sigma \in \mathcal{I}$, on a $p(\rho, \sigma) \prec p(\rho', \sigma)$.*

Des méthodes de Bellaïche-Chenevier [BCh] §6.5 et §7.8 permettent de démontrer l'assertion suivante de compatibilité locale-globale aux places premières à p :

Théorèm 3.5. *Soit $\Pi \in \mathcal{A}_0$, $\pi = \mathrm{BC}(\Pi)$. Pour toute place finie v de $\mathcal{K}$ première à p,*

$$(\rho_\Pi)_{|\Gamma_{\mathcal{K}_v}} \prec \iota_p \iota_\infty{}^{-1} \mathcal{L}(\pi_v \otimes | \bullet |^{\frac{1-n}{2}}).$$

3.6 Choix de la variété de Hecke

En vue de démontrer ces deux résultats, fixons dorénavant un $\Pi_0 \in \mathcal{A}_0$. Considérons la variété de Hecke *minimale* contenant Π_0 (voir [BCh, Example 7.5.1]), c'est à dire que nous faisons les choix suivants pour $(W_\infty, \mathcal{H}, e)$:

- $W_\infty := \bigotimes_v (\Pi_0)_v$, le produit tensoriel portant sur toutes les places réelles v de F qui ne sont pas dans $\cup_{w \in S_p} \Sigma_\infty(w)$.
- S est l'ensemble des places finies de F telles que Π_v est ramifiée et $v \notin S_p$. Pour chaque $v \in S$ (nécessairement $U(F_v) \simeq \mathrm{GL}_n(F_v)$), on fixe une composante connexe de Bernstein $\mathcal{B}_v$ de la catégorie des représentations lisses de $U(F_v)$.
- D'après l'extension par Schneider et Zink [SZ, Prop. 6.2] de résultats de Bushnell-Kutzko, on peut choisir un idempotent e_v de l'algèbre de Hecke de $U(F_v)$ (en fait associé à un K-type) de sorte que :
    - $b_v e_v = e_v$ où b_v est l'idempotent central de Bernstein associé à $\mathcal{B}_v$,
    - $e_v((\Pi_0)_v) \neq 0$,
    - pour toute irréductible π de $\mathcal{B}_v$, si $e_v(\pi) \neq 0$ alors $\mathcal{L}(\pi) \prec_I \mathcal{L}((\Pi_0)_v)$.

  Nous renvoyons au § 3.10 pour la notation $\prec_I$, ainsi qu'à [BCh, §6.5] pour cette traduction des résultats de [SZ].

On choisit le corps de coefficients E de sorte que $\mathcal{B}_v$, son point base, e_v, ainsi que le centre de Bernstein $\mathfrak{z}_v$ de $\mathcal{B}_v$, soient tous définis sur E. On choisit enfin pour algèbre de Hecke la E-algèbre

$$\mathcal{H} = \mathcal{H}^{S \cup S_p} \otimes (\bigotimes_{v \in S} \mathfrak{z}_v)$$

(le premier produit tensoriel étant sur $\mathbb{Z}$ et le second sur E).

Définition 3.7. *(X, ψ, ν, Z) désignera la variété de Hecke de U associée aux données $(S_p, W_\infty, \mathcal{H}, e)$ ci-dessus.*

Sous les hypothèses ci-dessus, l'ensemble $\mathcal{A}$ associé est le sous-ensemble des représentations automorphes Π de $\mathcal{A}_0$ telles que pour tout $v \in S$, Π_v est dans la composante de Bernstein $\mathcal{B}_v$ et dominée par $(\Pi_0)_v$. Par construction, si ψ est un système de valeurs propres de $\mathcal{H}$ associé à Π alors ψ détermine $\Pi^{S \cup S_p}$, et pour tout $v \in S$ le centre de Bernstein agit sur Π_v par $\psi_{|\mathfrak{z}_v}$. De plus, si ν est attaché comme en § 1.4 au choix d'un raffinement accessible de Π (il en existe toujours au moins un), alors ν détermine Π_∞ et ainsi que le polynôme caractéristique de $\mathcal{L}_W(\Pi_v)(\mathrm{Frob}_v)$ pour tout $v \in S_p$. Un conséquence de tout cela est que (ψ, ν) *détermine le L-paramètre de Π_v restreint au groupe de Weil (plutôt qu'au Weil-Deligne) pour toute place v de F.*

Enfin, on dira que $\Pi \in \mathcal{A}$ est *régulière* si pour tout $w \in S_p$ et tout $v \in \Sigma_\infty(w)$ le poids extrémal de la représentation de dimension finie Π_v de $U(F_v)$ n'est dans aucun mur d'une chambre de Weyl. On notera $\mathcal{A}^{\mathrm{reg}} \subset \mathcal{A}$ le sous-ensemble formé des Π qui sont régulières, et

$$Z^{\mathrm{reg}} \subset Z$$

le sous-ensemble paramétrant des (ψ, ν) où ψ est associé à un Π dans $\mathcal{A}^{\mathrm{reg}}$. C'est un sous-ensemble Zariski-dense et d'accumulation dans X (voir par exemple [BCh, Lemma 7.5.3]).

3.8 Le pseudo-caractère galoisien sur X et dèfinition de ρ_Π

Une conséquence simple du théorème 3.2 (a) est qu'il permet de définir un pseudo-caractère galoisien de Γ_K sur toute la variété de Hecke X, plutôt que seulement aux points définis par $\mathcal{A}^{\mathrm{reg}}$. L'objectif dans ce qui suit sera de déterminer ses propriétés qui découlent "formellement" de ce théorème.

Pour tout $x = (\psi, \nu) \in Z^{\mathrm{reg}}$, associé à un $\Pi \in \mathcal{A}^{\mathrm{reg}}$, on note ρ_x la représentation galoisienne ρ_Π définie par ce théorème. Notons que si deux Π donnent lieu au même x, alors $\rho_\Pi \simeq \rho_{\Pi'}$ par le (a) du théorème et Cebotarev : ρ_x est bien définie. Soient Σ l'ensemble des places finies de $\mathcal{K}$ qui sont au dessus d'une place de S ou au dessus de p, et $\Gamma_{\mathcal{K},\Sigma}$ le groupe de Galois d'une extension algébrique maximale de $\mathcal{K}$ non ramifiée hors de Σ. On renvoie à [BCh, Ch. 1] pour les généralités sur les pseudo-caractères. Soient $w \notin \Sigma$ une place finie de $\mathcal{K}$ et v la place de F au dessous de w, l'isomorphisme de Satake nous fournit un élément $h_w \in \mathcal{H}$ dans l'algèbre de Hecke sphérique de $(U(F_v), K_v)$ tel que pour toute représentation sphérique Π_v de $U(F_v)$, de changement de base π_w à $\mathrm{GL}_n(\mathcal{K}_w)$, h_w agit sur la droite sphérique de π_v par multiplication par $\mathrm{tr}(\mathcal{L}(\pi_w| \bullet |^{\frac{1-n}{2}})(\mathrm{Frob}_w))$.

Corollaire 3.9. *Il existe un unique pseudo-caractère continu*

$$T : \Gamma_{\mathcal{K},\Sigma} \to \mathcal{O}(X),$$

de dimension n, tel que pour tout $z \in Z^{\mathrm{reg}}$, $T_z = \mathrm{tr}(\rho_z)$. Pour tout $w \notin \Sigma$,

$$T(\mathrm{Frob}_w) = \psi(h_w) \in \mathcal{O}(X).$$

(Voir [Ch, Cor. 7.1.1], ou encore [BCh, Cor. 7.5.4].) Dans ce corollaire, et par la suite, nous notons T_x l'évaluation de T en un $x \in X(\overline{\mathbb{Q}}_p)$, i.e. $T_x(g) := T(g)(x)$. D'après un résultat général de Taylor et Procesi sur les pseudo-caractères, il existe une unique représentation continue semi-simple

$$\rho_x : \Gamma_{\mathcal{K},\Sigma} \to \mathrm{GL}_n(\overline{\mathbb{Q}}_p)$$

telle que $\mathrm{tr}(\rho_x) = T_x$.

Soit maintenant $\Pi \in \mathcal{A}$ qui n'est plus nécessairement régulière. Choisissons un raffinement accessible de Π aux places dans S_p, un système de valeurs propres ψ de $\mathcal{H}$ associé à Π, et considérons un point $z = (\psi, \nu) \in Z$ défini par ces choix. On pose

$$\rho_\Pi := \rho_z.$$

Par le théorème de Cebotarev, ρ_Π ne dépend ni du choix de raffinement, ni du choix de ψ.

3.10 Propriétés aux places dans S ne divisant pas p

Nous devons commencer par un préliminaire sur le centre de Bernstein de GL_n ([Bern]). Soient ℓ un nombre premier, F_ℓ une extension finie de $\mathbb{Q}_\ell$, et $m \geq 1$ un entier. Fixons une composante de Bernstein $\mathcal{B}$ de la catégorie des représentations complexes lisses de $\mathrm{GL}_m(F_\ell)$, de centre $\mathfrak{z}$.

Si π est une représentation lisse irréductible de $\mathrm{GL}_m(F_\ell)$, rappelons que

$$\mathcal{L}_W(\pi) : W_{F_\ell} \to \mathrm{GL}_m(\mathbb{C})$$

désigne la représentation du groupe de Weil W_{F_ℓ} de F_ℓ associée à π par la correspondance de Langlands locale. D'après la théorie de Bernstein, pour π dans la composante $\mathcal{B}$ la représentation $\mathcal{L}_W(\pi)$ est uniquement déterminée par le caractère de $\mathfrak{z}$ sur π, que l'on notera $z(\pi) \in \mathcal{B}$ ("le support cuspidal de π").

La composante $\mathcal{B}$ est associée à une classe inertielle (M, ω) pour un certain sous-groupe de Levi M de $\mathrm{GL}_m(F_\ell)$ et ω une supercuspidale de M. On sait que quitte à remplacer ω par une torsion non ramifiée, on peut supposer qu'elle est définie sur $\overline{\mathbb{Q}}$. On choisit enfin un corps de coefficients $E \subset \overline{\mathbb{Q}}$ suffisament grand, fini et galoisien sur $\mathbb{Q}$, de sorte que ω, $\mathcal{L}_W(\omega)$, $\mathcal{B}$, et le centre de Bernstein $\mathfrak{z}$ de $\mathcal{B}$, soient tous définis sur E. Soit $E[\mathcal{B}]$ l'anneau des coordonnées affines de $\mathcal{B}$.

Proposition 3.11. *Il existe un (unique) pseudo-caractère de dimension m*

$$T^{\mathcal{B}} : W_{F_\ell} \longrightarrow E[\mathcal{B}]$$

tel que pour toute irréductible π dans $\mathcal{B}$, $T^{\mathcal{B}}_{z(\pi)} = \mathrm{tr}(\mathcal{L}_W(\pi))$.

(Ici encore, pour $x \in \mathcal{B}(\mathbb{C})$ et $g \in W_{F_\ell}$ on pose $T^{\mathcal{B}}_x(g) = T^{\mathcal{B}}(g)(x)$.)

Preuve — Supposons tout d'abord que $m = qd$, $M = \mathrm{GL}_d(F_\ell)^q$ et que ω est le produit de q supercuspidales identiques $r_1 = r_2 = \cdots = r_q$ de $\mathrm{GL}_d(F_\ell)$. Soit $Y = \mathbb{G}_m^q$ le tore des caractères non ramifiés de M, que l'on identifiera à ceux de $(F_\ell^*)^q$ via le déterminant. Le groupe symétrique $\mathfrak{S}_q$ agit sur Y par permutation des coordonnées, ainsi donc que sur $Y \rtimes \mathfrak{S}_q$ par

$$((\xi_1, \ldots, \xi_q), \sigma) \cdot (\chi_1, \ldots, \chi_q) := (\chi_{\sigma^{-1}(1)}\xi_1, \ldots, \chi_{\sigma^{-1}(q)}\xi_q).$$

On choisit $\omega = (r_1, r_2, \ldots, r_q)$ pour point base de $\mathcal{B}$. Cela permet d'identifier ce dernier au quotient de Y par le sous-groupe fini Δ des $(\xi, \sigma) \in Y(\overline{\mathbb{Q}}) \rtimes \mathfrak{S}_d$ tels que $r_i \otimes \xi_i \simeq r_{\sigma^{-1}(i)}$ pour tout i. On pose alors, pour $g \in W_{F_\ell}$ et $\chi = (\chi_1, \cdots, \chi_q) \in Y$,

$$T^{\mathcal{B}}(g)(\chi) := \sum_{i=1}^{q} \mathrm{tr}(\mathcal{L}_W(r_i(g)))\chi_i(\mathrm{rec}(g)) = \mathrm{tr}(\mathcal{L}_W(\oplus_{i=1}^q r_i \otimes \chi_i)(g))$$

où $\mathrm{rec} : W_{F_\ell} \to F_\ell^*$ est l'homomorphisme de réciprocité du corps de classes local. Par construction,

- $T^{\mathcal{B}}(g) \in E[Y]^\Delta = E[\mathcal{B}]$ pour tout $g \in W_{F_\ell}$,
- Si π est dans $\mathcal{B}$, on a $T^{\mathcal{B}}_{z(\pi)} = \mathrm{tr}(\mathcal{L}_W(\pi))$.

Cela démontre la proposition pour ce cas particulier de classe inertielle.

Dans le cas général, on écrit $M = M_1 \times M_2 \times \cdots \times M_s$ et $\omega = (\omega_1, \ldots, \omega_s)$ où chaque (M_j, ω_j) est une classe inertielle du type précédent : $M_j = \mathrm{GL}_{d_j}(F_\ell)^{q_j}$ et ω_j est le produit de q_j supercuspidales identiques, disons égales à une certaine r_j, de $\mathrm{GL}_{d_j}(F_\ell)$. Quitte à modifier ω dans sa classe inertielle, on peut supposer que cette écriture à été choisie de sorte que si $i \neq j$ et $d_i = d_j$, alors r_i n'est pas une torsion non ramifiée de r_j. Dans ce cas, la composante de Bernstein est de manière naturelle un produit $\mathcal{B}_1 \times \mathcal{B}_2 \times \cdots \times \mathcal{B}_s$ des composantes de Bernstein des $\mathrm{GL}_{q_j d_j}(F_\ell)$ de classes inertielles les (M_j, ω_j). Il est immédiat que

$$T_{\mathcal{B}}(g) := \sum_{j=1}^{s} T_{\mathcal{B}_j}(g) \in \left(\bigotimes_j E[\mathcal{B}_j]\right) = E[\mathcal{B}]$$

a les propriétés requises. □

Démontrons maintenant la propriété (a') du théorème 3.3. Fixons $v \in S$ et notons w la place de F au dessous de v, de sorte que $U(F_w)$ s'identifie à $\mathrm{GL}_n(\mathcal{K}_v) = \mathrm{GL}_n(F_w)$. Par construction, on dispose d'une composante de Bernstein $\mathcal{B}_v$ de $\mathrm{GL}_n(\mathcal{K}_v)$, dont une E-structure du centre est $\mathfrak{z}_v$, et un morphisme d'anneaux $\psi : \mathfrak{z}_v \to \mathcal{O}(X)$, on peut donc considérer

$$T' : W_{\mathcal{K}_v} \to \mathcal{O}(X)$$

la composée de $T^{\mathcal{B}_v} : W_{\mathcal{K}_v} \to \mathfrak{z}_v = E[\mathcal{B}_v]$ et de ψ : c'est un pseudo-caractère de dimension n sur $W_{\mathcal{K}_v}$.

Lemme 3.12. $T_{|W_{\mathcal{K}_v}} = T'$.

Preuve — Par Zariski-densité de Z^{reg} dans X il suffit de vérifier l'égalité du lemme en un $z \in Z^{\mathrm{reg}}$. Si $z \in Z$, il existe par définition une représentation $\Pi(z) \in \mathcal{A}$ associée à z telle que $\iota_p \iota_\infty^{-1} \Pi(z)_w$ est dans la composante de Bernstein $\mathcal{B}_v$, l'action du centre $\mathfrak{z}_v$ sur cette dernière représentation étant donnée par ψ_z. Ainsi,

$$\forall z \in Z, \quad T'_z = \iota_p \iota_\infty^{-1} \operatorname{tr}(\mathcal{L}_W(\Pi(z)_w)) \tag{3.5}$$

par définition de $T^{\mathcal{B}_v}$ (i.e. par la Proposition 3.11). D'autre part, si de plus $z \in Z^{\mathrm{reg}}$, alors z provient d'un $\Pi \in \mathcal{A}^{\mathrm{reg}}$. Le (a) du théorème 3.2 appliqué à ce Π assure que $(T_{|W_{F_\ell}})_z = \iota_p \iota_\infty^{-1} \operatorname{tr}(\mathcal{L}_W(\Pi(z)_w))$, ce qui conclut. □

Le (a') du théorème 3.3 découle alors du lemme 3.12 et de l'identité (3.5).

Remarque 3.13. Les arguments ci-dessus démontrent aussi que pour tout $x \in X(\overline{\mathbb{Q}}_p)$, si π est un constituant irréductible de la $\overline{\mathbb{Q}}_p$-représentation de $U(F_w)$ associée à x dans l'espace des formes automorphes p-adiques de U (cf. [BCh, Def. 7.4.4]), alors $(\rho_x)^{\mathrm{ss}}_{\Gamma_{\mathcal{K}_v}}$ et $\mathcal{L}_W(\pi)$ sont isomorphes.

Démontrons maintenant le théorème 3.5. Nous avons déjà démontré le résultat si $v \notin S$, et pour $v \in S$ (ne divisant pas v) nous avons déjà démontré que les deux représentations de l'énoncé ont même semi-simplifiée.

Nous devons pour cela faire un rappel de notations issues de [BCh, §6.5.1]. Soit v une place finie de $\mathcal{K}$, $W_{\mathcal{K}_v}$ son groupe de Weil et $I_{\mathcal{K}_v} \subset W_{\mathcal{K}_v}$ le sous-groupe d'inertie. Soit $\mathcal{J}$ l'ensemble des classes d'isomorphie de représentations irréductibles de $I_{\mathcal{K}_v}$. Si ρ est une représentation de Weil-Deligne de $\mathcal{K}_v$, et si $\tau \in \mathcal{J}$, alors nous avons encore une décomposition en composantes isotypiques

$$\rho = \bigoplus_{\tau \in \mathcal{J}} \rho[\tau]$$

qui sont préservées par $I_{\mathcal{K}_v}$ et par l'opérateur de monodromie. En particulier, on peut encore définir pour chaque $\tau \in \mathcal{J}$ une unique partition $p_I(\rho, \tau)$ de l'entier $\dim(\rho[\tau])/\dim(\tau)$ qui détermine la classe de conjugaison de l'opérateur de monodromie sur $\rho[\tau]$. Si ρ et ρ' sont deux représentations de Weil-Deligne, on note

$$\rho \prec_I \rho'$$

si les représentations de $W_{\mathcal{K}_v}$ associées à ρ et ρ' sont isomorphes restreintes au groupe d'inertie et si $p_I(\rho, \tau) \prec p_I(\rho', \tau)$ pour tout $\tau \in \mathcal{J}$. Si $\rho \prec_I \rho'$ et $\rho' \prec_I \rho$, on notera aussi $\rho \simeq_I \rho'$.

Si $\lambda = (\lambda_1 \geq \lambda_2 \geq \cdots)$ est une partition d'un entier n, et si $r \in \mathbb{N}$, on note $r\lambda$ la partition de rn constituée de r copies de λ_1, puis r copies de λ_2 etc. Le résultat suivant est bien connu (et sous-entendu dans la preuve de [BCh, Prop. 6.5.3]).

Lemme 3.14. *Soit ρ une représentation de Weil-Deligne de $\mathcal{K}_v$.*

(i) *Soit $\sigma \in \mathcal{I}$ et τ un constituant irréductible de $\sigma_{|I_{\mathcal{K}_v}}$. Alors $\rho[\sigma'] \cap \rho[\tau] = 0$ si $\sigma' \in \mathcal{I}$ n'est pas une torsion non ramifiée de σ. De plus, $p_I(\rho, \tau) = rp(\rho, \sigma)$ où $r = \dim(\sigma)/\dim(\tau) \in \mathbb{N}$.*

(ii) *Si ρ' est une représentation de Weil-Deligne de $\mathcal{K}_v$ telle que $\rho^{\rm ss} \simeq \rho'^{\rm ss}$, alors $\rho \prec \rho'$ si et seulement si $\rho \prec_I \rho'$.*

Preuve — Soit σ' une représentation irréductible de $W_{\mathcal{K}_v}$. On peut trouver un entier $m \in \mathbb{Z}$ tel que si φ est un Frobenius de $W_{\mathcal{K}_v}$ alors φ^m commute aux images de I dans σ et σ'. Par le lemme de Schur, φ^m agit donc par un scalaire dans ces représentations, et quitte à tordre σ' par un caractère non ramifié, on peut supposer que ces scalaires sont identiques. La représentation τ s'étend donc en une représentation $\widetilde{\tau}$ du groupe de Weil W_M de l'extension M non ramifiée de degré m de $\mathcal{K}_v$, telle que $\sigma_{|W_M}$ et $\sigma'_{|W_M}$ contiennent $\widetilde{\tau}$. La formule pour l'induite d'une restriction montre que

$$\mathrm{Ind}_{W_M}^{W_{\mathcal{K}_v}} \widetilde{\tau} = \sigma \otimes \mathbb{C}[W_{\mathcal{K}_v}/W_M]$$

est un somme directe finie de torsions non ramifiées de σ, ce qui conclut le premier point de (i). Le fait que $\dim(\widetilde{\tau})(= \dim(\tau))$ divise $\dim(\sigma)$ un résultat classique de Clifford, le reste du (i) s'en déduit. Le (ii) découle du (i) et de ce que pour $r \geq 1$ et λ, λ' deux partitions,

$$\lambda \prec \lambda' \Leftrightarrow r\lambda \prec r\lambda'. \qquad \square$$

Ce lemme fait notamment le lien entre la relation $\prec$ introduite plus haut et la relation $\prec_I$ étudiée dans [BCh], qu'il ne reste qu'à appliquer. Supposons que v est au dessus de S et ne divise pas p. Par choix de l'idempotent e_v, tout $z \in Z$ est associé à une représentation $\Pi(z) \in \mathcal{A}$ telle que

$$\mathcal{L}(\Pi(z)_v) \prec_I \rho_0.$$

D'après le théorème 3.2 (a), il vient que

$$\forall z \in Z^{\rm reg}, \quad (\rho_z)_{|\Gamma_{\mathcal{K}_v}} \prec_I \rho_0.$$

D'après [BCh, Prop. 7.8.19] appliqué à $T : \Gamma_{\mathcal{K}} \to \mathcal{O}(X)$,

$$\forall x \in X, \quad (\rho_x)_{|\Gamma_{\mathcal{K}_v}} \prec_I \rho_0.$$

Cela vaut donc en particulier pour ρ_Π, qui est de la forme ρ_z pour $z \in Z$. Cela entraîne $\rho_{\Pi\,|\Gamma_{\mathcal{K}_v}} \prec \rho_0$ d'après le lemme 3.14 (ii) et le théorème 3.3 (a'), QED.

3.15 Propriétés aux places divisant *p*

Démontrons maintenant les points (b) – (e) du théorème 3.3. Ils se déduisent essentiellement de résultats de Sen [Sen] et Berger-Colmez [BeCo] rappelés ci-dessous.

Fixons L et F_p deux extensions finies de $\mathbb{Q}_p$ (respectivement, les "coefficients" et le "corps de base"), avec disons $L \subset \overline{\mathbb{Q}}_p$ galoisienne contenant F_p. Soit Y un L-espace rigide réduit, disons quasi-compact et quasi-séparé, M un $\mathcal{O}_Y$-module localement libre de rang n et

$$\rho : \Gamma_{F_p} \to \mathrm{GL}_{\mathcal{O}_Y}(M)$$

une $\mathcal{O}_Y$-représentation continue. Pour $y \in Y$ on notera $\rho_y : \Gamma_{F_p} \to \mathrm{GL}_n(L(y))$ (la classe d'isomorphie de) l'évaluation de ρ en y. En particulier, la théorie de Sen associe à ρ_y un polynôme

$$P_y^{\mathrm{sen}} \in (L(y) \otimes_{\mathbb{Q}_p} F_p)[t]$$

unitaire de degré n dont les racines sont les poids de Hodge-Tate généralisés de ρ_y : il y en a donc n par choix de plongement $F_p \to \overline{\mathbb{Q}}_p$.

Proposition 3.16 ([Sen]). *Il existe un unique polynôme $P \in \mathcal{O}(Y)[t]$ tel que pour tout $y \in Y$, l'évaluation en y de P coïncide avec P_y^{sen}.*

Proposition 3.17 ([BeCo]). *Supposons qu'il existe un sous-ensemble Zariski-dense $Z \subset Y$, tel que pour tout $z \in Z$ la représentation ρ_z est de de Rham (resp. semi-stable, resp. cristalline), et que ses poids de Hodge-Tate sont bornés indépendamment de z.*

Alors pour tout $y \in Y$, ρ_y est de de Rham (resp. semi-stable, resp. cristalline). Dans le cas cristallin, il existe de plus un unique polynôme $Q \in \mathcal{O}(Y)[t]$ tel que pour tout $y \in Y$, son évaluation Q_y en y coïncide avec le polynôme caractéristique de la plus petite puissance du Frobenius cristallin de $D_{\mathrm{cris}}(\rho_y)$.

En fait, Berger et Colmez démontrent le (ii) sous l'hypothèse technique supplémentaire que Y est affinoide et que $\mathcal{O}(Y)$ admet un $\mathcal{O}_L$-modèle[14] A et un A-module libre de type fini N muni d'une action A-linéaire de Γ_{F_p} tels que $N[1/p] = M(Y)$ comme $\mathcal{O}(Y)[\Gamma_{F_p}]$-modules. Pour en déduire la version énoncée ci-dessus, nous aurons besoin du lemme suivant étendant le fait bien connu qu'un sous-groupe compact de $\mathrm{GL}_n(L)$ admet un $\mathcal{O}_L$-réseau stable. Dans ce lemme, G est un groupe topologique compact quelconque et $\mathcal{M}$ un $\mathcal{O}_X$-module localement libre de rang n muni d'une représentation $\mathcal{O}_Y$-linéaire continue de G.

Lemme 3.18. *Il existe un $\mathcal{O}_L$-schéma formel de type fini $\mathcal{Y}$, et un $\mathcal{O}_{\mathcal{Y}}$-module localement libre de rang n muni d'une action $\mathcal{O}_{\mathcal{Y}}$-linéaire continue de G, tels que $Y = \mathcal{Y}[1/p]$ et $M = \mathcal{M}[1/p]$.*

Preuve — D'après Raynaud, il existe un $\mathcal{O}_L$-modèle formel de type fini $\mathcal{Y}_0$ de X ainsi qu'un $\mathcal{O}_{\mathcal{Y}_0}$-module cohérent $\mathcal{N}_0$ tels que $\mathcal{N}_0[1/p] = M$ (et on peut les choisir sans p-torsion). Comme $\mathcal{N}_0$ est ouvert dans M et $\mathcal{Y}_0$ est de type fini, il existe par continuité un sous-groupe ouvert H de G qui préserve $\mathcal{N}_0$. Par compacité de G, $\mathcal{N} = \sum_{g \in G} g\mathcal{N}_0$ est alors une somme finie, c'est donc un faisceau cohérent sur $\mathcal{Y}_0$ muni d'une représentation $\mathcal{O}_{\mathcal{Y}_0}$-linéaire continue de G, et tel que $\mathcal{N}[1/p] = M$. Il reste à le rendre localement libre.

Soit $\mathcal{I}$ le n-ième idéal de Fitting de $\mathcal{N}$ et considérons l'éclaté de $\mathcal{I}$ (c'est un éclatement admissible car $\mathcal{N}[1/p]$ est libre de rang n) : c'est un $\mathcal{O}_L$-schéma formel de type fini $\mathcal{Y}$

[14] On rappelle qu'un $\mathcal{O}_L$-modèle d'une L-algèbre affinoide $\mathcal{O}(Y)$ est une $\mathcal{O}_L$-algèbre $A \subset \mathcal{O}(Y)$ qui est topologiquement de type fini et telle que $A[1/p] = \mathcal{O}(Y)$. De tels modèles existent toujours et sont ouverts dans $\mathcal{O}(Y)$.

dont la fibre générique est encore Y. Notons $f : \mathcal{Y} \to \mathcal{Y}_0$ cet éclatement. D'après [GR, Lemme 5.4.3], le transformé strict $\mathcal{M}$ de $\mathcal{N}$ par f est localement libre de rang n sur $\mathcal{Y}$. Par construction, l'action naturelle de G sur $f^*\mathcal{N}$ passe à son quotient $\mathcal{M}$, et $(\mathcal{Y}, \mathcal{M})$ a les propriétés de l'énoncé. □

On justifie alors la proposition 3.17 comme suit. Soit I un intervalle borné de $\mathbb{Z}$, $\mathcal{Y}$ et $\mathcal{M}$ comme dans le lemme ci-dessus. Pour chaque ouvert affine $\mathcal{U} \subset \mathcal{Y}$ sur lequel $\mathcal{M}$ est libre, [BeCo, Thm. B, C] assure que le lieu des $y \in U := \mathcal{U}[1/p]$ tels que M_y $(=\rho_y)$ est de de Rham (resp. cristalline) et à poids de Hodge-Tate dans I est un fermé de U, notons-le U^I. Ce fait s'étend immédiatement aux $\mathcal{U}$ non nécessairement affines mais quasi-compacts et sur lesquels $\mathcal{M}$ est libre. Si $\mathcal{U}$ et $\mathcal{V}$ sont deux tels ouverts, $\mathcal{U} \cap \mathcal{V}$ a la même propriété, et l'assertion $(U \cap V)^I = U^I \cap V^I$ a donc un sens et est évidente. Autrement dit, *le lieu $Y^I \subset Y$ des $y \in Y$ tels que ρ_y est de de Rham (resp. cristalline) à poids de Hodge-Tate dans I est un fermé analytique de Y*. Le premier point de la proposition 3.17 suit, ainsi que la précision dans le cas cristallin car les polynômes Q^I évidents se recollent trivialement. En vue d'utilisations éventuelles futures, mentionnons le corollaire suivant que nous venons de démontrer.

Corollaire 3.19. *Les théorèmes A, B et C de [BeCo] restent valables si $\mathcal{X}$ (resp. S) loc. cit. est remplacé par un L-espace rigide réduit quasi-compact quasi-séparé quelconque (resp. par $\mathcal{O}_{\mathcal{X}}$), et la représentation V par un $\mathcal{O}_{\mathcal{X}}$-module localement libre muni d'une action $\mathcal{O}_{\mathcal{X}}$-linéaire continue de G_K.*

Terminons la preuve du théorème 3.3. Soit $\Pi \in \mathcal{A}$ et $z_0 \in Z$ qui lui est associé. Par la propriété d'accumulation de Z^{reg} aux points de Z, on peut trouver un ouvert affinoide $\Omega \subset X$ contenant z_0 et tel que $Z^{\mathrm{reg}} \cap \Omega$ est Zariski-dense dans Ω. Notons T_Ω le pseudo-caractère galoisien $\Gamma_{\mathcal{K}} \to \mathcal{O}(\Omega)$ obtenu par restriction de T à Ω. D'après [BCh, Lemme 7.8.11], il existe un espace rigide réduit séparé quasi-compact Y et un morphisme $f : Y \to \Omega$ ayant les propriété suivantes :

(a) f est propre et surjectif,

(b) il existe un ouvert admissible $U \subset \Omega$ Zariski-dense tel que $f^{-1}(U) \to U$ soit étale fini,

(c) il existe un $\mathcal{O}_Y$-module M localement libre de rang n muni d'une action linéaire continue de $\Gamma_{\mathcal{K}}$ dont la trace est le pull-back de T_Ω sur Y par f,

(d) pour tout $y \in U$, la $\Gamma_{\mathcal{K}}$-représentation $M_{f^{-1}(y)}$ est semi-simple (c'est donc ρ_y).

En particulier, $Z' := f^{-1}(U \cap Z^{\mathrm{reg}})$ est Zariski-dense dans Y.

Fixons $v \in S'_p$. Pour tout $z \in Z'$, la représentation $M_z \simeq \rho_z$, vue comme représentation de $\Gamma_{\mathcal{K}_v}$, est donc de de Rham (resp. semi-stable si $\mathcal{B}_v$ est la composante de Bernstein non ramifiée, resp. cristalline si $v \notin S$). Par le (b) du théorème 3.2, les poids de Hodge-Tate de ρ_z sont indépendants de $z \in Z'$ car ils ne dépendent que de la représentation W_∞ fixée dans la construction de X. La Proposition 3.17 assure donc que pour tout $y \in Y$, M_y est de de Rham (resp. semi-stable, resp. cristalline) restreinte à $\Gamma_{\mathcal{K}_v}$, ainsi donc que sa $\Gamma_{\mathcal{K}}$-semi-simplifiée $(M_y)^{\Gamma_{\mathcal{K}}-\mathrm{ss}} \simeq \rho_{f(y)}$. De plus, le polynôme de Sen de M_y est indépendant de $y \in Y$. Cela conclut (b), (c) et le premier point de (d).

Pour vérifier le second, qui suppose notamment que v est au dessus d'une place w de F qui n'est pas dans S, notons que l'algèbre de Hecke $\mathcal{H}$ contient l'algèbre de Hecke sphérique en w par définition, il existe donc par l'isomorphisme de Satake un polynôme $P_w(t) \in \mathcal{O}(\Omega)[t]$ tel que pour tout $z \in Z$, et $\Pi \in \mathcal{A}$ associée à z (de changement de base noté π), on ait

$$P_w(z)(t) = \iota_p \iota_\infty^{-1} \det(t - \mathcal{L}(\pi_v | \bullet |^{\frac{1-n}{2}})(\mathrm{Frob}_v)). \tag{3.6}$$

En particulier, le polynôme Q donné par la proposition 3.17 coïncide avec $f^*(P_w)$ sur Z' par le théorème 3.2 (c), donc $f^*(P_w) = Q$ par Zariski-densité. Le second point de (d) découle alors de (3.6) et de cette dernière égalité appliquée en un $y \in Y$ tel que $f(y) = z_0$.

Fixons $v \in S_p$ et vérifions (e). Soit $\kappa^{\mathrm{univ}} : T_v \to \mathcal{O}(\mathcal{W})^*$ le caractère continu universel de T_v. Il est bien connu que ce caractère est $\mathbb{Q}_p$-différentiable en l'identité, de sorte que l'on dispose d'une application $\mathbb{Q}_p$-linéaire naturelle

$$d\kappa^{\mathrm{univ}} : \mathrm{Lie}(T_v) \longrightarrow \mathcal{O}(\mathcal{W}).$$

Rappelons que le corps L contient une clôture galoisienne de F_v. L'espace $\mathrm{Lie}(T_v) \otimes_{\mathbb{Q}_p} L$ contient donc les différentielles en l'identité de tous les co-caractères algébriques de T_v (vu par restriction comme $\mathbb{Q}_p$-tore). Une $\mathbb{Z}$-base (e_i), $i = 1, \cdots, [\mathcal{K}_w : \mathbb{Q}_p]n$, de ces co-caractères étant choisie on peut ainsi lui associer des éléments universels $\kappa_i \in \mathcal{O}(\mathcal{W}) \otimes_{\mathbb{Q}_p} L$.

Considérons maintenant le morphisme naturel $X \to \mathcal{W} \times_{\mathbb{Q}_p} L$. On peut alors considérer le pull-back des κ_i ci-dessus à X. La recette exacte pour les poids de Hodge-Tate $k_{i,\sigma}(z)$ des $(\rho_z)_{|\Gamma_{\mathcal{K}_w}}$ pour $z \in Z^{\mathrm{reg}}$ (w une place choisie de $\mathcal{K}$ au dessus de v), donnée par le théorème 3.3 (b), est une certaine combinaison $\mathbb{Z}$-linéaire fixée (i.e. indépendante de z) de ces κ_i, naturellement indexée par $\{1, \ldots, n\} \times \mathrm{Hom}(\mathcal{K}_w, \overline{\mathbb{Q}}_p)$. Quitte à changer les e_i par des combinaisons linéaires, et les ré-indexer par ce dernier ensemble, on peut donc supposer

$$\forall z \in Z^{\mathrm{reg}}, \ \forall (i,\sigma) \in \{1, \ldots, n\} \times \mathrm{Hom}(\mathcal{K}_w, \overline{\mathbb{Q}}_p), \ k_{i,\sigma}(z) = \kappa_{i,\sigma}(z) \tag{3.7}$$

Replaçons nous maintenant dans le contexte de Π, z_0, Ω, Y etc...plus haut. Soit $P^{\mathrm{sen}} \in \mathcal{O}(Y)[t]$ le polynôme associé à $M_{|\Gamma_{\mathcal{K}_w}}$ donné par la proposition 3.16. Il vient que

$$P^{\mathrm{sen}} = \left(\prod_{i=1}^{n} (T - f^*(\kappa_{i,\sigma})) \right)_{\sigma \in \mathrm{Hom}(\mathcal{K}_w, \overline{\mathbb{Q}}_p)}$$

car cela vaut sur l'ensemble Zariski-dense Z' par (3.7). La proposition 3.16 assure donc que pour tout $y \in Y$, les poids de Hodge-Tate généralisés de M_y restreinte à $\Gamma_{\mathcal{K}_w}$ sont les $\kappa_{i,\sigma}(f(y))$, ainsi donc que ceux de sa $\Gamma_{\mathcal{K}}$-semi-simplifiée $(M_y)^{\Gamma_{\mathcal{K}}-\mathrm{ss}} \simeq \rho_{f(y)}$. Si y est tel que $f(y) = z_0$, alors pour chaque $\sigma \in \mathrm{Hom}(\mathcal{K}_w, \overline{\mathbb{Q}}_p)$ les $\kappa_{i,\sigma}(y)$ sont des entiers distincts par construction, ce qui conclut le (e), et termine la preuve du théorème 3.3.

Remarque 3.20. (i) L'argument ci-dessus montre que pour tout $x \in X$ et $v \in S'_p$, $(\rho_x)_{|\Gamma_{\mathcal{K}_v}}$ est de de Rham avec les poids de Hodge-Tate attendus, et cristalline quand de plus $v \notin S$. Ceci est très faux en général si $v \in S_p$, auquel cas les propriétés galoisiennes des ρ_x sont beaucoup plus subtiles (voir les travaux de Kisin, Colmez, et Bellaïche-Chenevier [BCh]). Nous contournons ici cette difficulté en autorisant dans la construction de X des places divisant p qui ne sont pas dans S_p.

(ii) Les arguments ci-dessus (et [BeCo, Thm.C]) montrent de plus que si le théorème 3.2 (c) était étendu en :

(c')*Si* $\Pi \in \mathcal{A}^{\mathrm{reg}}$ *et* $v \in S'_p$, *la Frobenius semi-simplifiée de la représentation de Weil-Deligne sous-jacente à* $D_{\mathrm{pst}}((\rho_\Pi)_{|\Gamma_{\mathcal{K}_v}})$ *est isomorphe à* $\iota_p\iota_\infty{}^{-1}\mathcal{L}(\pi_v\otimes|\bullet|^{\frac{1-n}{2}})$, on déduirait que pour $\Pi \in \mathcal{A}$ et $v \in S'_p$, $D_{\mathrm{pst}}((\rho_\Pi)_{|\Gamma_{\mathcal{K}_v}}) \prec \iota_p\iota_\infty{}^{-1}\mathcal{L}(\pi_v \otimes |\bullet|^{\frac{1-n}{2}})$. Par rapport au dévissage [ChH], il suffirait même de se restreindre au cas où v est décomposée sur F.

References

[BCh] J. Bellaïche & G. Chenevier, *Families of Galois representations and Selmer groups*, Astérisque 342, Soc. Math. France, 314 p. (2009).

[BCh2] J. Bellaïche & G. Chenevier, *The sign of Galois representations of automorphic forms for unitary groups*, Compositio Math. 147, 1337-1352 (2011).

[BeCo] L. Berger & P. Colmez, *Familles de représentations de de Rham et monodromie p-adique* dans *Représentations p-adiques de groupes p-adiques I : représentations galoisiennes et* (φ,Γ)*-modules*, Astérisque 319 (2008).

[Bern] J.-N. Bernstein, *Le centre de Bernstein* (rédigé par Pierre Deligne), dans *Représentations des groupes réductifs sur un corps local*, Herman, collection "Travaux en cours" (1984).

[Bu] K. Buzzard, *Eigenvarieties*, proceedings of the London Math. Soc. Symp. on *L-functions and Galois Representations*, Cambridge University Press, Durham (2007), 59–120.

[Bo] A. Borel, *Some finiteness properties of adele groups over number fields*, Publ. Math. I. H.É.S. 16 (1963), 5–30.

[Ch] G. Chenevier, *Familles p-adiques de formes automorphes pour* GL(n), Journal für die reine und angewandte Mathematik 570 (2004), 143–217.

[ChH] G. Chenevier & M. Harris, *Construction of automorphic Galois representations*, Cambridge Math. Journal 1 (2013), 53–73.

[C1] R.Coleman, *Classical and overconvergent modular forms*, Invent. Math. 124 (1996), 214–241.

[C2] R. Coleman, *p-adic Banach spaces & families of modular forms*, Invent. Math. 127 (1997), 417–479.

[CM] R. Coleman & B. Mazur, *The eigencurve*, in Galois representations in Arithmetic Algebraic Geometry (Durham, 1996), London. Math. Soc. Lecture Notes **254**, Cambridge Univ. Press (1998).

[E] M. Emerton, *On the interpolation of systems of eigenvalues attached to automorphic Hecke eigenforms*, Invent. Math. **164** (2006), 1–84.

[GR] C. Gruson & M. Raynaud, *Critères de platitude et de projectivité*, Invent. Math. **13** (1971), 1–89.

[Mac] I.G. Macdonald, *Symmetric functions and hall polynomials*, Oxford Math. Monographs (1979).

[Lab] J.-P. Labesse, *Changement de base CM et séries discrètes*, dans *On the Stabilization of the Trace Formula*, International Press (2011).

[Loe] D. Loeffler, *Overconvergent algebraic automorphic forms*, Proc. London Math. Soc. 102 (2011), no. 2, 193–228.

[M] B. Mazur, *The theme of p-adic variation*, Math.: Frontiers and perspectives, V. Arnold, M. Atiyah, P. Lax & B. Mazur Ed., AMS (2000).

[SZ] P. Schneider & E.-W. Zink, *K-types for the tempered components of a p-adic general linear group*, Journal für die reine und angew. Math. **517** (1999), 161–208.

[Sen] S. Sen, *An infinite dimensional Hodge-Tate theory*, Bull. Soc. math. France 121 (1993), 13–34.

[Shin] S.W. Shin, *Galois representations arising from some compact Shimura varieties*, Ann. of Math. 173 (2011), no.3, 1645-1741.

[Y] A. Yamagami, *On p-adic families of Hilbert cusp forms of finite slope*, Journal of Number Theory 123 (2007), 363–387.

A PATCHING LEMMA

CLAUS M. SORENSEN

Department of Mathematics, UC San Diego, La Jolla, USA.
E-mail address: csorensen@ucsd.edu

Abstract

In this note, we discuss how to patch up certain collections of Galois representations over solvable extensions. We give a fairly elaborate proof, based on ideas from [HT]. The patching lemma, referred to in the title, is of importance in the construction of automorphic Galois representations, and is used in [CH]. Descent from quadratic extensions was used earlier by Blasius and Rogawski in the case of Hilbert modular forms [BRo].

1 Patching: Extensions of prime degree

We discuss a patching argument, used in various guises by other authors. For example, see Proposition 4.3.1 in [BRo], or section 4.3 in [BRa]. Here we will prove a variant of a Proposition in the *first* version (v.1) of [CH] by Harris, which in turn was based on the discussion on pages 230–231 in [HT]. The proof in the first version of [CH] was somewhat brief, and somewhat imprecise at the end, so we decided to write up the more detailed proof below.

Throughout the paper we use Γ_F as shorthand notation for the absolute Galois group $\mathrm{Gal}(\bar{F}/F)$. The setup is the following: We let $\mathcal{I} \neq \varnothing$ be a set of cyclic Galois extensions E, of a fixed number field F, of *prime* degree q_E. We allow the prime q_E to vary with $E \in \mathcal{I}$. For every $E \in \mathcal{I}$ we assume we are given an n-dimensional continuous semisimple ℓ-adic Galois representation

$$\rho_E : \Gamma_E \to \mathrm{GL}_n(\bar{\mathbb{Q}}_\ell).$$

Here ℓ is a fixed prime. The family of representations $\{\rho_E\}$ is assumed to satisfy:

(a) Galois invariance: $\rho_E^\sigma \simeq \rho_E, \quad \forall \sigma \in \mathrm{Gal}(E/F)$,

(b) Compatibility: $\rho_E|_{\Gamma_{EE'}} \simeq \rho_{E'}|_{\Gamma_{EE'}}$,

for any two $E, E' \in \mathcal{I}$. These conditions are certainly necessary for the ρ_E to be of the form $\rho|_{\Gamma_E}$ for a representation ρ of Γ_F. What we will show below is that in fact (a) and (b) are also sufficient conditions if the collection $\mathcal{I}$ is large enough in the following sense.

Definition 1. For a finite set S of places of F, we say that a non-empty collection $\mathcal{I}$ of cyclic extensions E/F of prime degree is S-general if for any finite place $v \notin S$ of F there exist infinitely many $E \in \mathcal{I}$ in which v splits completely.

We will extend this notion to collections of solvable extensions in Definition 3 below. Another characterization of S-general collections $\mathcal{I}$ is as follows.

Lemma 1. *A collection $\mathcal{I}$ as in Def. 1 is S-general if and only if for all $v \notin S$ and for every finite extension M/F there exists an $E \in \mathcal{I}$ such that $E \cap M = F$ and v splits in E. (If this holds there will be infinitely many such E.)*

Proof. Note that since $[E : F]$ is prime $E \cap M = F$ is equivalent to $E \not\subset M$. □

A slightly stronger condition is:

Definition 2. We say that $\mathcal{I}$ is strongly S-general if the following holds: For any finite set Σ of places of F, disjoint from S, there is an $E \in \mathcal{I}$ in which every $v \in \Sigma$ splits completely. (Here the infinitude of such E is not assumed.)

To see that this is indeed *stronger*, we use the characterization in Lemma 1: Fix a finite place $v \notin S$, and a finite extension M of F. Clearly we may assume $M \neq F$ is Galois. Let $\{M_i\}$ be the subfields of M, Galois over F, with a simple Galois group. For each i we then choose a place v_i of F, not in S, which does *not* split completely in M_i. We take Σ to be $\{v, v_i\}$ in the above definition, and get an $E \in \mathcal{I}$ in which v and every v_i splits. If E was contained in M, it would be one of the M_i, but this contradicts the choice of v_i. Thus E and M are linearly disjoint. (This argument is based on a similar proof from [CH, v.1].)

Remark. An anonymous referee suggested the following alternate argument, which circumvents Lemma 1: Let $\{M_i\}_{i \in I}$ be the set of fields in $\mathcal{I}$ that are totally split at v. Suppose I is finite. Choose for each M_i a place $v_i \notin S$ not split in M_i and let $\Sigma = \{v_i\}_{i \in I} \cup \{v\}$. Then this violates Definition 2.

Example. Let $\Sigma = \{p_i\}$ be a finite set of primes. As is well-known, for odd p_i,

$$p_i \text{ splits in } \mathbb{Q}(\sqrt{d}) \Longleftrightarrow p_i \nmid d \text{ and } \left(\frac{d}{p_i}\right) = 1.$$

Here d is any square-free integer. Moreover, 2 splits when $d \equiv 1 \bmod 8$. The set of *all* integers d satisfying the congruences $d \equiv 1 \bmod p_i$, for all i, form an arithmetic progression. By Dirichlet's Theorem, it contains infinitely many primes. Therefore, the following family of imaginary quadratic extensions

$$\mathcal{I} = \{\mathbb{Q}(\sqrt{-p}) \text{ for primes } p\}$$

is strongly $\varnothing$-general. This gives rise to a similar family of CM extensions of any given totally real field F, by taking the set of all the composite fields $F\mathcal{I}$.

The main result of this section, is a strengthening of a Proposition in [CH, v.1]:

Lemma 2. *Let $\mathcal{I}$ be an S-general set of cyclic extensions E/F of (possibly varying) prime degree q_E, and let ρ_E be a family of semisimple Galois representations satisfying the conditions (a) and (b) above. Then there is a continuous semisimple representation*

$$\rho : \Gamma_F \to \mathrm{GL}_n(\bar{\mathbb{Q}}_\ell) \quad \text{with} \quad \rho|_{\Gamma_E} \simeq \rho_E$$

for all $E \in \mathcal{I}$. This determines the representation ρ uniquely up to isomorphism.

Proof. The proof below is strongly influenced by the proof in [CH, v.1], and the proof of Theorem VII.1.9 in [HT]. We simply include more details and clarifications. The proof is quite long and technical, so we divide it into several steps. Before we construct ρ, we start off by noting that it is necessarily *unique*: Indeed, for any place $v \notin S$, we find an $E \in \mathcal{I}$ in which v splits. In particular, $E_w = F_v$ for all places w of E dividing v. Thus, all the restrictions $\rho|_{\Gamma_{F_v}}$ are uniquely determined by $\rho_E|_{\Gamma_{E_w}}$. We conclude that ρ is unique by the Cebotarev Density Theorem and the semisimplicity of ρ (by looking at $\mathrm{tr}\rho$).

For the construction of ρ, we first establish some notation used throughout the proof. We fix an arbitrary base point $E_0 \in \mathcal{I} \neq \emptyset$, and abbreviate

$$\rho_0 \overset{\mathrm{df}}{=} \rho_{E_0}, \quad \Gamma_0 \overset{\mathrm{df}}{=} \Gamma_{E_0}, \quad G_0 \overset{\mathrm{df}}{=} \mathrm{Gal}(E_0/F), \quad q_0 \overset{\mathrm{df}}{=} q_{E_0}.$$

We let H denote the Zariski closure of $\rho_0(\Gamma_0)$ inside $\mathrm{GL}_n(\bar{\mathbb{Q}}_\ell)$, and consider its identity component H°. Define M to be the finite Galois extension of E_0 with

$$\Gamma_M = \rho_0^{-1}(H^\circ), \quad \mathrm{Gal}(M/E_0) = \pi_0(H).$$

Let T be the set of *isomorphism classes* of irreducible constituents τ of ρ_0, ignoring multiplicities. By property (a), the group G_0 acts on T from the right. We note that τ and τ^σ occur in ρ_0 with the same multiplicity. We want to describe the G_0-orbits on T. First, we have the set P of fixed points $\tau = \tau^\sigma$ for all σ. The set of non-trivial orbits is denoted by C. Note that any $c \in C$ has prime cardinality q_0. For each such c, we pick a representative $\tau_c \in T$, and let C_0 be the set of all these representatives $\{\tau_c\}$. Each $\tau \in C_0$ obviously has a trivial stabilizer in G_0.

Step 1: *The extensions of ρ_0 to Γ_F.*

Firstly, a standard argument shows that each $\tau \in P$ has an extension $\tilde{\tau}$ to Γ_F. This uses the divisibility of $\bar{\mathbb{Q}}_\ell^*$, in order to find a suitable intertwining operator $\tau \simeq \tau^\sigma$. For each isomorphism class $\tau \in P$ we fix such a lift $\tilde{\tau}$ once and for all. All the other lifts are then obtained from $\tilde{\tau}$ as unique twists:

$$\tilde{\tau} \otimes \eta, \quad \eta \in \hat{G}_0.$$

Here $\hat{G}_0 = \mathrm{Hom}(G_0, \bar{\mathbb{Q}}_\ell^*)$ is the group of characters of G_0.

Secondly, for any $\tau \in c$ we introduce the induced representation

$$\tilde{\tau} \overset{\mathrm{df}}{=} \mathrm{Ind}_{\Gamma_0}^{\Gamma_F}(\tau).$$

Since τ is *not* Galois-invariant, this $\tilde{\tau}$ is irreducible (cf. [CR, Cor. 50.6]). It depends only on the orbit c containing τ, and it is invariant under twisting by $\hat{G}_0$. It has restriction

$$\tilde{\tau}|_{\Gamma_0} \simeq \bigoplus_{\sigma \in G_0} \tau^\sigma.$$

If we let m_τ denote the multiplicity with which $\tau \in T$ occurs in ρ_0, we get that

$$\{\bigoplus_{\tau \in C_0} m_\tau \cdot \tilde{\tau}\} \oplus \{\bigoplus_{\tau \in P} \bigoplus_{\eta \in \hat{G}_0} m_{\tilde{\tau},\eta} \cdot (\tilde{\tau} \otimes \eta)\}$$

is an extension of ρ_0 to Γ_F for all choices of $m_{\tilde{\tau},\eta} \in \mathbb{Z}_{\geq 0}$ such that

$$\sum\nolimits_{\eta \in \hat{G}_0} m_{\tilde{\tau},\eta} = m_\tau$$

for every fixed $\tau \in P$.

Step 2: *$\rho_0(\Gamma_{NE_0})$ is dense in H, when N is linearly disjoint from M over F.*

To see this, let us momentarily denote the Zariski closure of $\rho_0(\Gamma_{NE_0})$ by H_N. N is a finite extension, so H_N has finite index in H. Consequently, we deduce that $H_N^\circ = H^\circ$. Now, NE_0 and M are linearly disjoint over E_0, and therefore

$$\Gamma_0 = \Gamma_{NE_0} \cdot \Gamma_M \Longrightarrow \rho_0(\Gamma_0) \subset \rho_0(\Gamma_{NE_0}) \cdot H^\circ \subset H_N.$$

Taking the closure, we obtain that $H_N = H$. (The first equality above follows from linear disjointness since $\mathrm{Gal}(\Omega/NE_0) \xrightarrow{\sim} \mathrm{Gal}(M/E_0)$ with $\Omega := MNE_0$.)

Step 3: *If N is a finite extension of F, linearly disjoint from M over F. Then:*

(1) *$\tau|_{\Gamma_{NE_0}}$ is irreducible, for all $\tau \in T$.*

(2) *$\tilde{\tau}|_{\Gamma_N}$ is irreducible, for <u>all</u> $\tau \in T$.*

(3) *$\tau|_{\Gamma_{NE_0}} \simeq \tau'|_{\Gamma_{NE_0}} \Rightarrow \tau \simeq \tau'$, for all $\tau, \tau' \in T$.*

(4) *$\tilde{\tau}|_{\Gamma_N} \simeq (\tilde{\tau}' \otimes \eta)|_{\Gamma_N} \Rightarrow \tau \simeq \tau'$ and $\eta = 1$, for all $\tau, \tau' \in P$ and $\eta \in \hat{G}_0$.*

Part (1) follows immediately from Step 2 (if some $\tau|_{\Gamma_{NE_0}}$ is reducible, the image $\rho_0(\Gamma_{NE_0})$ can be conjugated into a proper parabolic of a Levi subgroup of GL_n). Part (3) follows from Step 2 by a similar argument: If $\tau|_{\Gamma_{NE_0}} \simeq \tau'|_{\Gamma_{NE_0}}$ the image $\rho_0(\Gamma_{NE_0})$ can be conjugated into a subgroup of block-diagonal matrices $\mathrm{diag}(A, B, \ldots)$ in GL_n with $A = B$, which a fortiori contains $\rho_0(\Gamma_0)$, and thus $\tau \simeq \tau'$. Obviously (1) implies (2) for $\tau \in P$ since $\tilde{\tau}|_{\Gamma_{NE_0}} = \tau|_{\Gamma_{NE_0}}$, using that $\tilde{\tau}$ extends τ. For $\tau \notin P$ the proof of (2) goes as follows: Since N and E_0 are linearly disjoint over F, we see that $\Gamma_F = \Gamma_E \cdot \Gamma_0$. Hence,

$$\tilde{\tau}|_{\Gamma_N} = \mathrm{Ind}_{\Gamma_0}^{\Gamma_F}(\tau)|_{\Gamma_N} \simeq \mathrm{Ind}_{\Gamma_{NE_0}}^{\Gamma_N}(\tau|_{\Gamma_{NE_0}}),$$

by Mackey theory. Now, $\tau|_{\Gamma_{NE_0}}$ is irreducible and *not* Galois-invariant by (3). Part (3) also implies that $\tau \simeq \tau'$ in (4). Now suppose $\eta \in \hat{G}_0$ satisfies:

$$\tilde{\tau}|_{\Gamma_N} \simeq \tilde{\tau}|_{\Gamma_N} \otimes \eta|_{\Gamma_N}$$

for some $\tau \in P$. In other words, $\eta|_{\Gamma_N}$ occurs in $\mathrm{End}_{\Gamma_{NE_0}}(\tilde{\tau}|_{\Gamma_N})$, which is trivial by part (2). So, η is trivial on Γ_N and on Γ_0. Hence, $\eta = 1$ by disjointness.

Step 4: *Suppose $E \in \mathcal{I}$ is linearly disjoint from M over F. Then, for a unique choice of non-negative integers $m_{\tilde{\tau},\eta,E}$ with η-sum m_τ, we have the formula:*

$$\rho_E \simeq \{\bigoplus_{\tau \in C_0} m_\tau \cdot \tilde{\tau}|_{\Gamma_E}\} \oplus \{\bigoplus_{\tau \in P} \bigoplus\nolimits_{\eta \in \hat{G}_0} m_{\tilde{\tau},\eta,E} \cdot (\tilde{\tau} \otimes \eta)|_{\Gamma_E}\}.$$

In particular, ρ_0 and ρ_E have a <u>common</u> extension to Γ_F.

The uniqueness of the $m_{\tilde{\tau},\eta,E}$ follows directly from part (4) in Step 3. Recall,

$$\rho_E|_{\Gamma_{EE_0}} \simeq \rho_0|_{\Gamma_{EE_0}} \simeq \{\bigoplus_{\tau\in C_0} m_\tau \cdot \bigoplus_{\sigma\in G_0} \tau|^{\sigma}_{\Gamma_{EE_0}}\} \oplus \{\bigoplus_{\tau\in P} m_\tau \cdot \tau|_{\Gamma_{EE_0}}\},$$

by the compatibility condition (b). Here all the $\tau|^{\sigma}_{\Gamma_{EE_0}}$ are distinct by (3). First, let us pick an arbitrary $\tau \in P$. As representations of G_0, viewed as the Galois group of EE_0 over E by disjointness, we have

$$\mathrm{Hom}_{\Gamma_{EE_0}}(\tilde{\tau}|_{\Gamma_E}, \rho_E) \simeq \bigoplus_{\eta\in\hat{G}_0} \dim_{\bar{\mathbb{Q}}_\ell} \mathrm{Hom}_{\Gamma_E}((\tilde{\tau}\otimes\eta)|_{\Gamma_E}, \rho_E) \cdot \eta.$$

The $\bar{\mathbb{Q}}_\ell$-dimension of the left-hand side clearly equals m_τ, and the right-hand side defines the partition $m_{\tilde{\tau},\eta,E}$ of m_τ. Next, let us pick an arbitrary $\tau \in C_0$. By the same argument, using that $\tilde{\tau}$ is invariant under twisting by $\hat{G}_0$, we get:

$$\mathrm{Hom}_{\Gamma_{EE_0}}(\tilde{\tau}|_{\Gamma_E}, \rho_E) \simeq \dim_{\bar{\mathbb{Q}}_\ell} \mathrm{Hom}_{\Gamma_E}(\tilde{\tau}|_{\Gamma_E}, \rho_E) \cdot \bigoplus_{\eta\in\hat{G}_0} \eta.$$

Now the left-hand side obviously has dimension $m_\tau q_0$. We deduce that $\tilde{\tau}|_{\Gamma_E}$ occurs in ρ_E with multiplicity m_τ. Counting dimensions, we obtain the desired decomposition of ρ_E. Note that we have not used the Galois invariance of ρ_E. In fact, it is a *consequence* of the above argument, assuming $E \cap M = F$.

Step 5: *Fix an $E_1 \in \mathcal{I}$ disjoint from M over F. Introduce the representation*

$$\rho \stackrel{\text{df}}{=} \{\bigoplus_{\tau\in C_0} m_\tau \cdot \tilde{\tau}\} \oplus \{\bigoplus_{\tau\in P} \bigoplus_{\eta\in\hat{G}_0} m_{\tilde{\tau},\eta,E_1} \cdot (\tilde{\tau}\otimes\eta)\}.$$

Then $\rho|_{\Gamma_E} \simeq \rho_E$ for all extensions $E \in \mathcal{I}$ linearly disjoint from ME_1 over F.

By definition, and Step 4, we have that $\rho|_{\Gamma_{E_1}} \simeq \rho_{E_1}$. Take $E \in \mathcal{I}$ to be any extension, disjoint from ME_1 over F. We compare the decomposition of $\rho|_{\Gamma_E}$,

$$\rho|_{\Gamma_E} = \{\bigoplus_{\tau\in C_0} m_\tau \cdot \tilde{\tau}|_{\Gamma_E}\} \oplus \{\bigoplus_{\tau\in P} \bigoplus_{\eta\in\hat{G}_0} m_{\tilde{\tau},\eta,E_1} \cdot (\tilde{\tau}\otimes\eta)|_{\Gamma_E}\},$$

to the decomposition of ρ_E in Step 4. We need to show the multiplicities match:

$$m_{\tilde{\tau},\eta,E} = m_{\tilde{\tau},\eta,E_1}, \quad \forall\tau \in P, \quad \forall\eta \in \hat{G}_0.$$

By property (b), for the pair $\{E, E_1\}$, we know that $\rho|_{\Gamma_E}$ and ρ_E become isomorphic after restriction to Γ_{EE_1}. Once we prove EE_1 is linearly disjoint from M over F, we are done by (2) and (4). The disjointness folllows immediately:

$$EE_1 \otimes_F M \simeq E \otimes_F E_1 \otimes_F M \simeq E \otimes_F ME_1 \simeq EE_1M.$$

Step 6: $\rho|_{\Gamma_E} \simeq \rho_E$ *for* **all** $E \in \mathcal{I}$.

By Step 5, we may assume $E \in \mathcal{I}$ is *contained* in ME_1. Now take an auxiliary extension $\mathcal{E} \in \mathcal{I}$ linearly disjoint from ME_1 over F. Consequently, using (b),

$$\rho|_{\Gamma_{\mathcal{E}}} \simeq \rho_{\mathcal{E}} \Rightarrow \rho|_{\Gamma_{E\mathcal{E}}} \simeq \rho_{\mathcal{E}}|_{\Gamma_{E\mathcal{E}}} \simeq \rho_E|_{\Gamma_{E\mathcal{E}}}.$$

Thus, $\rho|_{\Gamma_E}$ agrees with ρ_E when restricted to $\Gamma_{E\mathcal{E}}$. It suffices to show that the union of these subgroups $\Gamma_{E\mathcal{E}}$, as $\mathcal{E}$ varies, is dense in Γ_E. Again, we invoke the Cebotarev Density Theorem. Indeed, let w be a place of E, lying above $v \notin S$. It is then enough to find an $\mathcal{E} \in \mathcal{I}$, as above, such that w splits completely in $E\mathcal{E}$. Then Γ_{E_w} is contained in $\Gamma_{E\mathcal{E}}$. We *know*, by the S-generality of $\mathcal{I}$, that we can find an $\mathcal{E} \in \mathcal{I}$, not contained in ME_1, in which v splits completely. This $\mathcal{E}$ works: This follows from elementary splitting theory, as E and $\mathcal{E}$ are disjoint.

This finishes the proof of the patching lemma. □

Remark. From the proof above, we infer the following concrete description of the patch-up representation ρ. First fix *any* $E_0 \in \mathcal{I}$, and let P be the set of Galois-invariant constituents τ of ρ_{E_0}. For each such τ, we fix an extension $\tilde{\tau}$ to F once and for all. Furthermore, let C_0 be a set of representatives for the non-trivial Galois orbits of constituents of ρ_{E_0}. Then ρ is of the following form

$$\rho \simeq \{\bigoplus_{\tau \in C_0} m_\tau \cdot \mathrm{Ind}_{\Gamma_0}^{\Gamma_F}(\tau)\} \oplus \{\bigoplus_{\tau \in P} \bigoplus_{\eta \in \mathrm{Gal}(E_0/F)^\wedge} m_{\tilde{\tau},\eta} \cdot (\tilde{\tau} \otimes \eta)\}.$$

Here the $m_{\tilde{\tau},\eta}$ are *some* non-negative integers with η-sum m_τ, the multiplicity of τ in ρ_{E_0}. This fairly explicit description may be useful in deriving properties of ρ from those of ρ_{E_0}.

2 Patching: Solvable extensions

We finish with a discussion of the generalization of the patching lemma to *solvable* extensions. Thus, $\mathcal{I}$ now denotes a non-empty collection of solvable finite Galois extensions E/F, and we assume we are given, for each $E \in \mathcal{I}$, a Galois representation ρ_E, as above, satisfying (a) and (b).

For any finite extension L/F let us introduce

$$\mathcal{I}_L \overset{\mathrm{df}}{=} \{E \in \mathcal{I} : L \subset E\}.$$

Moreover, for a finite set S of places of F, we denote by $S(L)$ the set of places of L which lie above a place in S.

Loosely speaking, we say that $\mathcal{I}$ is S-general if it is S-general in prime layers:

Definition 3. For a finite set S of places of F we say that a collection $\mathcal{I} \neq \varnothing$ of finite solvable Galois extensions of F is S-general if the following holds: For every finite extension L/F such that $\mathcal{I}_L \neq \varnothing$,

- either $L \in \mathcal{I}$,
- or the set

$$\{\text{prime degree Galois extensions } K/L \text{ with } \mathcal{I}_K \neq \varnothing\}$$

is $S(L)$-general (in the sense of Def. 1).

(We leave it to the reader to verify that this recovers Def. 1 in case $\mathcal{I}$ consists of prime degree extensions; the point is that the condition $\mathcal{I}_L \neq \varnothing$ amounts to having either $L \in \mathcal{I}$ or $L = F$ since $[L : F]$ divides one of the primes q_E.)

From now on we will make the additional hypothesis that all the extensions $E \in \mathcal{I}$ have uniformly *bounded heights*. That is, there is an integer $H_{\mathcal{I}}$ such that every index $[E : F]$ has at most $H_{\mathcal{I}}$ prime divisors (not necessarily distinct). Note that what we call the *height* of a finite solvable Galois extension E/F is called the *solvable index* in [CH, p. 10]; it is the integer $m \geq 0$ for which there is a tower

$$E = E_m \supset E_{m-1} \supset \cdots \supset E_0 = F$$

with each E_{i+1}/E_i cyclic of prime degree. We are thus assuming that $m \leq H_{\mathcal{I}}$ for all $E \in \mathcal{I}$, which will enable use to proceed with patching by induction on the smallest uniform bound $H_{\mathcal{I}}$.

Lemma 3. *Assume the collection $\mathcal{I}$ has uniformly bounded heights. Then $\mathcal{I}$ is S-general if and only if the following condition holds for every L with $\mathcal{I}_L \neq \varnothing$: Given a finite place $w \notin S(L)$ and a finite extension M over L, there is an extension $E \in \mathcal{I}_L$ linearly disjoint from M over L, in which w splits completely.*

Proof. The *if* part follows immediately by unraveling the definitions (and using Lemma 1): Indeed, for given w and M the condition in Lemma 3 guarantees the existence of an $E \in \mathcal{I}_L$, linearly disjoint from M over L, in which w splits. Now take a subfield $K \subset E$ of prime degree over L.

The *only if* part is proved by induction on the maximal height of the collection $\mathcal{I}_L$ over L, the height *one* case being Lemma 1. Suppose $\mathcal{I}_L$ has maximal height H, and assume the lemma holds for smaller heights. Let w and M be as above. By S-generality, there is a prime degree extension K over L with $\mathcal{I}_K \neq \varnothing$, disjoint from M over L, in which w splits; *or* $L \in \mathcal{I}$, in which case we may take $E = L$. In the first case fix a place $\tilde{w}$ of K above w. Now, $\mathcal{I}_K$ clearly has maximal height *less* than H. By the induction hypothesis there is an $E \in \mathcal{I}_K$, disjoint from MK over K, in which $\tilde{w}$ splits. This E works. □

Under the above assumptions on $\mathcal{I}$, a given place $w \notin S(L)$ splits completely in infinitely many $E \in \mathcal{I}_L$, unless L belongs to $\mathcal{I}$. One has a stronger notion, extending Def. 2, which only plays a minor role in this paper.

Definition 4. We say that $\mathcal{I}$ is strongly S-general if the following holds: For any L such that $\mathcal{I}_L \neq \varnothing$, and any finite set Σ of places of L disjoint from $S(L)$, there is an $E \in \mathcal{I}_L$ in which every $v \in \Sigma$ splits completely.

As in the prime degree case, treated above, one shows that this is indeed a stronger condition. Our next goal is to prove the following generalization of the patching lemma (Lemma 2) to certain collections of solvable extensions.

Theorem 1. *Let $\mathcal{I}$ be an S-general collection of solvable Galois extensions E over F, with uniformly bounded heights, and let $\{\rho_E\}_{E \in \mathcal{I}}$ be a family of n-dimensional continuous* *semisimple ℓ-adic Galois representations satisfying the conditions (a) and (b) above. Then there is a continuous semisimple representation*

$$\rho : \Gamma_F \to \mathrm{GL}_n(\bar{\mathbb{Q}}_\ell) \quad \textit{with} \quad \rho|_{\Gamma_E} \simeq \rho_E$$

for all $E \in \mathcal{I}$. This determines the representation ρ uniquely up to isomorphism.

Proof. Uniqueness is proved by paraphrasing the argument in the prime degree situation (cf. the proof of Lemma 2). The existence of ρ is proved by induction on the maximal height of $\mathcal{I}$ over F, the height one case being the previous patching Lemma 2. Suppose $\mathcal{I}$ has maximal height H, and assume the Theorem holds for smaller heights. Take an arbitrary prime degree extension K over F, with $\mathcal{I}_K \neq \varnothing$. Clearly $\mathcal{I}_K$ is an $S(K)$-general set of solvable Galois extensions of K, of maximal height strictly smaller than H. Moreover, the subfamily $\{\rho_E\}_{E \in \mathcal{I}_K}$ obviously satisfies (a) and (b). By induction, we find a continuous semisimple ℓ-adic representation

$$\rho_K : \Gamma_K \to \mathrm{GL}_n(\bar{\mathbb{Q}}_\ell) \quad \text{with} \quad \rho_K|_{\Gamma_E} \simeq \rho_E$$

for all $E \in \mathcal{I}_K$. We then wish to apply the prime degree patching lemma to the family $\{\rho_K\}$, as K varies over extensions as above. Taking $L = F$ in Def. 3, such K do form an S-general collection over F; or $F \in \mathcal{I}$, in which case $\rho = \rho_F$ does the job. It remains to show that $\{\rho_K\}$ satisfies (a) and (b). To check property (a), take any $\sigma \in \Gamma_F$, and note that ρ_K^σ agrees with ρ_K after restriction to Γ_E for an arbitrary extension $E \in \mathcal{I}_K$. The union of these Γ_E is dense in Γ_K by the Cebotarev Density Theorem: Every place w of K, outside $S(K)$, splits in some $E \in \mathcal{I}_K$, so the union contains Γ_{K_w}. To check property (b), fix prime degree extensions K and K' as above. Note that

$$(\rho_K|_{\Gamma_{KK'}})|_{\Gamma_{EE'}} \simeq (\rho_{K'}|_{\Gamma_{KK'}})|_{\Gamma_{EE'}}, \quad \forall E \in \mathcal{I}_K, \quad \forall E' \in \mathcal{I}_{K'}.$$

We finish the proof by showing that the union of these $\Gamma_{EE'}$ is dense in $\Gamma_{KK'}$. Let w be an arbitrary place of KK' such that $w|_F$ does not lie in S. Choose an extension $E \in \mathcal{I}_K$ linearly disjoint from KK' over K, in which $w|_K$ splits. Then pick an extension $E' \in \mathcal{I}_{K'}$ linearly disjoint from EK' over K', in which $w|_{K'}$ splits. By elementary splitting theory, w splits in EK', and any place of EK' above w splits in EE'. Consequently, w splits in EE', see the diagram:

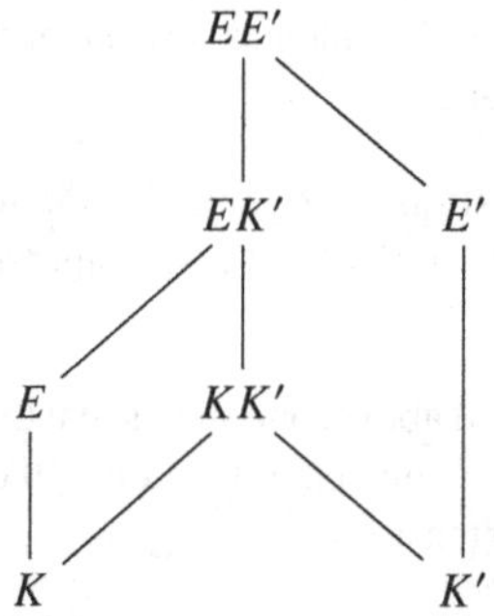

The union then contains the Galois group of $(KK')_w$. Done by Cebotarev. □

The previous result should be compared to the patching lemma in [CH, v. 1].

Acknowledgements

We would like to take this opportunity to thank D. Blasius, K. Buzzard, G. Chenevier, M. Harris, D. Ramakrishnan, S. W. Shin, and the anonymous referee for very useful feedback on an earlier draft of this paper which led to significant improvements.

Special thanks go to S. W. Shin, who pointed out some subtleties with Definition 3 and Lemma 3 in the initial draft (the possibility that $L \in \mathcal{I}$), and to M. Harris for allowing me to include many of his arguments in this note.

References

[BRa] D. Blasius and D. Ramakrishnan, *Maass forms and Galois representations*, Galois groups over $\mathbb{Q}$ (Berkeley, CA, 1987), 33–77, Math. Sci. Res. Inst. Publ., 16, Springer, New York, 1989.

[BRo] D. Blasius and J. Rogawski, *Motives for Hilbert modular forms*, Invent. Math. 114 (1993), no. 1, 55–87.

[CH] G. Chenevier and M. Harris, *Construction of automorphic Galois representations II*, Cambridge Math. Journal 1, 53–73 (2013).

[CR] C. Curtis and I. Reiner, *Representation theory of finite groups and associative algebras*. Reprint of the 1962 original. AMS Chelsea Publishing, Providence, RI, 2006.

[HT] M. Harris and R. Taylor, *The geometry and cohomology of some simple Shimura varieties*, With an appendix by Vladimir G. Berkovich. Annals of Mathematics Studies, 151. Princeton University Press, Princeton, NJ, 2001.

ON SUBQUOTIENTS OF THE ÉTALE COHOMOLOGY OF SHIMURA VARIETIES

CHRISTIAN JOHANSSON AND JACK A. THORNE

Department of Mathematical Sciences
Chalmers University of Technology and the University of Gothenburg
SE-412 96 Gothenburg, Sweden

Department of Pure Mathematics and Mathematical Statistics, Wilberforce Road, Cambridge CB3 0WB, United Kingdom

1 Introduction

Let L be a number field, and let π be a cuspidal automorphic representation of $\mathrm{GL}_n(\mathbb{A}_L)$. Suppose that π is L-algebraic and regular. By definition, this means that for each place $v|\infty$ of L, the Langlands parameter

$$\phi_v : W_{L_v} \to \mathrm{GL}_n(\mathbb{C})$$

of π_v has the property that, up to conjugation, $\phi_v|_{\mathbb{C}^\times}$ is of the form $z \mapsto z^\lambda \overline{z}^\mu$ for regular cocharacters λ, μ of the diagonal torus of GL_n. In this case we can make, following [Clo90] and [BG14], the following conjecture:

Conjecture 1.1. *For any prime p and any isomorphism $\iota : \overline{\mathbb{Q}}_p \cong \mathbb{C}$, there exists a continuous, semisimple representation $r_{p,\iota}(\pi) : \Gamma_L \to \mathrm{GL}_n(\overline{\mathbb{Q}}_p)$ satisfying the following property: for all but finitely many finite places v of L such that π_v is unramified, $r_{p,\iota}(\pi)|_{\Gamma_{L_v}}$ is unramified and the semisimple conjugacy class of $r_{p,\iota}(\pi)(\mathrm{Frob}_v)$ is equal to the Satake parameter of $\iota^{-1}\pi_v$.*

(We note that this condition characterizes $r_{p,\iota}(\pi)$ uniquely (up to isomorphism) if it exists, by the Chebotarev density theorem.) The condition that π is L-algebraic and regular implies that the Hecke eigenvalues of a twist of π appear in the cohomology of the arithmetic locally symmetric spaces attached to the group $\mathrm{GL}_{n,L}$. The first cases of Conjecture 1.1 to be proved were in the case $n = 2$ and $L = \mathbb{Q}$, in which case these arithmetic locally symmetric spaces arise as complex points of Shimura varieties (in fact, modular curves), and the representations $r_{p,\iota}(\pi)$ can be constructed directly as subquotients of the p-adic étale cohomology groups (see e.g. [Del71]). Similar techniques work in the case where $n = 2$ and L is a totally real field (see e.g. [Car86]).

The next cases of the conjecture to be established focused on the case where L is totally real or CM and π satisfies some kind of self-duality condition. When $n > 2$, or when L is not totally real, the arithmetic locally symmetric spaces attached to the group $\mathrm{GL}_{n,L}$ do not arise from Shimura varieties. However, the self-duality condition implies that π or one of its twists can be shown to descend to another reductive group G which does admit a Shimura variety. In this case the representations $r_{p,\iota}(\pi)$ can often be shown to occur as subquotients of the p-adic étale cohomology groups of the Shimura variety associated to some Shimura datum (G, X). The prototypical case is when L is a CM field and there is an isomorphism $\pi^c \cong \pi^\vee$, where $c \in \mathrm{Aut}(L)$ is complex conjugation. In this case π descends to a cuspidal automorphic representation Π of a unitary (or unitary similitude) group G such that Π_∞ is essentially square-integrable.

Going beyond the case where π satisfies a self-duality condition requires new ideas. The general case of Conjecture 1.1 where L is a totally real or CM field was established in [HLTT16] (another proof was given shortly afterwards in [Sch15a]). The difficulty in generalizing the above techniques to the case where π is not self-dual is summarized in [HLTT16] as follows:

According to unpublished computations of one of us (M.H.) and of Laurent Clozel, in the non-polarizable case the representation $r_{p,\iota}(\pi)$ will never occur in the cohomology of a Shimura variety. (1.1)

The purpose of this note is to expand on the meaning of this statement. According to [BLGGT14], an irreducible Galois representation is polarizable if it is conjugate self-dual up to twist. We first prove a negative result, showing that there are many Galois representations which are not conjugate self-dual up to twist and which do appear in the cohomology of Shimura varieties:

Theorem 1.2. *Let p be a prime, and fix an isomorphism $\iota : \overline{\mathbb{Q}}_p \cong \mathbb{C}$. Then there exist infinitely many pairs (L, π) satisfying the following conditions:*

1. *$L \subset \mathbb{C}$ is a CM number field and π is a regular L-algebraic cuspidal automorphic representation of $\mathrm{GL}_n(\mathbb{A}_L)$ such that $\pi^c \not\cong \pi^\vee \otimes \chi$ for any character $\chi : L^\times \backslash \mathbb{A}_L^\times \to \mathbb{C}^\times$.*
2. *There exists a Shimura datum (G, X) of reflex field L such that the associated Shimura varieties $\mathrm{Sh}_K(G, X)$ are proper and $r_{p,\iota}(\pi)$ appears as a subquotient of $H^*_{\acute{e}t}(\mathrm{Sh}_K(G, X)_{\overline{\mathbb{Q}}}, \mathcal{F}_{\tau,p})$ for some algebraic local system $\mathcal{F}_\tau$ and for some neat open compact subgroup $K \subset G(\mathbb{A}_{\mathbb{Q}}^\infty)$.*

(Here $\mathcal{F}_{\tau,p}$ denotes the lisse $\overline{\mathbb{Q}}_p$-sheaf on $\mathrm{Sh}_K(G, X)$ associated to $\mathcal{F}_\tau$; see § 1.2 below for the precise notation that we use.) It is therefore necessary to give a different interpretation to the assertion (1.1). The representations $r_{p,\iota}(\pi)$ appearing in Theorem 1.2 are necessarily special: in fact, the examples we construct are induced from cyclic CM extensions of L.

One subtlety here is that even if an irreducible representation $r : \Gamma_L \to \mathrm{GL}_n(\overline{\mathbb{Q}}_p)$ is conjugate self-dual up to twist (as one would expect e.g. for the n-dimensional representation attached to a RLACSDC[1] automorphic representation of $\mathrm{GL}_n(\mathbb{A}_L)$), it need not be the case that the irreducible subquotients of tensor products $r^{\otimes a} \otimes (r^\vee)^{\otimes b}$ are conjugate self-dual up to twist (and indeed, it is this possibility that we exploit in our proof of Theorem 1.2). This points to the need to phrase a condition in terms of the geometric monodromy group of r (i.e. the identity component of the Zariski closure of $r(\Gamma_L)$). The Galois representations that we construct in the proof of Theorem 1.2 are at least 'geometrically polarizable', in the sense that complex conjugation induces the duality involution on the geometric monodromy

[1] Regular L-algebraic, conjugate self-dual, cuspidal

group. The main point we make in this paper is that well-known conjectures imply that all Galois representations appearing in the cohomology of Shimura varieties are geometrically polarizable, using statements like our Principle 2.6 below. (In the body of the paper, we use the terminology 'odd' instead of 'geometrically polarizable'; see Definition 2.3.)

In order to fully address the question posed in (1.1), one must first answer the question of which kind of cohomology groups to consider. If $\mathrm{Sh}_K(G, X)$ is proper then ordinary étale cohomology with coefficients in an algebraic local system provides the only natural choice. In the non-compact case, one could consider ordinary cohomology, cohomology with compact support, or the intersection cohomology of the minimal compactification $\mathrm{Sh}_K^{\min}(G, X)$ of $\mathrm{Sh}_K(G, X)$. We first study the intersection cohomology, using its relation with discrete automorphic representations of $G(\mathbb{A}_\mathbb{Q})$. This leads, for example, to the following theorem.

Theorem 1.3. *Let L be an imaginary CM or totally real number field, and let $\rho : \Gamma_L \to \mathrm{GL}_n(\overline{\mathbb{Q}}_p)$ be a continuous representation which is strongly irreducible, in the sense that for any finite extension M/L, $\rho|_{\Gamma_M}$ is irreducible. Let (G, X) be a Shimura datum of reflex field L. Assume Conjecture 4.3 and Conjecture 5.1. Let $j : \mathrm{Sh}_K(G, X) \to \mathrm{Sh}_K^{min}(G, X)$ be the open immersion of the Shimura variety into its minimal compactification. If ρ appears as a subquotient of $H^*_{ét}(\mathrm{Sh}_K^{min}(G, X)_{\overline{\mathbb{Q}}}, j_{!*}\mathcal{F}_{\tau,p})$ for some algebraic local system $\mathcal{F}_\tau$, then ρ is conjugate self-dual up to twist.*

It is easy to construct examples of strongly irreducible Galois representations which are not conjugate self-dual up to twist (for example, arising from elliptic curves over an imaginary CM field L). Conjecturally, then, these Galois representations can never appear as subquotients of the intersection cohomology of Shimura varieties.

We can summarise the conjectures assumed in the statement of Theorem 1.3 as follows. Conjecture 4.3 asserts the existence of Galois representations attached to discrete cohomological automorphic representations π of $G(\mathbb{A}_\mathbb{Q})$, where G is a reductive group over $\mathbb{Q}$ such that $G(\mathbb{R})$ admits discrete series. It includes a rather precise formulation of local-global compatibility at infinity based on a connection with an A-parameter of π_∞. (For a closely related statement, see [Ser12, §8.2.3.4].) It would not be possible to formulate this using only the formalism of L-parameters (as opposed to A-parameters). Conjecture 5.1 is a weak consequence of Kottwitz's conjectural description of the intersection cohomology of the minimal compactification of a Shimura variety in terms of A-parameters, slightly reformulated here in a similar manner to [Joh13]. This focus on A-parameters is essential, since for a result like Theorem 1.3 the most interesting part of cohomology is indeed the part corresponding to non-tempered automorphic representations.

Since compactly supported cohomology is dual to ordinary cohomology, the other case to consider is that of the ordinary cohomology of non-proper Shimura varieties. In this case, Morel's theory of weight truncations can be used to reduce to the case of intersection cohomology. This leads, for example, to the following theorem.

Theorem 1.4. *Let (G, X) be a Shimura datum satisfying the assumptions of § 6, and let L be its reflex field. Let $\mathcal{F}_\tau$ be an algebraic local system on $\mathrm{Sh}_K(G, X)$, and let p be a prime. Then any irreducible subquotient $\overline{\mathbb{Q}}_p[\Gamma_L]$-module of $H^*_{ét}(\mathrm{Sh}_K(G, X)_{\overline{\mathbb{Q}}}, \mathcal{F}_{\tau,p})$ is isomorphic to a subquotient of $H^*(\mathrm{Sh}_{K'}^{min}(G', X')_{\overline{\mathbb{Q}}}, j_{!*}\mathcal{F}_{\tau',p})$ for some Shimura datum (G', X') of reflex field L.*

We note that the assumptions in § 6 hold in particular for the Shimura data associated to inner forms of unitary similitude and symplectic similitude groups. These are the groups used in [HLTT16] and [Sch15a], and which led us to be interested in these problems in the first place.

We now describe the organization of this note in more detail. In § 2 we review some principles from the representation theory of reductive groups, and consequences for what we call 'odd Galois representations'. In § 3 we prove Theorem 1.2 by explicitly constructing irreducible Galois representations in the cohomology of unitary Shimura varieties which are not conjugate self-dual up to twist. In § 4, we introduce the Langlands group and the formalism of L-parameters and A-parameters, and use this as a heuristic in order to justify Conjecture 4.3. In § 5 we combine this with Kottwitz's conjectural description of the intersection cohomology of Shimura varieties in order to state Conjecture 5.1 and then, using the groundwork done in § 2, to prove Theorem 1.3. Finally, in § 6 we sketch Morel's proof of Theorem 1.4.

1.1 Acknowledgments

We are very grateful to Laurent Clozel, Michael Harris, Frank Calegari, and Toby Gee for their comments on an earlier draft of this paper. We also thank the anonymous referee for their useful comments. This work was begun while JT served as a Clay Research Fellow. This project has received funding from the European Research Council (ERC) under the European Union's Horizon 2020 research and innovation programme (grant agreement No 714405).

1.2 Notation

A reductive group is not necessarily connected. If $G, H, \ldots$ are linear algebraic groups over a field Ω of characteristic 0, then use gothic letters $\mathfrak{g}, \mathfrak{h}, \ldots$ to denote their respective Lie algebras. We write $\mathfrak{sl}_2$ for the Lie algebra of SL_2; it has a basis of elements

$$x_0 = \begin{pmatrix} 0 & 1 \\ 0 & 0 \end{pmatrix}, t_0 = \begin{pmatrix} 1 & 0 \\ 0 & -1 \end{pmatrix}, y_0 = \begin{pmatrix} 0 & 0 \\ 1 & 0 \end{pmatrix}.$$

These satisfy the relations

$$[x_0, y_0] = t_0, [t_0, x_0] = 2x_0, [t_0, y_0] = -2y_0.$$

If $\mathfrak{g}$ is any Lie algebra and (x, t, y) is a tuple of elements of $\mathfrak{g}$ satisfying the same relations, then we call (x, t, y) an $\mathfrak{sl}_2$-triple in $\mathfrak{g}$.

If E is a field, then we write Γ_E for the absolute Galois group of E with respect to some fixed separable closure $\overline{E}$. If E is a number field and v is a place of E, then we write $\Gamma_{E_v} \subset \Gamma_E$ for the decomposition group at v, which is well-defined up to conjugation. If v is a non-archimedean place, then we also write $k(v)$ for the residue field of v, and q_v for the cardinality of $k(v)$. We write Frob_v for a geometric Frobenius element. We write $\mathbb{A}_E$ for the adèle ring of E, and $\mathbb{A}_E^\infty$ for its finite part. If p is a fixed prime, then we write $\epsilon : \Gamma_E \to \mathbb{Q}_p^\times$ for the p-adic cyclotomic character.

If G is a connected reductive group over $\mathbb{Q}$, then we write ${}^LG$ for its L-group, which we usually think of as a semi-direct product $\widehat{G} \rtimes \mathrm{Gal}(E/\mathbb{Q})$, where $\widehat{G}$ is the dual group (viewed as a split reductive group over $\mathbb{Q}$) and $E/\mathbb{Q}$ is the Galois extension over which G becomes an inner form of its split form. If p is a prime, then an L-homomorphism $\rho : \Gamma_{\mathbb{Q}} \to {}^LG(\overline{\mathbb{Q}}_p)$ is a homomorphism for which the projection $\Gamma_{\mathbb{Q}} \to \mathrm{Gal}(E/\mathbb{Q})$ is the canonical one.

If (G, X) is a Shimura datum, in the sense of [Del79b], and $K \subset G(\mathbb{A}_{\mathbb{Q}}^{\infty})$ is an open compact subgroup, then we write $\mathrm{Sh}_K(G, X)$ for the associated Shimura variety, which is an algebraic variety defined over the reflex field of the pair (G, X) (see [Mil83] for existence in the most general case). By an algebraic local system $\mathcal{F}_\tau$, we mean the local system of $\overline{\mathbb{Q}}$-vector spaces $\mathcal{F}_\tau$ on $\mathrm{Sh}_K(G, X)(\mathbb{C})$ associated to a finite-dimensional algebraic representation $\tau : G_{\overline{\mathbb{Q}}} \to \mathrm{GL}(V_\tau)$ such that the central character $\omega_\tau : Z(G)_{\overline{\mathbb{Q}}} \to \mathbb{G}_m$ is defined over $\mathbb{Q}$. If p is a prime and $\iota : \overline{\mathbb{Q}} \hookrightarrow \overline{\mathbb{Q}}_p$ is a fixed embedding, then we get a lisse étale sheaf $\mathcal{F}_{\tau,p}$ on $\mathrm{Sh}_K(G, X)$, which is the one considered in the introduction to this paper.

2 Odd Galois representations

In this section we discuss Galois representations $\rho : \Gamma_{\mathbb{Q}} \to H(\overline{\mathbb{Q}}_p)$, where H is a reductive group. We are particularly interested in representations which are odd, in the sense of Definition 2.3. If $H = {}^LG$, where G is a connected reductive group over $\mathbb{Q}$ such that $G^{\mathrm{ad}}(\mathbb{R})$ contains a compact maximal torus, then Definition 2.3 coincides with the one given in F. Calegari's note [Cal], but not otherwise (see also [Gro]).

Let Ω be a field of characteristic 0.

Definition 2.1. Let G be a connected reductive group over Ω. We say that an involution $\theta : G \to G$ is odd if $\mathrm{tr}(d\theta^{\mathrm{ad}} : \mathfrak{g}^{\mathrm{ad}} \to \mathfrak{g}^{\mathrm{ad}}) = -\operatorname{rank} G^{\mathrm{ad}}$, where θ^{ad} is the induced involution of the adjoint group G^{ad}.

If G is semisimple, then the class of odd involutions coincides with the class of Chevalley involutions (see e.g. [AV16]). In particular, they are all $G(\overline{\Omega})$-conjugate. However, in general we diverge from this class by allowing also involutions which e.g. act trivially on the centre of G.

Lemma 2.2. *Let G be a reductive group over Ω, and let $H \subset G$ be a closed reductive subgroup. Let θ be an involution of G which leaves H invariant and such that $\theta|_{G^\circ}$ is odd. Suppose that there exist cocharacters $\mu, w : \mathbb{G}_m \to H$ satisfying the following properties:*

1. *μ is regular in G°, and w takes values in $Z(G)$.*
2. *$\theta \circ \mu = \mu^{-1} w$.*

Then $\theta|_{H^\circ}$ is an odd involution.

Proof. After replacing G by its quotient by $Z(G^\circ)$, we can assume that G° is adjoint and $w = 1$. Let $T = Z_G(\mu)$, and consider the decomposition

$$\mathfrak{g} = \mathfrak{t} \oplus \mathfrak{g}^+ \oplus \mathfrak{g}^-$$

into zero, positive, and negative weight spaces for the cocharacter μ. Since $\theta \circ \mu = \mu^{-1}$, we see that θ swaps $\mathfrak{g}^+$ and $\mathfrak{g}^-$. Since $\operatorname{tr} d\theta = -\dim \mathfrak{t}$, we see that θ must act as -1 on $\mathfrak{t}$. Since μ factors through H, we find a similar decomposition

$$\mathfrak{h} = (\mathfrak{h} \cap \mathfrak{t}) \oplus \mathfrak{h}^+ \oplus \mathfrak{h}^-,$$

showing that $\operatorname{tr} d\theta|_{\mathfrak{h}} = -\operatorname{rank} H^\circ$, and hence that $\theta|_{H^\circ}$ is also odd. □

Definition 2.3. Let p be a prime and let G be a reductive group over $\overline{\mathbb{Q}}_p$, and let $\rho : \Gamma_{\mathbb{Q}} \to G(\overline{\mathbb{Q}}_p)$ be a continuous representation. We say that ρ is odd if $\theta = \operatorname{Ad} \rho(c)|_{G^\circ}$ is an odd involution.

Now let E be a number field, and let G be a reductive group over $\overline{\mathbb{Q}}_p$ for some prime p.

Definition 2.4. Let $\rho : \Gamma_E \to G(\overline{\mathbb{Q}}_p)$ be a continuous representation which is unramified almost everywhere. Then:

1. We say that ρ is mixed if there exists a cocharacter $w : \mathbb{G}_m \to G$ centralizing the image of ρ and such that for any representation $G \to \mathrm{GL}(V)$, $V \circ \rho$ is mixed with integer weights, and $V = \oplus_{i \in \mathbb{Z}} V^{w(t)=t^i}$ is its weight decomposition. In other words, there exists a finite set of finite places of E, containing the p-adic places, such that for any finite place $v \notin S$ of E, $\rho|_{\Gamma_{E_v}}$ is unramified and for any isomorphism $\iota : \overline{\mathbb{Q}}_p \cong \mathbb{C}$, any eigenvalue α of Frob_v on $V^{w(t)=t^i}$ satisfies $\iota(\alpha)\overline{\iota(\alpha)} = q_v^i$.
2. We say that ρ is pure if it is mixed and w takes values in $Z(G)$.
3. We say that ρ is geometric if for each place $v|p$ of E, $\rho|_{\Gamma_{E_v}}$ is de Rham. In other words, for any representation $G \to \mathrm{GL}(V)$, $V \circ \rho|_{\Gamma_{E_v}}$ is de Rham in the sense of p-adic Hodge theory.

Note that if ρ is mixed, then w is uniquely determined by ρ. When working with a mixed Galois representation, we will always write w for its corresponding weight cocharacter.

Let $\mathbb{C}_p$ denote the completion of $\overline{\mathbb{Q}}_p$. If ρ is de Rham then it is also Hodge–Tate, so there exists a cocharacter $\mu_{\mathrm{HT}} : \mathbb{G}_m \to G_{\mathbb{C}_p}$, again uniquely determined, such that $V_{\mathbb{C}_p} = \oplus_{i \in \mathbb{Z}} V_{\mathbb{C}_p}^{\mu_{\mathrm{HT}}(t)=t^i}$ is the Hodge–Tate decomposition of $V_{\mathbb{C}_p}$.

Definition 2.5. Let $\rho : \Gamma_E \to G(\overline{\mathbb{Q}}_p)$ be a geometric representation and let $v|p$ be a place of E. We call a Hodge–Tate cocharacter at v any cocharacter $\mu : \mathbb{G}_m \to G$ with the following properties:

1. μ takes values in the Zariski closure H of $\rho(\Gamma_E)$.
2. μ is $H(\mathbb{C}_p)$-conjugate to μ_{HT}.

Note that Hodge–Tate cocharacters always exist.

We conclude this section with a discussion of Hodge–Tate cocharacters satisfying special properties. This will be used as motivation in § 5. Let G be a reductive group over $\overline{\mathbb{Q}}_p$, and let $\rho : \Gamma_{\mathbb{Q}} \to G(\overline{\mathbb{Q}}_p)$ be a geometric representation of Zariski dense image. In this case, we expect that the following should be true:

- ρ is pure. Let $w : \mathbb{G}_m \to Z(G)$ denote the corresponding character.
- There exists a Hodge–Tate cocharacter $\mu : \mathbb{G}_m \to G$ and a complex conjugation $c \in \Gamma_{\mathbb{Q}}$ such that $\mathrm{Ad}(\rho(c))(\mu) = \mu^{-1} w$.

Indeed, let us suppose that we are in the "paradis motivique" described in [Ser94] (in other words, we assume the standard conjectures and the Tate conjecture). We are free to replace ρ by $\rho \times \epsilon$ and G by the Zariski closure of the image of Galois in $G \times \mathbb{G}_m$. According to the conjectures in [FM95], we should be able to find a faithful representation $R : G \to \mathrm{GL}(V)$ such that $R \circ \rho$ appears as a subquotient of the étale cohomology of a smooth projective variety X over $\mathbb{Q}$. Let $G_X \subset \mathrm{GL}(H^*(X_{\overline{\mathbb{Q}}}, \mathbb{Q}_p))$ denote the Zariski closure of the image of $\Gamma_{\mathbb{Q}}$; then G is isomorphic to a quotient of $G_{X,\overline{\mathbb{Q}}_p}$, so we just need to justify the existence of a Hodge–Tate cocharacter $\mu : \mathbb{G}_m \to G_{X,\overline{\mathbb{Q}}_p}$ and complex conjugation $c \in \Gamma_{\mathbb{Q}}$ satisfying the expected properties.

We will use the language of of Tannakian categories, as in [Ser94]. Let $\mathrm{Mot}_{\mathbb{Q}}$ denote the Tannakian category of motives over $\mathbb{Q}$, and let $\langle X \rangle$ denote its tensor subcategory generated by X. Let $\mathrm{Vec}_{\mathbb{Q}}$ denote the tensor category of finite-dimensional $\mathbb{Q}$-vector spaces. Then there are Hodge and Betti fibre functors

$$\omega_H : \langle X \rangle \to \mathrm{Vec}_{\mathbb{Q}}, X \mapsto \oplus_{i,j} H^i(X, \Omega^j_X)$$

and

$$\omega_B : \langle X \rangle \to \mathrm{Vec}_{\mathbb{Q}}, X \mapsto H^*(X(\mathbb{C}), \mathbb{Q}).$$

There is a Hodge–Betti comparison isomorphism $\alpha \in \mathrm{Isom}^{\otimes}(\omega_H, \omega_B)(\mathbb{C})$. If we fix a choice of isomorphism $\mathbb{Z}_p \cong \mathbb{Z}_p(1)$ of $\mathbb{Z}_p$-modules and an embedding $\overline{\mathbb{Q}} \hookrightarrow \mathbb{C}$, then there is determined an isomorphism $\beta \in \mathrm{Isom}^{\otimes}(\omega_H, \omega_B)(\mathbb{C}_p)$ (cf. [Fal88]). We write $M_{X,B} = \mathrm{Aut}^{\otimes}(\omega_B)$ for the usual motivic Galois group and $c \in M_{X,B}(\mathbb{Q})$ for the image of complex conjugation, and $M_{X,H} = \mathrm{Aut}^{\otimes}(\omega_H)$. Then the Hodge grading determines a Hodge cocharacter $\mu_H : \mathbb{G}_m \to M_{X,H}$ which satisfies

$$\mathrm{Ad}(c) \circ (\alpha \circ \mu_H \circ \alpha^{-1}) = w\, \mathrm{Ad}(c) \circ (\alpha \circ \mu_H^{-1} \circ \alpha^{-1}),$$

where w is the weight cocharacter, which is central and defined over $\mathbb{Q}$ (cf. [Del79a, §0.2.5]). Fix an isomorphism $\iota : \overline{\mathbb{Q}}_p \to \mathbb{C}$. We'll be done if we can show that $\iota^{-1}(\alpha \circ \mu_H \circ \alpha^{-1}) = (\iota^{-1}\alpha) \circ \mu_H \circ (\iota^{-1}\alpha)^{-1}$ is a Hodge–Tate cocharacter, when we identify $M_{X,B,\mathbb{Q}_p}$ with the group G_X above.

By definition, this means we must show that $(\iota^{-1}\alpha) \circ \mu_H \circ (\iota^{-1}\alpha)^{-1}$ is $M_{X,B}(\mathbb{C}_p)$-conjugate to the character $\beta \circ \mu_H \circ \beta^{-1}$. However, we have $\iota^{-1}\alpha \circ \beta^{-1} \in \mathrm{Isom}^{\otimes}(\omega_B, \omega_B)$ $(\mathbb{C}_p) = M_{X,B}(\mathbb{C}_p)$, so this is automatic.

Taking on board Lemma 2.2 and the above 'motivic' discussion, we arrive at the following unproven principle:

Principle 2.6. *Let G be a reductive group over $\overline{\mathbb{Q}}_p$, and let $i : H \to G$ be the embedding of a closed reductive subgroup. Let $\rho : \Gamma_{\mathbb{Q}} \to H(\overline{\mathbb{Q}}_p)$ be a geometric representation. Suppose the following:*

1. *$i \circ \rho$ is pure.*
2. *$i \circ \rho$ is odd.*
3. *The Hodge–Tate cocharacter of $i \circ \rho$ is regular in G°.*

Then ρ is odd.

It is instructive to discuss all of the above in a concrete example. Let us take the representation $\rho : \Gamma_{\mathbb{Q}} \to \mathrm{GL}_3(\overline{\mathbb{Q}}_p)$ constructed in [vGT94], and associated to a non-self dual cuspidal automorphic representation of $\mathrm{GL}_3(\mathbb{A}_{\mathbb{Q}})$ of level $\Gamma_0(128)$. More precisely, we consider the representation constructed there inside the étale cohomology of a surface; the computations of Frobenius traces in *op. cit.* support the hypothesis, but do not prove, that these representations are the same as the ones attached to the above-mentioned cuspidal automorphic representation.

The representation ρ is irreducible, by the argument on [vGT94, p. 400]. In fact, ρ has Zariski dense image in GL_3. (This can be established using some p-adic Hodge theory. Write H for the Zariski closure of $\rho(\Gamma_{\mathbb{Q}})$. Then H° contains the image of a Hodge–Tate cocharacter, which is regular in GL_3. If H° is abelian then there is an isomorphism $\rho \cong \mathrm{Ind}_{\Gamma_L}^{\Gamma_{\mathbb{Q}}} \chi$ for some degree 3 extension $L/\mathbb{Q}$ and geometric character $\chi : \Gamma_L \to \overline{\mathbb{Q}}_p^\times$. However, the infinity type of χ must be induced from the maximal CM subfield of L, which is totally real. This contradicts the fact that ρ is Hodge–Tate regular. Therefore H° is not abelian. If the derived group of H° has rank 1, then it is equal to PGL_2 in its 3-dimensional representation. The normalizer of PGL_2 in GL_3 is GO_3. Since ρ is not self-dual up to twist (see [vGT94, p. 400] again) this cannot happen. We see finally that the derived group of H° must have rank 2, and therefore that H is equal to GL_3.)

There exist a weight cocharacter w, a Hodge–Tate cocharacter μ, and a complex conjugation c of the form

$$w(t) = \mathrm{diag}(t^2, t^2, t^2),$$
$$\mu(t) = \mathrm{diag}(t^2, t, 1),$$
$$\rho(c) = \begin{pmatrix} 0 & 0 & 1 \\ 0 & -1 & 0 \\ 1 & 0 & 0 \end{pmatrix}.$$

Note in particular that $\mathrm{Ad}\,\rho(c) \circ \mu = w\mu^{-1}$. No twist of ρ is odd, because the odd involutions of GL_3 are outer. The representation $\rho \otimes \epsilon$ is pure of weight 0 and has trivial determinant.

Let $H = \mathrm{GL}_3$, and let G denote the special orthogonal group defined by the matrix

$$J = \begin{pmatrix} & & I_3 \\ & 1 & \\ I_3 & & \end{pmatrix}.$$

We write $R : H \to G$ for the embedding given by $g \mapsto \mathrm{diag}(g, 1, {}^t g^{-1})$. A calculation shows that if χ is an odd character, then $R \circ (\rho \otimes \chi)$ is odd, in the sense of Definition 2.3. Thus if χ is a geometric odd character, then $R \circ (\rho \otimes \chi)$ is geometric and odd. We note that arguing as in [HLTT16] or [Sch15b], we should be able to exhibit (the pseudocharacter of) any twist $R \circ (\rho \otimes \chi)$ by an odd geometric character of sufficiently large Hodge–Tate weight as a p-adic limit of (pseudocharacters of) G-valued representations of Zariski dense image attached to cusp forms on Sp_6 with square-integrable archimedean component. Since passing to a p-adic limit preserves the conjugacy class of complex conjugation, the oddness of $R \circ (\rho \otimes \chi)$ is a necessary condition for this to be possible. (We note that the oddness of the Galois representations attached to regular L-algebraic cusp forms on Sp_6, which is a consequence of the conjectures formulated in [BG14], follows from the results of Taïbi [Tb16].)

However, we cannot conclude that $\rho \otimes \chi$ is odd using Principle 2.6, because any such twist of ρ will fail one of the conditions there. If χ is not pure of weight -2, then $R \circ (\rho \otimes \chi)$ will not be pure. If χ is pure of weight -2 (for example, if $\chi = \epsilon$), then $R \circ (\rho \otimes \chi)$ will be pure of weight 0, but the Hodge–Tate cocharacter of $R \circ (\rho \otimes \chi)$ will not be regular.

3 Negative results

Let us fix a prime p and an isomorphism $\iota : \overline{\mathbb{Q}}_p \to \mathbb{C}$. In this section, we prove the following result (Theorem 1.2 of the introduction):

Theorem 3.1. *There exist infinitely many pairs (L, Π) satisfying the following conditions:*

1. *$L \subset \mathbb{C}$ is a CM number field and Π is a regular L-algebraic cuspidal automorphic representation of $\mathrm{GL}_n(\mathbb{A}_L)$ such that $\Pi^c \not\cong \Pi^\vee \otimes \chi$ for any character $\chi : L^\times \backslash \mathbb{A}_L^\times \to \mathbb{C}^\times$.*
2. *There exists a Shimura datum (G, X) of reflex field L such that the associated Shimura varieties $\mathrm{Sh}_K(G, X)$ are proper and $r_{p,\iota}(\Pi)$ appears as a subquotient of $H^*_{\acute{e}t}(\mathrm{Sh}_K(G, X)_{\overline{\mathbb{Q}}}, \mathcal{F}_{\tau,p})$ for some algebraic local system $\mathcal{F}_\tau$ and for some neat open compact subgroup $K \subset G(\mathbb{A}_\mathbb{Q}^\infty)$.*

Let q be an odd prime, and let K be a CM number field containing an imaginary quadratic field. Fix a CM type Φ_K of K. If K'/K is any CM extension, then we write $\Phi_{K'}$ for the induced CM type. Let E/K be a cyclic CM extension of degree q^2, and let E_0 denote the unique intermediate subfield of E/K.

Lemma 3.2. *Fix integers $(n_\tau)_{\tau \in \Phi_E}$. Then we can find a character $\psi : E^\times \backslash \mathbb{A}_E^\times \to \mathbb{C}^\times$ and a finite place v of K split over K^+ and inert in E, all satisfying the following conditions:*

1. *$\psi \circ \mathrm{N}_{E/E^+} = \| \cdot \|^{1-q^2}$ and $\psi|_{(E \otimes_{E,\tau} \mathbb{C})^\times}(z) = z^{n_\tau} \overline{z}^{1-q^2-n_\tau}$ for all $\tau \in \Phi_E$.*
2. *Let w denote the unique place of E lying above v. Then for each $g \in \mathrm{Gal}(E/K)$, we have $\psi|_{E_w^\times} \neq \psi^g|_{E_w^\times}$.*

Proof. This is a special case of [BLGGT14, Lemma A.2.5]. □

We now fix a tuple of integers $(n_\tau)_{\tau\in\Phi_E}$ with the following properties:

- For all $\tau, \tau' \in \Phi_E$ such that $\tau \neq \tau'$, we have $|n_\tau - n_{\tau'}| > 1$.
- For all $\tau_0, \tau_0' \in \Phi_{E_0}$ such that $\tau_0 \neq \tau_0'$, we have
$$\sum_{\substack{\tau\in\Phi_E\\ \tau|_{E_0}=\tau_0}} n_\tau \neq \sum_{\substack{\tau'\in\Phi_E\\ \tau'|_{E_0}=\tau_0'}} n_{\tau'}.$$
- There exists $\tau \in \Phi_E$ such that the matrix $(n_{\tau gh})_{g,h\in \mathrm{Gal}(E/K)}$ has non-zero determinant (note that this is a circulant matrix).

Fix ψ as in Lemma 3.2. Let $\psi_p = r_{p,\iota}(\psi) : \Gamma_E \to \overline{\mathbb{Q}}_p^\times$.

Lemma 3.3. *1. The representation $\rho_p = \mathrm{Ind}_{\Gamma_E}^{\Gamma_K} \psi_p$ is absolutely irreducible, and there is a regular L-algebraic cuspidal automorphic representation σ of $\mathrm{GL}_{q^2}(\mathbb{A}_K)$ such that $r_{p,\iota}(\sigma) \cong \rho_p$.*

2. The representation $\rho_p|_{\Gamma_{K_v}}$ is absolutely irreducible, and σ_v is a supercuspidal representation of $\mathrm{GL}_{q^2}(K_v)$.

Proof. The irreducibility of ρ_p is equivalent to the following statement: for all $g, h \in \mathrm{Gal}(E/K)$ such that $g \neq h$, $\psi_p^g \neq \psi_p^h$; or for all $g, h \in \mathrm{Gal}(E/K)$ such that $g \neq h$, $\psi^g \neq \psi^h$. This statement is true because it is true after restricting ψ to $(E\otimes_{\mathbb{Q}}\mathbb{R})^\times \subset \mathbb{A}_E^\times$. The existence of σ follows from the results of [AC89, Theorem 4.2], and σ is cuspidal for the same reason that ρ_p is irreducible: see [AC89, Corollary 6.5]. The second part is similar. □

Let H denote the Zariski closure of $\rho_p(\Gamma_K)$ in $\mathrm{GL}_{q^2}(\overline{\mathbb{Q}}_p)$, and let $\rho_H : \Gamma_K \to H(\overline{\mathbb{Q}}_p)$ denote the tautological representation. The group H sits in a short exact sequence

$$1 \longrightarrow \mathbb{G}_m^{\mathrm{Gal}(E/K)} \longrightarrow H \longrightarrow \mathrm{Gal}(E/K) \longrightarrow 1.$$

(To ensure that the image of ρ_p is Zariski dense in this group, we are using the condition imposed above that the matrix $(n_{\tau gh})_{g,h\in\mathrm{Gal}(E/K)}$ has non-zero determinant.) Recall that E_0 is the unique intermediate subfield of E/K, and consider the Hecke character $\chi = \psi|_{\mathbb{A}_{E_0}^\times}$. Let $\chi_p = r_{p,\iota}(\chi) : \Gamma_{E_0} \to \overline{\mathbb{Q}}_p^\times$. Let H_0 denote the pre-image in H of $\mathrm{Gal}(E/E_0)$. We can find a character $x : H_0 \to \mathbb{G}_m$ such that $x \circ \rho_H|_{\Gamma_{E_0}} = \chi_p$. We can find another character $y : H_0 \to \mathbb{G}_m$ such that $y \circ \rho_H|_{\Gamma_{E_0}} = \varphi_p$ is a non-trivial character $\mathrm{Gal}(E/E_0) \to \overline{\mathbb{Q}}_p^\times$. Let $R = \mathrm{Ind}_{H_0}^H(x \otimes y)$. Then R is a q-dimensional representation of the group H and we have $R \circ \rho_H \cong \mathrm{Ind}_{\Gamma_{E_0}}^{\Gamma_K}(\chi_p \otimes \varphi_p)$.

Proposition 3.4. *With notation as above, the representation $r_p = R \circ \rho_H$ has the following properties:*

1. *It is absolutely irreducible and Hodge–Tate regular.*
2. *There exists a cuspidal, regular L-algebraic automorphic representation π of $\mathrm{GL}_q(\mathbb{A}_K)$ such that $r_{p,\iota}(\pi) \cong r_p$.*
3. *There does not exist a character $\lambda : \Gamma_K \to \overline{\mathbb{Q}}_p^\times$ such that $r_p^c \cong r_p^\vee \otimes \lambda$.*

Proof. If $\tau_0 : E_0 \hookrightarrow \overline{\mathbb{Q}}_p$ is an embedding, let $m_{\tau_0} = \mathrm{HT}_{\tau_0}(\chi)$. Then we have

$$m_{\tau_0} = \sum_{\substack{\tau : E \hookrightarrow \overline{\mathbb{Q}}_p \\ \tau|_{E_0} = \tau_0}} n_\tau .$$

In particular, we see that the m_{τ_0}, $\tau_0 \in \Phi_{E_0}$, are pairwise distinct, and that the representation r_p is Hodge–Tate regular. This representation is irreducible because the conjugates $(\chi_p \otimes \varphi_p)^g$ are pairwise distinct as $g \in \mathrm{Gal}(E_0/K)$ varies: in fact, these characters already have distinct Hodge–Tate weights. The existence of π is again a consequence of [AC89, Theorem 4.2].

It remains to show that r_p is not conjugate self-dual up to twist. Let $\lambda : \Gamma_K \to \overline{\mathbb{Q}}_p^\times$ be a character, and suppose that $r_p^c \cong r_p^\vee \otimes \lambda$. Looking at determinants, we see that for each embedding $\tau : K \hookrightarrow \overline{\mathbb{Q}}_p$, we have $\mathrm{HT}_\tau(\lambda) = q(1-q)^2$. Restricting to Γ_{E_0}, we see that there exists $g \in \mathrm{Gal}(E_0/K)$ such that $\chi_p^c \otimes \varphi_p^c = (\chi_p^\vee \otimes \varphi_p^\vee)^g \otimes \lambda|_{\Gamma_{E_0}}$. Passing to Hodge–Tate weights, this gives for any $\tau_0 : E_0 \hookrightarrow \overline{\mathbb{Q}}_p$:

$$m_{\tau_0 c} + m_{\tau_0 g} = q(1-q^2).$$

Since we also have $m_{\tau_0} + m_{\tau_0 c} = q(1-q^2)$, we find $m_{\tau_0} = m_{\tau_0 g}$, hence $g = 1$ (using again Hodge–Tate regularity of r_p). This forces

$$\lambda|_{\Gamma_{E_0}} = \chi_p \chi_p^c \varphi_p \varphi_p^c = \epsilon^{q(q^2-1)} \varphi_p \varphi_p^c = \epsilon^{q(q^2-1)} \varphi_p^2.$$

(Note that $\varphi = \varphi^c$ because φ factors through the Galois group of a CM extension of the CM field K.) However, the character φ_p^2 does not extend to Γ_K (otherwise E/K would have Galois group $(\mathbb{Z}/q\mathbb{Z})^2$). This contradiction completes the proof. □

We now apply the following general result.

Proposition 3.5. *Let Γ be a profinite group, and let $\rho : \Gamma \to \mathrm{GL}_n(\overline{\mathbb{Q}}_p)$ be a continuous semisimple representation. Let H denote the Zariski closure of $\rho(\Gamma)$, and let $R : H \to \mathrm{GL}(V)$ be a finite-dimensional irreducible representation. Then there exist integers $a, b \geq 0$ such that $R \circ \rho$ occurs as a subquotient of $\rho^{\otimes a} \otimes (\rho^\vee)^{\otimes b}$.*

Proof. The group H is reductive as it has a faithful semisimple representation (because ρ is semisimple). Let r denote the tautological faithful representation of H on V. It then suffices to find integers $a, b \geq 0$ such that R occurs in $r^{\otimes a} \otimes (r^\vee)^{\otimes b}$. This is presumably standard, see e.g. [Del82, Proposition 3.1(a)]. □

By Proposition 3.5, we can find integers $a, b \geq 0$ such that r_p appears as a subquotient of $\rho_p^{\otimes a} \otimes (\rho_p^\vee)^{\otimes b}$. Let ℓ be the residue characteristic of the place v. We now fix a cyclic totally real extension $L_0/\mathbb{Q}$ of prime degree $d > a + b$ and in which ℓ splits, and set $L = K \cdot L_0$. We observe that this implies the following:

1. The base change π_L is cuspidal, and $r_p|_{\Gamma_L}$ is irreducible and still not conjugate self-dual up to twist. Indeed, $\mathrm{Gal}(E/K)$ is linearly disjoint from $\mathrm{Gal}(L/K)$, so we can just run the above arguments again with $\psi_p|_{\Gamma_L}$ instead of ψ_p.
2. The place v splits in L, so that if w is a place of L dividing v, then $\sigma_{L,w}$ is supercuspidal.

Let $\Sigma = \sigma_L$ and $\Pi = \pi_L$. We have now almost completed the proof of Theorem 3.1: we have constructed, from the data of the extension E/K and the character ψ, an automorphic representation Π which is regular L-algebraic and cuspidal but *not* conjugate self-dual up to twist. In order to complete the proof, we must show that there exists $\tau \in \Phi_L$ and a Shimura datum (G, X) of reflex field $\tau(L)$ such that a twist of $r_{p,\iota}(\Pi)$ by a geometric character appears as a subquotient of $H^*_{\text{ét}}(\text{Sh}_K(G, X)_{\overline{\mathbb{Q}}}, \mathcal{F}_p)$ for some choice of algebraic local system $\mathcal{F}$.

Fix an embedding $\tau_0 \in \Phi_K$ and disjoint subsets $\Sigma_0, \Sigma_1 \subset \Phi_L$ of embeddings extending τ_0, such that $|\Sigma_0| = a$ and $|\Sigma_1| = b$. Suppose given the following data:

1. A division algebra D over L of rank $n = q^2$ and centre L, together with an involution $* : D \to D$ such that $*|_L = c$.
2. A homomorphism $h_0 : \mathbb{C} \to D \otimes_{\mathbb{Q}} \mathbb{R}$ of $\mathbb{R}$-algebras such that $h_0(z)^* = h_0(\overline{z})$ for all $z \in \mathbb{C}$.

Then we can associate to $(D, *, h_0)$ a unitary group G_{00} over L^+, its restriction of scalars $G_0 = \text{Res}_{\mathbb{Q}}^{L^+} G_{00}$, a unitary similitude group G over $\mathbb{Q}$ containing G_0, and a Shimura datum (G, X) (see [Kot92, § 1] for details). We can choose this data so that the following conditions are satisfied (cf. [Clo91, § 2]):

1. At each place $w|v$ of L, D_w is a division algebra of invariant $1/n$. At each place $w \nmid vv^c$ of L, D is split and the group $G_{0,w|_{L^+}}$ is quasi-split.
2. For each $\tau \in \Sigma_0$, we have $n(\tau) = n - 1$. For each $\tau \in \Sigma_1$, we have $n(\tau) = 1$. For every other $\tau \in \Phi_L$, we have $n(\tau) = 0$.

The integers $n(\tau)$ here are as on [Kot92, p. 655]; the second condition here means that we have an isomorphism

$$G_{0,\mathbb{R}} \cong U(n-1,1)^a \times U(1,n-1)^b \times U(0,n)^{[L^+:\mathbb{Q}]-a-b}.$$

We note that the reflex field of (G, X) is equal to $\tau(L)$, for any $\tau \in \Sigma_0$.

The automorphic representation $\Sigma \otimes \| \det \|^{\frac{1-q^2}{2}}$ is RACSDC[2] and descends to an automorphic representation Σ_{G_0} of $G_0(\mathbb{A})$ with $\Sigma_{G_0,\infty}$ essentially square integrable and of strictly regular infinitesimal character. (This follows from e.g. the main theorems of [KMSW]. Since we are dealing here with a 'simple' Shimura variety, it is possible to prove the existence of this descent much more easily, along the same lines as in the proof of [Clo93, Proposition 2.3], making appropriate changes to deal with the presence of more than one non-compact factor at infinity.) Arguing as in the proofs of [HT01, Theorem VI.2.9] and [HT01, Lemma VI.2.10], we can extend Σ_{G_0} to a representation Σ_G of $G(\mathbb{A}_{\mathbb{Q}})$ such that the integer $a(\Sigma_G^\infty)$ of [Kot92] is non-zero. We can therefore apply [Kot92, Theorem 1] to conclude that there is an algebraic local system $\mathcal{F}$ such that the Σ_G^∞-part of $H^*_{\text{ét}}(\text{Sh}_K(G, X)_{\overline{\mathbb{Q}}}, \mathcal{F}_p)$ is isomorphic to a character twist of the representation

$$(\rho_p^{\otimes a} \otimes (\rho_p^\vee)^{\otimes b})^{|a(\Sigma_G^\infty)|}|_{\Gamma_L}.$$

[2] Regular algebraic, conjugate self-dual, cuspidal, cf. [CHT08]

In particular, it admits a twist of the representation $r_p|_{\Gamma_L} = r_{p,\iota}(\Pi)$ by a geometric character as a subquotient. This completes the proof of Theorem 3.1. (It is clear that we can generate infinitely many pairs (L, Π) just by varying our initial choices.)

4 Conjectures on Galois representations

Let G be a reductive group over $\mathbb{Q}$, and let ${}^LG$ be its L-group. In order to avoid a proliferation of subscripts, we will in this section fix a prime p and an isomorphism $\iota : \overline{\mathbb{Q}}_p \to \mathbb{C}$, and write ${}^LG$ also for ${}^LG(\mathbb{C})$, ${}^LG(\overline{\mathbb{Q}}_p)$, ${}^LG_{\mathbb{C}}$ and ${}^LG_{\overline{\mathbb{Q}}_p}$. We hope that in each case it will be clear from the context exactly which of these groups is intended. In order to analyse the cohomology of Shimura varieties in the next section, we introduce the formalism of the Langlands group, local and global L-parameters, and finally local and global A-parameters. We will make predictions using these ideas, and state precise conjectures which are independent of the existence of the Langlands group.

Following [Art02], the global Langlands group should be a locally compact topological group which is an extension

$$1 \longrightarrow K_{\mathbb{Q}} \longrightarrow L_{\mathbb{Q}} \longrightarrow W_{\mathbb{Q}} \longrightarrow 1,$$

where $W_{\mathbb{Q}}$ is the Weil group of $\mathbb{Q}$. For each place v of $\mathbb{Q}$, there should be a continuous embedding $L_{\mathbb{Q}_v} \to L_{\mathbb{Q}}$, defined up to conjugacy, where $L_{\mathbb{Q}_v}$ is the local Langlands group:

$$L_{\mathbb{Q}_v} = \begin{cases} W_{\mathbb{Q}_v} \times \mathrm{SU}_2(\mathbb{R}) & v \text{ non-archimedean;} \\ W_{\mathbb{Q}_v} & v \text{ archimedean.} \end{cases}$$

The irreducible n-dimensional continuous complex representations of the group $L_{\mathbb{Q}}$ should be in bijection with the cuspidal automorphic representations of $\mathrm{GL}_n(\mathbb{A}_{\mathbb{Q}})$. More generally, if π is an essentially tempered automorphic representation of $G(\mathbb{A}_{\mathbb{Q}})$, then one expects that there should be a corresponding continuous homomorphism $\phi : L_{\mathbb{Q}} \to {}^LG$ with the property that for each place v of $\mathbb{Q}$, π_v is in the L-packet corresponding to $\phi|_{L_{\mathbb{Q}_v}}$. (To formulate this statement supposes that the local Langlands correspondence for $G(\mathbb{Q}_v)$ is known. It thus has an unconditional sense at least if either v is archimedean, or v is non-archimedean and π_v is unramified.) The homomorphism ϕ should be an L-parameter, i.e. it should be semisimple, and the projection $L_{\mathbb{Q}} \to \pi_0({}^LG)$ should factor through the canonical surjection $L_{\mathbb{Q}} \to \Gamma_{\mathbb{Q}} \to \pi_0({}^LG)$. The condition that π is essentially tempered should imply that the image of ϕ is essentially bounded, i.e. bounded modulo the centre of ${}^LG$.

Let $\underline{W} = (\mathbb{G}_m \times \mathbb{G}_m) \rtimes \{1, c\}$, where c acts by swapping factors. Then there is an embedding $L_{\mathbb{R}} \to \underline{W}(\mathbb{C})$, which sends z to $(z, \overline{z}) \rtimes 1$ and j to $(-i, -i) \rtimes c$. We say that a homomorphism $\phi_\infty : L_{\mathbb{R}} \to {}^LG$ is L-algebraic if it is the restriction to $L_{\mathbb{R}}$ of a map $\underline{W}(\mathbb{C}) \to {}^LG(\mathbb{C})$ which comes from a morphism $\underline{W}_{\mathbb{C}} \to {}^LG_{\mathbb{C}}$ of algebraic groups. In this case we write $a_{\phi_\infty} : \underline{W}_{\mathbb{C}} \to {}^LG_{\mathbb{C}}$ for the corresponding morphism of algebraic groups, and call it the algebraic L-parameter corresponding to ϕ_∞. We say that an irreducible admissible representation of $G(\mathbb{R})$ is L-algebraic if its Langlands parameter is L-algebraic, and that an automorphic representation π of $G(\mathbb{A}_{\mathbb{Q}})$ is L-algebraic if π_∞ is.

We say that an L-parameter $\phi : L_{\mathbb{Q}} \to {}^L G$ is L-algebraic if $\phi|_{L_{\mathbb{R}}}$ is L-algebraic. Langlands has suggested that there should be a morphism from $L_{\mathbb{Q}}$ to the motivic Galois group of $\mathbb{Q}$ (with $\mathbb{C}$-coefficients). Based on the conjectures in [BG14] (see also [Art02, § 6]), one can guess that a morphism $\phi : L_{\mathbb{Q}} \to {}^L G$ factors through the motivic Galois group if and only if ϕ is L-algebraic. Passing to p-adic realizations, and bearing in mind the discussion at the end of § 2, this leads us to predict that for any L-algebraic morphism $\phi : L_{\mathbb{Q}} \to {}^L G$, there exists a continuous homomorphism $\rho_\phi : \Gamma_{\mathbb{Q}} \to {}^L G$ satisfying the following conditions:

1. ρ_ϕ is geometric and mixed. (We recall from Definition 2.4 that the weight cocharacter w is then defined.)
2. For each prime l, $\mathrm{WD}(\rho_\phi|_{\Gamma_{\mathbb{Q}_l}})$ is $\widehat{G}$-conjugate to $\phi|_{L_{\mathbb{Q}_l}}$.[3]
3. There exists a Hodge–Tate cocharacter μ of ρ_ϕ and a complex conjugation $c \in \Gamma_{\mathbb{Q}}$ such that $\mathrm{Ad}(\rho_\phi(c)) \circ \mu = \mu^{-1} w$ and the morphism $a_\rho : \underline{W} \to {}^L G$, $(z_1, z_2) \mapsto z_1^\mu z_2^{w-\mu}$, $c \mapsto \rho_\phi(c)$ is $\widehat{G}$-conjugate to $a_{\phi|_{L_{\mathbb{R}}}}$.

(Here, as in § 2, we write $w : \mathbb{G}_m \to Z({}^L G)$ for the weight cocharacter of the Galois representation ρ_ϕ.) This leads us to the following conjecture:

Conjecture 4.1. *Let π be an essentially tempered automorphic representation of $G(\mathbb{A}_{\mathbb{Q}})$ which is L-algebraic. Then there exists a continuous homomorphism $\rho_\pi : \Gamma_{\mathbb{Q}} \to {}^L G$ satisfying the following conditions:*

1. *ρ_π is geometric and pure.*
2. *For each prime $l \neq p$ such that π_l is unramified, $\rho_\pi|_{\Gamma_{\mathbb{Q}_l}}$ is unramified and $\rho_\pi(\mathrm{Frob}_l)$ is $\widehat{G}$-conjugate to the Satake parameter of π_l.*
3. *There exists a Hodge–Tate cocharacter μ of ρ_π and a complex conjugation $c \in \Gamma_{\mathbb{Q}}$ such that $\mathrm{Ad}\, \rho_\pi(c) \circ \mu = \mu^{-1} w$ and the morphism $a_\rho : \underline{W} \to {}^L G$, $(z_1, z_2) \mapsto z_1^\mu z_2^{w-\mu}$, $c \mapsto \rho_\pi(c)$ is $\widehat{G}$-conjugate to a_{π_∞}, the algebraic L-parameter of π_∞.*

We note that this conjecture makes no reference to the Langlands group. It is worth comparing this conjecture with those made in [BG14, § 3.2]. In *loc. cit.*, the authors do not restrict to essentially tempered automorphic representations, imposing instead only L-algebraicity. At infinity, they predict only the $\widehat{G}$-conjugacy class of $\rho_\pi(c)$ in ${}^L G$. By contrast, we are predicting both the conjugacy class of $\rho_\pi(c)$ and the existence of a Hodge–Tate cocharacter that is compatible with $\rho_\pi(c)$, in some sense. This is motivated by the discussion at the end of § 2. We note that this stronger prediction would be false without the restriction that π is essentially tempered, as one sees either by considering holomorphic Eisenstein series for GL_2 or holomorphic Saito–Kurakawa lifts on PSp_4 (cf. [Lan79, § 3]).

[3] Here we write WD for the Weil–Deligne representation associated to a p-adic representation of $\Gamma_{\mathbb{Q}_l}$, assumed to be de Rham if $l = p$. See for example [Tat79] in the case $l \neq p$, or [BM02, § 2.2] in the case $l = p$. We will soon restrict to unramified places in order to avoid any unnecessary complications.

Proposition 4.2. *Let π be an essentially tempered L-algebraic automorphic representation of $G(\mathbb{A}_\mathbb{Q})$, and suppose that π_∞ is essentially square-integrable. Suppose that Conjecture 4.1 holds for π. Then:*

1. $\operatorname{Ad}\rho_\pi(c)$ *is an odd involution of $\widehat{G}$, in the sense of Definition 2.1.*
2. *Let H_π denote the Zariski closure of the image of ρ_π. Then $\operatorname{Ad}\rho_\pi(c)$ is an odd involution of H_π°.*

Proof. We note that our definition of Hodge–Tate cocharacter implies that μ in fact factors through H_π. Therefore the second part of the proposition will follow from the first part and from Lemma 2.2 if we can establish the first part and at the same time show that μ is a regular cocharacter of $\widehat{G}$.

To prove the whole proposition, it therefore suffices to show that a_{π_∞} has the property that $a_{\pi_\infty}|_{\mathbb{G}_m\times 1}$ is a regular cocharacter and $a_{\pi_\infty}(c)$ acts as -1 on $\operatorname{Cent}(\widehat{G}, a_{\pi_\infty}(\mathbb{G}_m\times 1))/Z(\widehat{G})$. To see this, we just describe the L-parameters of the L-algebraic discrete series representations of $G(\mathbb{R})$. Fix a choice of maximal torus $T \subset G_\mathbb{R}$ which is compact mod centre. Let $B \subset G_\mathbb{C}$ be a Borel subgroup containing $T_\mathbb{C}$. Let $\widehat{T} \subset \widehat{B} \subset \widehat{G}$ be the corresponding maximal torus and Borel subgroups of the dual group. Let σ_T denote the L-action of complex conjugation on $\widehat{T}$ corresponding to the given real structure on T. Then the restriction of the L-action of $\widehat{G}$ to $\widehat{T}$ is given by $\operatorname{Ad} n_G \circ \sigma_T$, where $n_G \in N(\widehat{G}, \widehat{T})$ represents the longest element of $W(\widehat{G}, \widehat{T})$ with respect to the system of positive roots given by $\widehat{B}$.

If π_∞ is an L-algebraic essentially discrete series representation of $G(\mathbb{R})$, then, after possibly replacing a_{π_∞} by a $\widehat{G}$-conjugate, a_{π_∞} is given by the formula

$$(z_1, z_2) \mapsto z_1^\mu z_2^{w-\mu},$$
$$c \mapsto n_G z \rtimes c$$

for some element z of the centre $Z(\widehat{G})$ of $\widehat{G}$, and some cocharacters $\mu \in X_*(\widehat{T})$, $w \in X_*(Z(\widehat{G}))$, such that μ is regular and dominant with respect to $\widehat{B}$. In particular, $\operatorname{Ad} a_{\pi_\infty}(c)$ acts as σ_T on $\widehat{T}$. If $\widehat{T}^{\mathrm{ad}}$ denotes the image of $\widehat{T}$ in $\widehat{G}^{\mathrm{ad}}$, then σ_T acts as -1 on $\widehat{T}^{\mathrm{ad}}$. It follows that $\operatorname{Ad} a_{\pi_\infty}(c)$ is an odd involution of $\widehat{G}$, as desired. □

We note that a calculation of the type appearing in Proposition 4.2 has appeared already in the note of Gross [Gro]. We now turn to the question of generalizing this proposition to non-tempered automorphic representations. We will do this just for discrete automorphic representations, using Arthur's formalism of A-parameters. By definition, a global A-parameter is a continuous semisimple homomorphism

$$\psi : L_\mathbb{Q} \times \mathrm{SL}_2 \to {}^L G$$

such that the induced map $L_\mathbb{Q} \to \pi_0({}^L G)$ factors through the canonical one $\Gamma_\mathbb{Q} \to \pi_0({}^L G)$, and with the property that $\psi|_{L_\mathbb{Q}}$ is essentially bounded. To any A-parameter ψ we can associate an L-parameter ϕ_ψ, given by the formula

$$\phi_\psi(w) = \psi(w, \operatorname{diag}(\|w\|^{1/2}, \|w\|^{-1/2})),$$

where $\|\cdot\| : W_{\mathbb{Q}} \to \mathbb{R}_{>0}$ is the norm pulled back from the idele class group. We define a local A-parameter similarly to be a continuous semisimple homomorphism

$$\psi_v : L_{\mathbb{Q}_v} \times \mathrm{SL}_2 \to {}^L G$$

such that the induced map $L_{\mathbb{Q}_v} \to \pi_0({}^L G)$ factors through the canonical one $\Gamma_{\mathbb{Q}_v} \to \pi_0({}^L G)$, and with the property that $\psi_v|_{L_{\mathbb{Q}_v}}$ is essentially bounded. Arthur's conjectures predict that for any local A-parameter ψ_v, one should be able to define a set (called an A-packet) Π_{ψ_v} of representations of $G(\mathbb{Q}_v)$, containing the L-packet of ϕ_{ψ_v}. To any discrete automorphic representation π of $G(\mathbb{A}_{\mathbb{Q}})$, one should be able to associate a global A-parameter $\psi : L_{\mathbb{Q}} \times \mathrm{SL}_2 \to {}^L G$ with the property that for each place v of $\mathbb{Q}$, $\pi_v \in \Pi_{\psi_v}$.

We now discuss what this has to do with Galois representations. Let ψ be a global A-parameter such that ϕ_ψ is L-algebraic. Let (x, t, y) be the $\mathfrak{sl}_2$-triple in $\widehat{\mathfrak{g}}$ determined by $\psi|_{\mathrm{SL}_2}$, and let M_1 denote the centralizer in $\widehat{G}$ of this $\mathfrak{sl}_2$-triple. Let M_1' denote the centralizer in ${}^L G$ of this $\mathfrak{sl}_2$-triple. Let $M = M_1 \cdot \lambda_t(\mathbb{G}_m)$, where $\lambda_t : \mathbb{G}_m \to \mathrm{SL}_2 \to \widehat{G}$ is the cocharacter with derivative t, and let $M' = M_1 \cdot \lambda_t(\mathbb{G}_m)$. Then there are exact sequences

$$1 \longrightarrow M_1 \longrightarrow M_1' \longrightarrow \pi_0({}^L G) \longrightarrow 1$$

and

$$1 \longrightarrow M \longrightarrow M' \longrightarrow \pi_0({}^L G) \longrightarrow 1,$$

and ϕ_ψ factors through a homomorphism $\phi' : L_{\mathbb{Q}} \to M'$. This leads us to expect the existence of a representation $\rho_\psi : \Gamma_{\mathbb{Q}} \to {}^L G$ and an $\mathfrak{sl}_2$-triple (x, t, y) in $\widehat{\mathfrak{g}}$ satisfying the following conditions:

1. ρ_ψ is geometric and mixed. Moreover, $dw - t \in \mathfrak{z}(\widehat{\mathfrak{g}})$ and the following formulae hold: for each $\gamma \in \Gamma_{\mathbb{Q}}$,

$$\mathrm{Ad}\,\rho_\psi(\gamma)(x) = \epsilon^{-1}(\gamma)x, \mathrm{Ad}\,\rho_\psi(\gamma)(t) = t, \mathrm{Ad}\,\rho_\psi(\gamma)(y) = \epsilon(\gamma)y.$$

2. For almost all primes l, $\mathrm{WD}(\rho_\psi|_{\Gamma_{\mathbb{Q}_l}})$ is $\widehat{G}$-conjugate to $\phi_\psi|_{L_{\mathbb{Q}_l}}$.
3. There exists a Hodge–Tate cocharacter μ of ρ_ψ and a complex conjugation $c \in \Gamma_{\mathbb{Q}}$ with the following properties: we have $\mathrm{Ad}\,\rho_\psi(c) \circ \mu = \mu^{-1} w$. Define $a_\rho : \underline{W} \to {}^L G$, $(z_1, z_2) \mapsto z_1^\mu z_2^{w-\mu}$, $c \mapsto \rho_\psi(c)$. Then the pair $(a_\rho, (x, t, y))$ is $\widehat{G}$-conjugate to $(a_{\phi_\psi|_{L_{\mathbb{R}}}}, (d\psi(x_0), d\psi(t_0), d\psi(y_0)))$.

In contrast to the case of tempered representations, we do not see a way to phrase this prediction solely in terms of automorphic representations, without making reference to A-parameters. However, one can make a conjecture supposing only that the local A-packets at infinity have been defined. A definition has been given in [ABV92].

In order to avoid unnecessary complications here, we will now assume for the rest of §4 that $G_{\mathbb{R}}$ contains a maximal torus T which is compact modulo centre. Let $t_0 \in T(\mathbb{R})$ be such that $\mathrm{Ad}(t_0)$ induces a Cartan involution of $G_{\mathbb{R}}^{\mathrm{ad}}$. Let $K_\infty = G(\mathbb{R})^{t_0}$. We will state a conjecture only for automorphic representations which are (up to twist) cohomological, in the sense that there exists an irreducible algebraic representation τ of $G_{\mathbb{C}}$ such that

$H^*(\mathfrak{g}_{\mathbb{C}}, K_\infty; \pi_\infty \otimes \tau) \neq 0$. In this case we can use the A-parameters and packets described by Adams–Johnson (see [AJ87] and also [Art89, § 5]). The representations in these packets were later shown by Vogan and Zuckerman to be the unitary cohomological representations of $G(\mathbb{R})$ [VZ84, Vog84].

Conjecture 4.3. *Let π be a discrete L-algebraic automorphic representation of $G(\mathbb{A}_{\mathbb{Q}})$ such that a twist of π_∞ is cohomological. Then there exists an $\mathfrak{sl}_2$-triple (x, t, y) in $\widehat{\mathfrak{g}}$ and a continuous representation $\rho_\pi : \Gamma_{\mathbb{Q}} \to {}^L G$ satisfying the following conditions:*

1. *ρ_π is geometric and mixed. Moreover, $dw - t \in \mathfrak{z}(\widehat{\mathfrak{g}})$ and the following formulae hold: for each $\gamma \in \Gamma_{\mathbb{Q}}$,*

$$\mathrm{Ad}\, \rho_\pi(\gamma)(x) = \epsilon^{-1}(\gamma)x, \mathrm{Ad}\, \rho_\pi(\gamma)(t) = t, \mathrm{Ad}\, \rho_\pi(\gamma)(y) = \epsilon(\gamma)y.$$

2. *For almost all primes $l \neq p$ such that π_l is unramified, $\rho_\pi|_{\Gamma_{\mathbb{Q}_l}}$ is unramified and $\rho_\pi(\mathrm{Frob}_l)$ is $\widehat{G}$-conjugate to the Satake parameter of π_l.*

3. *There exists a Hodge–Tate cocharacter μ of ρ_π, a complex conjugation $c \in \Gamma_{\mathbb{Q}}$, and an A-parameter $\psi : L_{\mathbb{R}} \times \mathrm{SL}_2 \to {}^L G$ which is up to twist of the type described in [Art89, §5], with the following property: we have $\mathrm{Ad}\, \rho_\pi(c) \circ \mu = \mu^{-1} w$, and ϕ_ψ is L-algebraic. Define $a_\rho : \underline{W} \to {}^L G$, $(z_1, z_2) \mapsto z_1^\mu z_2^{w-\mu}$, $c \mapsto \rho_\pi(c)$. Then the pair $(a_\rho, (x, t, y))$ is $\widehat{G}$-conjugate to $(a_{\phi_\psi}, d\psi(x_0), d\psi(t_0), d\psi(y_0))$.*

This leads us to the following generalization of Proposition 4.2:

Proposition 4.4. *Let π be a discrete L-algebraic automorphic representation of $G(\mathbb{A}_{\mathbb{Q}})$ such that some twist of π_∞ is cohomological. Suppose that π satisfies Conjecture 4.3. Let $H_\pi \subset {}^L G$ denote the Zariski closure of the image of ρ_π. Then $\mathrm{Ad}\, \rho_\pi(c)$ is an odd involution of H_π°.*

This generalizes the second part of Proposition 4.2 because if σ is an essentially square-integrable representation of $G(\mathbb{R})$, then $G_{\mathbb{R}}$ contains a maximal torus which is compact mod centre and some twist of σ is cohomological. One can also generalize the first part of Proposition 4.2, although we don't prove this here as we don't need it.

Proof. By Lemma 2.2 and the final requirement of Conjecture 4.3, it will again suffice to show that if $\psi : L_{\mathbb{R}} \times \mathrm{SL}_2 \to {}^L G$ is a twist of one of the A-parameters considered in [Art89, §5] such that ϕ_ψ is L-algebraic, then the cocharacter $a_{\phi_\psi}|_{\mathbb{G}_m \times 1}$ is regular in the group $M_1^\circ \cdot \lambda_t(\mathbb{G}_m)$, where $M_1 = \mathrm{Cent}(\widehat{G}, \mathrm{SL}_2)$, and $\mathrm{Ad}\, a_{\phi_\psi}(c)$ induces an odd involution of M_1°.

Using Arthur's explicit description in [Art89, §5], we see that the pair $(a_{\phi_\psi}, (d\psi(x_0), d\psi(t_0), d\psi(y_0)))$ has the following form. First let $T \subset G_{\mathbb{R}}$ be a maximal torus which is compact mod centre, and let $B \subset G_{\mathbb{C}}$ be a Borel subgroup containing T. Let $\widehat{T} \subset \widehat{B} \subset \widehat{G}$ be the corresponding maximal torus and Borel subgroup of the dual group. Then there is a standard Levi subgroup $\widehat{L}$ such that (x, t, y) is a principal $\mathfrak{sl}_2$-triple in $\widehat{\mathfrak{l}}$, and a_{ϕ_ψ} is given by the formula

$$(z_1, z_2) \mapsto z_1^{\mu} z_2^{w-\mu},$$
$$c \mapsto w(-i)n_L^{-1}n_G z \rtimes c,$$

for some $z \in Z(\widehat{G})$ and some regular dominant cocharacter $\mu \in X_*(\widehat{T})$. (We note that this differs from [Art89, p. 30], where $n_L^{-1}n_G$ is replaced by $n_G n_L^{-1}$, but our choice appears to give a correct formula. We recall that n_G, n_L are elements of the derived groups of $\widehat{G}$ and $\widehat{L}$, respectively, which represent the longest elements of the respective Weyl groups with respect to the sets of positive roots determined by $\widehat{B}$.) There is a decomposition $\widehat{\mathfrak{g}} = \widehat{\mathfrak{l}} \oplus \widehat{\mathfrak{n}}^+ \oplus \widehat{\mathfrak{n}}^-$, where $\widehat{\mathfrak{n}}^+$ and $\widehat{\mathfrak{n}}^-$ are the Lie algebras of the unipotent radicals of, respectively, the standard parabolic subgroup containing $\widehat{L}$, and its opposite. This decomposition is $\widehat{\mathfrak{l}}$-invariant, hence $\mathfrak{sl}_2$-invarant. It follows that the Lie algebra of $\mathfrak{m}_1$ admits a similar decomposition

$$\mathfrak{m}_1 = \mathfrak{a} \oplus \mathfrak{m}_1^+ \oplus \mathfrak{m}_1^-,$$

where $\mathfrak{a}$ is the Lie algebra of the connected centre A of $\widehat{L}$. In fact A is a maximal torus of M_1: indeed, we have $\widehat{L} = \mathrm{Cent}(\widehat{G}, A)$, so any torus of M_1 containing A is necessarily contained in $\mathrm{Cent}(\widehat{L}, \psi(\mathrm{SL}_2))$, hence in A. The proof will be complete if we can show that $\mathrm{Ad}\, a_{\phi_\psi(c)}$ acts as -1 on the image of A in the adjoint group of M_1°. However, since $\mathrm{Ad}\, a_{\phi_\psi(c)} = \mathrm{Ad}(w(-i)n_L^{-1}n_G) \circ \sigma_G$, we see that $\mathrm{Ad}\, a_{\phi_\psi(c)}$ acts on $\widehat{T}$ as $\mathrm{Ad}(n_L^{-1}) \circ \sigma_T$. The action on A is therefore equal to $\sigma_T|_A$, which has the desired property (recall that σ_T acts as -1 on $\widehat{T}/Z(\widehat{G})$). □

We end this section by stating a variant of Conjecture 4.3 for discrete automorphic representations π such that π_∞ is cohomological (not just up to twist). This variant is based on the observation that π is then necessarily C-algebraic, in the sense of [BG14].

In [BG14, Proposition 4.3.1], Buzzard and Gee define a canonical extension of G by $\mathbb{G}_m$:

$$1 \longrightarrow \mathbb{G}_m \longrightarrow \widetilde{G} \longrightarrow G \longrightarrow 1.$$

They define the C-group ${}^C G$ to be the L-group ${}^C G = {}^L \widetilde{G}$. The virtue of the C-group is that if σ is an irreducible admissible representation of $G(\mathbb{R})$ which is cohomological, then its pullback σ' to $\widetilde{G}(\mathbb{R})$ has a canonical twist which is L-algebraic (see the proof of [BG14, Proposition 5.3.6]). Moreover, the group ${}^C G$ admits an explicit description: its dual group can be identified as the quotient

$$\widehat{\widetilde{G}} = (\mathbb{G}_m \times \widehat{G})/(-1, \chi(-1)), \tag{4.1}$$

where $\chi \in X_*(\widehat{T})$ is the sum of the positive roots with respect to some choice of Borel containing $\widehat{T}$ (the element $\chi(-1)$ is central and does not depend on the choice of Borel). This leads us to the following conjecture for automorphic forms on G, which follows from applying Conjecture 4.3 to suitable L-algebraic discrete automorphic representations of $\widetilde{G}$:

Conjecture 4.5. *Let π be a discrete automorphic representation of $G(\mathbb{A}_\mathbb{Q})$ such that π_∞ is cohomological. Then there exists an $\mathfrak{sl}_2$-triple (x, t, y) in $\widehat{\widetilde{\mathfrak{g}}}$ and a continuous representation $\rho'_\pi : \Gamma_\mathbb{Q} \to {}^C G$ satisfying the following conditions:*

1. *ρ'_π is geometric and mixed. Moreover, $dw - t \in \mathfrak{z}(\widehat{\mathfrak{g}})$ and the following formulae hold: for each $\gamma \in \Gamma_{\mathbb{Q}}$,*

$$\operatorname{Ad} \rho'_\pi(\gamma)(x) = \epsilon^{-1}(\gamma)x, \operatorname{Ad} \rho'_\pi(\gamma)(t) = t, \operatorname{Ad} \rho'_\pi(\gamma)(y) = \epsilon(\gamma)y.$$

2. *For almost all primes $l \neq p$ such that π_l is unramified, $\rho'_\pi|_{\Gamma_{\mathbb{Q}_l}}$ is unramified and $\rho'_\pi(\mathrm{Frob}_l)$ is $\widehat{\widehat{G}}$-conjugate to the image of $(l^{-1/2}, t(\pi_l))$, where $t(\pi_l)$ is the Satake parameter of π_l.*
3. *There exists a Hodge–Tate cocharacter μ of ρ'_π, a complex conjugation $c \in \Gamma_{\mathbb{Q}}$, and an A-parameter $\psi : L_{\mathbb{R}} \times \mathrm{SL}_2 \to {}^C G$ which is up to twist of the type described in [Art89, §5], with the following property: we have $\operatorname{Ad} \rho'_\psi(c) \circ \mu = \mu^{-1}w$. Define $a_\rho : \underline{W} \to {}^C G$, $(z_1, z_2) \mapsto z_1^\mu z_2^{w-\mu}$, $c \mapsto \rho'_\pi(c)$. Then the pair $(a_\rho, (x, t, y))$ is $\widehat{\widehat{G}}$-conjugate to $(a_{\phi_\psi}, (d\psi(x_0), d\psi(t_0), d\psi(y_0)))$.*

5 Intersection cohomology of Shimura varieties

In this section we will use conjectures of Arthur and Kottwitz to give a conjectural description of the Galois representations appearing in the intersection cohomology of Shimura varieties. Let (G, X) be a Shimura datum. We write E for the reflex field and $\mathrm{Sh}_K(G, X)$ for the associated Shimura variety over E with respect to some neat open compact subgroup $K \subset G(\mathbb{A}^\infty_{\mathbb{Q}})$. Let $\mathcal{F}_\tau$ be an algebraic local system. We fix a prime p and an isomorphism $\iota : \overline{\mathbb{Q}}_p \to \mathbb{C}$.

Let $j : \mathrm{Sh}_K(G, X) \to \mathrm{Sh}^{\min}_K(G, X)$ denote the open embedding of $\mathrm{Sh}_K(G, X)$ into its minimal compactification. The intersection cohomology groups

$$IH_{\tau,K} = H^*_{\text{ét}}(\mathrm{Sh}^{\min}_K(G, X)_{\overline{\mathbb{Q}}}, j_{!*}\mathcal{F}_{\tau,p})$$

are $\mathcal{H}(G(\mathbb{A}^\infty_{\mathbb{Q}}), K) \otimes \overline{\mathbb{Q}}_p[\Gamma_E]$-modules, where $\mathcal{H}$ denotes the usual convolution Hecke algebra, and are finite-dimensional as $\overline{\mathbb{Q}}_p$-vector spaces. Our aim is to understand the irreducible subquotient $\overline{\mathbb{Q}}_p[\Gamma_E]$-modules of $H^*_{\text{ét}}(\mathrm{Sh}^{\min}_K(G, X)_{\overline{\mathbb{Q}}}, j_{!*}\mathcal{F}_{\tau,p})$. If π^∞ is an irreducible admissible $\mathbb{C}[G(\mathbb{A}^\infty)]$-module, then we define

$$W_p(\pi^\infty) = \mathrm{Hom}_{G(\mathbb{A}^\infty_{\mathbb{Q}})}(\iota^{-1}\pi_\infty, \lim_{\substack{\to \\ K}} IH_{\tau,K}).$$

Then $W_p(\pi^\infty)$ is a finite-dimensional $\overline{\mathbb{Q}}_p[\Gamma_E]$-module and each irreducible subquotient of $H^*_{\text{ét}}(\mathrm{Sh}^{\min}_K(G, X)_{\overline{\mathbb{Q}}}, j_{!*}\mathcal{F}_{\tau,p})$ is isomorphic to an irreducible subquotient of some $W_p(\pi^\infty)$.

The action of $G(\mathbb{A}^\infty_{\mathbb{Q}})$ on intersection cohomology can be understood in terms of discrete automorphic representations. Indeed, the Zucker conjecture (proved independently by Looijenga and Saper–Stern) states that $IH_{\tau,K}$ can be identified with the L^2-cohomology $H^*_{(2)}(\mathrm{Sh}_K(G, X)(\mathbb{C}), \mathcal{F}_\tau \otimes_{\overline{\mathbb{Q}}} \mathbb{C})$, which admits a decomposition

$$H^*_{(2)}(\mathrm{Sh}_K(G, X)(\mathbb{C}), \mathcal{F}_\tau \otimes_{\overline{\mathbb{Q}}} \mathbb{C}) = \bigoplus_\pi m(\pi)(\pi^\infty)^K \otimes H^*(\mathfrak{g}_{\mathbb{C}}, K_1; \pi_\infty \otimes \tau),$$

where $K_1 = \mathrm{Cent}(G(\mathbb{R}), h)$ for some choice of $h \in X$. This led Kottwitz to give a conjectural description of the space $W_p(\pi^\infty)$ in terms of the A-parameters giving rise to π^∞ [Kot90]. The paper [Joh13] gave a reformulation of Kottwitz's conjecture in terms of the C-group. We now discuss some consequences of the reformulated conjectures in [Joh13] for the spaces $W_p(\pi^\infty)$.

In order to describe the contribution of π to cohomology, we recall ([Joh13, §1]) that the Shimura datum (G, X) determines an irreducible representation $r_C : {}^C(G_E) \to \mathrm{GL}(V)$, where V is a finite-dimensional complex vector space and G_E denotes base extension to the reflex field E. Then (see [Joh13, Conjecture 8]) Kottwitz[4] predicts the existence of an isomorphism

$$W_p(\pi^\infty) = \bigoplus_{\psi:\pi^\infty \in \Pi_{\psi^\infty}} (U_\pi \otimes (r_C \circ \rho'_\pi|_{\Gamma_E}))_{\epsilon_\psi},$$

where the direct sum runs over the set of A-parameters giving rise to π^∞. The precise meaning of the symbols U_π and ϵ_ψ is not important for us here; rather, we need only the following weaker consequence of this prediction, which we can again phrase independently of the existence of the Langlands group. (We note that, according to [Art89, p. 59], if $W_p(\pi^\infty) \neq 0$ then there should exist π_∞ such that $\pi = \pi^\infty \otimes \pi_\infty$ is a discrete automorphic representation of $G(\mathbb{A}_\mathbb{Q})$ and a twist of π_∞ is cohomological, in the sense of § 4.)

Conjecture 5.1. *Let π be a discrete automorphic representation of $G(\mathbb{A}_\mathbb{Q})$ such that $W_p(\pi^\infty) \neq 0$. Then there exists an $\mathfrak{sl}_2$-triple (x, t, y) in $\widehat{\widehat{\mathfrak{g}}}$ and a continuous representation $\rho'_\pi : \Gamma_\mathbb{Q} \to {}^C G(\overline{\mathbb{Q}}_p)$ as in Conjecture 4.5, and each irreducible subquotient $\overline{\mathbb{Q}}_p[\Gamma_E]$-module of $W_p(\pi^\infty)$ is isomorphic to an irreducible subquotient of the representation $r_C \circ \rho'_\pi|_{\Gamma_E}$.*

Proposition 5.2. *Let π be a discrete automorphic representation of $G(\mathbb{A}_\mathbb{Q})$ such that $W_p(\pi^\infty) \neq 0$. Suppose that Conjecture 5.1 holds for π, and let V be an irreducible subquotient $\overline{\mathbb{Q}}_p[\Gamma_E]$-module of $W_p(\pi^\infty)$. Let $H_V \subset \mathrm{GL}(V)$ denote the Zariski closure of the image of Γ_E and $\rho_V : \Gamma_E \to H_V(\overline{\mathbb{Q}}_p)$ the tautological representation. Let θ be an odd involution of H_V°. Let $c \in \Gamma_\mathbb{Q}$ be a choice of complex conjugation. Then there exists a finite Galois extension M/E such that $\rho_V(\Gamma_M) \subset H_V^\circ(\overline{\mathbb{Q}}_p)$ and the representations $\rho_V^c|_{\Gamma_M}$, $\theta \circ \rho_V|_{\Gamma_M}$ are $H_V^\circ(\overline{\mathbb{Q}}_p)$-conjugate modulo $Z_{H_V^\circ}$.*

Proof. Let $f : H_V^\circ \to H_V^{\circ,\mathrm{ad}}$ denote the projection to the adjoint group. We must show that there exists a finite Galois extension M/E such that $\rho_V(\Gamma_M) \subset H_V^\circ(\overline{\mathbb{Q}}_p)$ and the representations $f \circ \rho_V^c|_{\Gamma_M}$, $f \circ \theta \circ \rho_V|_{\Gamma_M}$ are $H_V^{\circ,\mathrm{ad}}$-conjugate.

Let $H_\mathbb{Q}$ denote the Zariski closure of $\rho'_\pi(\Gamma_\mathbb{Q})$ in ${}^C G$, and let H_E denote the Zariski closure of $\rho'_\pi(\Gamma_E)$. Then $H_E \subset H_\mathbb{Q}$ and $H_E^\circ = H_\mathbb{Q}^\circ$. There is a map $H_E \to \mathrm{GL}(V)$, and H_V is equal to the image of this map. Let $r_E : \Gamma_E \to H_E(\overline{\mathbb{Q}}_p)$ be the tautological representation. According

[4] Kottwitz makes this prediction under additional assumptions on G, namely that the derived group G^{der} is simply connected and the maximal $\mathbb{R}$-split subtorus of $Z(G)$ is $\mathbb{Q}$-split. However, it seems natural to predict this in general, and since our discussion is conjectural we will do this in order not to impose further conditions on G.

to Proposition 4.4, $\mathrm{Ad}(\rho'_\pi(c))$ induces an odd involution of H_E°, which we denote by θ'. Since θ' is odd, it must leave invariant the simple factors of $H_E^{\circ,\mathrm{ad}}$, and therefore the kernel of the map $H_E^{\circ,\mathrm{ad}} \to H_V^{\circ,\mathrm{ad}}$. Replacing θ with a conjugate, we can therefore assume without loss generality that the map $H_E^{\circ,\mathrm{ad}} \to H_V^{\circ,\mathrm{ad}}$ intertwines $(\theta')^{\mathrm{ad}}$ and θ^{ad}.

We have $r_E^c = \mathrm{Ad}(\rho'_\pi(c)) \circ r_E$. Let M/E be any Galois extension such that $r_E(\Gamma_M) \subset H_E^\circ$. Then $r_E^c|_{\Gamma_M} = \theta' \circ r_E|_{\Gamma_M}$. Pushing this identity down to $H_V^{\circ,\mathrm{ad}}$ gives the identity $f \circ \rho_V^c|_{\Gamma_M} = \theta^{\mathrm{ad}} \circ f \circ \rho_V|_{\Gamma_M} = f \circ \theta \circ \rho_V|_{\Gamma_M}$, which is what we needed to prove. □

This leads to the following result.

Theorem 5.3. *Let $\rho : \Gamma_E \to \mathrm{GL}_n(\overline{\mathbb{Q}}_p)$ be a continuous representation which is strongly irreducible, in the sense that for any finite extension M/E, $\rho|_{\Gamma_M}$ is irreducible. Let π be a discrete automorphic representation of $G(\mathbb{A}_\mathbb{Q})$, and suppose that Conjecture 5.1 holds for π. Then if ρ is isomorphic to a subquotient of $W_p(\pi^\infty)$, then ρ is conjugate self-dual up to twist.*

Proof. Let H_V denote the Zariski closure of the image of ρ in $\mathrm{GL}_n = \mathrm{GL}(V)$, and let θ be a Chevalley involution of H_V°. Let $\rho_V : \Gamma_E \to H_V(\overline{\mathbb{Q}}_p)$ denote the tautological representation. Our hypotheses imply that V is an irreducible representation of H_V°, and $V \circ \theta \cong V^\vee$. Proposition 5.2 implies that there is a finite Galois extension M/E such that $\rho_V(\Gamma_M) \subset H_V^\circ(\overline{\mathbb{Q}}_p)$ and $\rho_V^c|_{\Gamma_M}$ and $\theta \circ \rho_V|_{\Gamma_M}$ are $H_V^\circ(\overline{\mathbb{Q}}_p)$-conjugate modulo the centre of H_V°.

It follows that there is a continuous character $\chi_M : \Gamma_M \to \overline{\mathbb{Q}}_p^\times$ such that $\rho^c|_{\Gamma_M}$ and $\rho^\vee|_{\Gamma_M} \otimes \chi_M$ are $\mathrm{GL}_n(\overline{\mathbb{Q}}_p)$-conjugate. Lemma 5.4 implies that χ_M extends to a character $\chi : \Gamma_E \to \overline{\mathbb{Q}}_p^\times$ such that $\rho^c \cong \rho^\vee \otimes \chi$. This completes the proof. □

Lemma 5.4. *Let Γ be a group, let $\Delta \subset \Gamma$ be a finite index normal subgroup, and let $r_1, r_2 : \Gamma \to \mathrm{GL}_n(\overline{\mathbb{Q}}_p)$ be representations such that $r_1|_\Delta = r_2|_\Delta \otimes \chi$ for some character $\chi : \Delta \to \overline{\mathbb{Q}}_p^\times$. Suppose that for any finite index subgroup $\Delta' \subset \Gamma$, $r_1|_{\Delta'}$ is irreducible. Then χ extends to a character $\chi' : \Gamma \to \overline{\mathbb{Q}}_p$ such that $r_1 = r_2 \otimes \chi'$.*

Proof. If $\sigma \in \Gamma, \tau \in \Delta$, we have

$$r_1(\sigma\tau\sigma^{-1}) = r_1(\sigma)r_1(\tau)r_1(\sigma^{-1}) = r_1(\sigma)r_2(\tau)\chi(\tau)r_1(\sigma^{-1})$$

and

$$r_1(\sigma\tau\sigma^{-1}) = r_2(\sigma)r_2(\tau)r_2(\sigma^{-1})\chi(\sigma\tau\sigma^{-1}).$$

It follows that if $g_\sigma = r_2(\sigma)^{-1}r_1(\sigma)$, then for all $\tau \in \Delta$ we have

$$g_\sigma r_2(\tau) g_\sigma^{-1} = r_2(\tau)\chi^\sigma(\tau)\chi(\tau)^{-1}.$$

There are two cases. If $\chi^\sigma = \chi$ for all $\sigma \in \Gamma$, then the irreducibility of $r_2|_\Delta$ implies that $g_\sigma = \lambda_\sigma \in \overline{\mathbb{Q}}_p^\times$, and we are done on defining $\chi'(\sigma) = \lambda_\sigma$.

Otherwise, there exists $\sigma \in \Gamma$ such that $\chi^\sigma \neq \chi$. Then $r_2 \cong r_2 \otimes \chi^\sigma \otimes \chi^{-1}$. The character χ^σ/χ has order dividing n, and if $\Delta' = \ker \chi^\sigma/\chi$ then $r_2|_{\Delta'}$ must be reducible, contradicting our assumptions. This completes the proof. □

6 Cohomology of open Shimura varieties

The previous section deals with the question of Galois representations appearing as subquotients in the intersection cohomology of minimal compactifications of Shimura varieties. It is of course natural to ask what, if anything, changes if one considers other types of cohomology groups. The most natural choices here are cohomology and compactly supported cohomology (with values in an automorphic $\overline{\mathbb{Q}}_p$-local system) of the open Shimura varieties. These two are related by Poincaré duality, so it suffices to consider ordinary cohomology. The expectation is then that the Galois representations occurring as subquotients of the cohomology of open Shimura varieties are exactly the same as those occurring in the intersection cohomology of minimal compactifications.

At least under some assumptions on the Shimura datum, this is known and due to Morel. Since the precise statement in this form is not (as far as we know) in the literature, we sketch one way of deducing it from work of Morel [Mor08, Mor10, Mor], which we learnt from a talk of Morel at the Institute for Advanced Study [Mor11]. Any mistakes below are entirely due to the authors. See also work of Nair for another approach [Nai].

The proof relies of Morel's theory of weight truncations and Pink's formula. In [Hub97], Huber constructs a triangulated category, which we will denote by $D_m(Y, \overline{\mathbb{Q}}_p)$, of so-called horizontal mixed complexes of constructible $\overline{\mathbb{Q}}_p$-sheaves on any separated scheme Y of finite type over any number field F. Informally speaking, 'horizontal' means that the complexes extend to complexes of constructible $\overline{\mathbb{Q}}_p$-sheaves on some flat model of Y over some open subset U of $\operatorname{Spec} \mathcal{O}_F$. A horizontal complex is pure if its specializations to all but finitely many closed points of U are pure (with constant weight), and mixed complexes are those arising as extensions of pure complexes. $D_m(Y, \overline{\mathbb{Q}}_p)$ admits a perverse t-structure, whose core $Perv_m(Y)$ is the so-called category of horizontal mixed perverse sheaves on Y. We refer to [Hub97, §1-3] for precise definitions. In [Mor], Morel considers the full subcategory $\mathcal{M}(Y)$ of $Perv_m(Y)$ consisting of objects admitting a weight filtration; see [Hub97, Definition 3.7] for the definition of a weight filtration on objects of $Perv_m(Y)$. A weight filtration is unique if it exists and morphisms between two objects admitting weight filtrations are strict [Hub97, Lemma 3.8]. Morel proves that $\mathcal{M}(Y)$ and its bounded derived category $D^b(\mathcal{M}(Y))$ satisfies a long list of properties (cf. [Mor] Théorème 3.2 and Proposition 6.1), including stability under six operations and Tate twists and that it contains $\overline{\mathbb{Q}}_p[d]$ if Y is smooth and pure of dimension d. Moreover, all objects in $D^b(\mathcal{M}(Y))$ admit weight filtrations and one may define weight truncations $w_{\leq a}$, $w_{>a}$ ($a \in \mathbb{Z} \cup \{\pm\infty\}$) as in [Mor08]; see [Mor, §8].

We recall some notation for Shimura varieties and their minimal compactifications used in the previous section, and then set up some more notation. We let G be a connected reductive group over $\mathbb{Q}$ and (G, X) a Shimura datum where we allow G^{ad} to have simple factors of compact type over $\mathbb{Q}$ (to allow zero-dimensional Shimura varieties), and we write E for the reflex field of (G, X). If $K \subseteq G(\mathbb{A}_f)$ is a neat compact open subgroup, we let $\mathrm{Sh}_K = \mathrm{Sh}_K(G,X)$ denote the canonical model (over E) of the complex Shimura variety of level K. We let $\mathrm{Sh}_K^{\min} = \mathrm{Sh}_K^{\min}(G,X)$ denote its minimal compactification and we write $j : \mathrm{Sh}_K \to \mathrm{Sh}_K^{\min}$ for the open embedding. Recall that a parabolic subgroup $P \subseteq G$ defined over $\mathbb{Q}$ is called admissible if its image in every simple fact G' of G^{ad} is either G' or a maximal proper parabolic of G'. We write N_P for the unipotent radical of P, U_P for the center of N_P, and $M_P = P/N_P$ for the Levi quotient. We let X_P be the boundary component corresponding to P. Following [Mor10, p. 2–3], we make the following additional assumptions on (G,X): First, assume that G^{ad} is simple. Second, for every admissible parabolic P of G, there exist connected reductive subgroups L_P and G_P of M_P such that

- M_P is the direct product of L_P and G_P;
- G_P contains a certain normal subgroup G_1 of M_P defined by Pink [Pin92, (3.6)], and the quotient $G_P/Z(G_P)G_1$ is $\mathbb{R}$-anisotropic;
- $L_P \subseteq Cent_{M_P}(U_P) \subseteq Z(M_P)L_P$;
- $G_P(\mathbb{R})$ acts transitively on X_P, and $L_P(\mathbb{R})$ acts trivially on X_P;
- for every neat compact open subgroup $K_M \subseteq M_P(\mathbb{A}_f)$, $K_M \cap L_P(\mathbb{Q}) = K_M \cap Cent_{M_P(\mathbb{Q})}(X_P)$.

If G satisfies these assumptions, then G_P satisfies these assumptions for any admissible parabolic P of G [Mor10, Remark 1.1.1]. These assumptions are satisfied if G is an inner form of a unitary similitude or symplectic simlitude group [Mor10, Example 1.1.2]. Let us briefly describe the stratification on $\mathrm{Sh}_K^{\min}$ (under these assumptions). Let P be an admissible parabolic of G and let $Q_P \subseteq P$ be the preimage of G_P. There is a Shimura datum (G_P, X_P) with reflex field E. Let $g \in G(\mathbb{A}_f)$. Set $H_P = gKg^{-1} \cap P(\mathbb{Q})Q_P(\mathbb{A}_f)$, $H_L = gKg^{-1} \cap L_P(\mathbb{Q})N_P(\mathbb{A}_f)$, $K_Q = gKg^{-1} \cap Q_P(\mathbb{A}_f)$, and $K_N = gKg^{-1} \cap N_P(\mathbb{A}_f)$. Then there is a morphism

$$\mathrm{Sh}_{K_Q/K_N}(G_P, X_P) \to \mathrm{Sh}_K^{\min}$$

which is finite over its image. The group H_P acts on $\mathrm{Sh}_{K_Q/K_N}(G_P, X_P)$ and the action is trivial on the normal subgroup $H_L K_Q$; moreover, $H_P/H_L K_Q$ is finite. Quotienting out by the action of H_P gives a locally closed immersion

$$i_{P,g} : \mathrm{Sh}_{K_Q/K_N}(G_P, X_P)/H_P \to \mathrm{Sh}_K^{\min},$$

which extends to a finite morphism

$$\bar{i}_{P,g} : \mathrm{Sh}^{\min}_{K_Q/K_N}(G_P, X_P)/H_P \to \mathrm{Sh}_K^{\min}.$$

The boundary of $\mathrm{Sh}_K^{\min}$ is the union of the images of the $i_{P,g}$ for proper admissible parabolics P and elements $g \in G(\mathbb{A}_f)$. If P, P' are admissible parabolics and $g, g' \in G(\mathbb{A})_f$, then the images of $i_{P,g}$ and $i_{P',g'}$ are either equal or disjoint, and they are equal if and only if there exists a $\gamma \in G(\mathbb{Q})$ such that $P' = \gamma P \gamma^{-1}$ and $P(\mathbb{Q})Q_P(\mathbb{A}_f)gK = P(\mathbb{Q})Q_P(\mathbb{A}_f)\gamma^{-1}g'K$.

In view of this, we fix a minimal parabolic subgroup P_0 of G and let $P_1, \ldots, P_n$ be the admissible parabolics of G containing P_0, where the order is defined by $r \leq s$ if and only if $U_{P_r} \subseteq U_{P_s}$. We simplify the notation and write $N_r = N_{P_r}$, $i_{r,g} = i_{P_r,g}$ etc. Then, the boundary of $\mathrm{Sh}_K^{\min}$ is the union of the images of the $i_{r,g}$ for $r = 1, \ldots, n$ and $g \in G(\mathbb{A}_f)$, and $i_{r,g}$ and $i_{s,h}$ have the same image if and only if $r = s$ and there is a $\gamma \in G(\mathbb{Q})$ such that $P_r(\mathbb{Q})Q_r(\mathbb{A}_f)gK = P_r(\mathbb{Q})Q_r(\mathbb{A}_f)\gamma^{-1}hK$. For a fixed r, put

$$\mathrm{Sh}_{K,r} = \bigcup_{g \in G(\mathbb{A}_f)} \mathrm{Im}(i_{r,g}).$$

The $\mathrm{Sh}_{K,r}$ are locally closed subvarieties of $\mathrm{Sh}_K^{\min}$ and the collection $\mathrm{Sh}_{K,1}, \ldots, \mathrm{Sh}_{K,n}$ defines a stratification of $\mathrm{Sh}_K^{\min}$ in the sense of [Mor08, Définition 3.3.1]. We let $i_r : \mathrm{Sh}_{K,r} \to \mathrm{Sh}_K^{\min}$ denote the inclusion. By [Mor, Corollaire 8.1.4], all results of [Mor08, §3] go through in the setting of [Mor, §7]. In particular, following [Mor08, Proposition 3.3.4], we may define ${}^wD^{\leq \underline{a}} = {}^wD^{\leq \underline{a}}(\mathcal{M}(\mathrm{Sh}_K^{\min}))$ (resp. ${}^wD^{> \underline{a}} = {}^wD^{> \underline{a}}(\mathcal{M}(\mathrm{Sh}_K^{\min}))$) for any $\underline{a} = (a_1, \ldots, a_n) \in (\mathbb{Z} \cup \{\pm\infty\})^n$ to be the full subcategory of $D^b(\mathcal{M}(\mathrm{Sh}_K^{\min}))$ of objects C such that $i_r^* C \in D^b(\mathcal{M}(\mathrm{Sh}_{K,r}))$ (resp. $i_r^! C \in D^b(\mathcal{M}(\mathrm{Sh}_{K,r}))$) has weights $\leq a_r$ (resp. $> a_r$) for all $r \in \{1, \ldots, n\}$. Then $({}^wD^{\leq \underline{a}}, {}^wD^{> \underline{a}})$ defines a t-structure on $D^b(\mathcal{M}(\mathrm{Sh}_K^{\min}))$ and we get weight truncation functors $w_{\leq \underline{a}}$ and $w_{> \underline{a}}$ for all $\underline{a} \in (\mathbb{Z} \cup \{\pm\infty\})^n$. In particular, for any $r \in \{1, \ldots, n\}$ and $a \in \mathbb{Z} \cup \{\pm\infty\}$, we get weight truncation functors

$$w^r_{\leq a} := w_{\leq (+\infty, \ldots, +\infty, a, +\infty, \ldots, +\infty)};$$
$$w^r_{> a} := w_{> (-\infty, \ldots, -\infty, a, -\infty, \ldots, -\infty)},$$

where a is in the r-th place.

The final piece of notation and terminology we need concerns automorphic lisse $\overline{\mathbb{Q}}_p$-sheaves on Sh_K. Let Rep_G be the (semisimple) abelian category of algebraic representations of G over $\overline{\mathbb{Q}}_p$ and let $D^b(Rep_G)$ be its bounded derived category. There is an additive triangulated functor

$$\mathcal{F}^K : D^b(Rep_G) \to D^b_c(\mathrm{Sh}_K, \overline{\mathbb{Q}}_p)$$

to the derived category of constructible $\overline{\mathbb{Q}}_p$-sheaves on Sh_K [Pin92, (1.10)], [Mor06, 2.1.4]. If $V \in Ob(D^b(Rep_G))$, then all cohomology sheaves of $\mathcal{F}^K V$ are lisse and in particular perverse up to shift.

Assume now that (G, X) is of abelian type. We may then find a finite set of primes Σ of E, containing all primes above p, such that all objects above extend to $\mathrm{Spec}\, \mathcal{O}_E \setminus \Sigma$. More precisely, by [Mor10, Proposition 1.3.4] we may choose Σ such that conditions (1)–(7) on [Mor10, p. 8] are satisfied. In particular, conditions (5) and (7) imply that the functor $\mathcal{F}^K$ may be naturally viewed as a functor

$$\mathcal{F}^K : D^b(Rep_G) \to D^b(\mathcal{M}(\mathrm{Sh}_K));$$

similar remarks apply for all strata of the minimal compactifications. We remark that Rep_G has a notion of weight coming from the Shimura datum (see [Mor10, p.7]) and condition (7) says, in particular, that if V is pure then $\mathcal{F}^K V$ is pure. Note also that condition (6) says that Pink's formula holds for our integral model with the extended (complexes of) sheaves. Let us now state and prove the main theorem of this section.

Theorem 6.1. *Consider the collection of all Shimura data (G, X) of abelian type which satisfies the list of conditions in the bullet points above, with G^{ad} simple. Then the intersection cohomology groups $H^*_{ét}(\mathrm{Sh}_K^{\min}(G,X)_{\overline{\mathbb{Q}}}, j_{!*}\mathcal{F}^K V)$, for all (G, X) as above and all levels K and $V \in D^b(Rep_G)$, contain the same irreducible Galois representation as subquotients as the ordinary cohomology groups $H^*_{ét}(\mathrm{Sh}_K(G,X)_{\overline{\mathbb{Q}}}, \mathcal{F}^K V)$.*

Proof. We will prove the following claim by induction on d:

Claim: For any $d \in \mathbb{Z}_{\geq 0}$, the intersection cohomology groups $H^*_{ét}(\mathrm{Sh}_K^{\min}(G,X)_{\overline{\mathbb{Q}}}, j_{!*}\mathcal{F}^K V)$, for all (G, X) as in the theorem with $\dim \mathrm{Sh}_K(G,X) \leq d$ (and all levels K and $V \in D^b(Rep_G)$), contain the same irreducible Galois representation as subquotients as the ordinary cohomology groups $H^*_{ét}(\mathrm{Sh}_K(G,X)_{\overline{\mathbb{Q}}}, \mathcal{F}^K V)$ for all (G, X) as in theorem with $\dim \mathrm{Sh}_K(G,X) \leq d$ (and all levels K and $V \in D^b(Rep_G)$).

This would clearly give us the theorem. If $d = 0$, then $H^*(\mathrm{Sh}_K^{\min}, j_{!*}\mathcal{F}^K V) = H^*(\mathrm{Sh}_K, \mathcal{F}^K V)$ and the assertion is clear. For the induction step, assume $d = \dim \mathrm{Sh}_K \geq 1$. Without loss of generality assume that V is concentrated in a single degree and that it is pure. Let a be the weight of $\mathcal{V} := (\mathcal{F}^K V)[d]$; this is a pure perverse sheaf in $\mathcal{M}(\mathrm{Sh}_K)$. By [Mor08, Proposition 3.3.4] (which holds in our situation by [Mor, Corollaire 8.1.4]) we have a distinguished triangle

$$w^2_{\leq a} Rj_*\mathcal{V} \to Rj_*\mathcal{V} \to Ri_{2*}w_{>a}i_2^* Rj_*\mathcal{V} \to$$

in $D^b(\mathcal{M}(\mathrm{Sh}_K^{\min}))$. Applying [Mor08, Proposition 3.3.4] again to this triangle, we get a square

$$\begin{array}{ccccccc}
w^3_{\leq a} w^2_{\leq a} Rj_*\mathcal{V} & \longrightarrow & w^3_{\leq a} Rj_*\mathcal{V} & \longrightarrow & w^3_{\leq a} Ri_{2*}w_{>a}i_2^* Rj_*\mathcal{V} & \longrightarrow \\
\downarrow & & \downarrow & & \downarrow & \\
w^2_{\leq a} Rj_*\mathcal{V} & \longrightarrow & Rj_*\mathcal{V} & \longrightarrow & Ri_{2*}w_{>a}i_2^* Rj_*\mathcal{V} & \longrightarrow \\
\downarrow & & \downarrow & & \downarrow & \\
Ri_{3*}w_{>a}i_3^* w^2_{\leq a} Rj_*\mathcal{V} & \longrightarrow & Ri_{3*}w_{>a}i_3^* Rj_*\mathcal{V} & \longrightarrow & Ri_{3*}w_{>a}i_3^* Ri_{2*}w_{>a}i_2^* Rj_*\mathcal{V} & \longrightarrow \\
\downarrow & & \downarrow & & \downarrow &
\end{array}$$

of triangles. Continuing in this way, we get an $(n-1)$-dimensional hypercube of triangles, with 'top left corner' $w^n_{\leq a} \dots w^2_{\leq a} Rj_*\mathcal{V} = w_{\leq a} Rj_*\mathcal{V}$.[5] Since $\mathcal{V}$ is a pure perverse sheaf of weight a, $w_{\leq a} Rj_*\mathcal{V} = j_{!*}\mathcal{V}$ by [Mor08, Théorème 3.1.4]. By [Mor10, Proposition 1.4.5] (whose proof relies of Pink's formula and the theory of weight truncations, so is valid in our setting), the cohomology of the complexes $Ri_{n_r*}w_{>a}i^*_{n_r} \dots Ri_{n_1*}w_{>a}i^*_{n_1} Rj_*\mathcal{V}$ for

[5] See [Mor08, Théorème 3.3.5] for the equality in the Grothendieck group that results from this hypercube.

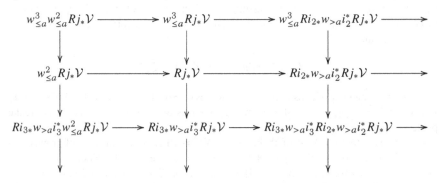

$n_1 < \ldots < n_r$, for any sequence in $\{2, \ldots, n\}$, are subquotients of direct sums of ordinary cohomology groups of automorphic complexes for boundary strata of $\mathrm{Sh}_K^{\min}$ (subquotients as the boundary strata are finite Galois quotients of Shimura varieties). As the complexes $Ri_{n_r*}w_{>a}i_{n_r}^* \ldots Ri_{n_1*}w_{>a}i_{n_1}^* Rj_*\mathcal{V}$ for $n_1 < \ldots < n_r$ together with $Rj_*\mathcal{V}$ make up the 'lower right' $2 \times \cdots \times 2$ hypercube and $j_{!*}\mathcal{V}$ sits in the top left corner, the induction step follows by taking long exact sequences of the triangles in the hypercube. This finishes the proof of the claim. □

To conclude, let us note that we would optimistically expect the theorem to hold with no assumptions on the (G, X), and that the proof would proceed along the same lines in an ideal world; we note that the Hodge-theoretic analogue holds by work of Nair [Nai]. At present, we do not know how one could try to remove the assumption of abelian type (unless one replaces $D^b(Rep(G))$ with a smaller subcategory, cf. [Mor10, Proposition 1.3.4]), since the link to geometry seems necessary to prove that the sheaves $\mathcal{F}^K V$ are pure of the expected weight when V is a pure representation. We would naively suspect that it might be possible to remove the other assumptions, but we have not looked into this.

References

[ABV92] Jeffrey Adams, Dan Barbasch, and David A. Vogan, Jr. *The Langlands classification and irreducible characters for real reductive groups*, volume 104 of *Progress in Mathematics*. Birkhäuser Boston, Inc., Boston, MA, 1992.

[AC89] James Arthur and Laurent Clozel. *Simple algebras, base change, and the advanced theory of the trace formula*, volume 120 of *Annals of Mathematics Studies*. Princeton University Press, Princeton, NJ, 1989.

[AJ87] Jeffrey Adams and Joseph F. Johnson. Endoscopic groups and packets of nontempered representations. *Compositio Math.*, 64(3):271–309, 1987.

[Art89] James Arthur. Unipotent automorphic representations: conjectures. *Astérisque*, (171-172):13–71, 1989. Orbites unipotentes et représentations, II.

[Art02] James Arthur. A note on the automorphic Langlands group. *Canad. Math. Bull.*, 45(4):466–482, 2002. Dedicated to Robert V. Moody.

[AV16] Jeffrey Adams and David A. Vogan, Jr. Contragredient representations and characterizing the local Langlands correspondence. *Amer. J. Math.*, 138(3):657–682, 2016.

[BG14] Kevin Buzzard and Toby Gee. The conjectural connections between automorphic representations and Galois representations. In *Automorphic forms and Galois representations. Vol. 1*, volume 414 of *London Math. Soc. Lecture Note Ser.*, pages 135–187. Cambridge Univ. Press, Cambridge, 2014.

[BLGGT14] Thomas Barnet-Lamb, Toby Gee, David Geraghty, and Richard Taylor. Potential automorphy and change of weight. *Ann. of Math. (2)*, 179(2):501–609, 2014.

[BM02] Christophe Breuil and Ariane Mézard. Multiplicités modulaires et représentations de $\mathrm{GL}_2(\mathbf{Z}_p)$ et de $\mathrm{Gal}(\overline{\mathbf{Q}}_p/\mathbf{Q}_p)$ en $l = p$. *Duke Math. J.*, 115(2):205–310, 2002. With an appendix by Guy Henniart.

[Cal] Frank Calegari. Even Galois representations. Unpublished lecture notes, available at `http://www.math.uchicago.edu/~fcale/papers/FontaineTalk-Adjusted.pdf`.

[Car86] Henri Carayol. Sur les représentations l-adiques associées aux formes modulaires de Hilbert. *Ann. Sci. École Norm. Sup. (4)*, 19(3):409–468, 1986.

[CHT08] Laurent Clozel, Michael Harris, and Richard Taylor. Automorphy for some l-adic lifts of automorphic mod l Galois representations. *Publ. Math. Inst. Hautes Études*

Sci., (108):1–181, 2008. With Appendix A, summarizing unpublished work of Russ Mann, and Appendix B by Marie-France Vignéras.

[Clo90] Laurent Clozel. Motifs et formes automorphes: applications du principe de fonctorialité. In *Automorphic forms, Shimura varieties, and L-functions, Vol. I (Ann Arbor, MI, 1988)*, volume 10 of *Perspect. Math.*, pages 77–159. Academic Press, Boston, MA, 1990.

[Clo91] Laurent Clozel. Représentations galoisiennes associées aux représentations automorphes autoduales de $\mathrm{GL}(n)$. *Inst. Hautes Études Sci. Publ. Math.*, (73):97–145, 1991.

[Clo93] Laurent Clozel. On the cohomology of Kottwitz's arithmetic varieties. *Duke Math. J.*, 72(3):757–795, 1993.

[Del71] Pierre Deligne. Formes modulaires et représentations l-adiques. In *Séminaire Bourbaki. Vol. 1968/69: Exposés 347–363*, volume 175 of *Lecture Notes in Math.*, pages Exp. No. 355, 139–172. Springer, Berlin, 1971.

[Del79a] P. Deligne. Valeurs de fonctions L et périodes d'intégrales. In *Automorphic forms, representations and L-functions (Proc. Sympos. Pure Math., Oregon State Univ., Corvallis, Ore., 1977), Part 2*, Proc. Sympos. Pure Math., XXXIII, pages 313–346. Amer. Math. Soc., Providence, R.I., 1979. With an appendix by N. Koblitz and A. Ogus.

[Del79b] Pierre Deligne. Variétés de Shimura: interprétation modulaire, et techniques de construction de modèles canoniques. In *Automorphic forms, representations and L-functions (Proc. Sympos. Pure Math., Oregon State Univ., Corvallis, Ore., 1977), Part 2*, Proc. Sympos. Pure Math., XXXIII, pages 247–289. Amer. Math. Soc., Providence, R.I., 1979.

[Del82] Pierre Deligne. Hodge cycles on abelian varieties. In *Hodge cycles, motives and Shimura varieties*, volume 900 of *Lecture Notes in Math.*, pages 9–100. Springer-Verlag, Berlin, 1982.

[Fal88] Gerd Faltings. p-adic Hodge theory. *J. Amer. Math. Soc.*, 1(1):255–299, 1988.

[FM95] Jean-Marc Fontaine and Barry Mazur. Geometric Galois representations. In *Elliptic curves, modular forms, & Fermat's last theorem (Hong Kong, 1993)*, Ser. Number Theory, I, pages 41–78. Int. Press, Cambridge, MA, 1995.

[Gro] Benedict Gross. Odd galois representations. Unpublished note, available online at `http://www.math.harvard.edu/~gross/preprints/Galois_Rep.pdf`.

[HLTT16] Michael Harris, Kai-Wen Lan, Richard Taylor, and Jack Thorne. On the rigid cohomology of certain Shimura varieties. *Res. Math. Sci.*, 3:Paper No. 37, 308, 2016.

[HT01] Michael Harris and Richard Taylor. *The geometry and cohomology of some simple Shimura varieties*, volume 151 of *Annals of Mathematics Studies*. Princeton University Press, Princeton, NJ, 2001. With an appendix by Vladimir G. Berkovich.

[Hub97] Annette Huber. Mixed pervese sheaves for schemes over number fields. *Compositio Math.*, 108(1):107–121, 1997.

[Joh13] Christian Johansson. A remark on a conjecture of Buzzard-Gee and the cohomology of Shimura varieties. *Math. Res. Lett.*, 20(2):279–288, 2013.

[KMSW] Tasho Kaletha, Alberto Minguez, Sug-Woo Shin, and Paul-James White. Endoscopic classification of representations: Inner forms of unitary groups. Preprint, available online at `https://arxiv.org/abs/1409.3731`.

[Kot90] Robert E. Kottwitz. Shimura varieties and λ-adic representations. In *Automorphic forms, Shimura varieties, and L-functions, Vol. I (Ann Arbor, MI, 1988)*, volume 10 of *Perspect. Math.*, pages 161–209. Academic Press, Boston, MA, 1990.

[Kot92] Robert E. Kottwitz. On the λ-adic representations associated to some simple Shimura varieties. *Invent. Math.*, 108(3):653–665, 1992.

[Lan79] R. P. Langlands. Automorphic representations, Shimura varieties, and motives. Ein Märchen. In *Automorphic forms, representations and L-functions (Proc. Sympos. Pure Math., Oregon State Univ., Corvallis, Ore., 1977), Part 2*, Proc. Sympos. Pure Math., XXXIII, pages 205–246. Amer. Math. Soc., Providence, R.I., 1979.

[Mil83] J. S. Milne. The action of an automorphism of **C** on a Shimura variety and its special points. In *Arithmetic and geometry, Vol. I*, volume 35 of *Progr. Math.*, pages 239–265. Birkhäuser Boston, Boston, MA, 1983.

[Mor] Sophie Morel. Complexes mixtes sure un schema de type fini sur Q. *Available at* `https://web.math.princeton.edu/~smorel/`.

[Mor06] Sophie Morel. *Complexes d'intersection des compactifications de Baily-Borel. Le cas des groupes unitaires sur* $\mathbb{Q}$. PhD thesis, Université Paris-Sud XI - Orsay, 2006.

[Mor08] Sophie Morel. Complexes pondérés sur les compactifications de Baily–Borel: le cas de variétiés de Siegel. *J. Amer. Math. Soc.*, 21(1):23–61, 2008.

[Mor10] Sophie Morel. *On the cohomology of certain noncompact Shimura varieties. With an appendix by Robert Kottwitz.*, volume 173 of *Annals of Mathematics Studies*. Princeton University Press, Princeton, NJ, 2010.

[Mor11] Sophie Morel. Intersection cohomology is useless. *Talk at the Institute for Advanced Study,* `https://video.ias.edu/graf/2011/03/22/morel`, 2011.

[Nai] Arvind Nair. Mixed structures in Shimura varieties and automorphic forms. *Available at* `http://www.math.tifr.res.in/~arvind/`.

[Pin92] Richard Pink. On ℓ-adic sheaves on Shimura varieties and their higher direct images in the Baily–Borel compactification. *Math. Ann.*, 292(2):197–240, 1992.

[Sch15a] Peter Scholze. On torsion in the cohomology of locally symmetric varieties. *Ann. of Math. (2)*, 182(3):945–1066, 2015.

[Sch15b] Peter Scholze. On torsion in the cohomology of locally symmetric varieties. *Ann. of Math. (2)*, 182(3):945–1066, 2015.

[Ser94] Jean-Pierre Serre. Propriétés conjecturales des groupes de Galois motiviques et des représentations l-adiques. In *Motives (Seattle, WA, 1991)*, volume 55 of *Proc. Sympos. Pure Math.*, pages 377–400. Amer. Math. Soc., Providence, RI, 1994.

[Ser12] Jean-Pierre Serre. *Lectures on* $N_X(p)$, volume 11 of *Chapman & Hall/CRC Research Notes in Mathematics*. CRC Press, Boca Raton, FL, 2012.

[Tat79] J. Tate. Number theoretic background. In *Automorphic forms, representations and L-functions (Proc. Sympos. Pure Math., Oregon State Univ., Corvallis, Ore., 1977), Part 2*, Proc. Sympos. Pure Math., XXXIII, pages 3–26. Amer. Math. Soc., Providence, R.I., 1979.

[Tb16] Olivier Taïbi. Eigenvarieties for classical groups and complex conjugations in Galois representations. *Math. Res. Lett.*, 23(4):1167–1220, 2016.

[vGT94] Bert van Geemen and Jaap Top. A non-selfdual automorphic representation of GL_3 and a Galois representation. *Invent. Math.*, 117(3):391–401, 1994.

[Vog84] David A. Vogan, Jr. Unitarizability of certain series of representations. *Ann. of Math. (2)*, 120(1):141–187, 1984.

[VZ84] David A. Vogan, Jr. and Gregg J. Zuckerman. Unitary representations with nonzero cohomology. *Compositio Math.*, 53(1):51–90, 1984.

Printed in the United States
by Baker & Taylor Publisher Services

Printed in the United States
by Baker & Taylor Publisher Services